D0762295

NANOSCALE

THE WILEY BICENTENNIAL—KNOWLEDGE FOR GENERATIONS

*E*ach generation has its unique needs and aspirations. When Charles Wiley first opened his small printing shop in lower Manhattan in 1807, it was a generation of boundless potential searching for an identity. And we were there, helping to define a new American literary tradition. Over half a century later, in the midst of the Second Industrial Revolution, it was a generation focused on building the future. Once again, we were there, supplying the critical scientific, technical, and engineering knowledge that helped frame the world. Throughout the 20th Century, and into the new millennium, nations began to reach out beyond their own borders and a new international community was born. Wiley was there, expanding its operations around the world to enable a global exchange of ideas, opinions, and know-how.

For 200 years, Wiley has been an integral part of each generation's journey, enabling the flow of information and understanding necessary to meet their needs and fulfill their aspirations. Today, bold new technologies are changing the way we live and learn. Wiley will be there, providing you the must-have knowledge you need to imagine new worlds, new possibilities, and new opportunities.

Generations come and go, but you can always count on Wiley to provide you the knowledge you need, when and where you need it!

WILLIAM J. PESCE
PRESIDENT AND CHIEF EXECUTIVE OFFICER

PETER BOOTH WILEY
CHAIRMAN OF THE BOARD

NANOSCALE
ISSUES AND PERSPECTIVES
FOR THE NANO CENTURY

Edited by

Nigel M. de S. Cameron
Center on Nanotechnology and Society
Illinois Institute of Technology
Chicago, Illinois

M. Ellen Mitchell
Institute of Psychology
Illinois Institute of Technology
Chicago, Illinois

WILEY-INTERSCIENCE
A John Wiley & Sons, Inc., Publication

Published by John Wiley & Sons, Inc., Hoboken, New Jersey
Published simultaneously in Canada

For general information on our other products and services or for technical support, please contact
our Customer Care Department within the United States at (800) 762-2974, outside the United States at
(317) 572-3993 or fax (317) 572-4002.

Wiley also publishes its books in a variety of electronic formats. Some content that appears in print
may not be available in electronic formats. For more information about Wiley products, visit our
web site at www.wiley.com.

Wiley Bicentennial Logo: Richard J. Pacifico

Library of Congress Cataloging-in-Publication Data:

Nanoscale: issues and perspectives for the nano century/editors, Nigel M. de S. Cameron,
    M. Ellen Mitchell.
        p. cm.
    Includes bibliographical references and index.
    ISBN 978-0-470-08419-9 (cloth)
    1.   Nanotechnology—Moral and ethical aspects.   2. Nanotechnology—Social aspects.
    I.   Cameron, Nigel M. de S.   II.   Mitchell, M. Ellen.
    T174.7.N3575 2007
    620'.5—dc22

                                                                          2007006004

Printed in the United States of America

10   9   8   7   6   5   4   3   2   1

"Now nanotechnology had made nearly everything possible, and so the cultural role in deciding what should be done with it had become far more important than imagining what could be done with it."

—Neal Stephenson, *The Diamond Age or a Young Lady's Primer* (1995)

"Each new power won *by* man is a power *over* man as well. Each advance leaves him weaker as well as stronger. In every victory, besides the general who triumphs, he is a prisoner who follows the triumphal car . . . *Human* nature will be the last part of Nature to surrender to Man. The battle will then be won. We shall have "taken the thread out of the hands of Clotho" and be free henceforth to make of our species whatever we wish it to be. The battle will indeed be won. But who, precisely, will have won it?"

—C. S. Lewis, *The Abolition of Man* (1943)

"[T]he discoverer of an art is not the best judge of the good or harm which will accrue to those who practice it."

—Plato, *Phaedrus* (c. 370 BC)

"Science Finds, Industry Applies, Man Conforms"

—Motto of Chicago World's Fair, 1933–34 (Century of Progress Exposition)

CONTENTS

Most Americans have not yet heard of nanotechnology, and many of those who have cannot offer a working definition of the term. This low profile is anomalous, disconcerting, and destined, before long, for a correction that could be dramatic in nature. It can, perhaps, be explained by a combination of low public interest in science and science policy in general, the recent dominance of the science space by the stem-cell and cloning debates, the wide variety of applications of nanoscale research, and the fact that there is not—yet—a significant political constituency with an interest in critiquing, or at least monitoring, the very extensive federal funding of work on the nanoscale.

Nevertheless, the broad social implications of this new wave of technology have been recognized in the funding process. When President Bush signed the 21st Century Nanotechnology Research and Development Act (the Act) in December of 2003, a sum of $3.7 billion was designated for nanoscale research over a period of 4 years. This federal largesse, now running in excess of $1 billion a year, is being distributed across more than 20 different agencies, with the National Science Foundation (NSF) as lead. The National Nanotechnology Initiative (NNI) is monitored by congressional reporting requirements and a supervisory committee designated by the President—a role that has been assigned to the President's Council of Advisors on Science and Technology (PCAST) in the White House Office of Science and Technology Policy.

The Act specifies the need to fund nano-related ethical, legal, and societal issues (NELSI) research in addition to work on the technology itself, in a manner that parallels the ELSI (ethical, legal, and societal issues) program established under the human genome project, the last major publicly funded science venture in the United States.

The human genome project was developed with the awareness that issues of science and technology cannot be pursued in isolation from their broader implications for society. The ethical, legal, and social issues raised by new technologies must be addressed in parallel, both to ensure that pitfalls unforeseen by scientists will be addressed in good time, and to help build public confidence in the technologies themselves. Alongside the NELSI issues, questions of environment, health, and safety (EHS) have also been singled out for research, as well as the need to review workforce implications and permeate the educational system with an understanding of this emerging technology and training of tomorrow's scientists.

What, then, are the fundamental questions raised by nanotechnology? At least three distinct areas of concern can be identified.

First, there are concerns about its safety. A recent report by Swiss Re, the world's largest reinsurance company, draws attention to substantial risk issues involved in this new technology that have yet to be assessed.[1]

Second, there are concerns about the impact on the way we lead our lives. For example, one prospect is of miniaturized RFID (radio frequency ID) transponders that would enable the location of each of us to be pinpointed. Technologies that have many beneficial applications can also pose new threats to social values like privacy, and, while not requiring their development, may suggest new directions for the culture. Another aspect of ethical concern is the so-called nano-divide, in that the new capacities that this technology may be expected to provide (e.g., in healthcare and many other fields) will not come without costs that could deepen economic divisions within and between nations.

Third, there are concerns about the capacity of nanotechnology to reshape human nature itself. Early NSF documents have framed development of nanotechnology in the context of the "convergence" of nanotechnology, biotechnology, information technology, and cognitive science (together referred to as NBIC), with a view to the "improvement" of "human performance." While some in the nano community downplay these capacities and others have exaggerated their significance, there is no doubt that a major strand of social concern relates to the potential employment of nanoscale products to effect changes to basic human capacities. The 2003 Act singles out the development of artificial intelligence and the enhancement of human intelligence as key issues of concern.

In 2000, the same year as the NNI was established, Bill Joy, cofounder and for many years chief technologist at Sun Microsystems, emerged as an early cultural critic of nanotechnology in his essay, "Why the Future Doesn't Need Us," published in the premier new technology monthly *Wired*.[2] Joy's argument was that nano, together with genomics and robotics, has the potential to eclipse human nature—either through an accident that destroys the species, or through human choices that lead to the supremacy of a nonhuman form of life. While his remarks may represent far-fetched projections of the future ungrounded in current data, they accurately reflect that nanotechnology can be applied to virtually anything because it refers only to scale and it may have the potential to transform every aspect of life, perhaps even the nature of *Homo sapiens* itself, at some fundamental level. Sifting the truth from the hype is difficult. Mihail C. Roco of the NSF, who has been the most influential voice in U.S. nano policy, has written:

> The vision of the NNI includes a path to discoveries of new properties and phenomena at the nanoscale, working directly at the building blocks of matter with cross-cutting approaches and tools applicable to almost all man-made objects, and development of highly efficient manufacturing. This is completed by the promise of better

[1]Annabelle Hett. 2004. Nanotechnology: Small Matter Many Unknowns. Swiss Re.
[2]Bill Joy. April 2000. Why the Future Doesn't Need Us. Wired Magazine 8.04.

comprehension of nature, increased wealth, better healthcare and long-term sustainable development.[3]

Perhaps the greatest challenges facing our society lie in our assessment of these projections, our management of the expectations they create, and our development of judicious policy approaches to the technology options that may result.

The essays that follow have been selected with the purpose of contributing to what we believe will be one of the greatest of all public debates. A debate that will benefit from full discourse that includes both information and opinion. While there is naturally some overlap between the two, they fall broadly into complementary categories: opinion pieces by visionaries, boosters and critics; and reviews of key areas of ethical, legal, and societal questions. These chapters are rife with strong opinion and new knowledge, and we invite you to use this volume to fuel the conversation.

NIGEL M. DE S. CAMERON

M. ELLEN MITCHELL

Chicago, Illinois

---

[3]Mihail C. Roco. Based on a presentation made at Cornell Nanofabrication Center, September 15, 2000. Available at: http://www.nsf.gov/crssprgm/nano/reports/roco_vision.jsp.

ACKNOWLEDGMENTS

We would like to acknowledge the aid of many colleagues in the planning and compiling of this volume, especially Michele Mekel, J.D., Associate Director and Legal Fellow of the Center on Nanotechnology and Society at Illinois Institute of Technology, who has played a major role in the editorial task; Dawn Willow, J.D., also a Legal Fellow; our Administrative Associate Joseph P. Oldaker; and Christine Sackmann and her team of student assistants who have also lent their willing energies. It has been a delight to work with the colleagues at our university and further afield who have gladly contributed the chapters herein.

NIGEL M. DE S. CAMERON

M. ELLEN MITCHELL

Chicago, Illinois

CONTRIBUTORS

Lori B. Andrews, J.D., is Distinguished Professor of Law at Chicago-Kent College of Law; the Director of the Institute of Science, Law, and Technology at Chicago-Kent College of Law, and Associate Vice President of Illinois Institute of Technology. She served as chair of the federal Working Group on the Ethical, Legal, and Social Implications of the Human Genome Project.

Debra Bennett-Woods, Ed.D., is Director and Associate Professor in the Department of Health Care Ethics in the Rueckert-Hartman School for Health Professions at Regis University, and a member of the Task Force on Nano-Ethics and Societal Impacts of the Colorado Nanotechnology Institute. She is also a Fellow of the Center on Nanotechnology and Society at Chicago-Kent College of Law/Illinois Institute of Technology.

Brent Blackwelder, Ph.D., is President of Friends of the Earth, an international group that lobbies for environmental causes. He is also a Fellow of the Institute on Biotechnology and the Human future.

Nick Bostrom, Ph.D., is Director of the Future of Humanities Institute at the University of Oxford. He is also co-founder and chair of the World Transhumanist Association.

Julie A. Burger, J.D., is the Assistant Director and Legal Fellow of the Institute for Science, Law, and Technology at Chicago-Kent College of Law/Illinois Institute of Technology. She previously practiced law at a Chicago-area firm.

Nigel M. de S. Cameron, Ph.D., is Director of the Center on Nanotechnology and Society, President of the Institute on Biotechnology and the Human Future, Research Professor of Bioethics and Associate Dean at Chicago-Kent College of Law/Illinois Institute of Technology. He founded the journal *Ethics and Medicine* in 1983 and has represented the United States at United Nations meetings on issues of technology policy.

William P. Cheshire, Jr., M.D., is Consultant in Neurology at the Mayo Clinic in Jacksonville, Florida; Associate Professor of Neurology at the Mayo Clinic College of Medicine; Director of the Mayo Autonomic Reflex Laboratory; and

Past Chair of the Autonomic Nervous System Section of the American Academy of Neurology. He is also a Fellow of the Center on Nanotechnology and Society.

Jerry C. Collins, Ph.D., is Research Associate Professor of Biomedical Engineering at Vanderbilt University.

Jessica K. Fender, M.S., is a student at Chicago-Kent College of Law, where she is an associate editor for both the *Chicago-Kent Law Review* and the *Chicago-Kent Journal of Intellectual Property*. She has also worked as a research assistant for the Institute for Science, Law, Technology, and the Institute on Biotechnology and the Human Future.

Ruthanna Gordon, Ph.D., is an Assistant Professor in the Institute of Psychology at Illinois Institute of Technology. She is also a Member of the Center on Nanotechnology and Society Advisory Panel.

David Guston, Ph.D., is Professor of Political Science at Arizona State University; associate director of the Consortium for Science, Policy and Outcomes. He is Principal Investigator and Director of the Center for Nanotechnology in Society at Arizona State University.

Jacob Heller is Policy Associate at the Foresight Nanotech Institute, and Founder and Director of A Computer in Every Home, a community organization that provides free computers and technical training to underprivileged students. He was selected as a Harry S Truman Scholar for his commitment to technology policy and public service.

Annabelle Hett, Ph.D., is Head of Emerging Risk Management at Swiss Re. Based in Zurich, she is responsible for the systematic identification, assessment, and evaluation of emerging risks on Group level and also has responsibility for screening existing exposures arising from novel, unprecedented scenarios and accumulations.

U.S. Congressman Mike Honda (D), M.Ed., is the U.S. Representative for the 15th Congressional District of California. He joined Science Committee Chairman Sherwood Boehlert in introducing the 21st Century Nanotechnology Research and Development Act, which was ultimately signed into law by President Bush on December 3, 2003.

C. Christopher Hook, M.D., is Consultant in Hematology, Special Coagulation and the Comprehensive Hemophilia Center, and Assistant Professor of Medicine at the Mayo Clinic in Rochester, Minnesota; and Director of Ethics Education at Mayo Clinic Graduate School of Medicine. He created and chairs the Mayo Reproductive Medicine Advisory Board, the DNA Research Committee, the Ethics Consultation Service, and the Mayo Clinical Ethics Council. Additionally, he is a Fellow of the Center on Nanotechnology and Society.

James Hughes, Ph.D., is Professor of Health Policy at Trinity College in Hartford Connecticut, and serves as Trinity's Associate Director of Institutional Research and Planning. He also serves as the Executive Director of the World Transhumanist Association and its affiliated Institute for Ethics and Emerging Technologies.

George A. Kimbrell, J.D., is Staff Attorney at the International Center for Technology Assessment in Washington, D.C. He works on legal developments in biotechnology, nanotechnology, and climate changes.

Kristen M. Kulinowski, Ph.D., is Executive Director for Education and Public Policy at the Center for Biological and Environmental Nanotechnology and for the International Council on Nanotechnology at Rice University.

Michele Mekel, J.D., M.H.A., M.B.A., is Associate Director of the Center on Nanotechnology and Society and Executive Director and Fellow of the Institute on Biotechnology and the Human Future—both at Chicago-Kent College of Law/Illinois Institute of Technology. A former Fulbright Fellow, she is also on the Board of the Converging Technologies Bar Association, and is an Associate Editor of *Nanotechnology, Law and Business*.

Sonia E. Miller, J.D., M.B.A., M.S.Ed., is a practicing attorney and principal of S.E. MILLER LAW FIRM, a boutique law firm that specialize in the implications of emerging and converging technologies. She is founder of the Converging Technologies Bar Association.

M. Ellen Mitchell, Ph.D., is the Director of the Institute of Psychology at Illinois Institute of Technology. She is Senior Fellow of the Center on Nanotechnology and Society, and a Fellow of the Institute on Biotechnology and the Human Future.

Christine Peterson is Founder and Vice President of Foresight Nanotech Institute. She serves on the Steering Committee of the International Council on Nanotechnology, the Editorial Advisory Board of NASA's *Nanotech Briefs*, and California's Blue Ribbon Task Force on Nanotechnology.

Trudy A. Phelps, Ph.D., is Standards Director at the Association of British Healthcare Industries (ABHI), Chairman of the ABHI Natural Rubber Latex Working Group, and Secretary to the European Medical Technology Industry Association (Eucomed) Standards Focus Group. She has chaired the European Commission's Committee on nanotechnology standards.

U.S. Congressman Brad Sherman (D), J.D., is the U.S. Representative for the 27th Congressional District of California.

Marty Spitzer, J.D., Ph.D., is a former Professional Staff Member of the House Committee on Science, Subcommittee on Environment, Technology, and Standards.

Marianne R. Timm, J.D., is a Patent Attorney at Suiter Swantz. She previously worked with the Institute on Science, Law, and Technology.

Vivian Weil, Ph.D., is Director of the Center for the Study of Ethics in the Professions and Professor of Ethics at Illinois Institute of Technology. She is a Member of the Advisory Panel of the Center on Nanotechnology and Society.

U.S. Congressman David Weldon (R), M.D., is the U.S. Representative for the 15th Congressional District of Florida.

PART I

POLICY AND PERSPECTIVES

This section opens with perspectives from members of the U.S. Congress, and includes some of those who wrote the 2003 21st Century Nanotechnology Research and Development Act that established the National Nanotechnology Initiative. It is the product of a roundtable at the Center on Nanotechnology and Society's first annual conference on nanopolicy (in 2006). Keynotes had been delivered by Mihail C. Roco, nanotechnology advisor at the National Science Foundation and the most influential figure in U.S. nanotechnology, and Sean Murdoch, who directs the trade group the NanoBusiness Alliance.

Central to the concerns of policymakers, technologists, and business leaders is the question of risk. This is discussed by Annabelle Hett, head of emerging technology risk at Swiss Re, now the world's largest reinsurance company and publisher of the influential report she authored on risk and nanotechnology. Risk covers many issues; one plainly lies in environmental hazards and toxicology concerns. Brent Blackwelder, U.S. President of the international environmentalist group Friends of the Earth, offers a somewhat different perspective, focused on issues of consumer safety. Looking more broadly at the need to maximize benefits and minimize risks, Jacob Heller and Christine Peterson write from the Foresight Nanotech Institute (of which Peterson was co-founder with K. Eric Drexler), the nano think tank that has long promoted the nano vision, including a special focus on "molecular" nanotechnology.

But the implications of a new technology range more broadly than quantifiable issues of safety and broader risk. Two psychologists, M. Ellen Mitchell and Ruthanna Gordon, tackle wider questions with one eye on the human dimension and another on the claims made for technological promise.

What of the purpose for which nanotechnology is being developed, and the wider policy context? Nick Bostrom from Oxford and James Hughes from Trinity College, Hartford, Connecticut, both leaders of the World Transhumanist Association, make their respective cases for a vision of the future in which "human nature" may have become a thing of the past, and yet in which technology enables persons to thrive in conditions that stretch our imagination. On the same theme, Nigel Cameron reviews

Nanoscale: Issues and Perspectives for the Nano Century. Edited by Nigel M. de S. Cameron and M. Ellen Mitchell
Copyright © 2007 John Wiley & Sons, Inc.

the European response to the National Science Foundation's first report on *Converging Technologies for Improving Human Performance*, which was seen as favoring the transhumanist vision (by enthusiasts and critics alike), and misunderstood by many as a statement of U.S. policy.

Taken together, these chapters set the scene for the cultural politics of the twenty-first century, setting out the promises and the perils of nanotechnology and sampling arguments that will be heard for many years to come.

CHAPTER 1

The View from Congress:
A Roundtable on Nanopolicy[1]

U.S. CONGRESSMAN MIKE HONDA, U.S. CONGRESSMAN BRAD SHERMAN, U.S. CONGRESSMAN DAVID WELDON, and MARTY SPITZER[2]

MARTY SPITZER

I am here representing the Chairman of the Science Committee, Mr. Boehlert, who has other obligations in the District today.

What I would like to do today is to: provide a little bit of an overview of three things; tell you a little bit about what the Science Committee has been doing and is doing in the area of the societal and environmental implications of nanotech; provide a quick overview of what the 21st Century Nanotechnology Research and Development Act, which created the federal apparatus to implement and carry out the Act, requires; and then talk to you about one element of the societal implications that everyone has some interest in and that is pressing upon us today—the environmental implications.

When the Act was passed in 2003, basically, it authorized almost $4 billion over 5 years to be spent by the federal government. It provided the statutory framework for what a lot of the federal government is doing today. So, it set up interagency committees, required annual reports to Congress, and set up an advisory panel that would report every 2 years. One of the things that some of the other panel members can speak to is that it specifically called for an emphasis on research as it relates to the societal implications of nanotechnology in order to understand the impact of these products on health and the environment. It also includes a study that's almost done, on the responsible development of nanotech. This is part of

---

[1]Edited transcripts from the policy panel at the Center on Nanotechnology and Society's Nanopolicy Conference at the National Press Club, Washington, DC, on April 28, 2006.
[2]Representing U.S. Representative Sherwood Boehlert, Chairman of the House Science Committee.

Nanoscale: Issues and Perspectives for the Nano Century. Edited by Nigel M. de S. Cameron and M. Ellen Mitchell
Copyright © 2007 John Wiley & Sons, Inc.

the triennial review that the National Academies of Science, and it is a one-time assessment to look at standards, guidelines, and strategies for ensuring the responsible development of nanotech. It is supposed to look at including issues of self-replicating nanoscale machines and devices, as well as the release of such machines into the environment.

The other piece of the societal implications is the environmental aspects of nanotech. There are really two components of that. There are the environmental applications, that is, technologies that are actually going to help to make the environment cleaner and better. Then there are the environmental health and safety implications, those things that may actually cause problems for humans, the environment, and our ecosystems. What I am going to do is spend some time talking more about these implications.

Now, the federal investment is approximately $1.1 billion, up from almost $500 million just a few years ago. That is matched with a lot of private-sector funding that is on the order of $2 billion, close to half a billion dollars per year in state funding, and then international investments in the arena of nanotech; the numbers are quite significant. The context for talking about environment—and really all the societal implications—is the potential growth of this sector, in terms of the entire economy. The numbers being tossed around are enormous. Lux Research predicts that, in 10 years or so, $2.6 trillion worth of products in the global marketplace will contain nanomaterials and 15% of manufacturing output will include some nanomaterials. This is enormous, and it puts a great deal of pressure on all of the systems (societal, business, infrastructure, and governmental) to deal with the changes that this new set of technologies will bring about. So it's in that context that we have been, as a Committee, thinking about the environmental implications. The initial concern is: Are we spending enough, and are we doing enough in this arena? So far, in Fiscal Year 2006, approximately $80 million of the federal investment was devoted to societal implications of nanotech. About 4%, or $38 million, of that was devoted to environmental implications, and most of that came under the oversight of the National Science Foundation, where they do investigator-driven research. The rest of the funding has been spent on economic workforce, educational, and other ethical issues. We are still waiting to hear what the 2007 proposed budget figures will be. Hopefully, we will have those pretty soon. And there is lots of private-sector money being spent, as well.

We are talking about a suite of technologies that are going to revolutionize the way we do things and how we live. And the questions are How will that happen? And what will we do as this unfolds? Do we have systems in place that are capable of keeping up with the rapid change of the technology?

In the environmental arena, we are dealing with two kinds of problems regarding nanotech. I am assuming, for a moment, that everyone agrees that there are probably some beneficial uses of nanotech, and that there are very good things that we are going to be able to do with them. Yet we have both real and perceived risks with which to deal. Among the real risks, it is probably true that most nanomaterials aren't going to be very harmful. Although there is early evidence that some of them are. So, the question is How do we distinguish them? It is not hard to imagine that the very elements that make the nanomaterials so beneficial and so exciting in their applications, their small size and their unique behavioral characteristics, are the same things that potentially make them dangerous to humans and the environment.

When the Woodrow Wilson Center did its study last year looking at the public perceptions of nanotech, it found a couple of things that were important and should guide us as we are thinking about how to move forward. The first is that the public generally does not know much about nanotech, but, when it does learn more about it, it is actually pretty optimistic. But, at the same time, when there is a void and the public does not have information, people tend to fill that void with other experiences (e.g., dioxin, nuclear power, and asbestos): think superfund and hazardous waste problems. And, if we leave that void there for too long, you can be sure that it will be filled with negative experiences. So, on the positive side, the public is reasonably positive toward nanotech when it does not know much about it. But, it also wants to make sure that there is a strong governmental role and a strong regulatory framework that ensure protections are in place. And the public just generally doesn't trust voluntary approaches to solve those kinds of problems. So the lesson, I guess, for the short run, is that there is time to shape public opinion, but, at the same time, we have to provide the public with the information that it really needs.

Many businesses have learned the lessons of not doing that, and I think that's one of the major reasons why there's so much consensus among the business community, the environmental community, and many regulators and government officials about the need to move and as quickly as possible. So, the common ground that we need for moving forward, in some sensible way, is really in place. The problem is that we actually do not have the information we need to make those decisions. Some of the things we are missing include a standard nomenclature; basic tools for measurement, toxicity screening, and risk characterization; and tested personal protective equipment for workers. And, even if we had all those things, we would have to think about them in a lifecycle framework. We cannot just think about workers; we have to think about products and their uses, and what happens to them at the end of their lives. We're seeing more and more examples of things that may give us some concern. When the Wilson Center did an inventory of the nanoproducts that are out there, it discovered more products than people thought are actually in the marketplace. The recent case in Germany regarding the "nano" product recall raises a whole bunch of questions that, actually, we cannot answer yet and that we cannot even get the basic facts about.

What has become clear from the Science Committee's standpoint is that we need a comprehensive research strategy. And we believe that was called for in the Act, and the federal government is in the process of putting together its version of what it thinks that ought to be. The Committee is going to hold a hearing when that report comes out. It's a little delayed from what we were hoping, and we're hoping to see something soon. That's going to give us something to look at and decide: if we are on the right track; if we are doing this at the right scale; and if we need to do more.

The Environmental Protection Agency (EPA) recently published a white paper that started to look at its regulatory framework. This is very important, not just for the EPA, but for the other agencies involved, the Federal Trade Commission and a variety of agencies.

As we think about those questions, I will just leave you with two last thoughts about some of the public policy and the bigger picture policy questions that must be addressed. One is, of course, as we deal with this industry, we are dealing with a lot of small companies. These are the start-ups that are developing these materials. They are not the large chemical companies that are used to regulation and that are used to having systems in place to deal with problems. Thus, we are going to have to make some special arrangements for and give some special attention to this sector. And there are a number of products that we are actually talking about using whose purpose is to be dispersed into the environment. We must grapple with what we are going to do and our assumptions about the appropriateness of that before we actually have the answers to the scientific questions that we must answer.

So, in closing, I would like to reiterate that we are making progress on these issues. In fact, in this arena, my experience suggests that we are making progress faster than we have ever made in any other area. That is a positive sign. And the community of interest shares common ground about the need to move forward. So the questions are: What do we do with that goodwill? How do we make the most of it? How do we direct our science and research effort to answer those questions as quickly as we can?

U.S. CONGRESSMAN DAVID WELDON

I practiced medicine for many years. Indeed, I still see patients about once a month in my congressional district. My undergraduate studies were in biochemistry, and I did some basic science research. I have always been very interested in issues of science, and I have actually been quite interested in the emerging field of nanotechnology, really from its infancy. I began reading about it more than 10 years ago in some of the science publications that I study. And I was certainly delighted when Congressman Honda helped move forward the legislation that got funding going for the technology. My particular areas of interest, as you can imagine, are nano's medical applications, as well its aerospace applications, as I represent the area of Florida that includes Cape Canaveral and Kennedy Space Center. I think there is a tremendous amount of potential for applications in our space program, and, obviously, we have all been talking for years about applications in medical technologies.

I think the Congress, when it originally funded this program, envisioned—at least based on my discussions with colleagues—a robust discussion of the ethical issues, as well beyond the toxicology and the environmental issues. And, specifically, I believe that a percentage of the funding should be devoted to the ethical, legal, and social implications. Now, what I mean by that is that I believe there should be a discussion of some of the fundamental issues associated with human dignity and the development of ethical guidelines that can set practical boundaries as we apply nanotechnology in the United States. In addition, there should be the development of a process in which these ethical guidelines can shape our funding decisions as the Congress moves forward in the years ahead to continue the funding.

My interest really got piqued tremendously when I read a federal publication talking about a whole host of potential human enhancements—impacts that could enhance memory, muscle strength, and coordination. When you start moving into a discussion beyond helping the blind to see or helping the crippled to walk, you are talking about the potential capabilities of applying these technologies to create human enhancements. I believe an ethical discussion needs to be conducted *now*—not when there is a private company, the licensing has already moved forward, and the funding has occurred 10 years prior.

When we had Dr. Marburger in front of the Science and Commerce Subcommittee that I am on, he shared a very interesting little vignette about the iPOD. The basic science funding that went into making that device possible is fascinating. I have one, but I did not know its background. It was funding not only from the National Science Foundation; you could trace back funding to the National Institutes of Health, the Department of Defense, and the DOE, which allowed the development of the technologies that went into creating this device. I say all of this just to make the point that we do not know where all this is going to go: We really do not. There could be some really wonderful breakthroughs that not only help people with problems, our defense department, our national security, and the war on terror, but they could, at least as has been the case with the iPOD, create whole new industries that employ thousands of people. So it is really an exciting field.

And it is great to be part of this at the ground level. I think some great things are going to come out of it. But I want to begin the ethical discussion, particularly of human-enhancing technologies, today. Let's have a vigorous debate or discussion on how these technologies could or should be applied. Here are some of the questions that I think we should be asking, and I would like to see them addressed in the near term:

- Should a distinction be made between treatment and enhancement? If so, what limits, if any, should be placed on research on human enhancement?
- In light of the President's concern that we not go down the route toward a Brave New World, and repeated statements by policymakers about the importance of safeguarding the human condition, to what extent could nanotechnology impact human dignity and integrity? And how can we best ensure that the development of nanotechnology in the United States supports existing bipartisan commitment to human rights and human dignity?
- What unique privacy concerns arise with the advent of nanotechnology? And what kinds of protections are necessary individually and societally to ensure that nanotechnology proceeds in society's best interests?
- How are ethical issues related to nanotechnology being addressed globally?

- And, finally, what are the policy implications of the emerging ethical issues related to nanotechnology? In other words, how does this bounce back to us? Do we need laws? Do we need regulations? Do we need congressional action?

Don't ask me to answer all of these questions. That's your job, and I am looking forward to hearing your thoughts.

U.S. CONGRESSMAN BRAD SHERMAN

I am Brad Sherman from California's best named city, Sherman Oaks. The year 2006 is my tenth year in Congress, my fourth year on the Science Committee, and my eighth year of worrying about something I call engineered intelligence, for which I will get to a definition in just a second.

First, a few observations. Nanotechnology is the hip new term for, really, all cutting-edge science. We owe a great debt of gratitude to the GMO-phobics who have illustrated for the scientific world why it is so important to discuss societal issues in a broadly based way. Science is not just for the scientists and the venture capitalists. Nanotechnology raises consumer safety issues; it raises environmental safety issues.

Others are discussing those issues. Congressman Weldon and I, I think, are focused on issues outside of that realm. My focus is on engineered intelligence, by which I mean either computer engineers developing a level of self-awareness and intelligence that surpasses human intelligence, or biological engineers creating new types of human beings or new types of mammals with superhuman or beyond-human intelligence. So whether it is the computer engineers using what could be called "dry nanotechnology" to give us HAL, or the bioengineers using "wet nanotechnology" to give us a 2000-lb mammal with two 50-lb brains capable of beating your kids on the LSAT, nanotechnology raises the question of whether humans, as we know them, will be the most intelligent species on this planet by the end of the twenty-first century.

I used to say that the last decision humans would make is whether our successors are carbon based or silicon based: Whether we will invent a superspecies through a computer or through biological engineering. Since then, I have learned a little bit more about science, and I have a couple of corrections. First, the future of computers is probably not a silicon substrate. And we are probably, before we face completely nonhuman and superhuman levels of intelligence, going to face the enhanced human: The chip in the brain that Dave alluded to. I call this "damp nanotechnology": A combination of DNA on the one hand and computer engineering on the other.

Mike Honda really played the key role in getting the 21st Century Nanotechnology Research and Development Act passed. My focus was on making sure that, whether it is the creation of self-aware machines, the enhancement of human intelligence, or other deoxyribonucleic acid (DNA)-based forms of intelligent life, that these be included in what is studied when we study the societal implications of nanotechnology. Since then, I have been working with the Defense Advanced Research Projects Agency (DARPA) on this, and a report should be issued soon. The NSF has decided to fund the Center for Nanotechnology in Society there at Arizona State University. And I have got to commend Congressman David Weldon for getting money in the appropriations bill specifying that a percentage of funds, in this case 3%, ought to go into looking at societal implications.

Now, let's put this into context in terms of the history of science. The twentieth century was a century of enormous scientific development. It allowed us to quintuple the human population in one century: Pretty good for a large land animal, even Darwin would have to admit. The most important of those scientific developments, or certainly the most explosive, was the development of nuclear weapons and nuclear technology. And I believe it is just about the only analogy we have to what will be the import of developing engineered intelligence. Einstein wrote a letter in 1939 to Roosevelt, and it is the first evidence I have seen of a top decision maker focusing on the implications of nuclear weapons. Six years later—with almost no societal thought, some thought among the scientists, and no involvement of society as a whole, or theologians, or philosophers—we had Hiroshima. Now, when nuclear weapons came to the fore, we saw them in the big form. We can imagine what the history on nonproliferation could have been if the first nuclear weapon had been half a kiloton, like the briefcase bombs that they talk about. But instead, humankind was confronted with nuclear weapons in their obvious import as we began to catch up—as the diplomats, the theologians, the philosophers, and society at large—tried to wrestle with the issue of how do we deal with nuclear technology.

When it comes to engineered intelligence, there are some substantial differences. One is that I expect that this technology will creep rather than explode. That is to say, I think we are going to see the chip in the brain before we see HAL or an existential elephant—meaning a super-large mammal with a super-large DNA brain. That prospect will make it harder to get society to concentrate on these potential issues. On the other hand, we have got a lot more lead time. It is not just 6 years from when decision makers and society as a whole become aware of the issue, and when the technology presents itself in all of its glory.

Today, the good news is We have got about 150 people here at this Center on Nanotechnology and Society Conference. Likewise, the bad news is We have about 150 people here.

However, I do not think that it was until the late 1940s that you could get a conference like this one to discuss the implications of nuclear weapons. That was well after Hiroshima. We had a panel before the Science Committee saying we are about 25 years away from engineered intelligence through computer engineering. But, it is probably a lot less until we face the chip-in-the-brain issues.

Now, like my colleagues, I do not have any answers. Rather, I hope to identify some of the questions. I know that the right time to start thinking about these questions is now. Do we want to create self-aware machines? And if so, what societal rights will those self-aware machines have? What is the definition of a human? As David Weldon points out, What level of chip enhancement do we find acceptable? Will computers that are superintelligent be self-aware?

I have asked DARPA to take a look at what steps we would take to engineer the maximum possible intelligence, while preventing (or seeking to prevent) self-awareness. If we have a self-aware computer, will it be ambitious? We are used to DNA-driven devices, life that is inherently ambitious, whether it is human or the smallest, least sophisticated creature. These are creatures that wish to survive and seem to wish to propagate—and may even wish to control. In contrast, my washing machine does not care if I turn it off, and I am not sure that a computer capable of an existential crisis will care whether that crisis is interrupted by an off switch. In contrast, the DNA work starts with raw material that is inherently ambitious, and I do not know if we understand it enough to deprogram such ambition, even if we wanted and decided to do so.

Then, we raise the issue of whether we start with human DNA, which raises all the stem cell issues and all those politics, or whether we start with chimp DNA. All I know is that the last time a new level of intelligence was on this planet, it was when our ancestors said Hello to Neanderthal. It did not work out well for Neanderthal.

We have got to address these questions, and it is going to take a lot longer than 6 years to address them. So, thank God we've got longer than 6 years. It will take better minds than mine to figure them out. But my hope is that these questions can be worked out with merely human intelligence.

U.S. CONGRESSMAN MIKE HONDA

Good afternoon. This is a fun place to be because we get to listen to all these questions that both Congressmen David Weldon and Brad Sherman bring up.

I think these are the kind of questions we thought of and we struggled over as we developed the bill that was the 21st Century Nanotechnology Research and Development Act. That bill would not have been successful without the guidance of Chairman Sherwood Boehlert. When we got together to do the bill, I asked him: "Do you think this bill will get through?" He said, "I am the chairman." So, here we have the formula for success for bills in the areas that we care about. We have a bipartisan approach, we have someone on appropriations who understands this stuff, and so things can work together.

I think that what we can agree upon today is one of the pieces that was important to me in forming that bill: the issue of education in ethics. That is really the issue of the public as we move along in this particular science. And as Brad has very well pointed out, we take lessons from the past, and we must be aware of what can happen when we do not listen and do not discuss things in an open, public way. Sometimes science can get ahead of us, in terms of our sense of propriety. This whole issue of nano goes to what I call the "Mork and Mindy" approach; you recall that on that show Robin Williams' character, Mork, often said: "Nano, nano." That was the first time you probably heard the term nano in a public realm. Since that time, we have moved on to higher expectations from what we all now understand nano stands for. The media has a great role, in terms of public education and all these things we care about. So, as policy makers and Americans, we should be having more public discussions, such as this one, and we should also be bringing along members of the media, who can accurately share the information that we are talking about and engage the rest of the public in this discussion.

Everybody who comes to my office to talk about nanotechnology talks about what their organization does and their efforts on the science side of the technology, and I always ask them what they and their organization is doing in the area of ethics. I spoke at speak a conference in 2004, where I addressed the responsibility that scientists and policy makers have to engage with the public in a discussion about this technology early on, before problems are upon us. At that time, I made a point about how the debate about stem cell research might have proceeded differently had such a conversation taken place and that I want to prevent the same thing happening in nanotechnology.

Today, I want to make a point about a difference I see between stem cell research and nanotechnology: one which I think bodes well for our chances of making more progress on reaching more of a consensus about nanotechnology. In comparing the ethics of stem cell research to those of nanotech, I think we should look at two separate sets of ethical questions that arise from stem cell research. One set of issues is the fairly intractable one that arises from the use of embryonic stem cells. Those of us in this room, like people across the nation, have fundamentally different views on when life begins. What this means is that some of us feel that it is acceptable to use stem cells from embryos, which destroys them in the process, while others of us believe that destroying embryos is destroying life, which is wrong. These are two very different, polar positions, and resolving them is quite challenging. The standoff has led to different policies at the state and federal levels, depending on which mindset has the majority in that jurisdiction. I am not going to try to solve this problem here. Congressmen Weldon and Sherman have quite aptly already stated that we are not here to solve these problems. But, rather, we know how tough the issues can be. As a result, I would rather look at the other set of ethical questions that comes with stem cell research. And, in this case, we can benefit from thinking about the kind of stem cell research that is more widely accepted, research that uses adult stem cells in which no embryos are destroyed, thus eliminating that issue away. I think those of us in this room, regardless of our position on embryonic stem cell research, can agree about the potential benefits of adult stem cell research, which might bring about new treatments and cures for disease. But there is still a set of ethical questions that goes along with even the use of adult stem cells. A big question in particular being How far should we go?

When I was listening to the discussion here today, I started to think about the movie, *The Island.* In that movie, there was a colony of people growing up and being well cared for, and, upon reflection, herded. It turns out that these people were being grown for purpose of providing replacement of organs for those people who live in the world beyond the Island who need such organs and have matching DNA. Those on the Island thought that when they left the Island that they were going to paradise and that they were being selected to go via a lottery system. But, actually, it was all by design. This movie represents the fact that we have the ability to bring up and explore a set of issues and questions through media, such as film.

For example, if I needed a new kidney and we could use adult stem cells to grow a new one, I think that we would all agree that this would be a good idea. But in the movie, the only way to get new kidneys was to grow a new me and then cut out those kidneys. I think we can all agree that that is not such a good idea, especially for the other me, who is a sentient being. And, so, we look at cloning human beings as probably a bad idea because it raises all sorts of ethical questions, especially with regard to harvesting organs.

There is a place to draw the line on things like that, and, similarly, I believe that we can find the lines to draw in the area of nanotechnology. I also think that it is essential that the conversation not just take place among scientists or policy makers, or even between those two groups. The general public has to live with this stuff. They are the ones who need to find it acceptable. So, we need to engage with the public as we try to figure out where and how the lines are to be drawn.

There are going to be extremes on both sides. Some will argue for no limits, maybe envisioning a world in which our minds and spirits might be separated from our physical bodies and able to live forever in a machine. Others will say that we should allow no biological applications of nanotech, and rather that nature should be allowed to run its course. Where we will end up is somewhere in the middle. We will find manufactured vaccines or treatments for diseases acceptable using human-grown biologics. Is an artificial nanoantibiotic that can kill bacteria really so far from that idea? I think that what we will find is that this will depend on how the nano antibiotic works. Some of the examples we might use include whether we can replace the eyeglasses that we use now to correct vision for those who have imperfect sight. The question is Is an artificial nano retina that improves eyesight really so far from that concept? Again, it will depend on whether it can provide performance that sees like a regular human eye or whether it lets you see new wavelengths and greater distances, in which case we are talking about something very different than just correcting a defect to a normal level. Again, if you are using the technology to restore hearing to the normal level of function that may well be accepted, whereas it is a whole new question if you are able to increase the frequencies one can hear, thereby enabling someone to eavesdrop on very quiet conversations, which we may not want. It might be fun as a school teacher to understand what is going on in the rest of the classroom. But it is still an ethical question. So we are going to have to look very carefully at the applications where nanotechnology is combined with biological systems and decide if we are willing to allow applications to improve human performance or not.

Such potential capabilities beg the question whether those applications would change what it is to be human? I refer back to two other movies, I, Robot, and A.I. The movie A.I. is about a young person who found out that he was a robot, and so he went on a quest to become more human. So the questions really are:

- Do we really want to allow these changes to take place?
- Is this something that people should have individual choices over?
- Is it really possible to have individual choice anyway?

- If one person does something like implanting a memory chip or math processor that would be implanted in the human brain using nanotechnology, does that mean that everyone else has to keep up?
- Who can afford it?
- Will there be a division among the people who can afford it and those who cannot?

These are weighty questions, and I do not pretend to have the answers. They are big questions and ones that we have to begin to pursue. So, what I believe we need to do is to educate the public as we move along and go point by point so that fear and anxiety does not overtake sound policy making. But educated input and good, robust debate must occur in this arena; that is what this democracy is all about. I believe that as thinking, feeling, and compassionate human beings, we will come to answers to these pressing questions by having these conversations, and, thus, I think that we will be able to guide technology. And, in the words of *Star Trek*, we will be able to go where no other man has gone before.

CHAPTER 2

Nanotechnology and the Two Faces of Risk from a Reinsurance Perspective

ANNABELLE HETT

INTRODUCTION

Until recently, the phrase "technology cuts both ways" was rarely associated with nanotechnology. However, its commercial utilization has triggered a heated debate in specialist circles, particularly because there is no universal assessment of nanotechnology's risks or of its hazards and opportunities. To reduce uncertainties and ensure a sustainable introduction of nanotechnology, efforts must be made to establish a common discussion platform that facilitates an open dialogue on risk analysis, risk management, and acceptable options for risk transfer.

The core business of the insurance industry is the transfer of risk. Thus, the insurance business identifies, analyzes, evaluates, and diversifies risk in order to minimize the total capital cost of carrying it. Traditional means of underwriting and diversification, however, reach their limits when it is no longer possible to assess the probability and severity of risks—especially if many companies, industry sectors, and geographical regions could be affected simultaneously.

By way of introduction, insurability depends on the following principles:

- Accessibility (probability and severity of losses must be quantifiable to allow underwriting).
- Randomness (time of the insured event must be unpredictable and occurrence independent of the will of the insured).
- Mutuality (exposed parties must join together to build a community in which the risk is shared and diversified).

Nanoscale: Issues and Perspectives for the Nano Century. Edited by Nigel M. de S. Cameron and M. Ellen Mitchell
Copyright © 2007 John Wiley & Sons, Inc.

- Economic feasibility (private insurers must be able to charge a premium that is commensurate with the risk, giving them a fair chance to write the business profitably over the long run).

Nanotechnology challenges the insurance industry because of the high level of uncertainty in terms of potential nanotoxicity or nanopollution, the ubiquitous presence of nanoproducts in the near future (across industry sectors, companies, and countries), and the possibility of long-latent, unforeseen claims. The insurance industry is concerned by the ubiquitous presence of nanotechnology because scientific evaluations of potential risks for human health and the environment are few and remain inconclusive. Scientists have been unable to draw upon toxicological studies or long-term experience. Sponsors tend to show greater interest in scientific progress or promising patents, and the studies needed for purposes of risk assessment often fail to materialize because of lack of research funding.

Moreover, an agreed upon framework is needed within which both publicly financed projects and the insurance industry's own risk analysis can be conducted. At present, such a framework exists only in partial form. Matters are complicated by the fact that there is no common terminology for the great variety of nanotechnological substances, products, or applications. Any structured scientific approach toward evaluating potential risks would require a standardization of these materials and applications. Further, a common language would allow comparison of scientific knowledge across industries and countries, and would also allow for labeling requirements.

In light of these shortcomings, stakeholders agree that the only way to prevent a polarized debate about nanotechnology, which may slow down future research and economic growth in this promising field, is to work toward a common approach to reducing the uncertainty and offer answers for pressing questions, particularly those concerning potential nanotoxicity and nanopollution. It is essential to intensify a risk dialogue among regulators, representatives from business and science, and the insurance industry.

Furthermore, risk communication efforts should include the broader public. In contrast to the debates on nuclear power and genetic engineering, the public is only beginning to view nanotechnology as a noteworthy hazard. The increase in media interest since the beginning of 2003 could change that, however, and lead to more lively debate on the two sides of the coin: the risks and opportunities that nanotechnology may represent. Whether the public accepts the new technology and sees in it advantages, or rejects it, will largely depend on how well informed it is and to what degree it is able to make objective judgments.

THE DIFFERENT APPROACHES TO RISK

The role of insurance is to "put a price tag on risk" by promoting awareness for new risks and giving incentives for precautionary measures that may help to mitigate large losses. For insurers, risk does not imply prevailing negative connotations.

Provided that they are carefully examined and thoroughly assessable, risks spell opportunity from the insurer's perspective, as risk is its business.

Widely different approaches range from the pole (left) where risk is perceived as a threat and a danger (an attitude that may be typified as "survival-driven" and focused on *loss* potential) to a pole (right) where risk is perceived as the realization of opportunity, which may be typified as "incentive-driven" and which focuses on *profit* potential. Whereas the former approach sees more risk than opportunity, the latter points to the opportunity in taking a risk. Whereas the former is a precautionary attitude that perceives risk as a possible consequence of an adverse situation to be avoided or reduced, the latter is an entrepreneurial attitude that looks at risk as a possible deviation from the expected, but which focuses on the positive prospects of chance and innovation.

Risk acceptance results from the risk assessment process, and the continuous balancing of the aforementioned aspects. The more complex an issue, the more difficult it is to achieve a balance. This equals an ambiguous risk acceptance that is not fixed to a definite "value" on the "risk–opportunity" axis.

A NEW KIND OF RISK

Technological leaps have been known in the past: riveting was succeeded by welding; natural fibers by artificial fibers; the piston by the turbine; the horse-drawn carriage by the automobile; and the electric conductor by optical cable. The flintstone that prehistoric humans used gave way to an improved implement, which was adapted to changing needs, improving the cutting tools' performance continuously over time.

Technological changes were traditionally accepted by insurers (e.g., the application of surgical laser technology), as little in the way of unforeseeable problems was feared in those contexts. As far as risk was concerned, those were *evolutionary* developments, with which insurance companies are generally prepared to cope— even if they do so reactively. Yet, historically, the technological risk landscape never saw such categorically drastic change; that there had been uncertainties defying the estimation or assessment of loss potentials to the extent of seriously threatening the risk-bearers.

The situation is different in relation to developments that, in terms of risk, are *revolutionary*, and whose potential for damage cannot be assessed. Nanotechnology has no long history of adaptation and gradual refinement; critical applications are applied without time-tested performance. Such revolutionary developments come in two different forms: first, potential risks related to events attributable to a cause, or the so-called "real risks"; and, second, those whose causality merely cannot be excluded, (i.e., the so-called "phantom risks" or phenomena *perceived* by the population as a threat, although no scientifically demonstrable causal connection can be established).

What makes nanotechnology completely *new* from the point of view of insuring against risk is the unforeseeable nature of the risks it entails and the recurrent and cumulative losses it could lead to, given the new properties—hence, different behavior—of nanotechnologically manufactured products.

NO FUTURE WITHOUT RISK

New technologies have always harbored new risks. Yet two things have changed. First, the dangers per se are becoming more difficult to understand. Technical systems are becoming increasingly complex, and their components are constantly being reduced in size. Whether, for example, the weak electromagnetic fields of mobile phones, genetically modified foodstuffs, or nanoparticles pose any real danger is still highly uncertain. Second, not only is innovation achieved and produced at ever-greater speed, but today's technology and business networks also disseminate that innovation faster and over wider areas. In other words, rather than prompting gradual and local damage, hidden risks may trigger widespread loss accumulations.

It would therefore be careless to insure the risks associated with new technologies before more is known about them. Insurance is a promise to compensate the insured for losses incurred in the future. If their dimensions are unknown, adequate risk financing is virtually impossible, and the insurer can, at best, limit the coverage or, in the extreme case, refuse to offer coverage altogether. This is unsatisfactory for all concerned, however. For the insurance industry, because it sees its task and business opportunity in contributing to the management of risks, including those that are hard to assess in individual cases; and for the policyholders, because they are left with any residual risks.

The lack of available insurance cover also has negative effects for society as a whole. To benefit from the opportunities offered by progress, the risks accompanying every technical or economic change must be acceptable. A necessary (albeit not the only) prerequisite for this is the assurance of financial coverage for possible claims. Risk management is primarily interested, therefore, in knowing what preconditions must be established to manage tomorrow's losses, which arise out of today's risks.

RISK IS KNOWLEDGE OF POSSIBLE LOSSES

The chief prerequisite for successful risk management is readiness to address questions, even if some are highly unsettling. What would happen, for example, if the Gulf Stream were to lose strength or even suddenly change its course? What would it mean if nanoparticles actually penetrated the human brain directly via the olfactory nerve? Who bears what responsibility if machines start making more and more decisions? What risks are created by the broad rejection of genetically modified foods? What social conflicts loom if, as a result of rising unemployment and increasing life expectancy, fewer and fewer people with earning power have to cover the costs of more and more pensioners?

It would be wrong not to examine such scenarios on the assumption that experience suggests that they are "improbable." In fact, predictions about the likelihood of multicausal losses actually depend either on sound understanding of cause-and-effect relationships or on a detailed loss history, and the risks of the future have neither of the two.

The immediate purpose of discussing such scenarios therefore is to differentiate between the possible and the impossible. Only risks that are identified can be systematically analyzed. Only then, on the basis of sound knowledge, can the extent and the likelihood of such risks be determined. In reality, however, the public debate about risks of the future is often dominated by equally irresponsible scaremongering and trivializing reassurance, both of which hamper any attempt at effective risk management.

INSURANCE IS NO SUBSTITUTE FOR SAFETY

If new risks cannot be thoroughly understood, it is all the more important to fully utilize all reasonable possibilities of risk prevention and/or reduction. While in specific, individual, cases the notion of what is reasonable may well be negotiable, the principle itself is not. Insurers are obliged on ethical grounds, if for no other reason, to limit coverage to those losses that cannot be prevented by any justifiable means.

A great deal of potential remains untapped in this respect. Risk reduction, for example, is still too often limited to efforts at reducing the probability of occurrence, even though many future risks indicate a strong trend toward ever-greater potential consequences. The failure of an information or power network can already today paralyze thousands of companies in a matter of minutes. By systematically limiting the consequences, many critical risks could be reduced at little cost.

This also applies to liability risks, which are becoming increasingly difficult to calculate due to the rapid changes in societal values. The state is called upon here to launch the public debate on risk as early as possible and to bring about consensual decisions before new technologies are introduced on a large scale. How the risks of new technologies are assessed in detail is secondary from the viewpoint of liability insurance; what matters more is a set of viable ground rules.

WARN EARLIER, REACT FASTER

The earlier changes in the risk landscape are recognized, the more time remains to analyze and react to them. This is where insurers can function as an early warning system because, by their very nature, they have more loss data than any other institution, and hence are the first ones to be able to detect deviations from existing empirical values. Such early warning systems are still used far too rarely, however. Indeed, many far-reaching loss developments of the past (e.g., the asbestos losses) could have been contained in their initial stages, if not avoided altogether.

In light of that experience, Swiss Re has built up expert teams to cover all aspects of risk, from questions of risk perception through the social and economic sciences to the development of future technologies. Adopting an interdisciplinary approach, they observe the changing contours of the risk landscape by holding, for example, workshops with experts from all conceivable fields in an effort to uncover new risks or changes in existing ones.

Given the vast differences in perspectives and goals among insurers, policyholders, and governmental organizations, tracking down changes in the risk landscape and developing the appropriate solutions requires close cooperation among all of the players. Swiss Re brings to this dialogue its technical competence in the field of risk financing and risk management, and a high degree of sensitivity regarding both the special needs of the industry and the concerns of society as a whole. In return, government agencies and business partners should involve Swiss Re in new developments from an early stage.

In the case of industry, this means providing the insurer with all risk-related information so that it can be jointly analyzed, evaluated, and used toward constructive solutions. Discussions with government representatives and political leaders should help fund urgently needed public research, and establish what legal and fiscal conditions should be created or maintained to handle future risks optimally. Absolute security is an illusion. The future is risky by its very nature, because any change entails new risks. Even so, as our communities do not evolve by chance, the risk landscape associated with them can be shaped accordingly. To do so, however, we must grasp the opportunity to identify and influence risks early on, and make adequate provision for the event that a loss occurs.

PROPHECIES OF DOOM ARE OF LITTLE USE

For the purposes of risk management, prophecies of doom are of little practical use; they describe developments that, although conceivable in detail with a great deal of imagination, we do not yet know whether they will become reality. Further, the future entails more than just new technologies and knowledge; the assessment of risks will also change. To date, for example, robots have been thought to be machines assembled from cog wheels, chips, and lengths of cable. One day, they may actually grow of their own accord out of programmed germ cells. The exact dangers of "electrosmog" will only be clarified with research methods whose details are still beyond our imagination. As for genetic engineering, it remains to be seen what stance future generations will take. It is conceivable that genetic authenticity will be seen as a flaw rather than something natural that deserves protection.

As we are only able to *imagine* the distant future, but not to predict it with any certainty, risk management is confined to the more immediate future, and hence to those risks that we are already able to influence selectively today. To do so, we need not assess what the distant future will hold. Rather, we must identify the drivers and mechanisms of change, because they determine the changes of the coming months and years.

All future technologies, for example, have one motive in common, namely, to technically reproduce and "optimize" natural processes. The aim of nanotechnology is to manipulate atomic and subatomic particles by means of special tools and arrange them at will. Artificial intelligence is intended to empower machines with thought capability. Genetic engineering is aimed at selectively changing the genetic code.

In fact, experience shows that the initial phases of innovation cycles are particularly risky because we, as part of the innovation process, must first learn how to cope with the new hazards. As a case in point, one of the most costly errors in the history of technology dates from the beginnings of computer science when, in order to save what was then extremely expensive memory space, year dates were entered only with their last two digits. The conversion to Y2K-compatible systems at the end of the last century cost billions of U.S. dollars.

RISKS ARE A MATTER OF DEFINITION

Risk implies possible loss. As we can only "lose" what is of value to us, risks are a direct reflection of religious, social, political, and economic perceptions of value, and, as such, the risk landscapes of various cultures differ considerably. Whereas most industrialized countries are investing ever-greater sums in environmental protection, Nature in many of the so-called developing countries is still regarded as threatening and not at all deserving of protection. While not even basic medical care is ensured in poor countries, services, such as psychotherapy for household pets, are being offered in leading industrialized countries.

Value perceptions also determine the distribution of risk or loss burdens within society. In continental Europe, liability claims are largely limited to compensation for medical expenses and loss of income. Physical and mental pain is assigned only a low monetary value. By contrast, the Anglo-Saxon legal system provides for punitive damages, which may grant satisfaction to the injured party far in excess of any measurable losses.

Both systems have specific advantages and drawbacks, which will not be addressed in further detail here: different countries have different customs and different risks. What is much more important from the perspective of practical risk management is to recognize that the increasing interdependence among different social systems is, itself, triggering new risks. Still uncertain is how the various national legal systems will be aligned as globalization progresses.

MANY CAUSES, MANY PERPETRATORS, NO LIABILITY?

Even within a given culture, perceptions of value can change rapidly (e.g., as a result of scientific progress). The natural sciences traditionally defined a cause as something that invariably produced a given effect. What could not be explained as a cause in this sense was considered to be accidental (e.g., most diseases). Yet modern science sees effects as the result of the complex interplay of many individual factors and circumstances, none of which is the sole cause in the classical sense, but just makes a greater or lesser contribution to the overall effect.

However, our scientific knowledge is not yet sufficient for us to assign the parts of a given overall effect to the individual contributory causes. The multicausal and multiconditional weave of cause and effect is only partially clear. For example, some epidemiological studies have found a higher incidence of leukaemia among children living in the vicinity of high-tension power lines. As most of the exposed children do not contract the disease, however, it remains unclear whether electromagnetic fields are involved at all and, if so, in what way. Further, as cases of leukemia had been recorded prior to the introduction of electronics, electrosmog can, if at all, only be one of many influencing factors.

WHAT TO BELIEVE—OR WHOM TO BELIEVE

Only science can prove how dangerous genetically engineered maize, mobile phones, nanoparticles, and pork sausages really are. But the "common sense" of democratic societies determines what is considered to be acceptable. Risk acceptance is not only a question of objective measurement, but of individual and collective perceptions of value. No civil technology kills and maims more people or emits more pollutants and noise to the environment than road traffic. Yet, automobility continues to be widely accepted.

As a side effect of this development, reputational risks are gaining importance. As consumers are hardly able to assess technical data indicating the efficiency and reliability of products on their own, confidence in brands becomes the decisive criterion in deciding what to buy, and the loss of a sound reputation looms large as one of the greatest corporate risks. It takes far less time to shake confidence in persons, companies, or institutions than to acquire a stock of factual knowledge. This phenomenon also explains the increasingly frequent, sudden shifts in the public perception of risk, which are mostly triggered by external events. On September 21, 2001, a chemical factory in Toulouse, France, was the scene of one of the most devastating explosions in the history of the chemical industry. The detonation tore a crater 10-m deep and 50 m wide into the factory premises. Window panes within a radius of 5 km were shattered. Some 30 people were killed, and more than 2400 were injured.

Had it not been for the events in New York and Washington only a few days before, this chemical disaster would certainly have generated a wide-ranging discussion on the risks related to the chemical industry, just as similar events in Bhopal, Seveso, and at Sandoz in Basle had done earlier. As it was, however, this explosion was not even registered by many international media. The risk topics of climate change and genetic engineering that had dominated the debate up to September 11, 2001, vanished from the headlines for many months.

For simple, linear systems, loss events can be predicted precisely if all cause-and-effect relationships are known and all relevant variables are measurable with sufficient accuracy. That is why we can, for example, calculate how many hours an aircraft propeller can operate before becoming critically warped through the centrifugal forces that cause the metal molecules to migrate gradually to the tips of the propeller blades.

For complex systems, however, accurate predictions are extremely difficult. No one can calculate when, where, and with what consequences the next aviation accident will occur. Yet for example, it is possible to estimate the risk, or average frequency and severity of air accidents to be expected over the next few months. Determining the behavior of complex systems by extrapolation from the past is only reliable if the system under observation has remained unchanged. In the case of nanotechnology, however, the system has been changed dramatically.

In conclusion, it may be true that many future risks cannot be reliably quantified at present. But they could be understood better if the parts of the puzzle we have were pieced together to form an initial picture. That would make clearer what parts are still missing. The sooner the risk assessment begins, the more time will be available for the time-consuming process of learning.

FAINT SIGNALS?

There is no lack of details. What is lacking is the constructive and integrated view that cannot be achieved by a monologue of alternately alarming and reassuring statements and appraisals. What is needed is the interdisciplinary and, in the age of globalization, increasingly intercultural "polylogue."

Swiss Re set up numerous communications platforms in the past few years to promote direct exchange beyond the limits of tradition, interest, and competence. It may not be a novel idea, but it works. At the Swiss Re Centre for Global Dialogue in Rüschlikon, Switzerland, discussions of the future go well beyond purely commercial interests. Swiss Re accepts responsibility as a globally active company, promoting the examination of long-term changes and the risks and opportunities associated with them. For that reason, Swiss Re is interested in exchanging ideas about the future with all stakeholders.

THE CHALLENGE OF RISK ASSESSMENT

Risks arising out of the introduction of new products or innovative technologies need not reveal themselves immediately and may occur after an interval of years. Insurers are aware that nanotechnology, however, is set to spread—to such a wide range of industries and in such a large number of applications and at such speed—that the individual claims conceivable on the basis of experience and resulting from the development, design, product, and application defects can hardly be expected to be long delayed. Things will become critical if systemic defects only emerge over time, or if a systematic change in behavior remains undetected for a long time. In that case, an unforeseeably large loss potential could accumulate, for example, in the field of health impairment, while the experts agree that the greatest potential benefits at present are in the fields of medicine and pharmaceuticals.

An assessment of risk or loss potential will only be possible if, and when, the first health impairments demonstrably manifest themselves. As there are indications that certain nanoparticles might only be recognized as harmful after a considerable time, the insurer must be familiar with the various application areas and the dissemination of potentially suspect nanoproducts, as well as the precautionary and loss-prevention measures being taken.

PUBLIC PERCEPTION OF RISK

Many are still unfamiliar with nanotechnology. What the approach actually is, what special qualities nanoproducts may have, and what the possible associated risks are is often unclear to the layperson. The approach is not just a question of an extremely multifaceted technology; the manufacturing processes and operating mechanisms of nanotechnological products remain largely inscrutable to observers, users, and consumers. This may lead to uncertainty in and skepticism by society at large, especially if the various risk aspects become the subject of public discussion.

However, there is no question that nanotechnology is passing through the stages of published and public opinion to take its place on a political agenda; its governance will be incorporated into legislation is only a matter of time. After the advent of genetic engineering, we know that public protest can brake the development of an emerging technology. So, it is clearly in the interest of the stakeholders of nanotechnology to take the misgivings and needs of society seriously, making allowance for them in subsequent stages of development.

With regard to public perception of risk, it is not a question of how dangerous a new technology is, but how dangerous it is *perceived* to be. The so-called "fright factor" tells us whether an issue has the potential to create panic, and hence is perceived as a threat. This includes the origin of a risk. Is it a new manmade technology, and hence a "homemade problem," or a natural occurrence with which people have been familiar for generations? If the damage anticipated turns out to be irreversible, far more fear will be generated than if people believe that countermeasures can be taken. The risk dialogue is particularly crucial here, because once a certain opinion has become socially established, it is extremely difficult, tedious, and costly to persuade people of the contrary.

FRIGHT FACTORS

Why do some risks trigger so much more alarm, anxiety, or outrage than others, seemingly regardless of scientific estimates of their seriousness? Research over many years in the so-called "psychometric" tradition has sought to find answers to this question. Some rules of thumb have emerged. Risks are generally more worrying and less acceptable if perceived:[1]

- To be involuntary (e.g., exposure to pollution) rather than voluntary.
- As inequitably distributed (some benefit whole others suffer the consequences).

[1]Bennett, P. and Calman, K. (1999). *Risk Communication and Public Health*. Oxford: Oxford University Press, p. 6.

- As inescapable by taking personal precautions.
- To arise from an unfamiliar or novel source.
- To result from manmade rather than natural sources.
- To cause hidden and irreversible damage (e.g., through onset of illness many years after exposure).
- To pose some particular danger to small children or pregnant women or more generally to further generations.
- To threaten a form of death (illness/injury) arousing particular dread.
- To damage identifiable rather than anonymous victims.
- To be poorly understood by science.
- As subject to contradictory statements from responsible sources (or even worse, from the same source).

Fear is also aroused by risks that are forced upon the consumer, in which case, he cannot take an independent decision. This underscores the importance of the product declaration, which enables the consumer either to accept a risk voluntarily or reject it. In the case of nanotechnology risk, contradictory statements made by scientists, on the one hand, and the public authorities, on the other, generate public mistrust. Disagreement among scientists on the question of what the most important factor for the potential toxicity of nanoparticles also threatens a decline in confidence, as the controversy once before (over the risks of BSE in the early 1990s) clearly demonstrated.

What will society decide with regard to the spread of nanotechnology? Will it support nanotechnology because it expects the technology to yield a number of definite advantages? Or will it take a sceptical view of nanotechnology's further development, given that the questions regarding possible risks have not been satisfactorily answered? Further, is the layperson able to distinguish between potentially harmful nanoparticles and the usually harmless nanotechnological applications and products?

BETTER SAFE THAN SORRY

In light of these questions, responsible authorities face the task of ensuring the safe handling of nanotechnological products and applications, due to their rapid dissemination into the marketplace. A sensible pursuit of technological research and development, on the one hand, while offering people and the environment the best possible protection against possible hazards, on the other, must be found. For more than 20 years, this challenge has prevented the introduction of the so-called "precautionary principle" in relation to new technologies. The precautionary principle demands the proactive introduction of protective measures in the face of possible risks, which science at present (in the absence of knowledge) can neither confirm nor deny.

This "better safe than sorry" conviction is an ethical approach prescribing that necessary measures to protect people and the environment should be introduced at an early stage, even if the scientific uncertainties regarding the risks have not yet been finally clarified. Whether and at what stage of development such measures should be adopted is equally difficult to determine. While one does not want to take costly protective measures that are unnecessary, especially if they might have a negative effect on continuing economic development, neither people nor the environment should be burdened with dangers that could have been avoided. The dilemma surrounding the precautionary principle is at present the subject of numerous public discussions—in no small part the one dealing with mobile phones.

In view of the dangers to society that could conceivably arise from nanotechnology, and given the uncertainty currently prevailing in scientific circles, the precautionary principle should be applied whatever the difficulties. The handling of nanotechnologically manufactured substances should be carefully assessed and accompanied by appropriate protective measures. This finding is particularly important for individuals whose jobs expose them to nanoparticles on a regular basis.

TOWARD SUSTAINABILITY

In sum, the optimal commercial use of nanotechnology is crucially dependent on cross-disciplinary dialogue, which should address the full scope of the two sides of the risk: potential hazards and inherent opportunities. As unexpected losses can destroy economic investments, far-sighted thinking is necessary. We would suggest that the main precondition for successful risk assessment in a technology as complex as nanotechnology is finding a *consensus* among the industry representatives, legislators, and research institutes concerned. It is one that must extend across national borders, regulatory inconsistencies, and different perceptions of risks and benefits. With new technologies, where neither the probability nor the extent of potential losses in these areas can be calculated precisely, pooling specialist knowledge and promoting dialogue will foster clarity that no one party can expect to achieve alone.

As an effective and competent risk carrier, insurers continue to identify, analyze and measure risks associated with new technologies in an effort to render them as transparent and quantifiable as possible. With this acknowledged strength deployed, and by fostering responsible behavior in light of emerging technologies, such as nanotechnology, the insurance industry gives foundation to harvesting sustainable benefits, both for its clients and society at large.

BIBLIOGRAPHY

Swiss Reinsurance Company. (2004). *Nanotechnology—Small Matter, Many Unknowns.* Zurich: Hett, A. et al.

Swiss Reinsurance Company. (2004). *The Risk Landscape of the Future.* Zurich: Brauner, C. et al.

CHAPTER 3

Ethics, Policy, and the Nanotechnology Initiative: The Transatlantic Debate on "Converging Technologies"

NIGEL M. DE S. CAMERON

ROOTS OF CONTROVERSY

In December 2001, the first National Science Foundation (NSF) conference on "Converging Technologies" galvanized policy discussion on nanotechnology and related technologies in the United States. The following summer, the NSF together with the Department of Commerce (DOC) published a report, entitled *Converging Technologies for Improving Human Performance Nanotechnology: Biotechnology, Information Technology, and Cognitive Sciences* (NBIC Report and NSF Report),[1] proposing a vision for the advancement of science through the convergence of nanotechnology (i.e., N for nano), biotechnology (i.e., B for bio), information technology (i.e., I for info), and cognitive technologies (i.e., C for cogno). This report was met with concern in many quarters, especially in Europe.

In fact, the European Commission (EC), executive of the now 25-member European Union, convened a High Level Expert Group (HLEG), to assess the U.S. document and its implications. It concluded that they were "striking" and assumed "strongly positivistic and individualistic" underlying values.[2] In contrast, the EU group sought to focus on the concept of moral pluralism, explaining that "one of the hallmarks of the European identity is the way [the EU] accept[s] and

---

[1] Roco, M. and Bainbridge, W. (eds.) (2002). *Converging Technologies for Improving Human Performance: Nanotechnology, Biotechnology, Information Technology, and Cognitive Sciences.* Available at http://www.tec.org/Converging Technologies/1/NBIC report.pdf. (Retrieved October 19, 2006).

[2] European Commission, Special Interest Group II. (2004). *Converging Technologies and the Natural, Social and Cultural World* 4.

---

respect[s] moral pluralism in the Member States of the EU."[3] The EU's CTs agenda should aim to accomplish cohesion between societal values and scientific advancement, and recognizes the "need to evaluate and to articulate our values once again not only at the individual level of human and constitutional rights, but also at the level of the ideals about our European societies."[3] This chapter examines the European CTs policy response as it endeavors to balance transformative potentials with societal concerns. Its basic thrust is that prospective ethical, legal, and social analysis will prevent "after the fact" regulation and allow policy to determine outcomes, rather than allowing outcomes to drive the process.

Until the 2001 NSF CTs conference, Converging Technologies for Improving Human Performance, from which the NBIC Report resulted, the terminology of converging technologies (i.e., CTs) had been used in various generic ways within the scientific community. In one earlier reference, the concept of "technological convergence" was described as "increasingly extend[ing] to growing interdependence between the biological and microelectronics revolutions, both materially and methodologically . . . ."[4] In the clearest parallel usage, which followed from the NSF CTs conference, a Canadian National Research Council foresight report added a fifth element to the NBIC list: ecology.[5] More commonly, the idea of CTs has been used in relation to a variety of specific disciplines and applications, such as CTs for therapy delivery or converging CTs for communication systems.[5] From Norway, for example, we have "converging technologies for salmon-productive aquatic environments—bioinformatics, environmental science, systems theory, salmon genomics, production biology, and economics."[6] The rapid advancement of nanotechnology research and development has generated growing interest in the nano-enabled convergence of various technologies and has led to several uses of CTs terminology.

"Converging Technologies" Terminology as a Reflection of Policy

That first, 2001 NSF CTs conference and the resulting NBIC Report had the effect of defining a distinctive set of approaches to the question of CTs, although other more generic uses continue outside of the United States. A powerful point is made in the cover design of the European HLEG report, which reads "*nano-bio-info-cogno-socio-anthro-philo- . . . geo-eco-urbo-orbo-macro-micro-nano.*" The expert group seeks to expand the idea of converging technologies to include cognitive science

---

[3]European Commission, Special Interest Group II. (2004). *Foresighting the New Technology Wave Expert Group SIG 2 Final Report*, V3.7 (11.7) 1, *Report on the Ethical, Legal, and Societal Aspects of the Converging Technologies* 5. Available at http://ec.europa.eu/research/conferences/2004/ntw/pdf/sig2_en.pdf (retrieved on October 19, 2006).

[4]Castells, M. (2000). *The Rise of the Network Society*. Oxford: Blackwell, p. 72.

[5]European Commission, High Level Expert Group. (2004). *Foresighting the New Technology Wave: Converging Technologies—Shaping the Future of European Societies* 13. Available at http://www.ntnu.no/2020/final_report_en.pdf (retrieved on October 19, 2006).

[6]See footnote 5, p. 14.

environmental science, systems theory, evolutionary anthropology, and social science, including philosophy, economics and the law:

> The European approach to CTs assumes that nano-, bio-, and info-technologies are not the only enabling technologies capable of enabling each other. This assumption affords a fresh look at extant disciplines and their knowledge systems. Enabling technologies and knowledge systems in the engineering, natural, social, and human sciences ....[7]

In broadening the notion of CTs, the EC moved away from the NBIC branding, as set forth in its response.

European Commission Response

On the national level in Europe, the United Kingdom and Germany voiced serious concern about the U.S. approach to CTs. The U.K.'s Royal Society and Royal Academy of Engineering called the NSF report an example of a "proposal[s] for radical human enhancement."[8] The German Parliament's 2003 report on nanotechnology frankly expressed a similar view on the United States' NBIC orientation toward CTs policy:

> The visions of Drexler, Joy and other extreme futurists—and also some of those developed in the environment of the "NNI"—are based extensively on assumptions about the future interactions between a number of new technologies. Such visions of the convergence of different technologies are the drivers of hopes of extensive and far-reaching changes to the conditions of human existence. ... The enthusiasm which optimistic futuristic visions can evoke is being deliberately utilized in the USA as a means of promoting technology development. However, such a "hope and hype" strategy is always precarious. Besides the positive effects of this strategy (e.g. incentives for young scientists, or arousing and sustaining political and business interest), there are conceivable adverse effects. First, there is the danger that expectations of nanotechnology will be set too high, making disappointment inevitable. Second, it may popularize the reverse of the optimistic futurism—a pessimistic futurism involving apocalyptic fears and visions of horror.[9]

The EC itself first called attention to United States' NBIC approach to CTs in the June 2003 issue of the *Foresighting Europe* newsletter,[10] which featured a report on the NSF conferences addressing *Converging Technologies for Improving Human Performance* and the integration of NBIC. The EC considered it necessary to "deal with the questions" raised by the NSF Report. In outlining the context of the European response in a projected report from the HLEG, it is interesting to note the use of the unbranded term "convergent technologies" and the listing of three, rather than

---

[7]See footnote 5, p. 39.
[8]Royal Society and Royal Academy of Engineering. (2004). *Nanoscience and Nanotechnologies: Opportunities and Uncertainties*. London: RS policy document 19/04.
[9]Büro für Technikfolgen-Abschätzung beim Deutschen Bundestag (TAB), Office of Technology Assessment at the German Parliament. (2003). *Summary of TAB Working Report No. 92: Nanotechnology*. Available at http://www.tab.fzk.de/en/projekt/zusammenfassung/ab92.htm (retrieved on October 19, 2006).
[10]Available at ftp://ftp.cordis.lu/pub/foresight/docs/for_newsletter2.pdf (retrieved October 19, 2006).

four, technologies (i.e., nano, bio, and info), omitting the most controversial component in the NSF NBIC agenda, cognitive science, from this initial statement of the question:

> In order to deal with the questions developed in the U.S. NBIC Report, the EC envisaged the establishment of a high-level expert group (i.e., the HLEG) on convergent technologies aimed at improving the understanding of human knowledge and cognition at large, *Foresighting the New Technology Wave: Converging Nano, Bio and Info-technologies and Their Social and Competitive Impact on Europe.*[11]

The EC proceeded to initiate the process, which has resulted in a series of documents outlining a science policy agenda for the coevolution of society and convergent technologies. That process is embodied by the EC-established, 25-member HLEG on *Foresighting the New Technology Wave,*[12] which was created in December 2003. Six key documents have resulted from the process: (1) a statement of mission for the HLEG; (2–5) four Special Interest Group (SIG) working documents in specific areas; and (6) the final report to the EC.[13]

Broadly, the HLEG aims to create a supportive climate in which ethics provides an orientation for the prudent integration of convergent technologies into society. Recognizing that it is "all too easy" to reject new technologies, the final report explains that "[t]o the extent that public concerns are included in the process, researchers and investors can proceed without fear of finding their work over-regulated or rejected."[14]

DEFINING CONVERGING TECHNOLOGIES

Although the NSF Report has popularized the NBIC acronym as *the* model for CTs, note that neither term has been incorporated into the NNI Strategic Plan[15] (December 2004), nor has either term been included in the language of the 21st Century Nanotechnology Research and Development Act, which passed the House (H.R. 766) and the Senate (S. 189) in December 2004.[16]

---

[11]European Commission. (2004). *Communications from the Commission Towards a European Strategy for Nanotechnology.* 11. Available at http://www.aver.e2/data/vav/vav.cu/nano.com.en.pdf (retrieved October 1, 2006).

[12]A list of members was retrieved on October 19, 2006. Available at http://europa.eu.int/comm/research/conferences/2004/ntw/pdf/experts_en.pdf.

[13]See footnotes 2, 3, 6; Special Interest Group I. (2004). *Foresighting the New Technology Wave—Quality of Life.* Available at http://ec.europa.eu/research/conferences/2004/ntw/pdf/sig1_en.pdf (retrieved on October 19, 2006); Special Interest Group III. (2004). *New Technology Wave: Transformational Effect of NBIC Technologies on the Economy, SIG 3 Report on Economic Effect* Version 3.2. Available at http://ec.europa.eu/research/conferences/2004/ntw/pdf/sig3_en.pdf (retrieved on October 19, 2006); European Commission. (2004). *Mission and Objectives of the High Level Expert Group: Foresighting the New Technology Wave* 3. Available at http://ec.europa.eu/research/conferences/2004/ntw/pdf/hleg-tor_en.pdf (retrieved October 19, 2006).

[14]See footnote 5, p. 8.

The HLEG documents identify three distinct meanings for CTs. The first is purely reductionist and speaks simply of the convergence that results from the scale of research, at which conventional scientific disciplines cease to be differentiated and come together in "a single engineering paradigm."[17] Thus, "[t]he field of nano-technology can be said, by itself, to bring about a convergence of domains."[17] The two further definitions, however, are of more interest and are described in one of four SIG documents commissioned by the HLEG. Specifically, the July 2004 SIG II report, which sets out two different notions of CTs—one construing CTs in terms of the "growth of knowledge and new technological perspectives," and the other, in the context of the "ideal of enhancement" and the "enhancement of man and nature," as "the challenge of modern society"[18]:

> The first one, the neutral one, is that nanoscience—by its nature and heuristics—will deeply influence the other disciplines. Interdisciplinarity will emerge just due to the heuristics of nanotechnology and this is exactly the reason why convergence should be stimulated . . . .[18]

This first view looks to scientific facts and the current state of affairs, and the enabling role of the various disciplines in relation to each other. Thus, "[t]he Expert Group defines CTs as 'enabling technologies and knowledge systems that enable each other in the pursuit of a common goal.'"[19] By contrast, the second under-standing of CTs is essentially ideological in nature, open to a culture of enhancement, and indeed a convergence between humankind, nature, and technology itself:

> The second view on convergence is one that does not consider the heuristics solely as an intrinsic and neutral feature of nanosciences. In this view convergence refers to a technological concept of human and nature. The heuristics do not refer solely to the intrinsic good of growth of knowledge. Convergence is explicitly given a (moral) value loaded content. The concept implies that nanosciences and convergence break (should break) through the boundaries of man, nature and technological artifacts.[20]

Conscious of this second conceptualization, the EC opts for the former view of "enabling technologies," stating:

> [T]he term "converging technologies" has taken on a new, specific meaning through nanotechnology and the subsequent formulation of "NBIC convergence." The field of nanotechnology can be said to bring about, by itself, a convergence of domains . . . . However, this unification of domains has not been called convergent and is not the

---

[15]National Science and Technology Council, Committee on Technology, Nanoscale Science, Engineering and Technology Subcommittee. (2004). National Nanotechnology Strategic Plan. Available at http://www.nano.gov/NNI_Strategic_Plan_2004.pdf (retrieved on October 19, 2006).

[16]15 U.S.C. §§ 7510, *et seq.*

[17]See footnote 5, p. 12.

[18]See footnote 3, p. 2.

[19]See footnote 5, p. 12.

[20]See footnote 3, p. 2.

sense in which we are here concerned with CTs. When referring to the potential of nanotechnology one speaks of it instead as a key or *enabling technology*. An enabling technology enables technological development on a broad front. It is not dedicated to a specific goal or limited to a particular set of applications. If nanotechnology is an enabling technology, so are information technology and biotechnology. An important step in the history of CTs was the realization that, aside from nanotechnology, there are other enabling technologies and knowledge systems that are open to new R&D challenges and ready to enable one another.[21]

In other words, "[c]onverging technologies are enabling technologies and knowledge systems that enable each other in the pursuit of a common goal."[22]

The convergence of nanotechnology and other emerging technologies in specific engineering projects expresses an underlying philosophical agenda, namely, the total constructability of humanity and nature.[23] The most direct and profound effect of CTs, therefore, to alter traditional boundaries among the self, nature, and the social environment, where the social environment includes people, groups of people, and informal and formal institutions; it also includes arenas and places, both physical and informational, where goods and beliefs are traded and transformed.

CONVERGING TECHNOLOGIES AND THE SOCIAL ORDER

The HLEG's terms of reference are careful to point out that its work does not constitute a critique of or response to the NSF Report, but it takes the NSF Report as a "starting-point"—rather than a "focal point."[24] Despite this diplomatic nicety, it is impossible to read these documents except as a response to NBIC; if one focus is European policy development, the other is clearly a critique of what was seen as the United States' development of CT policy. Concern is expressed about the role of the cognitive sciences in the NBIC formulation; the other three technologies are specified as "enabling technologies." Indeed, not only "questions" but "*sometimes profound reservations*" need to be voiced.[25] The broad concern of the HLEG in seeking to shape a distinctive EC CT policy is to embed CT policy development within existing European social norms and structures: "It is a priority to clarify the civil and societal benefits of this research to give them a new legitimacy and to put them firmly in the context of positive social dynamics."[25] Insofar as CTs entail potential developments with transformative social implications, the EC encourages "upstream" participation by the public and policymakers in the setting of science policy to ensure an organic connection between emerging technologies and social norms.

[21]See footnote 5, pp. 12–13.
[22]See footnote 5, p. 14.
[23]See footnote 5, p. 32.
[24]European Commission. (2004). *Mission and Objectives of the High Level Expert Group: Foresighting the New Technology Wave* 3. Available at http://ec.europa.eu/research/conferences/2004/ntw/pdf/hleg-tor_en.pdf (retrieved October 19, 2006).
[25]See footnote 24, p. 1.

Sharing the HLEG's concern with transhumanist tendencies, the German Parliament's 2003 review described the NSF Report as "serv[ing] as a vehicle for some highly idiosyncratic ideas, exhibits many biases and overly opinionated views, and suffers from a lack of forthrightness with regard to its proximity to 'transhumanist' and other radically futuristic thinkers."[26] The German Parliament went on to critique the U.S. policy laid out in the NSF Report:

> Overall, the initiative is technology-driven, seems to be heavily influenced by new governmental perspectives on national security after 9/11, and conceals that many of the assumed technical breakthroughs presuppose scientific knowledge and technological capabilities that will very likely not be available in the foreseeable future. Cognitive science is crucial for achieving most of the technological visions but its opportunities and limits are least addressed. Discussions of ethical, legal or social issues related to NBIC are largely avoided. Assessments of hazards and risks as well as the discussion of values and moral boundaries are missing. Among the most serious flaws are the technocratic understanding of society and culture, the dubious evocation of the renaissance, the vision of a perfect future, the carefree siding with the proponents of a neural turn in social sciences and humanities, the alarmingly deep fascination with man–machine-symbiosis, and a certain degree of disregard for diversity and for relevant research findings of other scientists and scholars.[26]

Furthermore, the German Parliament's assessment points out two particularly concerning notions found in the NSF Report:

> In one contribution (Canton 2004),[27] possible misuses of CT by autocratic regimes and the "specter of eugenics" are mentioned, but it is also deterministically stated that human enhancement and designed evolution will inevitably be future tools for shaping societies. In another paper,[28] a rather bizarre and polemical piece, the author predicts that a biology-inspired approach to social sciences "will allow us to engineer culture."[28]

A cognate concern has been voiced by my colleague Vivian Weil, in her essay, "Ethical Issues in Nanotechnology" in the NSF volume on *Societal Implications of Nanotechnology*.[29] She issued a careful warning that: "[w]hile trying to stay alert to unintended consequences, we should also try to avoid taking it for granted that there is wide agreement on the desirable consequences of various nanotechnology

[26]Coenen, C. et al. (2004). *Report on the Conference: Converging Technologies for a Diverse Europe.* Available at http://www.itas.fzk.de/tatup 1043/cova04htm#back 1.

[27]Canton, J. *Designing the Future: NBIC Technologies and Human Performance Enhancement* (paper for the New York Academy of Sciences, retrieved on October 19, 2006). Available at http://www.futureguru.com/docs/Final-NY-Academy-Paper-3-11-03.pdf, in Roco, M.H. and Montemagno, C.D., eds. (2004). *The Coevolution of Human Potential and Converging Technologies*, op. cit., 186–198.

[28]Bainbridge, W.S. (2004). *The Evolution of Semantic Systems*, in Roco, M.H. and Montemagno, C.D., eds. (2004). *The Coevolution of Human Potential and Converging Technologies*, op. cit., 150–177 (M.H. Roco and C.D. Montemagno eds.).

[29]National Science Foundation. Roco, M.H. and Montemagno, C.D., eds. (2004). *Societal Implications of Nanoscience and Nanotechnology* 244–251.

options."[30] This observation is mirrored in the key concerns reflected by the HLEG's four-stage approach to CTs. First, the HLEG seeks to embed CT policy within broader social and economic policy initiatives, such as the Lisbon Strategy. Second, the HLEG seeks "upstream" public participation in CT policymaking. Third, the HLEG sets out a series of high-level and specific initiatives to review the ethical and social implications of potential CT developments, one aim of which is plainly to secure the "upstream" public participation that it sees as key to policy and economic stability. Fourth, the traumatic effect on European science policy of the forces that doomed "genetically modified" (GM) agricultural products in Europe, an experience that has *inter alia* led to a determination to develop European technology policy in the context of risk management, is noted. In light of this history, the HLEG is approaching CTs with an interrelated strategy to prevent either the rejection or the imprudent application of new "enabling technologies."

Risk Management

The HLEG departs from the ambitions of U.S. policy by expressing, with greater emphasis, the needs of addressing inherent risks or "anxieties" implicit in technical uncertainties and of balancing the goals of improving human performance with the broader social interests. Another guiding principle, "Precaution, Anticipation, and Risk Management," considers the "precautionary measures one can reasonably anticipate to ameliorate risk, build trust, and offer scientists and society the safest way forward."[31]

For example, the HLEG's final report to the EC identifies the "dangerously high probability that toxic nanotubes or other nanoparticles with unknown effects will pervade the entire food chain and in this way possibly contaminate the environment at a large scale."[32]

> [T]his raises the question whether under these circumstances the uncontrolled tampering with nanomaterial is consistent with our ethic principles . . . . In fact the stability of nature seems to rest on an evolutionarily grown diversity which is balanced in a rather subtle way. . . . That is why ecologists warn against . . . the danger of allowing genetically manipulated organisms (GMOs) to spread out into the natural world. Faced with the unconceivable complexity of nature we have no idea what the consequences of such changes might be for the future of this wonderful world.[32]

By contrast, Jean-Pierre Dupuy expressed his view on the insufficiency of the NSF's consideration of the risks associated with CTs:

> [A]s regards potential threats, there is only one mention thereof in the whole report—a warning by M. Roco himself about the risk of wild self-replication of nanobots,

---

[30]See footnote 29, p. 195.
[31]See footnote 24, p. 3.
[32]See footnote 2, p. 54.

immediately followed by the usual caveat, "we all agree that while all possible risks should be considered, the need for economic and technological progress must be counted in the balance."[33]

Europe has learned that economic success depends on the management of risk—both real and perceived. Recounting Monsanto's GM food debacle, the EC recognizes that trust in public and private systems is crucial to economic success and that "trust only can be maintained when citizens continue to ask questions about risks and values in science and society."[34]

For example, there is the economic risk of investing in a technological promise that does not materialize. Inversely, there is a societal risk that consumer acceptance of new technologies outpaces the careful consideration of their consequences.... A third type of risk is inherited by CTs through the contributions from the various enabling technologies—these are the risks of nanotechnology, genetic engineering, pervasive communication technology, etc.[35]

Embedding Converging Technologies Policy

Seeking to develop a distinctively European approach to the prospect of CTs, the HLEG proposed placing CTs in the broader context of the Lisbon Strategy of "delivering stronger, lasting growth[,] and creating more and better jobs."[36] Accomplishing this goal is "key to unlocking the resources needed to meet [Europe's] wider economic, social and environmental ambitions,"[37] which entails:

[T]he sustainable development of Europe based on balanced economic growth and price stability, a highly competitive social market economy, aiming at full employment and social progress and a high level of protection and improvement of the quality of the environment.[38]

In order to ensure a viable European economy the EC lists three objectives: (1) to make Europe a more attractive place to invest and work; (2) to recognize knowledge and innovation as the beating heart of European growth; and (3) to shape the policies allowing European businesses to create more and better jobs.[39]

---

[33]Dupuy, J.P. (2004). Assumptions and Values of NSF Report, Converging Technologies for Human Performance (February 2004) *in* European Commission, *Foresighting the New Technology Wave—Expert Group: State of the Art Reviews and Related Papers* 134. Available at http://ec.europa.eu/research/conferences/2004/ntw/pdf/soa_en.pdf (retrieved on October 19, 2006).

[34]See footnote 3, p. 12.

[35]See footnote 5, p. 31.

[36]European Commission. (2005). *Working Together for Growth and Jobs: A New Start for the Lisbon Strategy.* Communication to the Spring European Council (COM (2005) 24) 5. Available at http://europa.eu.int/growthandjobs/pdf/COM2005_024_en.pdf (retrieved on October 19, 2006).

[37]See footnote 36, p. 7.

[38]See footnote 37, p. 3.

[39]See footnote 37, p. 4.

In place of the NSF concept embodied in "Converging Technologies for Improving Human Performance," the HLEG proposes an alternative "branded" approach that is deemed less ideological: "CTEKS: Converging Technologies for the European Knowledge Society."[40] Realizing the significance of a knowledge-based society, the EC intends to harness the novel transformative potential of CTs in order to advance the Lisbon Strategy regarding the EU's employment, international competitiveness, economic reform, and social cohesion. Accordingly, the CTEKS agenda is a policy tool to exploit scientific development for economic benefit. Specifically, CTEKS aim to:

> create greater European cohesion in research and to bring together the scientific communities, companies and researchers of Western and Eastern Europe, to stimulate young people's taste for research and careers in science, to improve the attraction of Europe for researchers from the rest of the world, to promote common social and ethical values in scientific and technological matters, to set up research along the lines of technological platforms which bring together public and private stakeholders in CTEKS initiatives on health, education, ICT infrastructure, environment, and energy.[41]

The HLEG has chosen to shape CTs policy according to particular applications, but under the brand of CTEKS. Similarly, the NSF Report outlines primary application areas of NBIC: expanding human cognition and communication; improving human health and physical capabilities; enhancing group and societal outcomes, national security; and unifying science and education.[1] However, the EC plans to structure its approach to CTs in a specialized, disciplinary framework, addressing more specific problems, which could include such focused research programs as: "CTs for natural language processing"; "CTs for the treatment of obesity"; or "CTs for intelligent dwelling."[42] The HLEG also intends to develop agendas on "contextualised technology" or "all technologies which improve productivity, competitiveness and working conditions, closely linked with identified needs of the society."[43] However, "contexualised technology" does not exclude CTs, "which can be suitable answers in certain cases, but that also stresses the improvement of competitiveness in economic sectors considered as 'traditional.'"[43]

Upstream Participation and Agenda Setting

Though not a formal policy document of the EC, the HLEG's final report is a direct response to U.S. CTs policy. In fact, in outlining one of the 10 guiding principles, "Realism," the final report notes that "[t]he US key report is repeatedly criticised for containing a very wide ranging set of technology development assumptions."[44]

---

[40]See footnote 5, p. 2.
[41]See footnote 5, p. 24.
[42]See footnote 5, p. 8.
[43]See footnote 5, p. 62.
[44]See footnote 5, p. 62.

Unrealistic projections and imprudent introduction can result in public anxiety and rejection of new technologies. Upstream participation among the consumer and investor communities can diffuse potential anxieties and provide guidance in policy development. Because of such potential risks, realizing the goals of CTs will depend on investor and consumer confidence, as well as upstream participation by stakeholders and the public:

> By making these visions subject to public debate, technology assessment moves upstream: Instead of considering the products that come out of the development, vision assessment addresses the hopes, dreams, and promises that go in and inform it.[45]

The HLEG recognized that while the prospective applications of CTs hold transformative potential, without an ethical agenda, certain applications may pose a threat "to culture and tradition, to human integrity and autonomy, perhaps to political and economic stability."[46] The HLEG regards U.S. policy objectives as stated in the NSF Report as a means for the United States to maintain global leadership in the fields of national security and scientific development. Moreover, the HLEG emphasized that ethics must be an *intrinsic* component to progress, where technological advancement must "harmonize with the values of diversity, social justice, international security, and environmental responsibility."[47] Whereas the NSF Report centers on the transformation of society, the EC focuses on cohesion between CTs and social values, and asks: "How can the enhancement of humans and Nature made possible by the challenge of convergence, be implemented to build up these values? How can it revitalise and enforce the basic values of our liberal democracies?"[48] The agenda-setting process itself is a central focus of the HLEG final report, which specifies that ethical considerations are "not external and purely reactive" but, rather, deliberately included in public policy considerations; otherwise, there is the risk "defin[ing] our society's problems ... and ourselves and our environment first, in terms of the machine metaphor."[48] Discourse must examine societal ideals of autonomy, dignity, environmental ethics, constitutional rights, and other fundamental human rights.

Ethics and Social Context

The HLEG response to the NSF Report brought to light the ethical imperative in CTs policy development. That is, as the HLEG approach insists, applications intended to improve the individual or to generate wealth must be compatible with societal values and conceptions of freedom, morality, and human nature. There must be societal critique of the vision behind CTs that will flush out the philosophical agenda that implicitly sustains its research practice.[49] Furthermore, the HLEG reminds

[45]See footnote 5, p. 49.
[46]See footnote 5, p. 3.
[47]See footnote 5, p. 9.
[48]See footnote 3, p. 4.
[49]European Commission, High Level Expert Group. (2004). *Foresighting the New Technology Wave: Converging Technologies—Shaping the Future of European Societies.*

policymakers that open discourse will "contribute to trust formation only when [science and technology] internalise the values of public moral[s]."[50] In the policy development process, public discourse will keep CTs from failing to address the significant ethical issues concerning quality of life, social cohesion, and solutions to humankind's main challenges. The HLEG's CTs agenda aims to accomplish cohesion between societal values and scientific advancement, and to recognize the "need to evaluate and to articulate our values once again not only at the individual level of human and constitutional rights, but also at the level of the ideals about our European societies."[51] Accordingly, Recommendation 12 states that:

> Upon advice from the European Group on Ethics (EGE), the mandate for the ethical review of European research proposals should be expanded to include ethical and social dimensions of CTs. Funding organizations in Member States are asked to take similar steps.[52]

As noted above, the HLEG described U.S. projections of CT applications as holding "strongly positivistic and individualistic" underlying values.[53] In contrast, the HLEG focuses on the concept of moral pluralism, explaining that "one of the hallmarks of the European identity is the way [the EU] accept[s] and respect[s] moral pluralism in the Member States of the EU."[54] At one point, the HLEG final Report notes that the NSF Report "says nothing about the rest of the world, the issues of poverty and deprivation, of sharing, of any benefits to the global challenges facing the 95% of the world's population who are not U.S."[55]; rather, U.S. policy seemingly emphasizes the acceleration of human efficiency and productivity. Furthermore, Dupuy comments on his view of the NSF Report's utilitarian bent:

> Technology is viewed as a means to an end; *i.e.* the approach is purely utilitarian . . . .
> What adds to the utilitarian frame is the resolute individualistic bias: "The right of each individual to use new knowledge and technologies in order to achieve personal goals, as well as the right to privacy and choice, are at the core of the envisioned developments."[56]

Instead, European policy endeavors to balance transformative potentials with social concerns, placing an equal emphasis on the social and economic dimensions of sustainable development and, as Europe plots a diverging course for CTs, it seeks to observe the "precautionary principle" by assessing risks and asking how these technologies will infringe upon social values. In contrast, Dupuy criticizes the

---

[50]See footnote 3, p. 15.
[51]See footnote 3, p. 4.
[52]See footnote 5, p. 5.
[53]See footnote 2, p. 6.
[54]See footnote 3, p. 5.
[55]European Commission, *Foresighting the New Technology Wave—Expert Group: State of the Art Reviews and Related Papers* 134. Available at http://ec.europa.eu/research/conferences/2004/ntw/pdf/soa_en.pdf (retrieved on October 19, 2006).
[56]See footnote 55, p. 134.

NSF Report for elevating economic superiority over social cohesion and empowerment:

> The socio-economic analysis is of an incredible poverty. There are no social dynamics or forces, only individual behaviour that can be predicted and corrected if need be—for instance, disruptive behaviour, terrorist acts, etc. The inflow of dollars is the only driver. The role of the government is to set the conditions for private initiatives to flourish while ensuring public acceptance.[57]

Clearly, the HLEG under sponsorship of the European Commission is seeking to distance European approaches to emerging technologies from the NBIC approach by putting clear blue water between the NBIC CTs model and CTEKS. One example that is discussed several times in the HLEG documents is the distinction between engineering "*of* the mind" and "*of* the body" and engineering "*for* the mind" and "*for* the body."[58] Instead of using CTs to pursue engineering "*of* the mind" by physically altering or enhancing the human brain, CTEKS is dedicated to engineering "*for* the mind" by striving to improve the cognitive environment.[59] That is, CTEKS "sets out to improve the environment in which humans sense, think communicate and decide" as opposed to physically altering or enhancing the human brain.[60] The final report notes that "early responses to a CT initiative in the United States raised alarms about transhumanist ambitions to 'improve human performance' by turning humans into machines."[61] To that end, Dupuy describes the NSF Report as steering away from "conservative and overcautious" ethics that would inhibit "the transformation of civilization, thanks to which 'the acceptance of brain implants, the role of robots in human society, and the ambiguity of death' will conform to new principles."[62]

INTERNATIONAL STANDARDS

In its final Report, the HLEG listed several recommendations dealing with the ethical review of research proposals and an overall approach to CTEKS that will focus the normative implications of freedom, morality, and human nature, including:

Recommendation 9:

That the Commission implement a "EuroSpecs" research process for the development of European design specifications for converging technologies, dealing with normative issues in preparation of an international "code of good conduct."[63]

[57]See footnote 55, p. 135.
[58]See footnote 5, p. 3.
[59]See footnote 5, p. 22.
[60]European Commission, High Level Expert Group. (2004). *Foresighting the New Technology Wave: Converging Technologies—Shaping the Future of European Societies—National Policy/Market Analysis, Summary of the U.S. Report on Converging Technologies and Improving Human Performance* 133.
[61]See footnote 60, p. 7.
[62]See footnote 55, p. 135.
[63]See footnote 5, p. 54.

Additionally, a proposed Societal Observatory of Converging Technologies would consist of "a standing committee for real-time monitoring and assessment of international CT research, including CTEKS ... with the primary mission to study social drivers, economic and social opportunities and effects, ethics and human rights dimensions. It also serves as a clearing house and platform for public debate."[64] Specifically, Recommendation 8 calls for a "permanent societal observatory [to] be established for real-time monitoring and assessment of international CT research, including CTEKS."[65] This recommendation explicitly provides that the Societal Observatory of Converging Technologies:

> [in b]uilding on existing models of European Observatories, ... should study social drivers, economic and social effects, ethics and human rights dimensions. Comparative studies on legal, regulatory, and normative frameworks should be commissioned by this Observatory as CTs pose novel challenges that escape traditional regulatory categories. Existing regulatory approaches in Member and Associated States, on the European level, and in the international arena should be canvassed for similarities and differences, conceptual gaps, and creative solutions—with a view towards a proposal of European standards especially for CTs that are developed outside the CTEKS research process.[65]

Moreover, the expert group addressed the legal challenges to CTEKS, recommending the appointment of a commission to study areas such as:

> Comparative studies on existing human rights, national legislation, international norms, professional codes, and standards as they might apply to converging technologies. With a view to creating international or, at least, common European standards, such studies should consider future regulatory management and risk management for converging technologies.[66]

CONCLUSIONS

A comparative assessment of these two documents sheds light on the central significance of the social context and assumed values within which technology policy is developed. It would seem that the EC, in initiating the HLEG process, may have assumed a policy status for the NBIC Report that—as a compilation of papers from a workshop—it never had. On the other hand, the fact that it begins with an extensive "executive summary" (not common in conference volumes, though a standard feature of policy documents), and that its editors included the most senior figure in U.S. nanotechnology policy, both explain why such a misunderstanding may have

---

[64]See footnote 5, p. 48.
[65]See footnote 5, p. 53.
[66]See footnote 5, p. 49.

been possible, and underline the significance of the document's central thrust as an expression of opinion on the part of influential federal science administrators. As we move toward the consideration and articulation of particular policy positions in the development of these technologies, the caveats in the European documents should be heeded.

CHAPTER 4

Scientific Promise: Reflections on Nano-Hype

M. ELLEN MITCHELL

INTRODUCTION

Barely a day passes in which the news fails to report on the latest incredible finding or scientific advance that promises to cure, eliminate, or otherwise eradicate some disease or condition; biomedicine is frequently the source of these claims. Such press releases and the barrage of announcements about burgeoning science create an environment in which anything seems possible. Indeed these declarations, particularly recent ones about nanotechnology, sometimes suggest that even immortality and an end to suffering may be delivered. They paint a picture in which our human worry about our fundamental vulnerability seems pointless because science will convey us from all ills.

Examples of extraordinary claims associated with advancing nanotechnology abound. One reads thus:

> Never before has any civilization had the unique opportunity to enhance human performance on the scale that we will face in the near future. The convergence of nanotechnology, biotechnology, information technology, and cognitive science (NBIC) is creating a set of powerful tools that have the potential to significantly enhance human performance as well as transform society, science, economics, and human evolution.[1]

[1]Canton J. Designing the future: NBIC technologies and human performance enhancement. In Roco M.C. and Montemagno C.D., eds. The coevolution of human potential and converging technologies. New York: Annals of the New York Academy of Sciences; 2004. pp. 186–198.

Nanoscale: Issues and Perspectives for the Nano Century. Edited by Nigel M. de S. Cameron and M. Ellen Mitchell
Copyright © 2007 John Wiley & Sons, Inc.

In a National Science Foundation (NSF)/Department of Commerce-sponsored report,[2] 20 ways that converging technologies could benefit humanity within the next couple of decades were identified. These included, but were not limited to, the following:

> The human body will be more durable, healthier, more energetic, easier to repair, and more resistant to many kinds of stress, biological threats, and aging processes . . . . A combination of technologies and treatments will compensate for many physical and mental disabilities and will eradicate altogether some handicaps that have plagued the lives of millions of people . . . . Anywhere in the world an individual will have instantaneous access to needed information, whether practical or scientific in nature . . . Engineers, artists, architects, and designers will experience tremendously expanded creative abilities . . . . The vast promise of outer space will finally be realized . . . . Average persons as well as policy makers will have a vastly improved awareness of cognitive, social, and biological forces operating in their lives.

In their overview about the National Nanotechnology Initiative, the same authors conclude:

> If we make the correct decisions . . . . The twenty-first century could end in world peace, universal prosperity, and evolution to a higher level of compassion and accomplishment.[2]

While the prospect of a healthy, worry (and suffering)-free life has great appeal, one has to wonder about the social and personal costs of the rollercoaster of expectancies that accompanies the cycle of hope and disappointment attendant to such claims, which seldom eventuate. Medical science has bestowed incredible advances upon us that have improved the quality of life for many, but there have also been failures and untoward effects. These frequently are reported as terrible consequences; thalidomide is an easy example to cite. The yo–yo of tall expectations followed by disappointment in turn gives rise to disbelief. It is the implications of states of alternating hope and skeptical disbelief and the cylcothymic mood shifts that accompany these states for individuals and the culture, and factors that add to their complexity, that are the focus of this chapter.

THE ROLE OF EXPECTATIONS IN PROCESSES AND OUTCOMES

Expectancies have long been shown to have a positive effect on outcomes.[3] The expectation for health improvement[4] and positive benefits from mental health treatment have been shown to be statistically correlated with better outcomes. For

---

[2]National Science Foundation. n.d. Large benefits from a small world. Available at http://www.nsf.gov/od/lpa/nsf50/discov/nanoadver.htm (accessed 2004 July 19).

[3]See, for example, Glass C.R. et al. Expectations and preferences. Psychotherapy 2001;38:455–461.

[4]See, for example, Mondloch M.V. et al. Does how you do depend on how you think you'll do? A systematic review of the evidence for a relation between patients' recovery expectations and health outcomes. Canadian Medical Association Journal 2001;165(2):174–181.

example, optimism accounts for 40% of the variance in outcomes for heart transplant patients, separate from the effects of preoperative physical health.[5] Similarly, positive expectations for outcomes in mental health treatment[6] have long been shown to be consistently associated with better outcomes. The literature in health and mental health is replete with studies of process and outcome that include an examination of expectations. More recently, some investigators[7] have studied unrealistic expectations for individual self-change that persist despite repeated failure. The cycle of failure and renewed effort is linked to unrealistic expectations. There appears to be a parallel cultural cycle in which the more we fail to deliver on the promise of invention, the more we seek the next new, exciting thing. Nanotechnology is our newest thrilling area on which people are beginning to turn the spotlight.

In some cases,[8] the focus of research on expectancies has been specifically on expectations of the individual and, in other cases, the research focus has been on the so-called placebo effect[9] associated with improvements that occur in response to presumably inert interventions. This rather unfortunate label conjures up a pejorative connotation implying that any positive effects are somehow phony or imagined; an image that does not do justice to the consistent scientific findings that people's expectations play a role in health outcomes. In general, attitude and expectations are well documented to have a relationship to experience, perceptions, and outcomes.[10] The integration of the body and mind, and the recognition that mental phenomena are also physical phenomena are capturing the attention of Americans as exemplified by the upsurge in interest in yoga, meditation, behavioral health, and the like.

Belief is a powerful motivator. One might recall that in the 1970s there was a great deal of press and activity about the drug Laetrile, a form of purified amygdalin found in apricot pits and raw almonds. People seeking a cancer cure were outraged that the U.S. Food and Drug Administration (FDA) would not allow the sale of Laetrile in the United States, and they traveled to Mexico to obtain the drug. Amygdalin, still not approved by the FDA, produces cyanide when ingested. In turn, the toxin does sometimes slow or stop tumor growth.[11] The point is not to question or

[5]Leedham B. et al. Positive expectations predict health after heart transplantation. Health Psychology 1995;14:74–79.

[6]See, for example, Safren S.A. et al. Clients' expectancies and their relationship to pretreatment symptomatology and outcome of cognitive behavioral group treatment for social phobia. Journal of Consulting and Clinical Psychology 1997;65:694–698.

[7]Polivy J. and Herman C.P. If at first you don't succeed: false hopes of self change. American Psychologist 2002;57:677–689.

[8]Kazdin A.E. and Wilcoxon L.A. Systematic desensitization and nonspecific treatment effects: a methodological evaluation. Psychological Bulletin 1976;83:729–758.

[9]Benson H. and Friedman R. Harnessing the power of the placebo effect and renaming it "remembered wellness." Annual Review of Medicine 1996;47:193–199.

[10]Garfield S.L. Research on client variables in psychotherapy. In Bergin A.E. and Garfield S.L., eds. Handbook of psychotherapy and behavior change. New York: John Wiley & Sons, Inc., 1994. pp. 190–228.

[11]See, for example, National Cancer Institute. n.d. Laetrile/amygdalin (PDQ®) Available at http://www.cancer.gov/cancer_information/doc.aspx?viewid=962BA852-6565-41CB-B4C1-23C00B3C5F9F (accessed 2006 Sept. 11).

debate the efficacy of Laetrile or the action of the FDA, but to point out the fact that people will go to great lengths to obtain and do things based on beliefs. In this case, people gave up conventional forms of treatment shown to have merit and traveled to another country in search of an alternative treatment with less scientific support.

Expectations for scientific advances in nanotechnology will play a role in the evolution of this science in two distinct ways. The first will be expectations related to the science itself and the sense of public trust or distrust in the endeavor. To the extent that people expect, or perhaps worry, that nanoscience will give rise to risk, regardless of the attendant benefits or facts, then nanoscience will likely suffer the fate of genetically modified foods in Europe, which were virtually eliminated by public opposition. To the extent that people think it will provide a magical cure, they will pursue it fervently. The second way that expectations will play a role in the evolution of nanoscience is related to the extent to which expectations for nanoscience are overinflated, then disappointment and disillusionment will ensue and, meanwhile, we may miss the opportunity to pursue other, perhaps less sexy, more conventional paths. Moreover, the degree to which promoters of products can manipulate expectations to increase sales, the more likely it is that disappointment will follow because the picture that will be painted in the name of marketing, while glitzy, will be unrealistic and only partially true at best. Kass writes:

> Entrepreneurs not only resist governmental limitation of their work or restrictions on the uses to which their products may be put. They also promote public demand. The success of enterprise often turns on anticipating ands stimulating consumer demand, sometimes even on creating it where none exists. Suitably stimulated, the demand of consumers for easier means to better-behaved children, more youthful or beautiful or potent bodies, keener or more focused minds, and steadier or more cheerful moods is potentially enormous . . . . By providing quick solutions for short term problems or prompt fulfillment of easily satisfied desires, the character of human longing itself could be altered with large aspirations for long term flourishing giving way before the immediate gratification of smaller desires. What to do about this is far from clear but its importance should not be underestimated.[12]

The subject of media hype is not new, and the manipulation of consumers is also old hat. However, what is new is the potential pervasiveness and penetration of nanotechnology across market sectors and populations. Referring only to size, nanoscience can be applied to virtually anything ranging from materials to circuitry, computing to engineering, cosmetics to biomedicine, the environment, and more. This seemingly unlimited potential fuels the excitement and generation of broad, sweeping, and somewhat arbitrary claims and projections of good and ill. Because the array of possibilities is so broad, the media can find something of interest for everyone and use that interest to increase viewer and consumer attention. Portrayals of advances in highly dramatic terms serve to draw more people to viewing.

[12]Kass L. Beyond therapy: biotechnology and the pursuit of happiness. Washington (DC): U.S. Government Printing Office; 2003. pp. 304–305.

The prominence of hyperbole in the nanotechnology discussion has been striking, prompting whole books on the subject. In a recent book on nano-hype, Berube[13] noted that nano-hype, or the promulgation of misguided promises that nanotechnology can fix everything, is being used by a host of people who serve to gain from the nano initiative, including, but not limited to, industrial leaders, bankers, investors, researchers, reporters visionaries, bureaucrats, and so on. Indeed the list is almost all-inclusive. He notes that the characterization of nanotechnology is in terms that are dramatic, drastic, fundamental, and significant. Further, he asserts that the rhetoric is itself problematic because of the effects that issue from it. A particularly troublesome risk that is difficult to measure is the risk of putting off alternative, needed actions, presumably with the hope that the problem will go away or be solved by a future that will fix itself through nanotechnology.

Berube describes the cycle of hope, hype, inflated expectations, disillusionment, and then a shake out period of enlightenment as a cycle that accompanies the advancement of most scientific discovery. He presents this cycle as a normal, or at least predictable, occurrence that is part of the natural course of evolution of advances that new trends follow. The question that remains unanswered, however, pertains to the cost of such cycles in terms of actions not taken, cultural mood, and individual responsibility. The manufacture of desire has an attendant corollary of dissatisfaction with the present. Chronic cycles of hope and disillusionment make for moods filled with dissatisfaction and longing. Thus, hype about the future also foments discontent with the present and gives rise to cycles of seeking that can never really be gratified.

In an article by Nisbet, et al.[14] on the role of media on public perceptions of science, it was found that women are less likely to believe in the promise of science and that there is a positive relationship between TV viewing and skepticism about science. Simultaneously, the portrayal of science as miraculous enhances the belief that science holds promise. It can be inferred from these findings that the entertainment aspect of TV with its emphasis on science fiction, fantasy, imagination, fear, and mystery may interfere with the development of realistic expectations about what science can actually deliver. The authors report that more realistic and balanced content is associated with material contained in newspapers and magazines. However, the average person now obtains the bulk of their knowledge about science from TV viewing,[15] and as Berube[16] notes, the science content generated by the media has little information or insight and is presented to a mostly scientifically unsophisticated audience, with flash and sparkle. As well, the glamour of TV portrayals contribute to a unitary view of people in which individual

---

[13]Berube D.M. Nano-hype: the truth behind the nanotechnology buzz. New York: Prometheus Books; 2006.

[14]Nisbet M.C. et al. Knowledge, reservations, or promise? A media effect model for public perceptions of science and technology. Communication Research 2002;29:584–608.

[15]National Science Board. 2006. Science and technology: public attitudes and understanding. In Science and engineering indicators 2006, Vol. 1. pp. 7.1–7.46. Report nr NSB 06-01. Available at http://www.nsf.gov/statistics/seind06/c7/c7 h.htm (accessed 2006 July 18).

[16]Reference 13 p. 7.

differences exist as superficial variations on an otherwise scientifically perfectible human body and life.

LINEAR CAUSAL MODELS

Genetic research is one focal area that has been associated with changing conceptualizations of behavior and individual differences. While the public has an increasing sense of genetic determinism, ironically, scientists are finding that there are multiple genetic and environmental influences that exist in an intricate interplay in which single-gene effects, and the concept of singular, linear effects, are less compelling than quantitative trait loci or gene systems.[17] Indeed, there is a risk of assigning genetic causal attributions in such a way that they overlook the fact that complex psychological traits are subject to environmental influences.[18] Work during the past two decades in the area of genetics and behavior has led to widespread acceptance by the public and by scientists of the view that genetics are a causal factor in individual differences, personality, and behavior.

The Human Genome Project raised the prospect of change at the most fundamental corporeal level. It ushered in the hope of securing changes that would endure, thus solving the problem of incomplete and impermanent treatments associated with most current medical interventions for chronic problems. Most medical interventions for chronic conditions are palliative rather than curative, or they are stabilizing but do little to change the underlying condition. For example, people can take insulin or thyroxine to treat diabetes or thyroid disease and experience a substantially improved quality of life. However, the diabetes or thyroid disease remains, and the treatment must be life long. Nanoscience is promising a methodology for undertaking genetic and cellular therapies to actually deliver on the promise of curing the underlying problem and restoring health. More aggressive future projections for the ability to end disease and deliver perfect health have also been among nanotechnology claims.[19]

A particularly deterministic and concrete causal model has accompanied the dream of human perfection. Arising from simplified scientific paradigms on the causes of illness and disorder, the public has developed a relatively unidimensional view of causes and effects in relation to disease and disorders. The evolution of causal models of diseases has followed a number of paradigmatic shifts, which have been embraced by the general populace in a fundamentally linear fashion, albeit the multivariate and recursive nature of problems is recognized in scientific

---

[17]Plomin R. and Crabbe J.C. DNA. Psychological Bulletin 2000;126:806–828.

[18]Plomin R. and Colledge E. Genetics and psychology: beyond heritability. European Psychologist 2001;6:229–240.

[19]Kurzweil R. Testimony of Ray Kurzweil on the societal implications of nanotechnology. U.S. House of Representatives Hearing to Examine the Societal Implications of Nanotechnology and Consider H.R. 766, The Nanotechnology Research and Development Act of 2003; 2003 April 9; Washington, DC. Available at http://www.kurzweilai.net/meme/frame.html?main = /articles/art0556.html (accessed 2006 Oct. 17).

circles. Briefly, the discovery of pathogens, for example, the spirochete for syphilis or streptococcus for throat infections, created a belief that exposure to pathogens resulted in disease, a finding supported by data with important public health implications because quarantine could be used to curb disease.

The evolution of health science required medicine to address cases of exposure in which no disease manifested itself and, thus, the model of immune function came to the fore. The discovery that the body could attack itself, and that disease could arise without exposure, but verily from within, then spawned an the investigation of illness inside the person. One answer was found in immunology and another in genetics. Thus, medicine and lay persons alike adopted a belief that testing for the presence of genetic causes might be useful to understand, treat, and prepare for various disorders. Nanotechnology not only promises a method for genetic engineering, but also for entering the body with targeted treatments. For a good overview of what is thought possible in medical nanotechnologies, the reader is referred to Roco and Montemagno.[20]

The advent of genetic testing for disease entities not necessarily yet in evidence, such as in the case of the breast cancer gene or the Alzheimer's gene in asymptomatic people, also promotes a linear causal view between specific, singular genomic components and disease and disorders. While different base genetic sequences have been identified as the culprits, the essential linearity of the conceptualization is the same as in the pathogenic model of disease. This deterministic view of the causes of disease and disorder obfuscated both the complex interplay with environment and the probabilistic nature of events. While it is likely that many Americans know about the so-called breast cancer and Alzheimer's genes, it is unlikely that most grasp the degree to which false positives, and potentially false negatives, are generated. It is also doubtful that most people have a solid understanding of concepts like marker variables, moderating variables, or mediating variables, nor should they because these are fairly complex constructs. Events like false positives, which occur at a high frequency in the case of genetic testing for Alzheimer's disease, for example, raise fundamental questions about what to believe, and they do not fit a unitary linear causal model. The conundrum for the average person is both what to believe and what to think when faced with a medical condition requiring a decision.

This oscillation between the euphoria that we have discovered a cause and therefore a cure must be around the corner, followed by the disillusionment associated with so many unknowns and false starts, engenders a cultural mood of skepticism, fatalism, and helplessness. In a 2001 book on the impact of genetic advances and the Human Genome Project, Andrews[21] devoted several chapters to the consequences of genetic testing for individuals. She noted that the availability of genetic testing has had an effect on self-concept, anxiety, identity, family

[20]Roco M.C. and Montemagno C.D., eds. The coevolution of human potential and converging technologies. New York: New York Academy of Sciences 2004.
[21]Andrews L.B. Future perfect: confronting decisions about genetics. New York: Columbia University Press; 2001.

relationships, use of services, proclivity to engage in preventive actions, and spousal relationships. Most interesting is the diminution of activities that contribute to health enhancement through prevention; activities like monitoring for early signs of disease. What is troublesome about this is that it suggests that the knowledge gleaned from scientific methods like genetic testing not only made some people more anxious and distressed, but it also contributed to self-defeating behavior. Most diseases are better treated early; even those that might be inevitable, as in the case of the clearly genetically based disorders. Most conditions, if identified early, allow people the opportunity to act to optimize health and quality of life. Yet, knowledge does not appear to lead to prudent action.

PERFECTION

The recognition that in-born genetic defects can give rise to problems ranging from disease through disorder neglects to acknowledge the fact that the average person has between five and 50 genetic mutations, and it promotes the belief of the value and possibility of being genetically perfect; verily defect free.[21] The idea that gene defects can be fixed in some process of nanogenetic engineering fails to take into account the elaborate systemic interactions that scientists are documenting.[22] Genetic science is forging ahead, driven by the idea that if people had perfect genes, then they would be problem free. Furthermore, the fantasy of perfect genes assuring perfect health shifts personal responsibility from the individual to an external higher or other authority. If, in fact, one's health were determined fully by genetics, it would hardly matter what one actually did with respect to lifestyle, diet, or healthcare. The notion that one could even have perfect genes is wishful thinking; an empirical question at least and not a foregone conclusion. It is not known if perfect genes would result in the perfect person or if some genetic defects actually serve a function that is not readily apparent. Nevertheless, the hyperbole surrounding nanoscience fuels the fantasy of perfection. As well, the media promote images of perfection that are both unrealistic and lacking in pluralistic perspective.

The view of a perfect body and a science that can fix anything and everything encourages the idea that perfection exists in some finite, definable fashion. If such a state or set of traits were defined, it would, by default, also delineate that which falls outside of perfection. Many people believe that it would be beneficial to eliminate whole classes of conditions or states (e.g., disabilities), which would, by many estimations, fall outside of the vision of perfection. However, once we diminish the variability of humankind, then the extraordinary (e.g., perfect) becomes ordinary.

Genetic science and nanoscience risk diminishing diversity by offering to eliminate disability and variability. It is important to bear in mind that many conditions of disability are social problems and perspectives, not physical conditions

---

[22]See, for example, Wong A.H.C. et al. Phenotypic differences in genetically identical organisms: the epigenetic perspective. Human Molecular Genetics 2005;14(1):1–18.

necessarily in need of fixing.[23] The point is not so much what we should keep, pursue, or undertake, but rather acknowledging the need for mindfulness about the attitudes, values, and culture we are engendering that arise in parallel with the hope of nano-based scientific discovery. Rather than hope for individual perfection, perhaps we should hope for better quality of life and community because there is a huge chasm between images of media perfection and the conditions under which millions of people live.

Canton suggests that baby boomers are the driving force in the human performance enhancement industry and asserts that, as they age, baby boomers will want longevity, personalized medicine, sensory enhancement, improved intelligence, mobility, and memory, and that they will demand enhancement as their right. He speculates with great conviction that enhancement will be a key lifestyle trend of the future. Moreover, Canton declares that future generations will want even more, and he forecasts that collective entitlement will be a formidable force in elections, research agendas, and the marketplace. The entitlement-to-perfection attitude is expanding, as we are more able to actually deliver many things, from vegetables to babies, all in perfect condition. Hence, expectations of perfection are forged. However, expectations for enhanced human performance and perfection remain juxtaposed against the backdrop of adverse conditions and uneven access to technology and healthcare that characterize less affluent baby boomers who are unemployed, without retirement or savings, and who struggle month to month to live.

Kass writes "the stupendous successes over the past century in all areas of technology, and especially in medicine, have revived the ancient dreams of perfection."[24] However, perfection is, at its core, an idea and not an object or a state. The psychological literature demonstrates that there are different components to people's perfectionistic strivings. In a study of the interpersonal expression of perfection, researchers[25] have identified several key aspects of perfectionism. These include perfectionistic self-promotion, nondisplay of imperfection, and nondisclosure of imperfection. As everyone has human foibles, this interpersonal stance essentially involves construction of a persona and active concealment of any and all qualities that are less than ideal. Perfectionistic self-promotion is the tendency to seek to display and declare one's perfection; nondisplay of imperfection is exactly as it sounds and also includes concealment of imperfection; nondisclosure of imperfection is associated with avoiding verbal acknowledgment or admission of imperfection. The authors suggest that this maladaptive orientation is closely linked with regulation of self-esteem and psychological distress. Yet, one can see that the culture promotes this form of impression management with the multiplicity of images of people as absolutely attractive, astoundingly wealthy, and unbounded by the constraints of the body or normal daily life.

---

[23]Tate D.G. and Pledger C. An integrative conceptual framework of disability; new directions for research. American Psychologist 2003;58:289–295.
[24]Reference 12, p. 46.
[25]Hewitt P.L. et al. The interpersonal expression of perfection: perfectionistic self-presentation and psychological distress. Journal of Personality and Social Psychology 2003;84:1303–1325.

The mere idea that people can and should seek to achieve perfection is also promoted by strong religious orientations that include striving for the achievement of a life that is in the image and vision of God. While this chapter is not about religious belief systems or religious striving, the upsurge of spiritual interest and religious conservatism appears parallel with technological advances. At a time when people are hard pressed to know what to believe because it is increasingly difficulty to sort out the real from the invented, the real from the hype, and the real from the virtual, religion provides something in which people can believe and which is not shifting.

ELUSIVE TRUTH

The recent decades have been fertile with findings and seeming facts that eventuate to be wrong, incomplete, or even dangerous. For example, millions of people who were smokers in the mid-twentieth century were assured that tobacco was safe, and yet, today, we know that it is patently unsafe.[26] Millions of women were given hormone replacement therapy, which was touted to protect against heart disease and cancer, as well as to ameliorate peri-menopausal symptoms, and now are face to face with data about profound risks of the drugs.[27] Millions of Americans ingested diet drugs now known to risk life[28] and accepted blood transfusions not knowing that fatal, but then-undetectable, diseases might be transmitted.[29] The recent spate of drugs (e.g., Vioxx), being withdrawn from the market are other examples of the abrupt turns of thought and so-called scientific fact confronting the American public.

A problem with scientific advances and the manner in which they are reported in real time is that new findings supplant old findings at a rate that changes so rapidly that it renders the audience cynical about accepting anything as truth because these so-called facts are apt to change tomorrow. While it remains an empirical question, it is likely that the problem is not that science has been wrong or made mistakes, but rather that science has been so far off, and the reversals so rapid: potential cures have later emerged as killers; benign activities have turned out to be deadly. These drastic shifts challenge all belief.

The attitudes of the public are influenced by tendencies and biases in estimations about the potential impact of future events. The magnitude of the gap between the over inflated promise and the severely contrasting reality contributes to greater amplitude in the emotional cycle. The alternating high hope and steep

[26]Shopland D.R. Tobacco use and its contribution to early cancer mortality with a special emphasis on cigarette smoking. Environmental Health Perspectives 1995;103:131–142.

[27]Humphrey L.L. et al. Postmenopausal hormone replacement therapy and the primary prevention of cardiovascular disease. Annals of Internal Medicine 2002;137:273–284.

[28]Mark E.J. et al. Fatal pulmonary hypertension associated with short-term use of fenfluramine and phentermine. New England Journal of Medicine 1997;337:602–606.

[29]AuBuchon J.P. et al. Safety of the blood supply in the United States: opportunities and controversies Annals of Internal Medicine 1997;127:904–909.

disillusionment cycles have an impact on internal states like mood, as well as on views of, and beliefs about, external events like scientific advances gone awry. Perceptions of science are thus colored by inflated hopes, dashed dreams, and shifting truths, as findings and corrections are reported in sound byte form by the media.

Recent literature[30] suggests that individual levels of happiness and conversely, unhappiness, are relatively stable. However, most people overestimate the degree to which potential future adverse events will disturb them, and also overestimate the anticipated positive impact that positive events will have.[31] For example, if one interviewed many people about their beliefs about the impact of winning a multimillion-dollar lottery, and then compared the projected consequences to the actual outcomes for winners, one would find that the receipt of money did not also deliver the glowing hopes of happiness.[32] Similarly, people sometimes dread adverse events, which do not turn out to be as catastrophic as anticipated. The combination of overinflated expectancies and shifting information about technological and medical advances is a particularly volatile combination. On the one hand, people are predisposed to exaggerate their hopes and fears of and for the future while they are also the recipients of alternating good and bad news. The propensity to swing from optimism to pessimism, anticipation to disappointment, is thus inflated, making it more difficult for individuals to adopt a solid and realistic stance about emerging science. This difficult combination is further compounded by the challenge of understanding and keeping current on science, and it is most pronounced with respect to nanoscience because it is an area about which the public knows very little.[33] Thus, in the absence of sound and reliable information, conjecture and speculation have as much meaning and value as anything else.

SCIENTIFIC KNOWLEDGE

While TV is the primary source of information by which most adults learn about science and technology, the Internet is increasing in its popularity as a source. The average adult American obtains the bulk of new science and technology learning from public media sources, such as TV, the Internet, and newspapers.[15] Some TV viewers watch more informationally based programs like those offered by the *Discovery* or *History* channels, but there is a great deal of content available via general TV viewing about science and medicine. Certainly, at any given time, it is easy to name multiple programs with science (e.g., *Numbers, CSI*) or medical (e.g., *Scrubs, ER*) thematic content. General TV viewing, in contrast to selective or directed TV viewing, is associated with lower science knowledge levels. Few

[30]Diener E. and Diener C. Most people are happy. Psychological Science 1996;7:181–185.

[31]Gertner J. The futile pursuit of happiness. New York Times Magazine. Published online: September 7, 2003. Available at http://query.nytimes.com/gst/fullpage.html?sec=health&res=9E0DEFD61538F 934A3575AC0A9659C8B63 (accessed 2006 Sept. 11).

[32]Gilbert D.T. Stumbling on happiness. New York: Knopf; 2006.

[33]Cobb M.D. and Macoubrie J. Public perceptions about nanotechnology: risks, benefits and trust. Journal of Nanoparticle Research 2004;6:395–405.

people read substantive scientific journals. Most of the information delivered by TV is presented as background to otherwise fictional entertainment programs. Hence, public knowledge about the use of DNA in forensic work, or medical procedures, arises from entertainment TV programming sources (e.g., *Law and Order, Grey's Anatomy, Dr. House, Extreme Make Over*). While the core information on disease and disorder may be accurate, the portrayal of outcomes is not necessarily based in reality.

It is common for popular TV programs to feature highly uncommon, verily rare, disorders and problems. It seems that the more novel or low frequency the problem, the more likely it is to be depicted on some of these programs. Characters, as they should in a good story, live or die in relation to the story line rather than their disorder per sc. Consequently, the views and expectations that people develop about science, scientific advances, and health outcomes are also not wholly realistic, but contain some mix of reality and fiction. Also noteworthy is the well-documented psychological phenomenon in which people are likely to commit errors of recall because of previous learning, interference, exposure to misinformation or related information, and practice.[34] Actually sorting out the facts and fictions of information that come through entertainment channels is particularly difficult because the contents are not identified with respect to level of veracity. All of this presumes that most people grasp the content that is conveyed, an assumption that may well be unfounded and incorrect.

The National Center for the Study of Adult Learning and Literacy (NCSALL) reports that, "40% working-age adults lack the skills and education needed to succeed in family, work, and community life today."[35] The functional literacy rates in America are so poor that a NCSALL report by Reder recommends that national education policy be shifted from achievement of the goal of high school equivalency to achievement of the status of college ready.[36] The NCSALL notes, in a summary of a chapter dedicated to findings from the 1992 National Literacy Survey: "Among Reder's findings, he shows that nearly one in four (22%) of the nation's college students seeking academic degrees lacks the literacy skills needed to meet the designated national benchmark for adult literacy."[37]

There is a distinction between general literacy required to navigate day-to-day exigencies of the world and medical literacy that is necessary for healthcare decision making. There are data demonstrating that the medical literacy of the general population is particularly problematic. Medical literacy refers to the ability to obtain,

[34]Norman K.A. and Schacter D.L. False recognition in younger and older adults: exploring the characteristics of illusory memories. Memory and Cognition 1997;25(6):838–848.
[35]National Center for the Study of Adult Learning and Literacy. About NCSALL: connecting research and practice to strengthening programs. Available at www.ncsall.net/index.php?id=17 (accessed 2006 Oct. 17).
[36]National Center for the Study of Adult Learning and Literacy. 2005. The GED and beyond. Focus on Policy 2003;1(1):1–8. Available at www.ncsall.net/?id=647 (accessed 2006 Oct. 17).
[37]Reder S. 2005. Adult literacy and postsecondary education students: overlapping populations and learning trajectories. The Annual Review of Adult Learning and Literacy 1999;1 Available at www.ncsall.net/?id=523 (accessed 2006 Oct. 17).

process, and understand basic health information and services needed to make appropriate health decisions."[38] Research[39] has demonstrated that education level is not a good index of medical literacy and, in a sample of adults utilizing services in public clinics, the average level of education was tenth grade, but the reading ability was found to be on a par with the fifth grade. Patients in private clinics had higher education and reading levels (college freshman and tenth grade levels, respectively), but most education materials are written at the eleventh grade to college sophomore levels. The data suggest that the level of reading competence necessary to understand medical matters is at or above the tenth grade level. This is in contrast to the sixth grade level of reading ability that is found when random samples of adults are assessed. Thus, a majority of people probably cannot understand the content of medically based literature. Science understanding is similarly poor. The complexity of nanoscience, based in quantum physics, concepts on a scale that is unfathomable, will demand even more sophisticated levels of understanding in order to apply that knowledge to purchases, investments, policy, and medical decision making.

In a press release by the Economic and Social Research Council,[40] it was noted that out of 12 basic scientific questions presented in a survey of 1000 people in England the average number of correct answers was four to five and that even among people with scientific educational qualifications only about five questions were answered correctly, on average. The authors note that people have opinions on issues and scientific matters even in the absence of knowledge.

With the advent of the Internet, ever more people are going on line to obtain health information. A 2006 report from Pew indicates that 60 million, or 45%, of Internet users cited Internet information as crucial or important in making at least one of eight decision points; and 40% used the Internet for coping with a major illness.[41] There are two issues associated with using the Internet as a source of health and science knowledge. The first problem pertains to source credibility. Information on the Internet can appear quite compelling, but be utterly false. Research has demonstrated that health-based information of the Internet is of uneven quality.[42] Almost anyone can launch a web page, and there is no systematic and effective method for evaluating the information that pops up.

---

[38]Zarcadoolas C. et al. Understanding health literacy: an expanded model. Health Promotion International 2005;20:195–203.

[39]Davis T.C. et al. The gap between patient reading comprehension and the readiability of patient education materials. Journal of Family Practice 1990;31:533–538.

[40]Economic and Social Research Council. Public strong on opinions, weaker on knowledge about science: main conduits of information not trusted. Published online: September 4, 2002. Available at http://www.esrcsocietytoday.ac.uk/ESRCInfoCentre/PO/releases/2002/september/public.aspx?ComponetId=2169&SourcePageId=1403 (accessed 2006 Oct. 17).

[41]Horrigan J. and Rainie L. When facing a tough decision, 60 million Americans now seek the Internet's help: the Internet's growing role in life's major moments. Published online: April 19, 2006. Available at http://pewresearch.org/obdeck/?ObDeckID=19 (accessed 2006 Oct. 17).

[42]Benigeri M. and Pluye P. Shortcomings of health information on the Internet. Health Promotion International 2003;18(4):381–386.

There is no way for the public to judge the accuracy of content they encounter on the Internet or TV without prior knowledge, which would preclude even needing the information. The acceptance of information as credible is influenced by personal variables, such as wishful thinking. Wishful thinking is a real phenomenon and, as Sjoberg[43] notes, people are pleasure seekers who assess general risk differently from personal risk and assign credibility or risk in accordance with a complex set of variables that are colored by our wishes.

A second problem with Internet-based information is related to the average reading ability level of the general public. While the Internet is particularly useful for the portrayal of information using graphics, images, and brief text, people with average levels of information processing and problem solving skills are likely to gravitate toward the most easily understood material, which might or might not be the most accurate material. These two factors, unclear credibility of sources colored by wishful thinking and literacy rates further complicate the hope and disillusionment cycle because they contribute to distortion. To the extent that advances like nanoscience are described as holding promise to help in all areas of life, coupled with a lack of understandable information, then imagination, both hope and fear based, will also have no boundaries.

THE ROLE OF BELIEFS

The confluence of intermingled fictional and nonfictional content, in tandem with truly amazing advances by science, also make it difficult for people to know what is normal, reasonable, or "good" in the scheme of things or to have guide posts for decision making. For the sake of simplicity and so as to not digress, good in this context will be taken to mean enhancing of health and quality of life. The need to make judgments based on information and a sense of norms is omnipresent. Daily decisions may range from relatively superficial or mundane questions about procedures like immunization, nutritional supplements, Lasix, or Botox to more complicated and potentially life-risking choices as in the case of organ transplants, cancer treatments, or genetic therapies. People also need to confront questions that are personal and interpersonal, including, for example, questions about the impact and meaning for identity associated with artificial parts or having another person's organs. Knowing what is the right thing to do is elusive and fluid at best.

Andrews noted that genetic testing has shifted the boundaries of normality and has the potential to change the culture by its affect on conceptualizations of individual and social responsibility, and by "challenging basic societal concepts, such as free will and equality."[44] If science allows one to potentially do anything, a promise that has been assigned to nanotechnology, then the questions of, "will we" and "should we" arise. These meta-level questions about what is right, what

[43]Sjoberg L. Neglecting the risks: the irrationality of health behavior and the quest for la *dolce vita*. European Psychologist 2003;87:266–278.
[44]Reference 21, p. 49.

is normal, and what is okay reach a higher pitch with the advent of nanoscience, which projects to offer methods and materials that may be able to give rise to life itself. The answers to these questions, in most cases, rest more upon fundamental beliefs and value systems than on facts.

Recent surveys conducted by the NSF[15] demonstrate that there is widespread belief among Americans in things like lucky numbers,[45] ESP, UFOs, and astrology.[46] It is reported[15] that more than 25% of Americans believe in astrology even though two-thirds agreed it was not scientific. Clancy cites statistics from polls indicating that 93% of Americans believe that extraterrestrials exist, 80% believe the U.S. government is concealing knowledge about extraterrestrials, and 27% of the population believes the earth has been visited by extraterrestrials.[47]

The scientific psychological knowledge about belief, memory, false memory,[48] and irrationality is enormous. False memories and irrationality abound right along side accurate memory and rationality. Moreover, the public appears to not only believe in undocumented phenomena, but seeks them out. The content of popular TV programs and movies (e.g., *Charmed*, *Supernatural*) and their success suggests a curiosity about, if not a fascination with, magic, science fiction, superstition, para-normal ideas, and the like. In his book on the subject of why people believe such things, Shermer concludes that: "More than any other, the reason people believe weird things is because they want to. It feels good. It is comforting. It is consoling."[49] He notes that skeptics and scientists are not immune to such beliefs. He suggests that the immediate gratification, simplicity, morality, and meaning of these beliefs are very compelling. While he does not cite much in the way of empirical findings to support this conceptualization, his presentation is thoughtful and thought provoking. Images of nanotechnologies are rife with possibility, spun off from and contributing to all sorts of beliefs: nanoparticles cannot be seen, so, in large measure, their pre-sence becomes almost a matter of faith; and nanotechnology purports to be the wave of the future with the potential to change everything limited only by the bounds of imagination.

Both Shermer and Clancy conclude that the need for meaning in human exist-ence may be more fundamental and compelling than is recognized by science; it is so strong that people invent meaning where perhaps none is apparent. What is striking is that some beliefs are so widespread (e.g., belief in extraterrestrials) that they also become intractable and impervious to reason or data. Long-time

[45]Losh S.C. et al. What does education really do? Educational dimensions and pseudoscience support in the American general public, 1979–2001. Skeptical Inquirer 2003;27(5):30–35.

[46]Moore D.W. Three in four Americans believe in paranormal: little change from similar results in 2001. Published online: June 16, 2005. Available at http://www.galluppoll.com/content/?CI=16915 (accessed 2006 Oct. 17).

[47]Clancy S.A. Abducted: how people come to believe they were kidnapped by aliens. Cambridge, MA: Harvard University Press; 2005.

[48]See, for example, Albarracin D. and Wyer R.S., Jr. The cognitive impact of past behavior: influences on beliefs, attitudes, and future behavioral decisions. Journal of Personality and Social Psychology 2000;79:5–22.

[49]Shermer M. Why people believe weird things: pseudoscience, superstitions, and other confusions of our time. New York: W.H. Freeman and Company; 1997. p. 275.

smokers who hold that smoking is harmless or that second-hand smoke is harmless, those who believe that the holocaust never happened, or proponents of rituals for revealing the gender of babies *in utero* exemplify widespread belief in fallacies.

To be sure, odd beliefs are also fostered by lore grounded in tradition (but not necessarily science) and by odd events. Nanotechnology, more than any other technology to date, will have the capacity to elicit odd beliefs because it purports to provide a method for changing the body and the world.

Most people have had an experience, a real and convincing experience, that defied logic, reason, or science. These apparent anomalies can create strong impressions and beliefs. Similarly, dreams are real phenomena with content arising from individual cognition that is idiosyncratic and fantastic. The content can be so persuasive that people will embrace the content as real rather than view it as the product the neural biochemical and electrical activity of a sleeping brain. Because nanotechnology is promising to lift all the boundaries, the prevalence and breadth of odd beliefs may well grow. Up until recent years, science has had boundaries. It has been circumscribed by measurement, the limits of what we could see, do, compute, and manipulate. The promise of nanotechnology is that anything is possible: once again, science and invention have become boundless.

The confluence of computers, nanoscale, biological systems, and chemical knowledge is ushering in a new science that purports to include creation, replication, and the promise to fix anything. This hope arises, in part, because of our tendency to overestimate the positive and the effects that positive events will have. As we look forward toward the prospect of scientific advances enabled by nanotechnology, we grow excited that the future will finally hold hope and cure because this is the human tendency. When we consider ill effects, we imagine total destruction, grey goo as it were. In fact, we cannot know the future, and we must navigate the present.

The challenge for the present is to develop an approach for emerging sciences to move forward quickly, but not so quickly as to engender irrational false hope and downward spiraling disappointment because our mood cycles interfere with stability. The frenetic quest for the next great thing risks compromising the present and more. Because it is difficult to discern what is real, it is more difficult to make informed rational decisions about how to act. Mainstream TV now features reality TV that, from all reports and description, is not real, but who knows for sure, and this is the point. Berube notes:

> The public is hearing hyperbole from both sides of the fence. As such, when a highly exaggerated interpretation is offered in the public culture, whether in print or in film, and it is sufficiently like the hype of the proponents and opponents, we get a problematic linkage. Differentiating fiction from reality becomes incredibly difficult when this occurs and the Hollywood blockbuster seems as real as anything found in *Science* or *Nature* to the rhetorically challenge.[50]

[50]Reference 13, p. 47.

When major public figures like Presidents lie, and major scientific findings like the safety of drugs turn out to be wrong, then questions of credibility and believability remain potent and unanswerable, and people may as well believe whatever "weird thing," to adopt Shermer's label, they wish. The problem arises not with the belief so much as with the actions that follow as a consequence of the beliefs. If people believe that nanotechnology will allow environmental pollution to be erased, as some suggest,[19] then there is no reason to exercise care for the environment in the ways that we live. But what if we are wrong? If people believe that nanotechnology will allow us to be reborn because all that is needed is a copy of our DNA sequence to reincarnate us, then why not commit patently destructive acts? The why-not question also has no answer or boundary. The prospect of being able to always have a "do-over" is comforting, and nanotechnology is promising to deliver the grand fix.

If cultures are organized largely by belief systems, as some[51] hold, and if the belief systems are struggling to be grounded about what is real and true, then human interactions are at risk of becoming more difficult and fractionated. Terrorism demonstrates that extreme beliefs can wreak havoc; extreme beliefs by their definition do not lend themselves to easy integration. It is unclear what beliefs will be spawned with the advent of nanotechnology, but gaps[52] in access to healthcare, technology, and literacy can only serve to widen and deepen differences in belief systems.

THE CASE FOR REASON, STABILITY, AND INTERDISCIPLINARITY

It is incumbent on scientific thinkers to strive to be the voice not only of reason, but also of reasonable belief. When scientists are given to hyperbole, as in the case of nano-hype, the risk of unrealistic, overinflated expectations is heightened. Scientists and policy makers are regarded as the authorities, a position that they enjoy and deserve, and with it comes responsibility to make measured remarks and projections. Nanotechnology hopes can blind us from funding research in other important areas of human problems, such as, violence, illiteracy, the environment, drug treatment, and conventional medicine in favor of the next great thing. The hope and the hype move us away from thinking about people and the quality of life in the present to considering abstract invention and an ephemeral future.

Bandura[53] notes the divestiture of different aspects of psychology to the disciplines of biology and asserts that there is a need to encompass the complex interplay between intrapersonal, biological, interpersonal, and sociocultural determinants of human functioning. This also holds for other disciplines than psychology. The

[51]Triandis H.C. The psychological measurement of cultural syndromes. American Psychologist 1996;51:407–415.
[52]Warschauer M. Technology and social inclusion: rethinking the digital divide. Cambridge, MA: MIT Press; 2004.
[53]Bandura A. The changing face of psychology at the dawning of a globalization era. Canadian Psychology 2001;42:12–24.

need to embrace a model of nanoscience that has room for the agentic capacities of people is as great as the need to continue to work on problems of the present rather than pursue an elusive future that, while filled with desire and promise, distracts us from the important realities before us. Understanding what is real, what is important, and what is possible is an ever-difficult and increasingly vexing problem. The many faces of nanoscience demand the involvement of many perspectives and disciplines. It would be short sighted to forge a head without the involvement of those many perspectives and voices to help focus the discourse and consider real possibilities. The cost of cycles in terms of actions not taken, cultural mood, and individual responsibility is immeasurable. Ironically, we seem to grow ever more skeptical of scientific fact while more convinced about the truth of individual belief and this renders us unable to determine what it true. Grounding nanotechnology in the present and in reality is a daunting, but important, task if nanoscience is to be directed to inventing and developing materials, devices, and methods that can make a difference in the quality of life for people.

CHAPTER 5

Beyond Human Nature: The Debate Over Nanotechnological Enhancement

JAMES HUGHES

NANOTECHNOLOGY THREATENS HUMANNESS?

Critics of the speculative future uses of nanotechnology sometimes suggest that there is some kind of "humanness" or "human dignity" threatened by nanotechnological enhancement. For example, in his theologically informed review of nanotechnology's promise, C. Christopher Hook, M.D., notes:

> It is one thing to use technology to repair an injury or to treat or heal an affliction, but it is quite another thing to use technology to engineer "better" human beings. Many who are healthy will likely be tempted to "enhance" themselves in various ways via cybernetics or to increase their longevity via nanotechnology.[1]

The Christian think tank opines:

> (Nanotechnology) has potentially dark sides to it. Repairing an injury or treating an illness is one thing, enhancing or engineering a "better" human being is another . . . . Unethical restoration is that which seeks to enhance, alter or improve the original design. The underlying question here could simply be: "what is this technology doing to human dignity?"[2]

Where does this anxiety about protecting humanness from nanotechnology come from?

[1]Hook, C. C. 2002. "In Whose Image?: Remaking Humanity Through Cybernetics and Nanotechnology," *Dignity*, Winter.
[2]Taylor, P. 2003. From Fiction to Fact: A Christian Perspective on Future Developments in Bioethics: Nanotechnology and Cybernetics. Center for Bioethics and Public Policy.

UNHELPFUL ONTOLOGICAL CONCRETENESS IN HUMAN COGNITION

One of the few things that may be unique to *Homo sapiens* among all animals is our facility for creating abstract concepts, and one of the earliest abstract concepts was probably the idea that there are spirits in the human body, in animals, and in things. This hypothesis of the abstract ontological unity and continuity of bearness, mountainness, or the human soul was a natural extension of the human self-awareness, the illusion of a unity and continuity in self-identity.[3] This attribution of an abstract ontology to things has its uses, as it allows us to make predictions about how creatures and things of a kind will behave.[4] The belief that others have an inner-self similar to our inner-self is the root of empathy. But these vitalist illusions can also trap us into positing identities that do not exist, of making inaccurate predictions, and persisting with dysfunctional and limiting beliefs.

Human nature is one such limiting, dysfunctional, illusory, and inaccurate belief, the inadequacy of which is revealed in the debates over the moral uses of human enhancement technologies. Take for instance Leon Kass's grounding of his opposition to human enhancement in the existence of Platonic ideal types, including a unitary and inviolate human nature:

> (Creatures) have their given species-specified natures: they are each and all of a given sort. Cockroaches and humans are equally bestowed but differently natured. To turn a man into a cockroach ... would be dehumanizing. To try to turn a man into more than a man might be so as well .... We need a particular regard and respect for the special gift that is our own given nature.[5]

Without ever clearly defining what this human nature is, Kass deploys the concept to both separate us from our continuity with other animals, and bar any improvement in our condition. When exactly does a human's evolution into a cockroach violate his or her human nature? Is it the loss of a skeleton, the growth of the carapace, the hairy legs, or the compound eyes? Can I have tiny antennae, but not big ones? Is it simply the obsessive compulsive fixation on the scent of food and avoiding light? When do humans become dehumanized in becoming more than human?

Few proponents of a distinctive and unitary human nature or soul attempt to answer these questions because they do not have a clear definition of human nature to begin with. They cannot specify when hominids obtained human nature or a soul, or which specific transhuman modifications would rob us of this vitalist essence. They can not agree which aspects of the mind and behavior are part of the soul or human nature, and which are unnatural, or how parts of human nature might also be shared by other animals. Only after we have deconstructed their illusory theory of a human nature, can we begin a serious discussion of the qualities of the human condition worth preserving.

[3]Dennett, D. 1991. *Consciousness Explained*. London. Little Brown.
[4]Dennett, D. 1987. *The Intentional Stance*. Cambridge, MA: Bradford Books.
[5]Kass, L. "Ageless Bodies, Happy Souls: Biotechnology and the Pursuit of Perfection," *The New Atlantis*, Number 1, Spring 2003, pp. 9–28.

HUMAN NATURE HAS NO CLEAR DEFINITION

One clear problem with the idea of human nature is that, despite thousands of years of investigation, and intimate access to the subject of investigation, there is no agreement about what human nature is. Are we innately good, compassionate, and altruistic, or evil, sinful, and selfish? Is moral striving a liberation of our true human nature from sinful influences or capitalism, or is moral behavior a persistent struggle of the good in human nature against its dark side? Or are we a blank slate, morally and behaviorally, or inscribed with all our personality traits, and even beliefs, at birth? Some writers identify human nature with the apparently distinctive human capacities for cognition, language, tool-use, and the creation of meaning and categories, while others include physiology that we share with other species, such as mortality, limbically-mediated emotions, and our genetic predispositions for altruism and aggression.

Cognitive neuroscience, ethology, and evolutionary psychology are attempting to specify the exact structure and epidemiology of human cognitive traits, and clarify which capacities and impulses are genetically innate and which are plastic or learned. These efforts continue to generate enormous insight, but they have also given some succor to advocates of human nature and natural law, even though these sciences simultaneously challenge the traditional understandings of free will, personal identity, and human exceptionalism. Nonetheless, the re-reification of human nature by some evolutionary psychologists has led them to echo Kass' pessimism about human enhancement. Leading evolutionary psychologist Stephen Pinker says, for example:

> After decades of exile, the concept of human nature is back. It has been rehabilitated both by scientific findings that the mind has a universal, genetically shaped organization, and by philosophical analyses that have dispelled the fear that the concept is morally and politically tainted. So if human nature exists, can it be changed? Attempts to redesign human nature ... are generally recognized as futile, dangerous, and unnecessary to achieve moral and political progress.[6]

This powerful desire to reconstruct the "Ought" on the genetic "Is" gives us reason to question the claim that evolutionary psychology has revealed a unitary and universal human nature. In *Adapting Minds*, David Buller's[7] careful deconstruction of evolutionary psychology, he argues that the field has generated little evidence that human beings have specific genetically driven modules adapted for Pleistocene existence. Rather, Buller believes the distinctive achievement of human evolution was the development of general cognitive plasticity, a dynamic adaptive intelligence that has allowed humans to invent and reinvent ourselves.

Of course, it is true that there are myriad genetic, hormonal, and physiological features that shape our desires, thoughts, and behavior, some of which we share with most other human beings. But this constellation of influences fails as a

[6]Pinker, S. 2003. "Can We Change Human Nature?" A Talk presented at The Future of Human Nature, April 11–12, 2003.
[7]Buller, David. 2005. Adapting Minds: Evolutionary Psychology and the Persistent Quest for Human Nature. Cambridge MA: MIT Press, Bradford Books, 2005.

theory of human nature on both analytical and normative grounds. It fails analytically because it posits a vague constellation of species-typical traits that had no clear beginning, are not actually species-specific, and are not clearly threatened by any specific enhancement. Normatively, the argument fails because we are not morally bound by our genes.

HUMAN NATURE: NO CLEAR BEGINNING AND NO CLEAR BOUNDARY WITH OTHER SPECIES

There is no clear beginning for human nature or the human species. There was, we can assume, no day when all our hominid precursors gave birth to modern humans with opposable thumbs, hidden estrus, upright posture, language ability, abstract cognition, and tool use. These traits may have emerged abruptly in evolutionary time, but the periods were still tens or hundreds of thousand of years. Which grandmothers or grandfathers would the defenders of human nature determine finally had "it," and were not just savage beasts like their parents? Our branch of the evolutionary tree shows continuous change, right up through the last 15,000 years.[8] Without specifying which traits confer membership in humanity, it is not clear whether our genetic differences from Pleistocene humans mean we share human nature with them or not. Did the recently discovered tool-using "hobbits" of Indonesia, *homo florensis*, have human nature?

Similarly, we share with primates almost all the qualities that allegedly make us special: self-awareness, culture, language, and tool use. No, they are not good at abstract reasoning or grammar, but then neither are small children, the demented, or the developmentally delayed, and yet they apparently have human nature.

Accepting that the things we value and attribute to human nature are actually shared continuously with nonhuman ancestors and contemporaneous species is not a devaluation of those traits, or of humanity. In fact, it is only by affirming the value of reason, language, compassion, and culture making that we can build an ethical framework to guide human enhancement technologies.

HUMAN NATURE HAS NO CLEAR ENDING

Without a clear definition of human nature, or specification of the things of value, the opponents of human enhancement technology flounder in defining which enhancements cross the line. Francis Fukuyama and the President's Council on Bioethics see the line being crossed with Ritalin, antidepressants, antitrauma drugs, and preimplantation genetic diagnosis, while others focus further along on the advent of superintelligent immortals and human–animal hybrids. David Reardon, an antiabortion

---

[8]Philips, M. L. 2006. "Many Human Genes Evolved Recently", *New Scientist*. March 7. Available at http://www.newscientist.com/channel/being-human/dn8812.html: Voight B. F., et al. 2006. "A Map of Recent Positive Selection in the Human Genome," *PLOS Biology* 4(3): e72. Available at http://biology.plosjournals.org/perlserv/?request=get-document&doi?=10.1371/journal.pbio.0040072.

activist who is promoting an amendment to the Missouri constitution to forbid human genetic engineering, cloning, and transhumanism, has said:

> Any ethic that fails to (1) define human nature, and (2) assign some value to protecting human nature, inherently lacks the ability to find any limits on the justifications that can be offered to alter or destroy human nature, human beings, or humanity.[9]

Reardon is, of course, completely wrong. Although human nature is being deployed to stop enhancement, the vague and chimerical concept provides no clear lines and policy conclusions. It is only when we let go of the notion of a unitary and inviolate human nature that we can turn to the challenge of delineating which features of embodied human existence are so important that we want to preserve them from technological modification, and which are so central that we want to encourage their enhancement and further evolution.

HUMAN NATURE IS NOT NORMATIVE

Even if we do have some clear set of evolved traits that were distinctively human, they are not normatively binding on us.[10] To the extent that we are born with impulses for aggression, racism, or selfishness, or limits on our capacity for wisdom, awe, or compassionate action, we may, in fact, be morally obliged to modify human nature.[11]

The boldest and most interesting defense of the naturalistic fallacy of a moral imperative of human nature comes from Francis Fukuyama. Fukuyama argues that human rights and social solidarity are grounded in a shared human nature. Any effort to tinker with human nature will erode social solidarity and lead to totalitarianism. But he explicitly refuses to define human nature, calling it simply "Factor X." He states:

> Factor X cannot be reduced to the possession of moral choice, or reason, or language, or sociability, or sentience, or emotions, or consciousness, or any other quality that has been put forth as a ground for human dignity. It is all these qualities coming together in a human whole that make up factor X.[12]

This argument for human nature as an ineffable *gestalt* is very convenient. If human nature were the sum of these features rather than their irreducible whole, then they might be individually improved, and human nature with them. If human

[9]Reardon, D. 2005. "Unenhanced Ethics," *PLOS Medicine*. Available at http://medicine.plosjournals.org/perlserv/?request=read-response&doi=10.1371/journal.pmed.0020121.
[10]Bayertz, K. 2003. "Human Nature: How Normative Might It Be?", *Journal of Medicine and Philosophy*. 2003 Apr 28(2):131–50.
[11]Savulescu, J. 2005. "New Breeds of Humans: The Moral Obligation to Enhance," *Ethics, Science and Moral Philosophy of Assisted Human Reproduction* (Volume 10, supplement 1) pp. 36–39.
[12]Fukuyama, F. 2002. Our Posthuman Future: Consequences of the Biotechnology Revolution. New York. Farrar, Strauss, and Giroux.

nature was self-awareness, empathy, and the ability for abstract thought, for example, then a green-skinned, four-armed transgenic could still be part of the Jeffersonian polity, and a superior citizen if it were smarter and more empathic.

But Fukuyama's Factor X is also a unique argument that the diversity of humanity must stay within its existing standard deviations from the mean of human traits. This allows Fukuyama to answer the challenge that a normative human nature excludes some existing humans who do not fit this ideal typical model, such as the disabled. Variation in intelligence, longevity, or morphology are acceptable, so long as we stay within our existing parameters of variation. Although our social unity can apparently still encompass conjoined twins, amputees, people born with fur or tails, and the developmentally delayed, mentally ill, and extremely smart, too many kids on Ritalin or too many 130-year olds would apparently break the bell-curved social contract.

But if people 4 feet tall can feel solidarity with people 7 feet tall, why can't the average person not be 6 feet tall instead of 5-and-one-half feet tall? Why would everyone enjoying the happiness or intelligence experienced by the luckiest 1% of the population fracture humanity into racial subgroups? Certainly, the sudden adoption by a minority of superintelligence, immortality, and uploading would challenge existing understandings of shared citizenship, just as shared citizenship had to be forged across racial differences in the past. But human enhancement technologies pose no challenge to Fukuyama's normative standard deviation if all members of a society become more intelligent, long-lived, and beautiful, and gradually move the bell curve to the right.

For Fukuyama and the other bioconservatives, this blurring of the line between unhumans and posthumans is even more horrifying than the emergence of an entirely separate posthuman species. As all good flows from the people of our race having pure Factor X, and race pride in the goodness of our shared Factor X, it must be protected from the complexities of a multiracial society and even more from race-mixing contamination.

THE INESCAPABLE RACISM OF THE HUMAN NATURE CONCEPT

The use of the concept of human nature today is, we see, inescapably racist, *human racist*, with the same consequences for tyranny, violence, and suppression of human diversity as the ideology of European racism before it. The human racists are more inclusive racists than their forebears, but racists nonetheless in their effort to ground solidarity in biological characteristics instead of shared recognition that another being has self-awareness, feelings, and thoughts like our own. We hear in the panicked demands to ban the mixing of human and animal deoxyribonucleic acid (DNA) striking echoes of the demands to protect the purity of the white race from mongrelization. The root of this racialist anxiety was laid bare in Mary Douglas' work;[13] it is the taboo on

---

[13]Mary Douglas 1966. Purity and Danger: An Analysis of the Concepts Pollution and Taboo. London: Routledge & Kegan Paul, 1970.

the violation of categories, the ritual taboos against blurring of lines between male and female, white and black, animal and human; and humans and the gods.

In fact, Yuval Levin, executive director of the President's Council on Bioethics, explicitly embraced Douglas' analysis as the mission of "conservative bioethics" in the inaugural issue of the bioconservative journal *The New Atlantis*. The goal of conservative bioethics, he says, is to defend the taboos that:

> ... stand guard at the border crossings between the realm of the properly human and those of the beasts and the gods. When the boundaries are breached, when degradation or hubris is given expression, our stomachs recoil.[14]

This alleged self-evident repugnance is the same rationale for bans on race-mixing given by all racialists.

The irony is that human-racism is being promoted by some progressives precisely as a means to unify humanity through "species consciousness," just as white American identity was used to meld together Poles, Irish, and Italians, and pan-Arabism and pan-Africanism was promoted to transcend nationalism and tribalism.

The doctrine of a unifying human nature has also become an unquestioned assumption in human rights discourse. For example, the United Nations *Universal Declaration on the Human Genome and Human Rights*[15] states:

> The human genome underlies the fundamental unity of all members of the human family, as well as the recognition of their inherent dignity and diversity.

As with bans on miscegenation, human racists demand bans on human enhancement technologies in order to protect the purity of the human race. President Bush called for a ban on human–animal hybrids in his 2006 State of the Union message, and Missouri has become the first U.S. state to consider a ban on human–animal hybrids, cloning, human–genetic modification, and transhumanism. Bioethicists George Annas and Lori Andrews have been working with an international network of opponents of human enhancement toward an international treaty to make human genetic modification a "crime against humanity." Genetic enhancement, they say:

> can alter the essence of humanity itself (and thus threaten to change the foundation of human rights) by taking human evolution into our own hands and directing it toward the development of a new species, sometimes termed the "posthuman." ... Membership in the human species is central to the meaning and enforcement of human rights.[16]

---

[14]Levin, Y. 2003. "The Paradox of Conservative Bioethics," *The New Atlantis*, Number 1, Spring 2003, pp. 53–65.

[15]United Nations General Assembly. 1998. *Universal Declaration on the Human Genome and Human Rights*. UN General Assembly.

[16]Annas, G. et al. 2002. "Protecting the Endangered Human: Toward an International Treaty Prohibiting Cloning and Inheritable Alterations," *American Journal of Law and Medicine* 2002; 28, 2&3: 151–178.

Again, like the white supremacists, Annas justifies the suppression of posthumanity on the grounds that they are destined to engage in race war to enslave or exterminate the pure humans:

> The posthuman will come to see us (the garden variety human) as an inferior subspecies without human rights to be enslaved or slaughtered preemptively. It is this potential for genocide based on genetic difference, that I have termed "genetic genocide," that makes species-altering genetic engineering a potential weapon of mass destruction.[16]

THE VIOLENT POTENTIAL OF THE HUMAN RACISTS

Is it mere hyperbole to point to the similarity between the race war apocalypticism of the white supremacists and the species-extermination apocalypticism of the bioconservatives? Unfortunately not. Beyond the violence that would be done to human life, longevity, and well being by attempts to ban any modification of our chimerical human nature, there is the actual violence that apocalyptic human-racism has already generated, and will generate. Theodore Kaczynski, aka "the Unabomber," waged a bombing campaign for *18 years* in the United States against scientists engaged in projects that he thought threatened human nature, principally through cybernetics and genetic engineering. He wrote:

> Human nature has in the past put certain limits on the development of societies. People could be pushed only so far and no farther. But today this may be changing, because modern technology is developing way of modifying human beings . . . getting rid of industrial society will accomplish a great deal. It will relieve the worst of the pressure on nature so that the scars can begin to heal. It will remove the capacity of organized society to keep increasing its control over nature (including human nature).[17]

Bombers of abortion clinics are also soldiers in the human-racist effort. In the embryo rights belief system, *all* bearers of the human genome have equal moral worth, just as *only* bearers of this human genome have worth. The Christian Right's "Manifesto on Biotechnology and Human Dignity," which calls for a ban on human genetic modification, makes clear the link it sees between defense of the unborn and bans on human enhancement, stating:

> The uniqueness of human nature is at stake. Human dignity is indivisible: the aged, the sick, the very young, those with genetic diseases—every human being is possessed of an equal dignity; any threat to the dignity of one is a threat to us all . . . at every stage of life and in every condition of dependency they are intrinsically valuable and deserving of full moral respect.[18]

---

[17]Kaczynski, T. 1995. "Unabomber Manifesto." Available at http://www.thecourier.com/manifest.htm.
[18]Anderson, C. et al. 2003. "Manifesto on Biotechnology and Human Dignity." Available at http://www.cbc-network.org/redesigned/manifesto.php.

It is, therefore, no surprise to see common cause being made between prochoice leftist opponents of human enhancement and antiabortion activists around their common ideology of human racism.

BEYOND HUMAN NATURE: THE NEED FOR A BROAD NORMATIVE RANGE FOR ACCEPTABLE HUMAN ENHANCEMENT

In conclusion, I am *not* arguing for a *laissez faire* approach to nanotechnology or human enhancement, unfettered by moral analysis and political regulation. It would be immoral, and perhaps suicidal, for liberal democracies to be indifferent to the directions in which human beings might evolve using human enhancement technologies. But the concept of a unitary and inviolate human nature is fundamentally the wrong place to start in the analysis of which aspects of human life we want to preserve, suppress, or extend. Rather, we need to make clear that it is our capacities for consciousness, feeling, reason, communication, growth, and empathy, all of which we share to a greater or lesser extent with other animals, that we are willing to use our technologies and the agencies of our collective suasion—legislation, regulation, social norms, and economic incentives—to encourage. It is greed, hatred, ignorance, violence, sickness, and death that we wish to discourage, whether part of human nature or not.

Yes, as a part of that project, we must take account of the insights of neuroscience and evolutionary psychology, even if the efforts to mold them into a natural law is wrong-headed and flawed. Understanding the way our genetic constitution shapes our thought and behavior is essential if we want to use human enhancement technology to improve the human condition, and pursue moral goals that were impossible before human enhancement. So, I will close with Peter Singer's closing thought in his essay *A Darwinian Left*, which argues that the Left must accept that utopian projects have indeed crashed on the shoals of intractable innate human characteristics. But, he says, "there may be a prospect for restoring more far-reaching ambitions of change. We do not know to what extent our capacity to reason can, in the long run, take us beyond the conventional Darwinian constraints on the degree of altruism that a society may be able to foster."[19] I hope, with universal access to human enhancement technologies, we will soon find out.

[19]Singer, P. 2000. A Darwinian Left: Politics, Evolution, and Cooperation. Yale University Press. New Haven, CT.

CHAPTER 6

Nanotechnology Jumps the Gun: Nanoparticles in Consumer Products

BRENT BLACKWELDER

INTRODUCTION

Friends of the Earth U.S. is part of Friends of the Earth International, the world's largest grassroots-based environmental advocacy network with member organizations in 71 countries. In partnership with Friends of the Earth Australia, we issued a report in May of 2006 entitled *Nanomaterials, Sunscreens and Cosmetics: Small Ingredients Big Risks*. The report highlighted the rush to put nanoingredients in consumer products, such as sunscreens and body lotions.

This chapter provides an in-depth presentation of the key findings of our report and makes the case for an immediate moratorium on the production of such consumer items.

Our report focused on the use of nanoparticles in the personal care industry because this sector is one of the primary early adopters of nanomaterials. We recognize that the impacts of nanotechnology reach much further than those associated with the toxicity of personal care products. Nanotechnology's broader impacts on the environment, risks for workers, socioeconomic impacts, and ethical problems are serious issues, but are not discussed in this chapter.

Friends of the Earth is concerned that the nanotechnology industry is rapidly introducing potentially hazardous nanomaterials into our bodies and into our environment without adequate scientific study to ensure that we understand its risks and can prevent harm occurring to people and environment. Friends of Earth is calling for a moratorium on the further commercial release of personal care products that contain nanomaterials, and the withdrawal of such products currently on the market, until adequate, publicly available, peer-reviewed safety studies have been completed, and adequate regulations have been put in place to protect the

Nanoscale: Issues and Perspectives for the Nano Century. Edited by Nigel M. de S. Cameron and M. Ellen Mitchell

general public, the workers manufacturing these products, and the environmental systems into which waste products will be released.

The central finding of our study is that the case of nanotechnology products represents one of the most shocking breakdowns of modern health regulation. It is all the more unpardonable given the tragic failure to learn lessons from asbestos, 1, 1, 1-trichloro-2, 2-bis (P-Chlorophenyl) ethane (DDT), and Polychlorobiphenyls (PCBs). We have found that corporations around the world are rapidly introducing thousands of tons[1] of nanomaterials into the environment and onto the faces and hands of hundreds of millions of people, despite the growing body of evidence indicating that nanomaterials can be toxic for humans and the environment.[2]

THE PROCEDURES USED TO DETERMINE PRODUCTS CONTAINING NANOINGREDIENTS

Nowhere are nanomaterials entering manufacturing and reaching the consumer faster than in personal care products and cosmetics. In 2004, the United Kingdom's (UK's) Royal Society noted that of the engineered nanomaterials in commercial production, the majority were being produced for use in the cosmetics industry.[3] The rush to incorporate nanomaterials in personal care products and cosmetics is especially concerning given the poorly understood risks of nanotoxicity.

As a result of our analysis, Friends of the Earth believes that there are at least several hundred cosmetics, sunscreens, and personal care products that contain nanomaterials now available in the global market. This figure is likely to be a conservative estimate.

In the absence of mandatory product labeling, it is difficult to estimate the number of cosmetics, sunscreens, and personal care products containing nanoparticles that are now commercially available. Estimates necessarily rely on information that product manufacturers—or the few government regulators collecting data on the use of nanomaterials—choose to make publicly available and readily accessible.

Publicly available websites list consumer products that are thought to contain nano-particles, but they rely on the accuracy of information provided by the manufacturer or product distributor. The work conducted by the Woodrow Wilson Center

---

[1]Woodrow Wilson International Center for Scholars (2006). A Nanotechnology Consumer Products Inventory. Available at: http://www.nanotechproject.org/index.php?id=44 Accessed 03.04.06; See also, Swiss Re (2004), Nanotechnology: Small matter, many unknowns. Available at http://www.swissre.com.

[2]For excellent overviews of the emerging field of nanotoxicology, see Oberdörster G. et al. (2005). "Nanotoxicology: an emerging discipline from studies of ultrafine particles." Environmental Health Perspectives 113(7):823–839; Hoet P. et al. (2004). "Nanoparticles—known and unknown health risks." Journal of Nanobiotechnology 2:12; and Oberdörster G. et al. (2005). "Principles for characterising the potential human health effects from exposure to nanomaterials: elements of a screening strategy." Particle and Fibre Toxicology 2:8.

[3]The Royal Society and The Royal Academy of Engineering, UK (2004). Nanoscience and nanotechnologies. Available at http://www.royalsoc.ac.uk.

for International Scholars Project on Emerging Nanotechnologies in its inventory of consumer products,[1] was of great help in the compilation of this appendix.

This appendix includes 116 products: 71 cosmetics, 23 sunscreens, and 22 personal care products that are now thought to incorporate nanomaterials. We recognize that this data represents a small fraction of personal care products containing nanomaterials that are currently on the market, and may not reflect the overall pattern of nanoparticle use across these sectors.

Products listed in this appendix include deodorants, soap, toothpastes, shampoos, hair conditioners, sunscreens, antiwrinkle creams, moisturizers, foundations, face powders, lipstick, blush, eye shadow, nail polish, perfumes, and after-shave lotions. Manufacturers include: L'Oréal; Estée Lauder; Proctor and Gamble; Shiseido; Chanel; Beyond Skin Science, LLC; Revlon; Dr. Brandt; SkinCeuticals; Dermazone Solutions; and many more.

The appendix shows that a wide range of nanomaterials is already being incorporated into personal care products. Nanoscale ingredients listed in the appendix include nanoparticles of titanium dioxide, zinc oxide, alumina, silver, silicon dioxide, calcium fluoride, and copper, as well as nanosomes, nanoemulsions, and nanoencapsulated delivery systems. Disturbingly, seven face creams list fullerenes as ingredients: a substance found to cause brain damage in fish[4] and toxic effects in human liver cells.[5]

The Australian Therapeutic Goods Administration (TGA) has stated that there are close to 400 sunscreen products alone that contain nanoparticle titanium dioxide and or nanoparticle zinc oxide that are currently commercially available in Australia.[6] However, the TGA has failed to disclose the names of these products, leaving the public to guess which of its sunscreens contain nanomaterials. There is no information available on the use of nanomaterials within the nontherapeutic cosmetics and personal care sectors in Australia.

The U.S. Food and Drug Administration (FDA) has not disclosed any relevant figures for the United States. Our report, based on preliminary web searches of publicly available information, contains the details of 116 cosmetics, personal care products, and sunscreens that now incorporate nanomaterials.

On its website,[7] the FDA notes that: "FDA is aware that a few cosmetic products claim to contain nanoparticles to increase the stability or modify release of ingredients." Our findings suggest that this estimate is seriously outdated; regulators in both Australia and the United States need to recognize the rapid market expansion of personal care products and cosmetics containing nanomaterials.

---

[4]Oberdörster E. (2004). "Manufactured nanomaterials (fullerenes, C60) induce oxidative stress in the brain of juvenile largemouth bass." Environmental Health Perspectives 112:1058–1062.

[5]Sayes C. et al. (2004). "The differential cytotoxicity of water-soluble fullerenes." Nanolett. 4:1881–1887.

[6]Australian TGA (2006). Safety of sunscreens containing nanoparticles of zinc oxide or titanium dioxide. Available at: http://www.tga.gov.au/npmeds/sunscreen-zotd.htm (accessed March 3, 2006).

[7]U.S. Food and Drug Administration (1999). HHS, Sunscreen Drug Products For Over-The-Counter Human Use; Final Monograph, 64 Fed. Reg. 27666-27693, 27671.

WHY SIZE MATTERS

The fundamental properties of matter change at the nanoscale. The properties of atoms and molecules are not governed by the same physical laws as larger objects or even larger particles, but by "quantum mechanics." The physical and chemical properties of nano-sized particles can therefore be quite different from those of larger particles of the same substance. Altered properties can include, but are not limited to, color, solubility, material strength, electrical conductivity, magnetic behavior, mobility (within the environment and within the human body), chemical reactivity, and biological activity.[2]

Nanotoxicology is an emerging field, with a small number of peer-reviewed studies published to date. It is often suggested by proponents of nanotechnology that we do not yet know enough about the behavior of nanoparticles to determine whether they pose enhanced risks to human health. However, the existing body of toxicological literature suggests clearly that nanoparticles have a greater risk of toxicity than larger particles.

This body of evidence has been sufficient for the world's oldest scientific organization to warn that we should not continue to release products containing nanomaterials until we have vastly improved requirements for safety testing.[3] There is a general relationship between particle size and toxicity; the smaller a particle, the greater its surface-area-to-volume ratio, and the more likely it is to prove toxic.[8] Toxicity is partly a result of the increased chemical reactivity that accompanies a greater surface area to volume ratio.[8]

The small size, greater surface area, and enhanced chemical reactivity of nanoparticles result in increased production of reactive oxygen species (ROS), including free radicals.[8] The ROS production has been found in a diverse range of nanomaterials, including carbon fullerenes, carbon nanotubes, and nanoparticle metal oxides.

ROS and free radical production is one of the primary mechanisms of nanoparticle toxicity; it may result in oxidative stress, inflammation, and consequent damage to proteins, membranes, and DNA.[8]

Size is therefore a key factor in determining the potential toxicity of a particle. Other factors influencing toxicity include shape, chemical composition, surface structure, surface charge, aggregation, and solubility. Because of their size, nanoparticles are more readily taken up by the human body than larger sized particles and are able to cross biological membranes and access cells, tissues, and organs that larger sized particles normally cannot.

Nanomaterials can gain access to the blood stream following inhalation or ingestion, and possibly also via skin absorption, especially if the skin is compromised.[9]

---

[8]Nel A. et al. (2006). "Toxic potential of materials at the nanolevel." Science Vol 311:622–627. Hoet P., Bruske-Holfeld I. and Salata O. (2004). "Nanoparticles – known and unknown health risks." Journal of Nanobiotechnology 2:12.

[9]Tan M. (1996). "A pilot study on the percutaneous absorption of microfine titanium dioxide from sunscreens." Australasian Journal of Dermatology 37(4):185–187; Lansdown, A., Taylor, A. (1997). "Zinc and titanium oxides: promising UV-absorbers but what influence do they have on the intact skin?" International Journal of Cosmetic Science 19:167–172.

Once in the blood stream, nanomaterials can be transported around the body and are taken up by organs and tissues, including the brain, heart, liver, kidneys, spleen, bone marrow, and nervous system.[10] Once in the blood stream, the major distribution sites for nanoparticles appear to be the liver, followed by the spleen. The length of time that nanoparticles may remain in vital organs and what dose may cause a harmful effect remains unknown.

Diseases of the liver suggest that the accumulation of even harmless foreign matter may impair its function and result in harm. Carbon nanotubes (nanoscale cylinders made of carbon atoms) have been shown to cause the death of kidney cells and to inhibit further cell growth.

Many types of nanoparticles have proven to be toxic to human tissue and cell cultures, resulting in increased oxidative stress, inflamatory cytokine production, DNA mutation, and even cell death. Unlike larger particles, nanoparticles may be transported within cells and be taken up by cell mitochondria and the cell nucleus, where they can induce major structural damage to mitochondria, cause DNA mutation, and even result in cell death.[11]

Nanoparticles of titanium dioxide and zinc oxide used in large numbers of cosmetics, sunscreens, and personal care products are photoactive, producing free radicals and causing DNA damage to human skin cells when exposed to ultraviolet (UV) light. Nanoparticle titanium dioxide has been shown to cause far greater damage to DNA than does titanium dioxide of larger particle size. Whereas 500-nm titanium dioxide particles have only a small ability to cause DNA strand breakage, 20-nm particles of titanium dioxide are capable of causing complete destruction of supercoiled DNA, even at low doses and in the absence of exposure to UV. The potential for sunscreens containing nanoparticles to result in harm is made greater as ROS and free radical production increases with exposure to light and UV.[12]

HEALTH RISKS

Use of personal care products poses clear risks of exposure to untested nanomaterials: they are used daily, are designed to be used directly on the skin, may be inhaled, and are often ingested.

Furthermore, many cosmetics, and personal care products contain ingredients that act as "penetration enhancers,"[13] raising concerns that they may increase the

---

[10]Geiser M. et al. (2005). "Ultrafine particles cross cellular membranes by non-phagocytic mechanisms in lungs and in cultured cells." Environmental Health Perspectives 113(11):1555–1560.

[11]Li N. et al. (2003). "Ultrafine particulate pollutants induce oxidative stress and mitochondrial damage." Environmental Health Perspectives 111(4):455–460; Savic R et al. (2003). "Micellar nanocontainers distribute to defined cytoplasmic organelles." Science 300:615–618.

[12]Donalson K. et al. (1996). "Free radical activity associated with the surface of particles: a unifying factor in determining biological activity?" Toxicology Letters 88:293–298. See footnotes 2, 4, 9–11.

[13]Environmental Working Group (2004). Skin Deep. Available at http://www.ewg.org/issues/cosmetics/FDA_warning/index.php Environmental Working Group (2006). Skin Deep: News about the safety of popular health and beauty brands. Available at http://www.ewg.org/reports/skindeep2/.

likelihood of skin uptake of nanomaterials and possible entry into the blood stream. In 2004, the world's oldest scientific organization, the Royal Society, warned that the risks of nanotoxicity were significantly serious as to warrant nanomaterials being assessed as new chemicals.[3] It warned that the toxicity of nanoparticles cannot be predicted from the known properties of larger-sized particles.

The Royal Society recommended that "ingredients in the form of nanoparticles should undergo a full safety assessment by the relevant scientific advisory body before they are permitted for use in products."[3] The Society also recommended that products containing nanoscale ingredients should be clearly labeled, to enable people to make an informed decision about using these products.[3]

But despite recognition at the highest scientific levels of the enhanced risks associated with nanomaterials used in cosmetics and personal care products, there are as of yet no regulations anywhere in the world that specifically cover their manufacture and marketing. Meanwhile, there is no requirement anywhere in the world for labeling of nanoscale ingredients to allow the public to make an informed choice about using nanoproducts.

Carbon fullerenes (buckyballs), currently being used in some face creams and moisturizers, have been found to cause brain damage in fish, kill water fleas, and have bactericidal properties.[4] Even low levels of exposure to fullerenes have been shown to be toxic to human liver cells.

Researchers are investigating the ability of surface coatings and modifications to make nanomaterials, such as fullerenes, safe. However, studies have shown that both surface coatings and modifications can be weathered over a 1- to 4-hour period by exposure to the oxygen in air or by UV irradiation, suggesting that the protective qualities of surface coatings can be short lived. There is also a concern that ingested coatings could be metabolized to expose the core harmful nanomaterial.[2]

Friends of the Earth challenges the ethics of regulators who would permit fullerenes (nanoparticles linked to brain damage and exhibiting toxicity) to be included in moisturizers and face creams in the absence of independent safety testing. Yet in an act of disturbing regulatory negligence, that is exactly what has happened.

The risks associated with this rash incorporation of fullerenes into cosmetics is underscored by the recent comment by Prof. Robert F. Curl, Jr., who shared the 1996 Nobel Prize in Chemistry for his codiscovery of fullerenes, that he would avoid using cosmetics containing fullerenes until their risks were better understood: "I would take the conservative path of avoiding using such cosmetics while withholding judgment on the actual merits or demerits of their use."[14]

In fact, when a scientist at an international nanotoxicology meeting recently asked her 200 colleagues present who would feel comfortable using face cream that contained fullerenes, fewer than ten indicated that they would.

The sobering reality is that, whereas these 200 scientists are in a position to understand the significance of the health risks posed by fullerenes, and are able to make a decision to avoid such products, most consumers lack this vital information,

[14]Halford B. (2006). "Fullerene For The Face: Cosmetics containing C60 nanoparticles are entering the market, even if their safety is unclear." Chemical & Engineering News. Vol. 84 (13):47.

and rely on government regulators to protect their safety by preventing such dangerous products from being released onto the market.

Cosmetics manufacturers, and even the Australian Therapeutic Goods Administration,[6] claim that the potential for nanoingredients in sunscreens and personal care products to be toxic to living cells and tissues is not a serious concern because nanoparticles remain in the outer layers of dead skin. The problem is that no one knows if this assertion is true. We do know that broken skin is an ineffective barrier and enables particles up to 7000 nm in size to reach living tissue. This suggests that the presence of acne, eczema, or shaving wounds is likely to enable the uptake of nanoparticles.

The Royal Society has called for additional research into the influence of skin condition, including sun burn, on the uptake of nanomaterials, especially in the assessment of nanomaterials found in sunscreens and cosmetics.[3] However, the fact that many cosmetics and personal care products are used on blemished skin or following shaving has been largely ignored in the discussion about skin uptake of nanomaterials found in personal care products to date. If nanoparticles are able to penetrate the outer layer of dead skin cells and gain access to the living cells within, they can join the blood stream and circulate around the body with uptake by cells, tissues, and organs. Other substances, for example, organic liquids, pharmaceuticals, and phthalate monoesters in personal care products,[15] are known to access the blood stream via skin uptake. However, there has been very little published research into skin uptake of nanomaterials in cosmetics and personal care products that are already commercially available. Penetration of intact skin is in part dependent on particle size, meaning that skin uptake of nanoparticles is comparably more likely than uptake of larger particles. The ability of 1000-nm particles to access the dermis when intact skin is flexed has been demonstrated. This suggests that uptake of 100-nm particles is possible in at least some circumstances. Preliminary study of the ability of zinc oxide and titanium oxide nanoparticles to cross the skin has produced conflicting results. Most studies found that these nanoparticles did not reach the living cells, while at least two pilot studies suggest that they did.[9] However, the few studies that have examined the ability of nanoparticles to cross the skin have generally been narrow in scope and have not adequately investigated the role of key variables that may influence skin uptake. It is especially important to investigate the role of base carriers that enhance skin uptake of nanoparticles by altering skin structure or increasing the solubility of the nanoparticle in the skin.

Skin Deep, a recent report by U.S.-based Environmental Working Group on the health risks of commercially available cosmetics and personal care products, found that more than one-half of all cosmetics contained ingredients that act as "penetration enhancers."[13] This suggests that testing of skin uptake of nanoparticle

[15]Duty S. et al. (2005). "Personal care use predicts urinary concentration of some phthalate monoesters." Environmental Health Perspectives 113(11):1530–1535. For example see Pflücker P. et al. (2001). "The Human Stratum corneum Layer: An Effective Barrier against Dermal Uptake of Topically Applied Titanium Dioxide." Skin Pharmacology and Applied Skin Physiology 14 (Suppl. 1): 92–97; Lademann J., et al. (1999). "Penetration of titanium dioxide microparticles in a sunscreen formulation into the horny layer and the follicular orifice." Skin Pharamacol Appl. Skin Physiol 12:247–256.

ingredients should be undertaken in the context of whole products, recognizing that other product ingredients may play a penetration enhancing role. Exposure to nano-materials in "real life" conditions must also be investigated given that flexing and massage have been demonstrated to increase skin uptake of larger particles, drugs, and dyes. Physical and chemical properties of nanoparticles that may influence skin uptake and that require investigation are particle size and shape; surface characteristics including the presence of coatings; electronic charge; and dose.

Publicly funded research into the interactions between nanomaterials and the skin is being undertaken currently by both the European Union (EU) and the United States. However, little of this information has yet been published in peer-reviewed, publicly accessible literature, and most studies are likely to continue for several years before publishing their results.

FAILURE TO CONDUCT OR PUBLICIZE HEALTH STUDIES

While we know very little about the toxicological effects of nanomaterials, such as titanium dioxide and zinc oxide on the human body, we know even less about a host of other nanomaterials currently being used in cosmetics, including carbon fullerenes (buckyballs), and iron, aluminum, zirconium, silicon, and manganese nano oxides.

One of the key problems is that we do not know how much safety research the sunscreen and cosmetics manufacturers are actually conducting. Some manufacturers claim that their products are "photostable" (i.e., do not produce ROS or free radicals when exposed to light or UV), or that their technology "helps to keep free radicals at bay." However, in the absence of peer-reviewed, publicly accessible information from cosmetics companies, it is impossible to know how adequate safety assessment has been.

As Sue Windebank, senior spokesperson for the UK Royal Society, said in 2005:

> it seems that there is really very little publicly funded research looking into the effects of nanoparticles being taken into the body through the skin.... The cosmetics companies may of course be doing their own research, but much of the information about what kind of safety assessments are being undertaken is not publicly listed.... Our concern is that manufacturers ensure that the toxicological tests that they use recognize that nanoparticles of a given chemical will often have different properties to the same chemical in its larger form and may have greater toxicity.... It is certainly not a cloak and dagger situation with the cosmetics companies, but it would help if they were more transparent about the results of their safety tests.[16]

This sense of frustration has been echoed by Dr. Bethany Halford, scientist and science journalist, writing in *Chemical & Engineering News* about the lack of safety data available for the face creams that contain fullerenes, for which she was assured by the manufacturer that (unpublished) safety testing had been carried

[16]Sue Windebank, senior spokesperson for the UK Royal Society, cited in Pitman, S. (2005). "Scientific body calls for more transparency on nanoparticles." Cosmetics Online-Europe.

out: "Why don't manufacturers make [safety] data readily available to their customers . . .? It doesn't seem that much to ask when you're paying about $250 for a jar of face cream."[17]

The UK Royal Society has made clear its view that greater safety testing of products that contain nanomaterials, and greater transparency in the conduct of safety testing, is required. Its 2004 joint report with the UK Royal Academy urged companies wishing to commercialize cosmetics containing nanomaterials to publish peer-reviewed, publicly accessible safety studies, and then label their products to allow consumers to make an informed choice:

> We recommend that ingredients in the form of nanoparticles undergo a full safety assessment by the relevant scientific advisory body before they are permitted for use in products. . . . We recommend that manufacturers publish details of the methodologies they have used in assessing the safety of their products containing nanoparticles that demonstrate how they have taken account that properties of nanoparticles may be different from larger forms . . . We recommend that the ingredients lists of consumer products should identify the fact that manufactured nanoparticulate material has been added.[3]

The call for new safety assessment of nanoingredients in cosmetics has even been echoed by some industry commentators, including Simon Pitman, editor of CosmeticsDesign.com and CosmeticsDesign-Europe.com, who warned in 2005: "Nanotechnology creates substances with new chemical properties that we do not yet understand. A science with such huge potential deserves closer attention to the possible risks, before it falls the wrong side of belated discoveries of toxicity."[18]

Mathew Nordan, vice president of research for nanotechnology research firm Lux Research, Inc., has also argued for (government-funded) toxicological testing of each nanomaterial to assess its threats to human and environmental health, stating: "It only takes one bad apple to spoil the bunch."[19]

STATUS OF REGULATIONS ON NANOTECHNOLOGY

Increasing numbers of cosmetics and personal care products contain nanomaterials and increasing numbers of scientific papers are demonstrating the general risks associated with nanotoxicity. Yet there has been little effort on the part of the regulators to slow the expansion of the nanocosmetics sector until proper safety testing is done; in particular, testing that ensures personal care products containing

---

[17]Halford, B. "Fullerene for the Face: Cosmetics Containing C_{60} Nanoparticles are Entering the Market, Even if Their Safety is Unclear." *Chemical & Engineering News* (*Science & Technology*) Vol. 84, No. 13 (March 27, 2006). Available at Chemical & Engineering News. Accessed on October 18, 2006.

[18]Pitman S. (2005). "The Evidence on Nanotechnology." Cosmetics Design.Com Europe. Available at http://www.cosmeticsdesign-europe.com/news/ng.asp?n=61511-l-oreal-estee-lauder (accessed April 3, 2006).

[19]Matthee N. Vice President of Research for Lux Research, Inc., cited in Service R. (2005). "Calls Rise for More Research on Toxicology Nanomaterials." *Science* Vol. 310:9.

nanomaterials are safe for the workers who manufacture them, the public who uses them and the environment into which waste nanoproducts are inevitably released.

In Australia, the National Industry Chemicals Notification and Assessment Scheme (NICNAS) regulates the safety of ingredients in cosmetics and personal care products and the Therapeutic Goods Administration (TGA) regulates sunscreens. However, these regulators fail to distinguish between nanoparticles and larger-sized particles. Manufacturers of cosmetics and personal care products are not required to seek approval from NICNAS for the use of nanoparticle ingredients where the use of larger-sized particles of the same substance has already been approved.

Manufacturers of all sunscreens must apply to the TGA for marketing approval, but current regulations do not require manufacturers to distinguish between larger-sized particles and nanoparticles.

Australian regulation of nanomaterials in personal care products therefore remains based on the flawed assumption that the toxicity of nanoparticles can be predicted from the known properties of larger-sized particles. This flies in the face of recommendations from the UK Royal Society for nanoparticles to be assessed as new chemicals.[3]

In the United States, manufacturers of sunscreens are required to seek premarket approval from the FDA if their products are "new drug" products. However, in 1999, the FDA made a decision to allow nanoparticle ingredients to be used in sunscreens without new safety assessments, based on previous safety assessment of larger-sized particles.[7] The FDA has virtually no authority over cosmetics and personal care products and cannot require manufacturers to conduct safety studies. Only 11% of the 10,500 ingredients used in cosmetics products have been assessed for safety by the industry-funded Cosmetics Industry Review Panel.

A recent report by the Woodrow Wilson Center's Project on Emerging Nanotechnologies strongly criticized the current approach to regulating cosmetics as wholly inadequate in dealing with the risks posed by nanotechnologies: "Although the FDCA [Food, Drug and Cosmetic Act] has a lot of language devoted to cosmetics, it is not too much of an exaggeration to say that cosmetics in the USA are essentially unregulated."[18]

RESEARCH AND FUNDING OF NANOTECHNOLOGY SAFETY

In one of the few concrete responses from governments to the Royal Society's recommendations, the EU requested last year that its Scientific Committee on Consumer Products to review previous decisions to allow nanoparticle titanium dioxide and zinc oxide to be permitted for use in sunscreens without new safety assessments.[20] However, there are as yet no specific regulations applying to the use or manufacture of nanoparticle ingredients in cosmetics and personal care products.

[20]European Commission (2005). Scientific Committee on Consumer Products: Request for a scientific opinion: Safety of nanomaterials in consumer products. Available at http://europa.eu.int/comm/health/ph_risk/committees/04_sccp/docs/sccp_nano_en.pdf (accessed May 3, 2006).

The emerging findings of the dangers of nanoparticles have rung alarm bells for eminent scientific bodies, including the Royal Society and the Science Council of Japan, both of which have called for greater public funding of the health risks posed by nanoparticles as a matter of urgency.[21]

Governments worldwide have invested billions of dollars of public money in nano research, but they have been more interested in supporting research into profitable commercial applications of nanotechnology, or military research, than health and safety testing. For example, in the $1.3 billion budget for the National Nanotechnology Initiative,[22] only $38.5 million (less than 4%) was earmarked for the study of the health, safety, and environmental impacts of nanotechnology. Conversely, the Department of Defense received $436 million (33.5% of the nanotechnology budget).

The growing evidence of the toxicological risks posed by nanomaterials has prompted increased (albeit inadequate) public funding of studies investigating nano-technology's threats to health, safety, and the environment, including the following:

- In the United States, government agencies, including the FDA, and the National Institute of Environmental Health Sciences, are cooperating through the NTP to study the skin absorption and phototoxicity of nanoparticles of titanium dioxide and zinc oxide preparations used in sunscreens and cosmetics. The NTP is also looking at the uptake and toxicity of fullerenes.

- The Australian government has not yet formally recognized the need to fund nanotechnology research into health and environmental risks of nano-materials. The Therapeutic Goods Administration recently published a lit-erature review of existing studies into the potential for nanomaterials in sunscreens to be absorbed through the skin.[6] However, that review failed to clearly recognize the inadequacies of studies conducted to date or the need for more thorough research.

- The EU has launched a research project called Nanoderm to investigate the quality of the skin as a barrier to formulations containing nanoparticles.

- Japan has launched a collaborative research initiative that includes an evalu-ation of nanomaterials' implications for risk assessment, health issues, environ-mental issues, ethical and social issues, and public acceptance.[20]

- The UK government has not earmarked any specific money for study of the health impacts of nanocosmetics and other consumer products (earning a sharp rebuke from the Royal Society),[3] but has invited research bids for areas it has identified as priorities for nanotechnology research, including the impacts of nanomaterials for human health and the environment. Most of these studies will take several years before publishing results, and much further work will then be required before reliable conclusions can be drawn.

[21]Royal Society-Science Council of Japan (2005). Report of workshop on impacts on nanotechnologies 11–12 July 2005. Available at http://www.royalsoc.ac.uk.
[22]U.S. National Nanotechnology Initiative (2006). What is nanotechnology? Available at http://www.na-no.gov/html/facts/whatIsNano.html See a supplement to the President's FY 2006 Budget.

CONCLUSIONS

Civil society groups, such as Friends of the Earth and others, have argued that the sensible response to a situation where the risks of nanotoxicity have been clearly identified, but remain poorly understood, is to place a moratorium on the commercialization of nanoproducts until the necessary safety research has been conducted.

CHAPTER 7

Nanotechnology: Maximizing Benefits, Minimizing Downsides

JACOB HELLER and CHRISTINE PETERSON

INTRODUCTION

Nanotechnology has immense potential. Within the next few years, nanotechnology is expected to be used in important and diverse applications, from localized cancer treatments to powerful new solar technologies to superstrong-yet-lightweight materials. In the longer term, productive nanosystems, or the capability to build macroscale objects from the molecular level up, should revolutionize the way our products are produced, making it possible to create any object with atomic precision. To advocates and observers of nanotechnology, many potential future benefits of nanotechnology are clear. It is also clear that certain nanotech inventions might also have negative consequences, such as endangering our health or triggering a nanoweapons arms race. Our goal should be to guide nanotechnology's development in order to maximize its potential benefits and minimize its downsides. This guiding process will necessarily involve the input of the technology and business communities, policymakers, nonprofit organizations, and individual citizens.

It is crucial that concerned constituents begin to consider the implications of certain aspects of nanotechnology and participate in informed debate now, as today's choices will necessarily determine tomorrow's technologies. If we were to judge, for example, that many nanotechnologies are potentially so militarily dangerous that exports should be prohibited, much nanotech development would halt immediately. On the other hand, if we were to judge that offensive military use of nanotechnologies is far in the future, obviating the need for export controls today, nanoproducts would almost certainly be exported broadly, whatever the risks may be. Choosing not to make judgments on matters like these is still a choice: it is a decision that the future of this important technology is, instead, left to a combination

Nanoscale: Issues and Perspectives for the Nano Century. Edited by Nigel M. de S. Cameron and M. Ellen Mitchell
Copyright © 2007 John Wiley & Sons, Inc.

of special-interest political motives and markets. For those who believe that these forces alone may not make the best choices for us as a society, it is important that we start considering these choices for ourselves.

This chapter is meant to bring out some of these choices: present and ongoing political and ethical debates that will bear greatly on the future of nanotechnology. What is presented here is neither exclusive nor exhaustive. Each issue, however, is of importance to the policy debates surrounding nanotechnology. We hope that bringing up these issues will generate discussion, thought, and action. Although we may come to tentative conclusions here about these policy matters, we encourage you to consider and come up with new alternatives. In cases where we present existing policy alternatives, we hope we present these alternatives fairly, even those with which we do not agree.

This chapter covers three broad policy areas: innovation, regulation, and implications. Innovation policies are the ways that governments can help ensure nanotechnology is developed, and determine which applications of nanotech are encouraged. Such policies include patent law and government financing of nanotech research and development. Regulatory policy concerns the areas of nanotech development that should arguably be controlled or outlawed, including any potentially dangerous nanoparticles, nanomedical "human enhancements" on those unable to choose for themselves, and the export of nanotechnologies that could be used as weapons. Finally, there is current policy discussion on the possible implications of nanotechnology, including its effects on poverty, inequality, and privacy. Although we discuss these issues through the lens of U.S. policy, most of these policy issues are applicable to the entire world. How governments all over the world approach the financing of nanotechnology, regulate its potential risks, and deal with its implications will shape the future of nanotech and its impact on humanity.

INNOVATION

Government Financing of Nanotech Research and Development

We begin with one of the less-controversial topics in nanotech policy: Should governments finance research and development of nanotechnology? Some answer No on the grounds that either nanotechnology will necessarily carry negative consequences or that governments should not finance scientific research, but those who hold either of these opinions are in the minority. Most believe that nanotechnology has at least the potential for good and believe that it will likely be greatly delayed without government backing, as most early-stage scientific research projects are too risky for the private sector to take on. The libertarian view that financially risky basic research, including in nanotechnology, is best funded only by companies, foundations, and individuals is held by a small minority in the United States today.

Due to this consensus, the U.S. federal government leads the world in nanotechnology research and development funding. The U.S. National Nanotechnology

Initiative (NNI), the research and development (R&D) program that coordinates multiagency efforts in nanotech and nanoscience, allocates more than $1 billion to 14 agencies. Since its inception in 2000, it has been largely regarded a success. Most recently, the President's Council of Advisors on Science and Technology (PCAST) found that the funding is "very well spent," and that "the program is well managed."[1] One criticism offered has been that the program is insufficiently ambitious, particularly in the area of molecular machine systems, the central requirement for productive nanosystems.[2] This issue was clarified in the National Research Council's 2006 review of the NNI, which recommended increases in experimental work toward atomically precise manufacturing.[3]

Based on its relatively near-term focus, the NNI has made valuable contributions to the development of near-term nanotechnologies. With NNI funding, researchers have been developing gold nanoshells that can target malignant cancer cells, low-cost hybrid solar cells, quantum dots that can open the door to much faster computing, and nanoscale iron particles that could reduce the costs of cleaning up contaminated groundwater.[4] Due largely to this high level of funding, the United States currently leads the world in nanotech patents, startups, and papers published.[4]

Some argue that more can and should be done. The United States is beginning to lose its lead in government-sponsored nanotech R&D relative to the rest of the world, and its level of spending is not keeping pace with nanotech development. When adjusted for purchasing-power-parity (a comparison of how much a dollar can buy in the United States relative to other countries), other governments are spending more per capita on nanotech research and development than the United States. By using this scale, the United States spent $5.42 per capita in government funding for nanotech R&D in 2004, while South Korea spent $5.62, Japan, $6.30, and Taiwan, $9.40.[5] China spends $611 million annually (after adjusting for purchasing-power-parity) on nanotech research, nearly 40% of U.S. federal funding.[6] Furthermore, the proposed US nanotech budget for FY 2006 was actually lower than FY 2005 funding when adjusted for inflation;[7] FY 2007 will likely have a similar decreases in funding.

Besides the amount of funding, the structure and duration of most NNI-funded research is problematic. The NNI budget pressures have made the peer review process more conservative, and most grants are given for only 1 year, so many

---

[1]Kvamme, E. F. *Hearing on: Nanotechnology: Where Does the US Stand? The Research Subcommittee of the Committee on Science of the United States House of Representatives.* 115th Cong., 2nd Sess. 1 (2005).
[2]Drexler, K. E. (2004). Nanotechnology: from Feynman to Funding. *Bulletin of Science, Technology and Society* 24, 21–27.
[3]National Materials Advisory Board (2006). *A Matter of Size: Triennial Review of the National Nanotechnology Initiative.* Washington, DC: National Academies Press.
[4]Lane, N. and Thomas K. (2005). The National Nanotechnology Initiative: Present at the Creation. *Issues in Science and Technology* 21(4), 51–52.
[5]Nordan, M. M. *Hearing on: Nanotechnology: Where Does the US Stand? The Research Subcommittee of the Committee on Science of the United States House of Representatives.* 115th Cong., 2nd Sess. 1 (2005).
[6]See footnote 5.
[7]Lane, N. and Kalil, T. (2005). The National Nanotechnology Initiative: Present at the Creation. *Issues in Science and Technology* 21(4), 51–52.

researchers "work ahead"—they essentially perform a given experiment before writing the associated grant proposal, thereby ensuring a successful appearance and increasing the odds of year-after-year funding. This necessarily constrains risk taking and creativity, both of which are essential for major breakthroughs in nanotech research. The U.S. federal government should consider following Japan's example and fund research projects for durations of 5 years, or preferably longer.

Intellectual Property Issues and Nanotech

The leading innovative industries of the twentieth century (computer hardware, software, the Internet, and biotechnology) began with very little patenting. By contrast, nanotechnology patenting has exploded from the outset, with more than 3700 nanotech patents filed between 2001 and 2003—a large number when compared to the relatively few marketed products.[8] The consequences of this boom in nanotech patenting are yet to be seen, but they could impact the level of innovation in nanotech-based industries for years to come.

Patents are traditionally thought to promote innovation. By granting a limited monopoly to the patent holder, patents allow innovators to obtain more profit from their inventions, increasing the incentive to innovate and helping the innovator recoup research costs. Further, a patent signals to potential investors that the innovator is producing tangible, possibly marketable, new results.

On the other hand, patents restrict access to new processes, which could slow down the pace of innovation in an industry. Patent holders can choose to charge licensing fees, or not license patented subject matter at all. As nanotech-based applications are just beginning, most major innovations are basic building blocks that will provide the foundation for future nanotech growth. Most of these building-block nanotechnology innovations have so far been patented, including semiconducting nanocrystals, light-emitting nanocrystals, carbon nanotubes, oxide nanorods, a method for self-assembling nanolayers, and atomic force microscopes.[9] It is likely that many nanotech innovations on the horizon will require the licensing of one or more of these patents. This will mean that nanotechnology research and development efforts will be required to pay hefty licensing fees, will be forced to "work around the patent," or worse, will not be able to operate because a patent holder refuses to license. Thus, patenting could impede growth in nanotechnology.

There are policy responses a government can employ to prevent nanotechnology patents from hampering innovation. It can impose a "strict utility" requirement on all nanotechnology patents, requiring that all patents filed are for a usable product, not a basic idea. This would prevent the patenting of building-block innovations, and only allow later, downstream inventions to be patented. Some argue, however, that

---

[8]Sampat, B. (2005). *Examining Patent Examination: An Analysis of Examiner and Applicant Generated Prior Art.* (Working Paper). Available at http://faculty.haas.berkeley.edu/wakeman/ba297tspring05/Sampat.pdf.

[9]Lemley, M. (2005, November). Patenting Nanotechnology. *Stanford Law Review* 58, 603–604.

nanotechnology's unique technology and industry structure make it a bad candidate for implementing a strict utility requirement.[10]

Another option is for government to mandate that all publicly funded research must be licensed nonexclusively (not just to one or a few parties). The U.S. government does have authority to take this action under the Bayh–Dole Act of 1980 (which originally granted the right to organizations to hold patents for publicly funded innovations), but this authority not yet been put to use. As most of the basic research happening today is being publicly funded at universities, this would help open up a lot of the basic building-block discoveries for further research and improvement. Further policy research is needed to determine whether such an action may be inadvisable in certain high-cost industries, such as pharmaceuticals, which currently require expensive regulatory approvals.

REGULATION

Nanoparticle Safety

Although the field of nanotechnology is relatively new, concerns are already arising that some nanoscale particles may prove toxic to humans or the environment. An emerging body of studies reveals that we are simply uncertain on what effects, if any, nanoparticles will have on the environment, health, and safety (EHS). More must be done to clear up this uncertainty, so that whatever impacts today's early nanomaterials can be appropriately dealt with in a proactive matter, before widespread problems arise.

Some recent studies on EHS effects reveal that nanoparticles have the potential to be unsafe. For example, one study, which has since been widely called into question, found that the introduction of fullerenes (buckyballs) into water with largemouth bass significantly increased cellular damage to the brain tissue.[11] This work has yet to be confirmed. In other studies, nanotubes were found to cause significant damage to the lungs when inhaled.[12] Other theoretical concerns have been raised, because nanoparticles have the unique ability to easily pass through cell walls and can permeate the blood–brain barrier.[13] Nanoparticles may also be bactericidal, which means that they can be highly damaging if introduced into ecosystems where bacteria are at the bottom of the food chain.[14]

---

[10]Almeling, D. S. (2004). Patenting Nanotechnology: Problems with the Utility Requirement. *Stanford Technology Law Review.*

[11]Holmes, B. (2004, April 3). Carbon "footballs" harm fish. *New Scientist* 182(11).

[12]Raloff, J. (2005, March 19). Nano Hazards: Exposure to minute particles harms lungs, circulatory system. *Science News Online* 167(12). Available at http://www.sciencenews.org/articles/20050319/fob1.asp (retrieved October 17, 2006).

[13]Kreuter, J. et al. (1995). Passage of peptides through the blood-brain barrier with colloidal polymer particles (nanoparticles). *Brain Research* 674, 171–174.

[14]Wiesner, M. et al. (2006). Assessing the Risks of Manufactured Nanomaterials. [Electronic version] *Environmental Science and Technology Online* 40(14).

These studies are preliminary, and concerns are mostly speculation. Other research has found that most nanoparticles do not pose a serious threat to human and environmental safety, and many industry leaders and researchers believe that, in theory, most nanoparticles should be benign.[15] Currently, we are simply uncertain as to what EHS problems these early nanomaterials may pose.[16]

Nanotechnologies are expected to offer many benefits for human health and the environment. The very properties of nanoparticles that make them potentially dangerous, such as their ability to cross the blood–brain barrier or destroy specific cells, make them excellent candidates for advanced medication and new drug delivery methods, and may hold the cure for diseases like cancer and AIDS.[17] Nanotechnologies that are under development today can also help monitor pollution, lower energy requirements, and reduce the use of harmful cleansing chemicals. Adopting a strict "precautionary principle," and halting all research in nanotechnology, would rob humanity of nanotech's many benefits indefinitely.

It would be beneficial to discover and appropriately deal with any EHS issues that nanotechnologies may present before they become widespread, while still allowing for nanotech innovation to progress. Not only would this approach help avoid human suffering and damage to the environment that may result from certain nanoparticles, but it would ensure that costly legal battles and a widespread backlash against nanotechnologies are avoided. A better understanding of the EHS impact of nanotechnologies would also clarify the regulatory landscape, which would help foster a favorable nanotech business climate: Nanotech businesses are already withholding some investments because they are afraid the "ground will shift underneath them".[18]

To further understanding of what risks nanoparticles may present, Foresight Nanotech Institute and others have recommended that the U.S. government allocate substantially more funding toward studying the EHS implications of nanomaterials.[19] In FY 2006, the NNI allocated around $38.5 million less than 4% of its budget) to studying the EHS implications. A substantial increase, to at least $100 million, will probably be necessary to properly research EHS implications.[16] Given past histories with substances such as asbestos and lead paint, significant spending on studying EHS risks is warranted to help avoid the huge sums potentially required for legal battles and cleanup costs. Most importantly, it would allow policymakers and government agencies to make proactive policies that manage and prevent possible harms to EHS.

[15]*Small Times Magazine.* "Nanotechnology: Are Safety Concerns Real Or Imagined? Experts Disagree." Available at http://www.smalltimes.com/document_display.cfm?document_id=11460 (retrieved May 3, 2006).

[16]Balbus, J. et al. (2005). Getting Nanotech Right the First Time. *Issues in Science and Technology* 21(4).

[17]de Grey, A. (2002). Increasing Health and Longevity of Human Life. *Foresight Nanotech Institute.* Available at http://foresight.org/challenges/health002.html (retrieved October 17, 2006).

[18]Nordan, M. *Hearing on: Nanotechnology: Where Does the US Stand? The Research Subcommittee of the Committee on Science of the United States House of Representatives.* 115th Cong., 2nd Sess. 1 (2005).

[19]Letter to House Appropriations Committee regarding Funding for Nanotech Health and Safety Research. Available at http://www.epa.gov/sab/pdf/gulledge_acc_np_letter_appropriations.pdf (retrieved February 17, 2006).

Beyond simple nanoparticles, tomorrow's nanodevices and nanosystems may present EHS risks of their own. An early effort to guide development in a safer direction can be seen in the Foresight Guidelines on Responsible Nanotechnology Development, which provide a series of scorecards for researchers, industry, and government to evaluate their progress.[20]

Human Enhancement

The long-term goal of nanotechnology is to be able to fully manipulate molecular and atomic structures. As humans are made of the same basic building blocks as the rest of the natural world, nanotechnology should eventually enable the ability to change human tissues and cells at the molecular level. This will open doors in medicine previously thought impossible, and it will enable us to extend the length and quality of human life. It will also open the door to "enhancements" of the body—in IQ, appearance, and capabilities. Such enhancements would be desired by many, but they also bring up important moral, ethical, and legal questions that society has not yet had to face.

Nanotechnology would likely allow for an enormous array of human enhancements and medical treatments. In the long run, nanotechnology should enable us to analyze and repair any physical ailment in the body. This would mean that nanotechnologies would be able to repair a patient who is damaged or diseased back to full health; an aged body and brain could be restored to the equivalent of a youthful state. The eventual implication could be the end of pain, disease, and aging. Yet, compared to more dramatic changes to the body, these innovations would be *relatively* less controversial, because they are extensions of modern medicine's pursuit of health.

The more controversial changes would likely be "unnatural" enhancements to human talent: much higher intelligence and memory capacity, significantly heightened sense of awareness, even entirely new senses, astonishing athletic capability and strength, and beauty enhancements are just a few examples. These types of enhancements do exist in some forms today, including steroids, Ritalin, and other Attention Deficit and Hyperactivity Disorder (ADHD) medications, Prozac, and plastic surgery, and they have already attracted significant amounts of controversy. We can expect that more dramatic nanotechnology-enabled enhancements will mirror the same controversies, but will be even more contentious.

There is a wide spectrum of positions in the human enhancement debate. On one end is the argument that humans should be allowed to alter themselves to whatever extent they choose. Rational adults are entrusted with decisions about their own health and level of bodily enhancement because only they have control over their own bodies. Under this position, children and the mentally handicapped would not be entitled to enhancement procedures without the consent of their guardians, because they are not considered legally competent, so they cannot make important

---

[20]Jacobstein, N. (2006, April). Foresight Guidelines for Responsible Nanotechnology Development. *Foresight Nanotech Institute and IMM*. Available at http://www.foresight.org/guidelines/current.html (retrieved October 17, 2006).

decisions about their own bodies. This closely mirrors the dominant ideology in the United States today over issues concerning Ritalin and plastic surgery.

Some worry, however, that human enhancements may create a "race to the bottom"—in this case, possibly more accurately termed a "race to the top." Would it be possible to compete for a job if everyone else was enhanced mentally, to compete in athletics if everyone else was strength enhanced, or compete for a spouse if you were not physically enhanced? Especially if nanoenhancement was not entirely safe or carried long-term repercussions, there could be an argument that allowing it in the first place forces people into an impossibly difficult position, and leaves everyone worse off in the long run. This argument, sometimes used in the current debate over steroids, has been proposed as a justification for government regulation of human enhancement.

Others worry that nanotech-enabled enhancements could exacerbate current economic and social disparities. Especially if the technology were expensive and limited, as is normally the case in the early days of any technology, only well-off people would be able to become enhanced. This would mean that the rich would become stronger, smarter, and more beautiful—and as a result, richer. The gap would significantly widen between those who could afford enhancements and those who could not, and the threat of creating a permanent, involuntarily unenhanced underclass would be real. There is the further concern that those controlling most resources, the enhanced class, would feel increasingly disconnected from the underclass, and, as a result, would disengage from efforts to improve their lot, potentially trapping the poor in their position.

Even those most opposed to controls on such procedures may be persuaded that society has a legitimate interest in protecting the well being of citizens unable to choose for themselves. For example, while the vast majority of parents try to do what is best for their children, a minority do not. Parents have been known to cripple their children to enhance their ability to beg; analogous efforts could result in bizarre body modifications with the sole goal of publicity and financial gain. We have a responsibility to protect children against such "enhancements," even when desired by parents.

There are practical considerations about placing limits on human enhancement. As soon as such technologies exist, some will want to enhance themselves, and some doctors will be willing to provide enhancements, regardless of their legal status. If enhancements are banned within a given country, people will simply go elsewhere to become enhanced. If they are banned worldwide, there would likely be an extensive black market. Permitting enhancement procedures may become an issue of necessity if governments prefer to maintain a minimal ability to protect those who need it.

Luckily for regulators, human enhancement of the magnitude discussed here will likely not come about for decades. This should allow time for discussion, reflection, and debate to occur.

Export Controls

Nanotechnology is expected to bring advanced drug-delivery methods, superstrong light-weight materials, and the ability to heal damaged environments. However, these same technologies could also be used to create potent weapons: drug-delivery systems could be turned into bioweapons, strong materials could radically improve fighter aircraft, and nanoenvironmental "healers" could wreak havoc on ecosystems if tweaked. There is a concern that potentially powerful nanotechnologies could fall into nondemocratic or terrorist hands and be used in offensive weapons. To deal with that prospect, policymakers are already considering placing export controls on certain nanotechnologies—attempting to prevent them from reaching unfriendly entities.

Export controls are not new, and there is already a legal and policy framework that deals with all weapons systems and technologies that would most likely apply to nanotechnology. For example, just as today the sale of nuclear weapons technology to "rogue states," such as North Korea, is not permitted, evolving export controls would similarly prohibit the sale of tomorrow's nanoweapons to such states.

The issue is complicated when technologies are "dual-use": they can be used both for civilian and military applications. Today, we do not allow the sale of certain pharmaceuticals to specific foreign countries, because, although they could be used to cure disease, they also contain compounds that can be used in the production of chemical or biological weapons. Even the video-game console PlayStation 2 was subject to Japanese export controls, because some feared that its processing power could be used to break encryption.[21] Some fear that the same could be true of many nanotechnologies, such as the ones mentioned above: Although they do have benign purposes, they could also be used with malignant intent and destructive force. Because the set of nanotechnologies is so broad and will continue to increase for decades, we simply do not know how often it will be possible to reverse-engineer nanotech products and weaponize a given technology.

Already, the prospect of export controls is having a chilling effect on nanotechnology-based companies. Nanotech businesses, afraid that they will not be able to sell their products at a global level, are considering moving their new efforts outside of the United States and, presumably, other countries that use export controls. If this were to happen in large numbers, the effect on the United States' and similar countries' lead in nanotechnology would be disastrous. It would also negate the effect of any export control regime, because companies operating beyond controls would presumably be able to sell their products to whomever they wish. It is for this reason that the President's Export Council (PEC), upon studying the issue of nanotech export controls, concluded that they would have to be "multilateral in order to be effective."[22]

Even if nanotech companies do not move, restrictions on exports could have a dampening effect and put U.S. nanotech companies at a competitive disadvantage. Export controls restrict trade and commerce, which means that U.S. companies

---

[21]C Net News, 2000. *Japan slaps export controls on PlayStation 2*. Available at http://news.com.com/2100-1040-239322.html (retrieved October 17, 2006).

[22]Marriott, Jr., J. W. (2006). *President's Export Council Letter to President Bush Concerning Export Controls on Nanotechnology*. Available at http://www.ita.doc.gov/TD/PEC/nanotech.html (retrieved October 17, 2006).

would not be able to profit fully from their innovations, discouraging further research and development. More troublesome is that they could prevent global collaboration and research sharing, especially with up-and-coming nanotechnology-based industries in countries, such as China. Policymakers must be aware of this cost to innovation and the economy, a concern that was also echoed by the PEC.[22]

Some specific export restraints on specific nanotechnologies will prove necessary. It would not be prudent to share with irresponsible entities technology that is already in a weaponized form, or can easily be used for destructive purposes. Significant research into what those technologies would be may be necessary, so nanotech companies can know in advance which types of products would be subject to restricted trade.

IMPLICATIONS

Poverty and Disparity

Nanotechnology has the potential to dramatically improve the position of the world's poorest people. Nanotech inventions could potentially help fight tropical disease, produce an abundance of food, provide for cleaner water, make the transport of goods easier and cheaper to people in remote areas, and provide clean and cheap energy sources. However, whether these inventions are created or made cheap enough for the poor in a timely fashion is questionable. The history of other technologies suggests that nanotech may not benefit those most in need, instead strengthening the positions of their richer competitors. Working to ensure that nanotech benefits all of humanity (rich and poor alike) will require thoughtful strategies. Deliberate action will need to be taken by noncommercial actors if the benefits from nanotechnology are to be widespread and not exacerbate already large disparities between the developed and developing worlds.

Nanotechnology has been singled out for its unique potential to alleviate the suffering of the poor in the developed and developing worlds. In the long run, productive nanosystems are expected to enable cheaper and more efficient production processes, which may lower the price of all generic necessities.[23] Historically, technology has been the driving force behind economic development and progress; different rates of technological progress (and the cultural factors underlying those rates) largely explain the disparity in incomes and quality of life between the developed and developing world.[24] Nanotechnology, if opened up to the developing world, could have profoundly positive effects on their potential for growth.

Parts of the developing world have already been relatively involved in nanotech research and development. As mentioned above, on a purchasing-power-parity-adjusted basis, China currently spends around $600 million annually ranking third in nanotech research after the United States and Japan. Countries including

[23]Foresight Nanotech Institute (2005). *The Technology Roadmap for Productive Nanosystems.* Available at http://www.foresight.org/roadmaps/index.html (retrieved October 17, 2006).
[24]Sachs, J. (2002, August 17). The essential ingredient. *The New Scientist* 52.

Brazil, India, Thailand, and South Africa have already allocated tens of millions of dollars for nanotechnology. One can hope that the fruits of this research will benefit the economies of these countries, which together represent almost one-half of the world's population, including some of the world's poorest people.

Nanotechnologies currently under development could also be applied to solve many of the problems faced by the developing world. A paper by Fabio Salamanca-Buentello et al.[25] found that nanotechnology could be applied to attain at least five of the eight United Nations Millennium Development Goals (MDG). For example, the production of hydrogen storage systems based on carbon nanotubes, photovoltaic cells based on quantum dot technology, and nanocrystals for hydrogen creation could greatly enhance environmentally sustainable growth, the seventh MDG. Nanotech scientists also widely agree that nanotechnology could be used to increase agricultural productivity, enable better and cheaper water enhancement and purification technologies, and aid in disease diagnosis and screening.[25]

Similar hopes were echoed for other technologies in the past, but often benefits to the poor have not yet materialized. In the 1980s, there was great hope that biotechnologies, especially genetically modified organisms (GMOs), would solve hunger problems in the developing world. More than one-quarter of a century later, it is apparent that most of the benefits of GMOs to date have accrued in the developed world, where they are most widely grown and consumed. Many of the innovations that were supposed to dramatically improve the quality of life in the developing world, such as plants that could grow in the arid deserts of Africa, have not been developed. The same is currently true with pharmaceuticals. Antiretroviral acquired immune deficiency syndrome (AIDS) medication can substantially increase the quality and length of life for those infected with AIDS, while significantly decreasing the rate of transmission of the deadly virus. However, most of these drugs are held under patents, which raise the prices of these drugs (because they are being produced by a monopoly) and do not allow developing countries to develop their own, cheaper generic drugs. Because of this, tens of millions of the world's poorest people with AIDS have no access to lifesaving medication. Despite this emergency, the World Trade Organization did not grant the needed humanitarian needs-based waiver until 2003. Even now, the required license procedure is claimed to be complex, cumbersome, and inefficient.[26]

Nanotechnology could easily go the way of GMOs and AIDS drugs, especially if policy action is not taken.[27] If it did, it might affect the developing world in an even more serious manner than GMOs, as not only would the benefits of nanotech innovations be denied to the developing world, but the economic gap between the developed and developing world could widen. Nanotechnology benefits come from both the production process and the final product. Without enough capital and know-how

[25]Salamanca-Buentello, F. (2005, May). Nanotechnology and the Developing World. *PLoS Medicine* 2(5). Available at http://www.pubmedcentral.nih.gov/articlerender.fcgi?tool=pubmed&pubmedid=15807631#pmed-0020097-b6 (retrieved October 17, 2006).
[26]Net Aid. (2005). *WTO Approves Change to Bring Cheaper Drugs to Poor*. Available at http://www.netaid.org/press/news/page.jsp?itemID=27393428 (retrieved October 17, 2006).
[27]Invernizzi, N. and Guillermo F. (2005). Nanotechnology and the Developing World: Will Nanotechnology Overcome Poverty or Widen Disparity?. *Nanotechnology Law and Business* 2(3).

to accumulate and use nanotech-enabled production technologies, the developing world would be even more strongly outcompeted by the developed world, possibly losing markets they now address successfully. The increased economic disparities created could be highly destabilizing, with associated increases in terrorism and warfare.

This prospect could be made less likely through conscious efforts in both the developed and developing world. Such efforts might include increased investment into encouraging world nanotechnology research and development, allowing developing countries an exemption on specific basic nanotech patents, or using an "open source" model for such inventions, and substantial resource and knowledge sharing between the developed and developing worlds. Without targeted action, it is likely that many of the benefits nanotechnology can provide to the developing world will be delayed by at least a generation or more—the 20-year term of a patent.

Surveillance

Nanotechnology will eventually enable supercomputing on a very small scale, detection of minute amounts of substances and genomes, and implantation of microchips into humans. Certain applications of these technologies could be beneficial in promoting economic progress, health, and environmental preservation. But such technologies come with a darker side: They can potentially open new opportunities for governments, individuals, and private interests to violate privacy. How nanotechnology will affect our privacy, and what actions governments should take, are important issues that need to be addressed before nanosurveillance becomes ubiquitous. Without careful consideration, debate, and possibly new policy action, "Big Brother" may end up being very, very small.

The most discussed form of nanotechnological surveillance is nanosensors. These sensors, already under development and production, can detect minute amounts of chemicals in the air or water. For example, Owlstone Nanotech, a New York based company, is producing dime-sized wireless sensors that can detect toxins and explosive materials in the air.[28] It should not be long before nanosensors are much smaller. Another possible nanosurveillance innovation might be extremely small cameras. Researchers at Hiroshima University and Nippon Hoso Kyokai have reportedly already been able to find a silicon nanocrystal film that is photoconductive, which is the first step in creating highly miniaturized cameras.[29] Human-implanted microchips could also become a tool of surveillance, as they would be able to track a person's location and possibly what that person consumes (e.g., illegal drugs or junk food).[30]

[28]Choi, C. (2006, May 23). Nano-loaded wireless sensors. *United Press International.* Available at http://www.physorg.com/news67611680.html (retrieved October 17, 2006).
[29]Mehta, M. (2002). On Nano-Panopticism: A Sociological Perspective. Available at http://chem4823.usask.ca/~cassidyr/OnNano-Panopticism-ASociologicalPerspective.htm (retrieved October 17, 2006).
[30]Gutierrez, E. (2004). *Privacy Implications of Nanotechnology.* Electronic Privacy Information Center. Available at http://www.epic.org/privacy/nano (retrieved October 17, 2006).

Advanced surveillance has some potentially positive uses. With sophisticated nanosensors properly placed, it would be very difficult or impossible to sneak a bomb into a secured location, such as an airport. If such sensing nanotechnologies were ubiquitous, producing a bomb undetected would also be very difficult. Human-implanted chips have already been proposed to be used to track the location of Alzheimer's patients.

However, there exists a substantial threat to personal privacy if such technologies are taken too far. For example, the same nanosensors that could "keep us safe" could also be used to track eating habits, smoking, or drug use—any activity that leaves a chemical trace. More ominously, it should be possible to track DNA, which would enable the tracking of individuals' locations over time.

Of course, such detailed surveillance would be regarded as Draconian today. However, security concerns, especially if the "war on terror" continues into the indefinite future as seems likely, may prompt the widespread use of nanosensors. Political pressure to deploy such sensors might increase over time, as nanotech-enabled weapons become technically feasible. Once such technology is ubiquitous, it may become politically feasible to suggest that it be used to monitor activities, such as drug use.

Privacy violations will not only come from governments. As mentioned above, nanotechnologies could enable other actors, including individuals and corporations, to observe where one has been and what one is consuming. Insurance companies have an incentive to want these technologies to be put in use, because they would allow such companies to more accurately calculate risk.

Not all such applications are necessarily bad; arguably, concerned caregivers should be able to know where their mentally disabled patients are, corporations could use tracking information to better develop and market products, and with better risk calculations, insurance companies could charge fairer rates, rewarding healthier behavior. However, they do raise serious questions about how much surveillance by certain actors is too much.

The current legal setting will already discourage many privacy violations. For example, the Supreme Court case *Kyllo v. United States* found that it was illegal for the police to use heat-sensing technology on Kyllo's house to determine whether or not Kyllo was using heat lamps to grow marijuana, without first obtaining a warrant.[31] The Court decided that the Fourth Amendment prohibition against unreasonable search and seizure also prohibits the government from using "a device that is not in general public use ... to explore details of a private home that would previously have been unknowable without physical intrusion."[31] The precedent in *Kyllo* could be extended to nanotechnology surveillance.[30] However, as the recent bout over the U.S. phone-tapping program shows, the letter of the law and its practice sometimes drastically differ, and as long as a surveillance technology exists, governments will be tempted to use it. Further, *Kyllo* would permit such surveillance once a given technology becomes generally available, as video cameras are today.

[31] *Kyllo v. United States*, 533 U.S. 27 (2001).

A practical compromise may arrive in the form of "sousveillance," or watching-from-below. In this model, citizens themselves voluntarily collect data, including on those in public office, discouraging governments from converting data on an individual into a civil liberties violation. In his book *The Transparent Society*, David Brin argues that this ability could enable society to preserve freedoms despite technological advances affecting privacy.[32]

Nanotechnology surveillance policies, too, should be worked out sooner rather than later. If policymakers and citizens begin to get comfortable with widespread use of surveillance nanotechnology, it may be too late to prevent abuses. Most privacy laws and legal precedents already in place, such as *Kyllo*, could apply to nanotechnology. However, new legislation may be necessary, as was the case after camera phones became widely used. Legislatures and citizens should be vigilant on nanotech surveillance issues, because the risk of abuse is so great.

CONCLUSIONS

The possibilities that nanotechnology holds for humanity, although in some cases decades off, are immense. In the next few years, we may benefit from inventions in fields as diverse as medicine and space travel. Over the next few decades, the advent of productive nanosystems is expected to reshape the way we live, making it possible to easily construct any object with atomic precision. Research and development on its own, however, will not necessarily produce outcomes that we find socially desirable. While nanotech cancer treatments, water purifiers, and solar panels are being researched in the labs, questions of policy—which will affect what is funded, what is regulated, and who will benefit from their production—remain unresolved. To maximize benefits from nanotechnology, these issues must be discussed, debated, and decided. There is still time for this process to occur, but, if we delay our decisions too long, they will either be decided for us, or we will be forced to make more difficult and costly decisions later. Our nanotech policy should not be an afterthought. Nanotechnology should be approached with the foresight it deserves.

---

[32]Brin, D. (1999). *The Transparent Society: Will Technology Force Us to Choose Between Privacy and Freedom?*. New York: Perseus Books Group.

CHAPTER 8

Reasoning About the Future of Nanotechnology

RUTHANNA GORDON

INTRODUCTION

Nanotechnology, along with the other developing technologies of the twenty-first century, will have been more closely examined in its infancy than any previous development. The discussion of ethical, legal, and social implications already covers not only the probable nano-related advancements of the next 10 years, but their possible successors during the next 50. Even in the short term, these speculations vary wildly both in expected impact on society and in optimism. Should we hope for better sunscreen, or neural enhancement, or immortality? Should we fear new cancers, or increased economic stratification, or destruction by out-of-control self-replicating devices? What trade-offs between benefit and danger will turn out to be worthwhile?

These are important questions, which this chapter does not address. Instead, as a psychologist, I want to examine the thought processes that lead those who do address them to their particular answers. One of the great strengths of humanity is our ability to speculate about things that are not directly before us. In fact, Jean Piaget framed his well-known theory of cognitive development around the child's increasing skill at hypothetical logic: at reasoning about that which might be, rather than only that which is.[1] However, even mature adults are not actually as good at this as one might hope. In order to avoid overtaxing our cognitive resources, we simplify our speculations, ignoring some factors and exaggerating the impact of others. We turn possibilities into likelihoods, and likelihoods into certainties. Our reason may be biased by fear or wishful thinking. None of these limitations is likely to disappear, even with the most advanced neural enhancements; indeed, they are the price we

[1]See Piaget J. 1962. Play Dreams and Imitation in Childhood. New York: Norton.

Nanoscale: Issues and Perspectives for the Nano Century. Edited by Nigel M. de S. Cameron and M. Ellen Mitchell
Copyright © 2007 John Wiley & Sons, Inc.

pay for creativity and mental flexibility. However, when we know they are present, and work to compensate for them, we can better prepare for the future that we speculate about.

Several areas of psychological research can provide insight into the ways we think about as-of-yet-undeveloped nanotechnologies. A number of researchers have looked at counterfactual reasoning: the ability to consider alternatives to present reality. Although much of the work in this area focuses on how we create possible alternatives to past events, it turns out that we use these simulations of the past to change our plans for the future, as well. Additionally, the factors that cause us to consider particular alternate pasts give insight into factors that change the way we think about events to come. Another line of study looks at time orientation: How focused particular individuals are on the past and the future, and how far from the present their focus extends. Although most people have preferences with regard to time orientation, it is also possible to temporarily induce a particular focus, improving one's ability to consider distant possibilities. Finally, many studies provide direct insight into the biases that influence our predictions and plans, and into the ways that we can minimize them. This chapter gives an overview of these areas of research, and applies their findings to our reasoning about nanotechnology and other developing technologies.

COUNTERFACTUAL REASONING

Counterfactual reasoning is primarily the skill of considering possibilities. It requires the ability to look at what *has* happened and to think about how it might have happened differently. It also allows that same flexibility with respect to the future: the ability to consider more than one possible outcome of current trends. In order to do this well, one must be willing to draw from many areas of knowledge, creating complex simulations of the interactions among many factors.[2] Inherent in this requirement is the danger that, when choosing which factors to focus on, one must inevitably leave out peripheral factors, which, in fact, may turn out to impact the scenario in question.

In illustrating the processes of counterfactual thought, I am going to use my own hypothetical scenario: A student with training in economic theory and sociology, attempting to predict the most likely future uses of nanotechnology. We will call her Jane. When creating her predictions, her first and most obvious bias will be a tendency to place the greatest weight on evidence from her own fields of expertise. This is not due to any undue belief in the importance of these areas relative to others, but simply to greater familiarity with possible factors because her knowledge base in these fields is richer. So, while she may ask herself, broadly "How will changes in computer science affect the development of nano-related technologies?," she will ask more and more detailed questions related to the effects of sociology and economics.

[2]Turner, 1996.

When people create counterfactual scenarios about the past, they follow certain predictable lines of reasoning. For example, we prefer to imagine specific "turning point" events, rather than broader changes.[3] Given a story in which a man leaves work 5 minutes early and gets into a car accident, most people will first think, "I wish he had left on time," rather than "I wish he had driven a safer car" or "I wish he worked some place else." This also illustrates another tendency: when seeking to avoid negative outcomes, we prefer changes toward normalcy rather than away from it. We want to explain unpleasant events in terms of something unusual that could have been avoided, not in terms of something that we do regularly and that we now must consider changing. We are also more likely to speculate about changes within our control than outside of it.[4]

Although there is considerably less research available, it is reasonable to speculate that similar preferences affect counterfactual reasoning about the future. Jane may be reluctant to imagine nano-related changes in those things that she considers most indicative of normal life, or miss the ways in which unexpected natural disasters might influence the priorities of researchers in the field of nanotechnology. Taking an example from the past, many people made predictions about the development of computerized networks. Few predicted the changes in dating patterns and other social behavior resulting from the ability to affiliate with people across the globe. Imagining the existence of improved communication was easier than imagining its effects on seemingly unrelated customs that were taken for granted. In the same way, we may unintentionally overlook the ways in which nanotechnology could alter those aspects of the present to which we are most devoted.

There is at least one way in which future-oriented, counterfactual reasoning seems to differ from past-oriented reasoning. Rather than imagining state-change turning points, we prefer to assume that trends will continue along their current trajectories.[5] In spite of any number of analyses indicating that technological development is proceeding on a logarithmic curve, most predictive scenarios follow a linear curve because it is easier to imagine. When predictions do follow the more ambitious curve, we often see undue certainty. There is an interesting paradox in the dual proposition that: (1) nano and other developing technologies will soon reach a technological breaking point beyond which daily life will bear little resemblance to its current form; and (2) we can confidently describe what life will be like beyond that point (at least whether life will be "better" or "worse" than it is now) (e.g., Kurzweil and other proponents of a coming "singularity").[6]

Counterfactual reasoning about the past supports a number of possible goals. Consideration of negative alternatives allows us to see how "things could have been worse," and improves our opinion of actual events. In the same way, the creation of negative future scenarios can be a form of mental self-protection.[7] Those

[3]Rescher, 1964; Roese and Olson, 1995.
[4]Miller et al., 1990; Girotto et al., 1991.
[5]Weber, 1996.
[6]Kurzweil, 2005.
[7]Shepperd et al., 1996; Shepperd et al., 2000; Sanna et al., 2005.

who expect the worst are not likely to be disappointed. Ideally, people who have thought carefully about the potential negative effects of nanotechnology will be well prepared to prevent them. However, once again, the human tendency toward certainty can be a trap, in this case leading to fatalism.[8] This sort of "catastrophizing" can cause negative predictions to become self-fulfilling.[9] Environmental activists have already discovered the difficulties of finding a balance between acknowledging unpleasant possibilities associated with nanotechnology and convincing people that they can be avoided. Too little evidence of impending disaster can lead to the public cheerfully ignoring the problem. Too much evidence causes people to feel helpless in the face of onrushing events and impending disaster. For those who favor particular nanotechnology-related scenarios, the same issues hold. Not only negative, but also positive, nanopredictions can be problematic if they come to seem inevitable. Positive futures, after all, must be worked toward if they are to come to pass. However, if people feel that positive futures are unavoidable, such effort will seem unnecessary. At the moment, there is little enough consensus surrounding predictions about nanotechnology that this is not an immediate concern. However, this may change as those predictions make their way into popular culture.

Other emotional factors can affect our speculations, as well. We crave predictability and controllability in the world, but these needs may conflict.[10] The balance described in the previous paragraph is one aspect of this conflict. Belief that some things cannot be avoided provides at least the comfort of predictability, but it comes at the cost of one's own input into events. For some, this may be a worthwhile cost, because another motivating factor is the desire to avoid blame. If, running through alternate pasts, we find many in which things get worse, we cannot be held responsible for the degree to which things are actually bad. Likewise, if nanorelated disaster is considered inevitable, then personal responsibility for such an outcome appears to be minimized.

We also have a need for cognitive consistency.[10] We are motivated to avoid suggesting that disliked causes might lead to good outcomes, or that good causes might be linked to bad outcomes. One might, for example, be reluctant to consider the possibility that, if one had never met a now-disliked ex, one would never have been introduced to a favorite restaurant. On a historical level, one might be reluctant to believe that a fascist political leader had been responsible for some positive economic impact, even while freely acknowledging that his or her negative impact far outweighed any benefit. It is easy to see how this could apply to future predictions about nanotechnology. Those who have become attached to a potential positive consequence of nanotechnology may be less able to appreciate prospective side effects of those same developments. Even before taking into account the desirability of future results, something or someone well liked in the present would be most easily seen as leading to positive future outcomes. If Jane is an early adopter,

[8]Vasey and Borkovec, 1992.
[9]Davey et al., 1996; Peterson et al., 1998.
[10]Tetlock and Belkin, 1996.

who always delights in having the latest technology, she may be reluctant to consider future scenarios in which nanotechnology has a strong negative effect. In this way, counterfactual reasoning is often as much about the present as about the timeframe ostensibly under discussion. It is one of many tools that we use to support our core beliefs about the way the world works.

Past- and future-oriented counterfactual thinking are related skills. By considering how the past might have turned out differently, we open ourselves to the idea that the future is not predetermined.[11] Unfortunately, this skill is not often taught, and for most people, it is not one that comes naturally beyond the immediate, personal past. American grade schools, for example, tend to teach history without mentioning possible courses of action that were avoided, or what the results might have been if different choices had been made.[12] The products of these classes learn that the past was a series of events, each progressing naturally and unavoidably from previous events. This encourages thinking of future options as similarly narrow, and encourages the already-described propensity for predictive overconfidence.

Any given future nano-related scenario is likely to be based on specific posited changes from the current situation. One basic constraint on the creation of these scenarios is the salience of changeable factors in the current situation.[13] If, out of the many areas of nanotechnology research, Jane knows quite a bit about medical developments, but very little about the design of new construction materials, then her scenarios are likely to feature the consequences of improved human health (or nano-mediated plagues, if she is feeling pessimistic). On the other hand, she may neglect the possibility of extensive changes in how houses are designed and built.

This variability in knowledge base is only one factor affecting the salience of relevant information. Right now, if the reader were asked to create a new scenario about nanotechnological development, medicine and construction would be likely to spring to mind first. The more often, and the more recently, a particular aspect of a situation has been mentioned, the more likely it is to enter your thoughts when you consider the issue. I have just made building design incrementally more likely to show up in the nano-related predictions of everyone who reads this chapter. Fortunately, my power over the course of more than a few minutes is minimal, as my discussion becomes integrated in your mind with everything else you have read and heard about nanotechnology, and becomes only a small part of the weight that each aspect receives. The changeability of an event also becomes more salient if it is visibly abnormal, or if an alternate possibility recognizably almost occurred (a near miss). Action is more salient than inaction—unless the inaction was abnormal; it is easy to see when someone did something that resulted in a problem, and slightly harder to see if he or she missed an opportunity.

Events that are closer in time are more salient as well, so that future nano-related scenarios are most likely to develop from changes close to one's present perspective.

[11]Weber, 1996.
[12]Loewen, 1996.
[13]Seelau et al., 1995.

As one moves into considering possibilities far into the future, it becomes more difficult to keep track of specific events that might change. Instead, we depend more on broad, schematic representations of possibilities.[14] So if Jane tries to make a prediction about nanotechnology based on changes within a year's time, she may posit that a specific funding measure gets passed, increasing nanotechnology research in a particular domain. If she were making predictions for 20 years in the future, she might more broadly suggest that support for the sciences, and nanotechnology research as a whole, will increase. We usually represent temporally distant events at this sort of high, general level, focusing on a few abstract features. Temporally close events are more likely to be represented with concrete and incidental details, and to have a rich contextual representation for their causes and results. This means that distant predictions are less accurate. This should be no surprise, and it is reasonable that this would be the case. However, these same predictions are made with greater confidence than the more accurate short-term predictions, because when we are aware of possible variation in intricate details that undermines our confidence. Because these details are less available for distant nanotechnological scenarios, the uncertainty involved becomes less apparent to us.

Is it possible to minimize this long-term overconfidence in our predictions? Fortunately, one of the best ways to make a factor salient (at least in the moment) is simply to bring it up. By asking people to explicitly consider details that could affect far-future nano-related scenarios, those scenarios become less abstract and confidence decreases.

TIME ORIENTATION

We live in the present, but define it in terms of both what we bring with us from the past, and our plans, expectations, and fears for the future. The degree to which we focus our cognitive resources on these two ends of the temporal spectrum, or on the present moment itself, varies across a number of dimensions.[15] The most obvious one is direction. Some people focus primarily on the past, a strength in writing a memoir, studying history, or simply being nostalgic. Others focus on future possibilities. Too much of the former can lead to difficulty improving one's current circumstances, while too much of the latter leads to the usual consequences for failing to learn from the past. Another axis of variation is extension, or depth of focus.[16] A student may be future focused in the sense of planning for graduation 2 years hence, while some of the contributors to the current volume speculate about changes over the course of decades—not only in nanotechnology, but in all the areas that may affect and be affected by it. Jane should have a strong future focus with a long extension. However, her economic and sociological training probably also give her at least a medium extension into the past. Past and future focus are

---

[14]Trope and Liberman, 2003.
[15]Lasane and O'Donnell, 2005.
[16]Bluedorne, 2005.

not mutually exclusive; in fact, under ideal circumstances, they can inform each other.

A more complex factor in time orientation is structural. Humans are essentially story-telling organisms, and we create narratives to connect the events, and potential events, of our lives. In cognitive psychology, we refer to these schemata as the over-arching patterns that we use to categorize and understand the world around us. We look for causal relationships not only because they permit us to predict new developments, but because they are emotionally satisfying. However, not everyone perceives themselves in the same sort of narrative.

Landman describes several basic narrative worldviews that vary in the optimism of their expectations, as well as in how meaningful one expects one's life, and the lives of humanity at large, to be.[17] These narratives are created from our own experiences and perceptions, but once in place, they also guide them. Because they are built up throughout people's lifespan, the evidence of something that does not fit into them must be quite strong in order to be considered. If Jane narrates her life in terms of progress and ultimate triumph, one would not expect her to easily make predictions at odds with that schema. In the domain of nanotechnology, she may focus on advanced medical and other applications that will benefit society, and either downplay potential dangers or believe that humanity's problem-solving abilities will be, at least, equal to the challenge. Evidence for this point of view will seem most plausible, because it fits in within the cognitive structures that she has developed over the course of several years. Another person, equally optimistic in his or her own way, might create a narrative of humanity triumphing over (rather than using) impersonal technology, so that evidence of nanotechnology's dangers would seem more salient.

Time orientation is heavily influenced by developmental experience, as well as by the availability of resources that permit effective future planning. High socio-economic status (SES) in childhood is correlated with the development of future perspective. As more resources become available, problem-solving becomes more effective, and, therefore, people become more willing to engage in it. Additionally, SES is related to educational opportunities and, therefore, to explicit training in this sort of planning. Zimbardo and Boyd[18] found that lessons involving mental simulations of the future increased measures of future time perspective in students who had previously lacked this orientation. In general, education leads to a greater ability to reason about things outside one's own experience.[19] Training can also be implicit: Books and movies that cover extended periods of time, or adults regularly discussing long-term prospects, all increase a child's ability to speculate and plan. Exposure to these experiences will improve the skills integral to thinking about future possibilities. It may be prudent, as development continues to speed up, to explicitly teach children to speculate about nano and other developing technologies, encouraging them to take a long-term perspective. The National

[17]Landman, 1995.
[18]Zimbardo and Boyd, 1999.
[19]Harris and Leevers, 2000.

Nanotechnology Infrastructure Network Education, a pilot program based out of the Georgia Institute of Technology, is one current attempt to do this.

Willingness to think about the future is also affected by the availability of temporal resources. People who feel that time is limited, whether because of illness, political upheaval, or any upcoming major life transition, tend to be oriented toward the present and toward short-term planning. People who feel that time is open ended are more likely to be future oriented.[20] These perceptions vary across the lifespan, with older adults tending toward a more time-limited perspective. However, external factors can overrule this. For example, researchers performed tests of temporal orientation in Hong Kong just before and just after the transition to Chinese rule.[21] They found that, on average, city residents were more present oriented before the transition and more future oriented afterward. The future-oriented, time-open thinkers showed a greater willingness to seek out new information, as well as better recall of what they had learned earlier on. Similar changes can be induced deliberately, by asking people to imagine themselves in a situation that would normally produce a particular temporal focus.

One area in which people may commonly be pushed to think about future possibilities is in the corporate world. Particularly among start-ups working on new technologies, such as nanotechnologies, a good balance between short- and long-term planning is vital to success. We can see this as the need for diversity in depths of extension. Research in this area suggests a very simple force encouraging depth of planning: In general, the longer someone must wait for feedback on his or her actions, the greater depth of perspective he or she gains.[22] Members of research and development departments, whose feedback is limited by the cycle of experimentation, show longer-term planning than members of marketing departments, who deal with swifter feedback loops. Successful executives, meanwhile, work with multiple time horizons, planning for results at multiple intervals; an ability that is correlated with financial success.[23] This strategy will be familiar to readers in the sciences, who are trained to always have multiple projects in different stages of development at any given time. These findings should be encouraging, as they suggest that the practicalities of nano-related research will naturally push those involved in it to use an appropriate range of time frames when considering their results.

As may be gathered from the results reported above, a strong future time perspective confers a number of benefits. Future orientation is associated with high achievement across domains.[24] It encourages not only personal success, but a willingness to engage in behaviors benefiting society at large.[25] Future orientation can also have emotionally beneficial effects, exemplified by someone who responds to a mistake by planning avoidance strategies for the next time will have more positive emotional

[20]Carstensen et al., 1999; Carstensen, Mikels, and Mather, 2006.
[21]Fung et al., 1999.
[22]Bluedorne, 2005.
[23]Judge and Spitzfaden, 1995.
[24]Lasane and O'Donnell, 2005.
[25]Joireman et al., 2001.

affect than someone who merely regrets the errors made.[26] These advantages, however, are not without cost. Future orientation comes with the ability to anticipate fearful outcomes, as well as hopeful ones, and can lead to increased stress over large-scale dangers, such as environmental degradation.[27] This fear can, in turn, lead to increased reluctance to consider evidence that conflicts with one's schemata and expectations.[28] We often compensate for anxiety with an increased need for certainty, and this inflexibility can cause us to miss options in our planning for the future. In an area, such as nanotechnology, where hopes and fears can run so strong, these effects are likely to be exacerbated.

Time perspective can provide some insight into why people think differently about nano and other developing technologies, and why they sometimes have trouble communicating these differences. Those who specialize in speculating about technology share a strong future orientation. Where they differ is in their narrative schemata. A lifetime of experience feeds into what we estimate as the likelihood of any given scenario. Nevertheless, most future-oriented thinkers will be willing to predict large-scale nano-related changes, even if they disagree about their direction.

However, the ultimate development of nanotechnology will be dependent on the attitudes that form toward it in society at large. Several studies have attempted to measure and predict these attitudes.[29] These measurements are limited by the fact that respondents are generally initially ignorant about nanotechnology, and the training necessary probably does not match the information that will be available when such knowledge is more pervasive in society.[30] Time perspective is an area of individual variation that may also be glossed over by these studies. Those who are past or present oriented are likely to underestimate potential nano-related changes, a difference that will be reflected in their concerns. Even for those who are future oriented, differences in depth of focus will strongly affect their input into nano policy.

BIASES IN PREDICTION AND PLANNING

The ability to plan for the long-term future has always been important, but it poses unique challenges in today's society, where change is rapid and the amount of information available outstrips anything that previous generations were ever able to access. However, the organism that tries to deal with this abundance is still essentially a plains ape. Our reasoning abilities have evolved to deal with concrete information immediately available to our senses, to make quick decisions on the assumption that useful data will be missing or ambiguous, and to learn about the world outside of our immediate vicinity from stories told around a campfire.

[26]Boninger et al., 1994.
[27]Slee and Cross, 1989; Nurmi, 2005.
[28]Simon et al., 1997; Routledge and Arndt, 2005.
[29]See, e.g., Macoubrie, 2005; NanoJury UK.
[30]Gordon, 2006.

These abilities can be great strengths, but they can interact strangely with a world of statistical analyses, one in which events on the other side of the world frequently affect our everyday lives. The disconnect becomes even stronger when thinking about nanotechnology, which has the potential to alter the ways that we perceive and use matter itself.

The most basic bias in human reasoning lies in the way we naturally build our models of the world. We have already explored how counterfactual reasoning is affected by how easy it is to think of particular changeable factors. Likewise, we judge the frequency or probability of events by how many past instances of such events are available in our memories.[31] For example, if you were asked, off the top of your head, what percentage of American children have braces, you would think about your or your children's classmates and how many of them you recalled wearing orthodontic equipment. Depending on the culture and socio-economic status of the people you have known, your answer is likely to be higher or lower than the actual figure. This inaccuracy, while difficult to avoid, will probably not have any detrimental effects on your life.

The problem becomes apparent when we make more complex judgments, ones that have large-scale impact and for which a great deal of contradictory information is available. For nanotechnology, it may often be impossible to study *all* of the available material. Our choices of what to read and listen to are likely to lean toward those who agree with us. The average politically involved citizen, for example, will pay the most attention to books and blogs that support their favored views. Most people prefer to spend time engaging those who consider them right than those who disagree with them. This is simply more comfortable and pleasant. After all, why waste time reading incorrect opinions? However, it has the added effect of making the arguments for one's own side seem more salient and common, and the arguments against less so, reinforcing one's beliefs and making a careful consideration of opposing opinions more difficult.[32] The same trap is all too easy to fall into with nanotechnology. There is a wide enough range of opinion available that one could spend all of one's time engaging with those who share one's expectations and beliefs.

Another problem arises when our tendency to judge probabilities based on mental availability is combined with our inclination toward narratives. If you are asked how likely someone is to be assaulted while walking through a city park, we move into a realm where we may have minimal personal experience. However, you have almost certainly read books and seen movies in which people get assaulted in parks. Add in news reports, and instances of this sort of crime become highly available, and your estimate of the danger goes up. In fact, the more often people are exposed to this type of news, the more danger they believe they are in, regardless of how safe their actual environment may be.[33] The tendency of news media to replay clips or descriptions of dramatic events plays into this, because each repeat acts as a separate instance in memory. A single event, given enough media focus, can make a one-time danger

[31]Tversky and Kahneman, 1973.
[32]Kahneman and Tversky, 1979; Sanna et al., 2005.
[33]Strange and Leung, 1999.

seem extraordinarily common. This focus may even be fictional (just ask anybody who refused to swim in the ocean after seeing *Jaws*). If anything, fiction containing specific details can lead to greater salience than a less detailed news report. Certainly, it carries more weight in the human mind than the abstract report of a study, even if the statistics of that study's findings are impeccable. This is the case even for those trained to think scientifically, thus the presence of Jane in my examples.

It is precisely this type of learning that is neglected when we discuss future societal attitudes toward nanotechnology. Where the participants in NanoJury UK learned from clear data carefully presented by experts, the public at large is more likely to read Michael Crichton's *Prey*, or the other popular presentations that will surely follow. So far, the great pro-nanotechnology bestseller remains to be written. However, people also draw attitudes toward technology as a whole from their preferred fictions, and then generalize to new situations. Such stories illustrate facts (or falsities) with vivid personalities and events. That vividness, in turn, leads to vivid cognitive representations of the ideas in question, which makes those ideas more salient and compelling.[34] The persuasive power of fiction has been well documented in the scientific literature,[35] but it is also clear to anyone who has read *Uncle Tom's Cabin* or *The Grapes of Wrath*. The right book or movie could well have greater potential than any logical argument to persuade people to take a particular stance on nanotechnology.

As mentioned earlier, we interpret the world in terms of our long-standing schemata and expectations. These grow from our life experiences, including the sort of stories described above. We find it easier to perceive and accept evidence if it supports these expectations, and to dismiss evidence that contradicts them. This is often known as the confirmation bias.[36] Humans like to be right, and we actively seek out information that justifies our prior beliefs. This increases our confidence in those same beliefs, continuing the cycle. Changing a closely held belief requires considerably more persuasion than coming to that same new belief from a position of uncertainty. However, that position of uncertainty, from which data can truly be considered impartially, is a difficult one to maintain. We have evolved to form initial judgments very quickly, without waiting for all the evidence to come in, because frequently, it never will.

We are not only biased to assume that our new experiences will be like our past experiences. We also have hopes and desires for what those new experiences will be. Jane, for example, wants to live a long and mentally healthy life, so that she can publish into her 90s (or retire at 65 and have a second career; she is flexible). She also enjoys hiking, and hopes that whatever else the future will bring, we will manage to preserve our natural resources for public use. This becomes relevant when she looks at possible developmental patterns for nanotechnology. She is likely to give extra weight to sources that underplay the difficulties with medical

[34]Gerrig, 1997.
[35]See, e.g., Prentice et al. 1997; Green and Brock, 2000.
[36]Wason, 1968; Woll, 2002.

nanotechnology, or to talk about how nano may be used to combat pollution. She may also seek reasons to dismiss the environmental hazards of nanoparticles. This type of desire-based reasoning has been documented under a wide variety of circumstances and is difficult to overcome.[37]

Most people who spend time thinking about nano-related developments have preferences about its eventual form, whether they hope to preserve their own lifespan, the environment, or just the status quo. If they perceive themselves as having a low level of control over events, they may be predisposed to find and believe evidence that events will play out as they wish. Alternatively, those with a higher perceived level of control may be motivated to come up with a clear causal chain leading to an outcome they wish to avoid, a chain that contains key links they can work to break.[38] This latter group may miss possibilities that would lead to the same outcome, but be less controllable.

The points of vulnerability in the reasoning process, at which these biases can have their effects, are many. The availability of information in memory, for example, need not be a constant. Recall is facilitated by cues, both in the world and in one's own thoughts, that bear some similarity to the information being sought.[39] The common experience of walking into the kitchen to get something, only to forget what you were looking for, illustrates this. Most people will go back to the room from whence they came, and this will be sufficient to remind them that they needed, say, a pair of scissors. The environment where you first thought about the scissors acts as a cue to recalling them. Desire can have the same kind of effect. If Jane is thinking about the environmental impact of nanotechnology, and has encountered information about both positive and negative potential effects, her wished-for conclusions may act as a memory cue, making the evidence for positive effects easier to bring to mind. That increased availability means that she will think of more examples of positive than negative impacts, naturally leading to the assumption that the bulk of the evidence is in its favor.[40]

When people are directly confronted by information that conflicts with their desired conclusions, there are avenues of defense that still remain open. They may judge the source of the new information as unreliable,[41] or misremember it as coming from a less credible source than it actually did.[42] They may seek out or recall specific arguments against the new information.[43] Although new information really can lead us to change our minds, we are often remarkably tenacious in clinging to our comfortable world-views.

There are also reasoning biases that interact directly with consideration of nanotechnological and other future developments. The most notable of these is the planning fallacy. Complex tasks appear easier, and likely to take less time, as their

---

[37]See, e.g., Kunda, 1990; Gordon et al., 2005.
[38]Roese and Olson, 1995.
[39]Tulving and Thompson, 1973; Balch and Lewis, 1996.
[40]Kunda, 1990.
[41]Ditto et al., 1988.
[42]Cooke et al., 2003; Gordon et al., 2005.
[43]Pyszczynski and Greenberg, 1987.

distance from the present increases.[44] This is illustrative of a larger bias that can be seen as similar to spatial perspective. As you look out toward the horizon, more distant objects have less detail and may be processed more abstractly (e.g., as "a woman walking toward me" rather than as "Aunt Hilda, wearing a raincoat"). Likewise, most people will judge the likely productivity of a week next month as greater than that of the current week. Up close, we see the dense schedules and inevitable distractions that are not apparent at a greater remove.[45] Similarly, our representation of future events becomes more prototypical (more like our schemata) with greater distance.[46] We are simply more aware of the unusual details that disrupt our routine today than those that may do so next week.

When thinking about nanotechnological development, a long-term version of the planning fallacy may increase the perceived ease of distant goals. Because they are a more immediate difficulty, the specific problems with research into carbon nanotubes may appear more salient than potential difficulties with medical nanobots. It is likely that we are vulnerable to underestimating the timetable for long-term nanotechnological developments, just as we are for more personal projects. Predictions, such as Kurzweil's, or plans, such as the National Cancer Institute's 2015 "challenge goal," may take longer to bring to fruition than their proponents expect or hope.[47] (Obviously, there are those who are less vulnerable to the planning fallacy for their own projects. It is an open question whether these people might make better technological forecasters, as well.)

For individuals, this bias can be ameliorated by asking people to think about the concrete details of future actions.[48] That is, "buying a new car," which might look simple from a year's remove, will be better planned for and more realistically scheduled if broken down into "visiting dealers," "comparing prices on-line," and the various other steps of the process. A similar exercise might well overcome the tendency to overconfidence in large-scale nano-related predictions. Concrete details help us to see where unexpected problems and unintended consequences could affect our timelines, but they are not something that we consistently include in distant scenarios without deliberate practice.

CONCLUSIONS

After reading this chapter, the reader may find him- or herself hoping for the rapid development of a nano-enabled neural prosthesis that will overcome these cognitive inadequacies. We seem to be a deeply imperfect species. We dismiss overwhelming statistics in favor of our own limited experience, and our own experience in favor of fiction. We let wishful thinking bias our reasoning. We leave important data out of

---

[44]Buehler et al., 1994.
[45]Liberman and Trope, 1998.
[46]Trope and Liberman, 2003.
[47]Kurzweil, 2005.
[48]Atance and O'Neill, 2005.

almost every attempt to model the true state of the world. However, it is important to bear in mind that all of these imperfections serve, or are the cost of, important mental functions. The same cognitive structures that cause us to weight selective personal experience over statistics, also allow us to separate hearsay from those things about which we have direct knowledge. The mental shortcuts that cause us to jump to conclusions also allow us, when necessary, to reach rapid decisions under stress. It is hard to imagine a Version 2.0 of human cognition that could leave out these "bugs" without replacing them with new and complimentary ones.

None of these limitations are absolutes. Most of them can be ameliorated to some degree by an awareness of their presence, and the deliberate cultivation of counter-acting thought processes. By actively seeking out data that contradicts our conclusions, or by considering concrete factors that may affect future scenarios, we can minimize the effects of many of these biases. Like any good habits, these compensations take training and practice. Fortunately, learning is one of the things that humans are good at.

Throughout this chapter I have given examples from the development of nano-technology. When predicting and planning for the advances of the coming decades, it is vital that we use our reasoning skills to their fullest capacity. However, my ultimate point is that no matter how much the world changes around us, we are likely to think about these new problems using the same basic methods that we once did to think about hunting mammoth. Our humanity is neither as easy to transcend as some hope, nor as fragile as some fear.

BIBLIOGRAPHY

Atance C.M. and O'Neill D.K. The emergence of episodic future thinking in humans. Learning and Motivation 2005;36:126–144.

Balch W.R. and Lewis B.S. Music-dependent memory: the roles of tempo change and mood mediation. Journal of Experimental Psychology: Learning, Memory, and Cognition 1996;22:1354–1363.

Berube D. 2005. Report on Nano Jury, Nanotechnology Implications and Interactions. Available at http://www.nanotechweb.org/articles/news/4/9/14/1.

Bluedorne A.C. Future focus and depth in organizations. In Strathman A. and Joireman J. (Eds.) Understanding Behavior in the Context of Time: Theory, Research, and Application. Mahwah, NJ: Erlbaum; 2005.

Boninger D.S. et al. Counterfactual thinking: from what might have been to what may be. Journal of Personality and Social Psychology 1994;67:297–307.

Buehler R. et al. Exploring the "planning fallacy": why people underestimate their task completion times. Journal of Personality and Social Psychology 1994;67:366–381.

Carstensen L.L. et al. Taking time seriously: a theory of socioemotional selectivity. American Psychologist 1999;54:165–181.

Carstensen L.L. et al. Aging and the intersection of cognition, motivation and emotion. In J. Birren and K.W. Schaie (Eds.), Handbook of the Psychology of Aging; 2006. pp. 343–362.

Cooke G.I. et al. Halo and devil effects demonstrate valenced-based influences on source-monitoring decisions. Consciousness & Cognition 2003;12:257–278.

Crichton M. Prey. New York: Harper Collins; 2002.

Davey G.C. et al. Catastrophic worrying as a function of changes in problem-solving confidence. Cognitive Therapy and Research 1996;20:333–344.

Ditto P.H. et al. Appraising the threat of illness: a mental representational approach. Health Psychology 1988;7:3–201.

Fung H.H. et al. Influence of time on social perspectives: implications for life-span development. Psychology and Aging 1999;14:595–604.

Gerrig R.J. Experiencing Narrative Worlds. New Haven, CT: Yale University Press; 1997.

Girotto V. et al. Event controllability in counterfactual thinking. Acta Psychological 1991;78:111–133.

Green M.C. and Brock T.C. The role of transportation in the persuasiveness of public narratives. Journal of Personality and Social Psychology, 2000;79:701–721.

Gordon R. Future hopes, future fears: predicting public attitudes toward nanotechnology. Nanologues, April 2006;2:4.

Gordon R. et al. Wishful thinking and source monitoring. Memory and Cognition 2005;33:418–429.

Harris P.L. and Leevers H.J. Reasoning from false premises. In Mitchell P. and Riggs, K.J. Children's Reasoning and the Mind. Sussex, UK: Psychology Press; 2000.

Joireman J.A. et al. Integrating social value orientation and the consideration of future consequences within the extended norm activation model of proenvironmental behaviour. British Journal of Social Psychology 2001;40:133–155.

Judge W.Q. and Spitzfaden M. The management of strategic time horizons within biotechnology firms: the impact of cognitive complexity on time horizon diversity. Journal of Management Inquiry 1995;4:179–196.

Kahneman D. and Tversky A. Intuitive prediction: biases and corrective procedures. Management Science 1979;12:313–327.

Kunda Z. The case for motivated reasoning. Psychological Bulletin 1990;108:480–498.

Kurzweil R. The singularity is near. New York: Viking; 2005.

Landman J. Through a glass darkly: worldviews, counterfactual thought, and emotion. In Roese N.J. and Olson J.M. What Might Have Been: The Social Psychology of Counterfactual Thinking. Mahwah, NJ: Erlbaum; 1995.

Lasane T.P and O'Donnell D.A. Time orientation measurement: a conceptual approach. In Strathman and Joireman (Eds.), Mahwah, NJ: Erlbaum; 2005.

Liberman N. and Trope Y. The role of feasibility and desirability considerations in near and distant future decisions: a test of temporal construal theory. Journal of Personality and Social Psychology 1998;75:5–18.

Loewen J.W. Lies My Teacher Told Me: Everything Your American History Textbook Got Wrong. New York: Touchstone; 1996.

Macoubrie J. Informed public perceptions of nanotechnology and trust in government. Woodrow Wilson Center for Scholars and Pew Charitable Trusts; September 2005.

March J.G. Exploration and exploitation in organizational learning. Organization Science 1991;2:71–87.

Miller D.T. et al. Counterfactual thinking and social perception: thinking about what might have been. In M.P. Zanna (Ed.). Advances in Experimental Social Psychology (Vol. 23). New York: Academic Press; 1990. Nanonation: What's Next. (n.d.) (Retrieved August 11, 2005.) Available at http://www.nanojury.org.

National Cancer Institute. NCI challenge goal 2015: Eliminating the suffering and death due to cancer (retrieved October 10, 2006). Available at http://www.cancer.gov/aboutnci/2015.

National Nanotechnology Infrastructure Network Education and Training. Available at http://www.nnin.org/nnin_education_training.html (retrieved September 13, 2006).

Nurmi J. Thinking about and acting on the future: the development of future orientation across the lifespan. In Strathman and Joireman (Eds.); Mahwah, NJ: Erlbaum; 2005.

Piaget J. The stages of intellectual development of the child. In Slater A. and Muir D. (Eds.) (1999). The Blackwell Reader in Developmental Psychology. Blackwell: Padstow, UK. 1962. pp. 35–42.

Peterson C. et al. Catastrophizing and untimely death. Psychological Science 1998;9: 127–130.

Prentice D.A. et al. What readers bring to the processing of fictional texts. Psychonomic Bulletin and Review 1997;4:416–420.

Pyszczynski T. and Greenberg J. Toward an integration of cognitive and motivational perspectives on social inference: a biased hypothesis-testing model. In Berkowitz L. (Ed.). Advances in Experimental Social Psychology (Vol. 20). New York: Academic Press; 1987.

Rescher N. Hypothetical Reasoning. Amsterdam: North-Holland; 1964.

Roese N.J. and Olson J.M. Counterfactual thinking: a critical overview. In Roese and Olson (Eds.); "What Might Have Been": The Social Psychology of Counterfactual Thinking (pp. 1–55). Mahwah, NJ: Erlbaum; 1995.

Routledge C. and Arndt J. Time and terror: managing temporal consciousness and awareness of mortality. In Strathman and Joireman (Eds.), Mahwah, NJ: Erlbaum; 2005.

Sanna L.J. et al. Yesterday, today, and tomorrow: counterfactual thinking and beyond. In Strathman and Joireman (Eds.), Mahwah, NJ: Erlbaum; 2005.

Seelau E.P. et al. Counterfactual constraints. In Roese and Olson (Eds.), Mahwah, NJ: Erlbaum; 1995.

Shepperd J.A. et al. Bracing for loss. Journal of Personality and Social Psychology 2000;78:620–634.

Shepperd J.A. et al. Abandoning unrealistic optimism: performance estimates and the temporal proximity of self-relevant feedback. Journal of Personality and Social Psychology 1996;70:844–855.

Simon L. et al. Terror management and cognitive-experiential self-theory: evidence that terror management occurs in the experiential system. Journal of Personality and Social Psychology 1997;72:1132–1146.

Slee P.T. and Cross D.G. Living in the nuclear age: an Australian study of children's and adolescents' fears. Child Psychiatry and Human Development, 1989;19:270–278.

Strange J.J. and Leung C.C. How anecdotal accounts in news and in fiction can influence judgments of a social problem's urgency, causes, and cures. Personality and Social Psychology Bulletin 1999;25:436–449.

Tetlock P.E. and Belkin A. Counterfactual thought experiments in world politics. In Tetlock P.E. and Belkin A. (Eds.). Counterfactual Thought Experiments in World Politics: Logical, Methodological, and Psychological Perspectives. Princeton, NJ: Princeton University Press; 1996.

Trope Y. and Liberman N. Temporal construal. Psychological Review 2003;110:403–421.

Tulving E. and Thompson D.M. Encoding specificity and retrieval processes in episodic memory. Psychological Review 1973;80:352–373.

Turner M. Conceptual blending and counterfactual argument in the social and behavioral sciences. In Tetlock P.E. and Belkin A. (Eds.), Princeton, NJ: Princeton University Press; 1996.

Tversky A. and Kahneman D. Availability: a heuristic for judging frequency and probability. Cognitive Psychology 1973;5:207–232.

Vasey M.W. and Borkovec T.D. A catastrophizing assessment of worrisome thoughts. Cognitive Therapy and Research 1992;16:505–520.

Wason P.C. Reasoning about a rule. Quarterly Journal of Experimental Psychology 1968;20:273–281.

Weber S. Counterfactuals, past and future. In Tetlock P.E. and Belkin A. (Eds.), Princeton, NJ: Princeton University Press; 1996.

Woll S. Everyday Thinking: Memory, Reasoning, and Judgment in the Real World. Mahwah, NJ: Erlbaum; 2002.

Zimbardo P.G. and Boyd J.N. Putting time in perspective: a valid, reliable individual differences metric. Journal of Personality and Social Psychology 1999;77:1271–1288.

CHAPTER 9

Nanotechnology and Society: A Call for Rational Dialogue

JERRY C. COLLINS

I've studied now Philosophy
And Jurisprudence, Medicine,
And even, alas! Theology
All through and through with ardour keen!
Here now I stand, poor fool, and see
I'm just as wise as formerly.
Am called a Master, even Doctor too,
And now I've nearly ten years through
Pulled my students by their noses to and fro
And up and down, across, about,
And see there's nothing we can know![1]

ANCIENT AND NEW NANOTECHNOLOGY

The root "nano" comes from the Greek word for "dwarf." The use of nanotechnology is not new. Thousands of years ago artists used nanoparticles to give paints spectacular hues and visual effects.[2] More recently, nanoparticles in solvents have been used to restore paintings to their original lustrous state.[3]

Similarly, ancient civilizations have been able to contribute astounding engineering and architectural feats. The magnificent temples of Greece, the aqueducts of

[1]Goethe, J. W. von. *Faust* (tr. George Madison Priest) (1808).
[2]Erhardt, D. (2003). Materials conservation: not-so-new technology. Nature Materials Vol. 2, 509–510. Available at www.ismn.cnr.it/Symp-O-NatureMaterials.pdf.
[3]Baglioni et al. Nanoparticle technology saves cultural relics: potential for a multimedia digital library. Available at http://wang.ist.psu.edu/MNET/pdf/baglioni_crete.pdf.

Nanoscale: Issues and Perspectives for the Nano Century. Edited by Nigel M. de S. Cameron and M. Ellen Mitchell

Rome, the tunnel of Hezekiah, the Seven Wonders of the Ancient World all attest to the ingenuity and understanding of humankind.

The term "nanotechnology" was popularized by K. Eric Drexler in his book *Engines of Creation: The Coming Era of Nanotechnology*.[4] Drexler raised the possibility of using molecular machinery for large-scale fabrication, with potentially uncontrollable outcomes. More recently, Michael Crichton drew upon fear of the dangers of self-organizing nanoparticles in his best-selling novel *Prey*.[5]

The famous physicist Richard Feynman, in his lecture "There's Plenty of Room at the Bottom,"[6] implied that one of the frontier areas of small particles was not physics, but biology. Feynman asked What are the most central and fundamental problems of biology today? They are questions like: What is the sequence of bases in the DNA? What happens when you have a mutation? How is the base order in the DNA connected to the order of amino acids in the protein? What is the structure of the RNA; is it single or double chain, and how is it related in its order of bases to the DNA?"[6] Feynman's lecture is widely regarded as one of the nucleation points of modern nanotechnology. A half century later, nanotechnology is tied through molecular medicine to the Human Genome Project and its successor, proteomic research, which seeks to understand molecular structure and function of both DNA and the thousands of expressed proteins that have been identified. Intervention and therapy at the molecular level for the individual person is one of the great aspirations of those who work in molecular medicine and victims of molecularly identifiable disorders, their families, friends, and caregivers.

SMALLER AND SMALLER

The Greeks also had a philosophical concept of the atom as the smallest building block of nature, as expressed by Democritus in the sixth century BC. It was not until the nineteenth century, however, that John Dalton enunciated the basis of atomic theory[7]:

- Every element is made of atoms.
- All atoms of any element are the same.
- Atoms of different elements are different (size, properties).
- Atoms of different elements can combine to form compounds.
- In chemical reactions, atoms are not made, destroyed, or changed.
- In any compound, the numbers and kinds of atoms remain the same.

These principles have been sufficient to sustain aspects of modern chemistry to the present time. However, other aspects have been subject to radical change; the process has often been painful (see below).

[4]Drexler, K. E. Engines of Creation: The Coming Era of Nanotechnology. Anchor Books, New York, 1990.
[5]Crichton, M. *Prey*. HarperCollins, 2002. New York.
[6]Feynman, R. P. There's plenty of room at the bottom. Eng Sci (CalTech) 1960; 23:22–36.
[7]Freudenrich, C. C. How atoms work. Available at http://science.howstuffworks.com/atom1.htm.

The Royal Society and the Royal Academy of Engineering in the United Kingdom have developed a comprehensive report on nanotechnology available on the Web.[8] In this report entitled *Nanoscience and Nanotechnologies: Opportunities and Uncertainties*, published on July 29, 2004, they state:

> The size range that holds so much interest is typically from 100 nm down to the atomic level (approximately 0.2 nm), because it is in this range (particularly at the lower end) that materials can have different or enhanced properties compared with the same materials at a larger size. The two main reasons for this change in behavior are an increased relative surface area, and the dominance of quantum effects. An increase in surface area (per unit mass) will result in a corresponding increase in chemical reactivity, making some nanomaterials useful as catalysts to improve the efficiency of fuel cells and batteries. As the size of matter is reduced to tens of nanometers or less, quantum effects can begin to play a role, and these can significantly change a material's optical, magnetic or electrical properties.[9]

In multiple ways, we are trying to understand that which we cannot see.

THE FAUSTIAN BARGAIN AND STEM-CELL RESEARCH

The nature of public debate over nanotechnology has been paralleled by, and is indeed related to, the debate over stem-cell research. The great promise of nano-medicine to approach and ameliorate disease has been made possible by genomic mapping and proteomic life science. With this knowledge has come the realization, however, that profound issues, such as the sanctity and nature of life itself, are at stake. The salient question becomes not "Can we do it?" but "Should we do it?"

The legend of Faust has fascinated imagination and informed discussion for centuries. Faust, in his desire for knowledge and the power knowledge would bring, bargained his soul to the devil. In the soliloquy at the first of this article, Goethe's words express Faust's anguish. Although he has obtained knowledge, Faust sees "there's nothing we can know." Nevertheless, he turns to the lore of the magician and dreams of a time when:

> no more with bitter sweat
> I need to talk of what I don't know yet,
> So that I may perceive whatever holds
> The world together in its inmost folds,
> See all its seeds, its working power....[10]

These words could well have been written about stem-cell research or nanotechnology.

[8]U.K. Nanotechnology Working Group. The Royal Society and the Royal Academy of Engineering. London. Available at http://www.nanotec.org.uk/workingGroup.htm. (July 2004).
[9]See footnote 8, p. 5.
[10]Goethe, J. W. von. *Faust* (tr. George Madison Priest) (1808).

The enigma of Faust has been echoed in the great biomedical ethical discussions of recent times. A month before President Bush's announcement of stem-cell policy, Lance Morrow expressed this ambivalence in *Time*.[11]

> It's the stem cells that are getting most of the moral and ethical debate at the moment. President Bush is considering whether taxpayers' money should be used to fund stem cell research. He is under pressure from both sides. Normally, stem cells for this research come from unneeded frozen embryos at fertility clinics, material that would routinely be discarded. The news from Norfolk was bad timing, since it adds a sinister implication of human life brought into being entirely for the purpose of being cannibalized for parts. Some see only the good in these Faustian quests and manipulations—the miracles of healing.
>
> Some see only the evil—monstrous possibilities, wicked abuse.
>
> The two possibilities are, in fact, twins—the dark and light sides of the Western intellectual quest. You see the twinning in the Faust legend. In the Medieval reading of it, Faust is damned to hell for his pact to obtain supernatural powers of knowledge from the devil—an act of human encroachment upon divine prerogatives. But (as Roger Shattuck points out in his splendid book "Forbidden Knowledge"), the Enlightenment gave Faust an opposite reading. The German dramatist G.E. Lessing's Faust, in the mid-eighteenth century, was not damned for his pact with the devil, but, on the contrary, saved, because of his now-admirable striving after knowledge.
>
> So when we look at stem cell research and other Faustian intrusions into the divine workmanship, we see, alternately, damnation or salvation in the exercise.
>
> I can see both, simultaneously. The jury is still out. And may always be out. This ambivalence is simply the dualism of the world, the secret of its magnetic fields, its gigantic plus and stupendous minus. We split the atom, and what was the moral meaning of Hiroshima? The lives saved? Or the lives incinerated?

The same sort of ambivalence was expressed in the Biomedical Engineering Society Bulletin editorial, "Tough Call," submitted the morning of President Bush's stem-cell research announcement.[12] Whether embryonic stem-cell research should proceed, and on what basis, is indeed a tough call. Responsible, caring people want to affirm the value and uniqueness of life and at the same time enhance the quality of life and wellbeing of the living.

LEARNING IS A PAINFUL PROCESS

The model of Figure 9.1 represents a general learning algorithm, not just for the progress of science, but for any process of understanding. On the left-hand side of this is a region depicted "Measurement," representing interaction with an outside world. In the arena of science, this interaction could be the outcome of a

---

[11]Morrow, L. The Faustian bargain of stem cell research. *Time*, June 12, 2001.
[12]Collins, J. C. Tough call. *Biomedical Engineering Society Bulletin* 25(3):1, 2001.

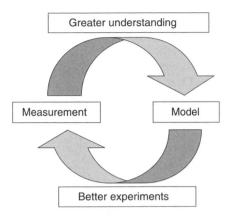

Figure 9.1. A universal learning model.

hypothesis-driven experiment, or an isolated observation. Such outcomes, when consistently interpreted, contribute to increased understanding of the process in question.

Accumulated knowledge of this sort leads the observer to organize what has been learned into a conceptual model that lends consistency to the observations. This process is not limited to science, but is ubiquitous. Conceptual models of every sort are continually created, updated, and discarded. For example, a person considering crossing a street looks at traffic. Based on what is seen, the person may decide to cross the street. If the crossing is a safe one, the person's conceptual model of "what-it-takes-to-cross-the-street-in-traffic" is reinforced. However, if the speed of an oncoming car is underestimated, if the car accelerates, or if the pedestrian slips or falls into its path, the model may change abruptly. In modern scientific terminology, this sort of change may be termed chaotic.

Scientific models have sometimes been developed in advance of measurements, such as was the case with Einstein's theory of relativity and the Michelson–Morley experiments some years later. However, the development of new models representing departures from widely held and staunchly defended years of experimentation and thought is generally more noteworthy and more often associated with chaotic change.

The history of advances in engineering science in the last 200 years is replete with such examples. The nineteenth and early twentieth centuries have been well described by P.V. Atkins in his monograph *The Second Law*.[13] In the late 1700s, Englishman James Watt developed the steam engine, which was refined for industrial use and used to great military advantage by the British. How did it work? In the early 1800s, scientific theory was still inadequate to explain what we know as the thermodynamics of the steam engine. Measurement dictated the need for a new model. (At other times, models, such as Einstein's theory of relativity, have preceded abilities to measure.)

The French scientist Sadi Carnot played an early role in the development of the new thermodynamic model, expressed in his paper "Reflections on the Motive

[13]Atkins, P. V. *The Second Law*. W. H. Freeman and Co., New York, 1986.

Power of Heat (or, in earlier versions, Fire)."[14] His work was the first to advance the concept of the heat engine and to relate heat to work in its analysis.

Over the following 40 years, the model of thermodynamic science was under constant change. Other scientists, such as James Joule, Rudolf Clausius, James Thomson (Lord Kelvin), and William Clerk Maxwell, advanced the science of thermodynamics. Their work culminated in a statement of the conservation of energy and in the recognition that the universe is tending toward a state of disorder, as evidenced by the increase of a defined thermodynamic property, entropy. These times, however, were characterized, as were those of the scientist Galileo, by bitter controversy. Passions ran high, and personal attacks often accompanied scientific criticism. Some of the personal rancor of the day has been captured by Irving Klotz.[15] In case after case, Klotz relates that some of the greatest anger was expressed by scientists and others whom history ultimately demonstrated were mistaken or in error in their opinions.

Rethinking a scientific model (Fig. 9.1) is always painful, particularly to those who have their work and reputations tied to a model being brought into question. The work of two scientists in the latter part of the nineteenth century, however, challenged not only the science but also the theological and philosophical underpinnings of the culture of the times. The identity of one scientist, Charles Darwin, is not difficult to surmise. His classic work *Origin of Species*[16] has been termed "the most important biology book ever written." Implications of this work, that random processes were at work in the development and sustenance of life, have been enthusiastically received by much of the scientific world, but are a threat to religious and other orthodoxies.

The other scientist, Stefan Boltzmann, is perhaps less well known to the public. He was a pioneer of statistical mechanics, but is perhaps best known scientifically for his insight that matter, like life, was at its core random in behavior. On his tombstone in Vienna is carved an equation containing the constant that bears his name, relating other thermodynamic quantities to entropy, or disorder. Boltzmann's career, like that of many others, was marked with controversy and bitter criticism from other scientists whose views contrasted with his own. In 1906, he took his own life. Whether his tragic outcome might have been altered had dialogue been kinder and gentler is a matter of conjecture. The bitterness and personal nature of the attacks against him, however, are a matter of public record.

The work of Darwin, Boltzmann, and other scientists of that era were of profound, unsettling significance that persists almost two centuries later. Not just the life process, but the universe itself, is random, not deterministic, in its basis. This knowledge is counterintuitive. We fail to notice it because evolutionary change is so gradual and because thermodynamic behavior of large systems (i.e., many molecules) is macroconsistent. Our instincts are betrayed by modern science. With Donne, we lament "Tis all in pieces, all coherence gone."[17] When that occurs,

[14]Carnot, S. *Reflections on the motive power of heat.* Translated by R. H. Thurston, American Society of Mechanical Engineers, New York, 1943.
[15]Klotz, I. *Diamond Dealers and Feather Merchants: Tales from the Sciences.* Birkhaeuser, Boston, 1986.
[16]Darwin, C. *On the Origin of Species by Means of Natural Selection, or The Preservation of Favoured Races in the Struggle for Life.* John Murray, London, 1859.
[17]Donne, J. *An Anatomie of the World* (1611).

baser instincts such as turf, ego, and personal control can arise, putting the possibility of rational discourse at risk.

CURRENT ISSUES IN NANOTECHNOLOGY

There are physical risks involved in dealing with nanoparticles. The Royal Society report cited earlier[18] provides the following procedures for dealing with such risks:

> The general approach to assessing and controlling risk involves identification of hazard (the potential of the substance in question to cause harm) and then a structured approach to determining the probability of exposure to the hazard and the associated consequences. Risk is usually controlled in practice by reducing the probability of exposure, although the first principle of risk management is to substitute less hazardous for more hazardous substances where possible. An appreciation of hazard (for example, toxicity or likelihood of explosion) is required to determine to what extent exposure should be controlled. Risk is controlled by limiting release of the material to air or water, and/or by interrupting the pathways by which the substance reaches the receptor where it could cause harm (for example an organ in the body), making an understanding of exposure pathways and likely quantities essential to risk management. In any new technology, foresight of possible risks depends on a consideration of the life cycle of the material being produced. This involves understanding the processes and materials used in manufacture, the likely interactions between the product and individuals or the environment during its manufacture and useful life, and the methods used in its eventual disposal.[19]

Other exposure risk concerns may not be peculiar to nanoscience, according to the report:

> The fact that nanoparticles are on the same scale as cellular components and larger proteins has led to the suggestion that they might evade the natural defenses of humans and other species and damage cells. It is important to set these concerns in context by noting that humans have always been exposed to some types of nanoparticles arising from natural sources such as atmospheric photochemistry and forest fires, and exposures to millions of pollutant nanoparticles per breath have been commonplace since the first use of fire.[20]

ETHICAL ARGUMENTS FOR NANOTECHNOLOGY AND BIOTECHNOLOGY

Outside of environmental, health, and safety issues associated with nanotechnology, there exists active debate among academics and policymakers regarding the

[18]U.K. Nanotechnology Working Group. The Royal Society and the Royal Academy of Engineering. London. (July 2004). Available at http://www.nanotec.org.uk/workingGroup.htm.
[19]See footnote 18, p. 36.
[20]See footnote 18, p. 35.

ethics of nanotechnology as applied to the life sciences and medicine. One argument in favor of medical applications of nanobiotechnology is the possibility of the development of personalized medicine at the molecular level. Alleviation of human suffering is a universal goal well worth the effort. Also, it is generally the case that the first delivery of a new technology is the most expensive. The first recipient of a technology is likely to be the recipient of millions of research and development expenditures. However, as technology becomes widespread the per-unit cost frequently drops dramatically, in a Moore's law scenario.

Another economic argument in favor of the development of nanotechnology is directed toward the worker. Many rural areas of the United States are economically challenged. Per capita family income is low. One factor is the loss of manufacturing base, a phenomenon occurring in many regions of the United States.[21] Such a loss leaves a local workforce at considerable economic risk as they become unemployed or increasingly dependent on low-paying jobs dependent on economic cycles.

There is an emerging dichotomy in the United States and other developed countries between the rich and the poor. As affluent people, who typically have high-paying, skilled jobs, acquire more disposable income, they travel more and buy more. This social development exacerbates the growing class of underpaid and unskilled workers, who tend to be engaged in the service industries relied upon by the wealthy. Clearly, the acquisition of high-technology jobs and job skills is of ethical importance from the standpoint of economics.

Other ethical implementation problems exist within the science itself. One problem that has emerged in potential therapeutic use of nanoparticles is the potential of their being toxic to cells. The quantum dot configuration is particularly useful as a tagged marker; however, certain sizes and materials of quantum dots have shown evidence of cytotoxicity. Investigators, such as colleague Todd Giorgio, are testing nanoparticles intended for delivery of anticancer therapeutics in *in vitro* systems[22] and will eventually initiate animal and later clinical testing. Geometrical configurations, such as nanotubes, are being tested for cytotoxicity,[23] and in either configuration, modifications are being sought to circumvent the problem.

One of the most comprehensive reviews of nanotechnology was commissioned by the United Kingdom's Office of Science and Technology in 2003. The primary goals of this review included determining the need for regulations of the control of nanotechnology, particularly in health, safety, toxicity, and ethics. This report was released in 2004 and resides on the Web at www.nanotec.org.uk/final report.htm.[8] Authors of the report cite the following social factors of potential importance: "specific technical and investment factors; consumer choice and wider public acceptability; the political and macro-economic decisions that contribute to the development of major technologies and outcomes that are viewed as

---

[21]Friedman, T. *The World is Flat: A Brief History of the Twenty-First Century.* Farrar Straus Giroux, New York, 2005.

[22]Kuhn, S. J. et al. Characterization of superparamagnetic nanoparticle interactions with extracellular matrix in an *in vitro* system. Ann Biomed Eng. 34(1):51–8, 2006.

[23]Brunner, T. J. et al. *In vitro* cytotoxicity of oxide nanoparticles: comparison to asbestos, silica, and the effect of particle solubility. Environ. Sci. Technol. 2006, 40:4374–81.

desirable; and legal and regulatory frameworks." It is obvious, however, that in nanotechnology, as in any rapidly developing field, that these factors are subject to sudden alteration." These and other issues need to be addressed publicly as nanoscience and nanotechnology continue to emerge.

IS RATIONAL PUBLIC DISCUSSION OF NANOTECHNOLOGY POSSIBLE?

Technological breakthroughs, such as stem-cell research, have been consistently characterized by social and ethical issues of misuse, misunderstanding, and misappropriation. These developments have generally occurred not because wrong conclusions were drawn as a result of public discussion, but because they were not understood and discussed enough. Public discussion after the fact often generates more heat than light.

Interestingly, technologies that offer real threats in the present tend not to be newer.

> There are certainly ethical concerns with any new technology that must be considered. The prospects of a run away technology as described in Michael Crichton's book *Prey* would be a sad outcome, but the current state of the art in nanotechnology in no way enables that outcome .... There are real dangers in the world, and those that concern us now are 50-year-old technologies, lethal in the hands of individuals and organizations that would choose to use them .... Unfortunately, the barrier between science and science fiction is only as high as the imagination of a talented novelist ... Even some professional colleagues lean over the line at times, seduced by the publicity of and the potential that this notoriety brings in terms of funding or other opportunities. Yet the practical reality of constructing self-assembling, autonomous machines smaller than a single bacterium that can scurry about like little fleas is still only the product of an artist's imagination. What we should be concerned with is the more mundane and the root causes for the growing desire to use them. Education is the key.[24]

Other examples of old technologies posing current threats exist. For example, nuclear science and technology has not been utilized to its full potential. There are reasons for this situation that are now publicly discussed, and need to be. Early development of nuclear technology culminating in use of the world's first nuclear weapon was kept highly secret, however, because the antagonists responsible for its use did not want either their enemies or the public to be aware of what was happening. The need for military secrecy trumped the public need to know.

Genetically, modified foods hold promise of feeding, and providing necessary but missing nutrients, to billions of hungry people. Realization of this promise has been damaged, however, by the manner in which earlier information was not disclosed.

[24]Waldron, A. M. et al. Too small to see: educating the next generation in nanoscale science and engineering. In *Nanofabrication Towards Biomedical Applications*, Wiley-VCH Verlag GmbH & Co. KGaA, Weinheim.

Economic factors prevented public discussion of not only the possibility, but also the implications of this development until the spectacular revelations of "StarLink Corn in Our Taco Bell" became the issue of the day for talk shows and tabloid journals.

Public news media are certainly culpable in public disinformation. This environment has developed for a number of reasons. First, the content and format of public television news, never particularly laudable, has been radically altered. In the insatiable drive for market share, networks have abandoned extended rational discussion in favor of sound bites and teasers that convey little or no useful information but keep the traumatized listener, it is hoped, tuned in for the next commercial or advertisement. The late Neil Postman articulated this problem admirably in *Amusing Ourselves to Death*.[25]

The irony of the culpability of television was captured in the recent highly acclaimed movie *The Insider*.[26] Scientist Jeffrey Wigand has risen to the rank of vice president of a large tobacco company. He is tortured, however, by the knowledge that companies have conspired to deny publicly the truth that tobacco products contain harmful addictive additives. A company agreement he has signed prohibits disclosure of that or any company information. When encouraged by a producer of CBS's *60 Minutes* to reveal this information, Wigand agrees. However, the trauma to himself and his family is almost unbearable. In the meantime, CBS is under threat from the tobacco company, who threaten to sue and thus drive down the price of a pending sale of CBS, which in turn cancels its plan to air the interview with Wigand. Eventually, the show is aired, but the personal toll on Wigand and others in the story is obvious. Wigand became a high school teacher and is now a national consultant on the dangers of smoking. In a 2006 interview with *60 Minutes*, he stated that his income is a tenth of what it was but he has recovered from his estrangement from some family members and is now respected by them.

Media and politics are particularly culpable in disinformation. Daniel Boorstin[27] described more than 40 years ago how media coverage of the Kennedy–Nixon televised debates may have contributed to Kennedy's election by portraying Kennedy's appearance in a more favorable way. Nixon's shadowed face contrasted with Kennedy's youthful charm, according to Boorstin, and political campaigns ever since have intensified and tried to capitalize on these effects. The current environment of ubiquitous negative campaign commercials in elections, driven by party politics and special interests at every level, diverts time, money and effort from more constructive endeavors and contributes to public disinformation.

Powerful economic forces that serve good purposes, but are also driven by problematic goals, also deserve mention. The research budget for a single large pharmaceutical company exceeds the entire budget of the U.S. National Science Foundation (NSF). Company research is driven by a single purpose: to make money for the shareholder. The number of Congressional lobbyists employed by the pharmaceutical industry exceeds the number of members of Congress. With

[25]Postman, N. *Amusing Ourselves to Death*. Penguin, 1985.
[26]*The Insider* (movie). Walt Disney Video, 1999.
[27]Boorstin, D. *The Image: Guide to Pseudo-Events in America*. Vintage, New York, 1961.

the recently enacted lifting of the ban on advertisements of drugs by pharmaceutical companies, the advertising budgets of these companies sometimes exceed their research budgets, although the companies may dispute this conclusion. As is the case with medium content in general, the problem is not that advertising is innately wrong but that it can be used excessively, and to mislead.

TEACH IT TO OUR CHILDREN

I have been aware much of my life about the term "nano." In my science textbooks, the prefix "nano" fell between "micro" and "pico." My first meaningful experience of the term, however, came during a lecture given at the University of Kentucky by Admiral Grace Hopper, co-author of the computer language COBOL, in the 1970s. She greeted us by handing each of us a length of electronic lead wire approximately 11.5 inches long. As some of you already realize, this is approximately the length of space light travels in a nanosecond. The content of her lecture was on the limitations spatial length would impose on the speed of computers. The size of computer circuitry has decreased by several orders of magnitude, and its speed increased commensurately, since then; we are now in nano dimensions of space, as well as time. Admiral Hopper's lecture was on the frontier and benefits of parallel processing. I have never forgotten Admiral Hopper or that lecture.

Nanotechnology should be taught to our children. There can be no doubt that nanotechnology is a part of our future. Application areas are already being developed, and estimates (although they may be excessive) are that the industry will generate more than a trillion dollars during the next decade.

One research group in New York has developed a systematic approach to help middle school students understand and appreciate nanotechnology.[24] These authors state:

> We have elected to focus our attention on the next generation of potential scientists and engineers. We are engaging young people who often do not view science as an educational opportunity let alone a career for them. *They do not see themselves as scientists*, and that is a significant barrier that we seek to overcome. We work in concert with their teachers, recognizing that this partnership will only work if we understand their world. With more and more mandated curricula, we must meet the needs of the schools rather than continue to offer content that has no relevance to the rest of the educational experience. Presently, we operate three middle-school science clubs for girls ... we also host three after-school science clubs for underrepresented minorities ... exercising our belief that these young students have all the potential in the world ... we make an honest effort to engage kids and have them hopefully begin to believe that science is a good think and learning about science can be exciting. Regardless of whether these kids go on to get their Ph.D. in nanotechnology, it is important for them to believe that they can do it.

> Furthermore, we offer events for the general public ... Finally, in April, 2003, a traveling museum exhibit, *It's a Nano World*, debuted ... we estimate that the exhibit will be

visited by more than one million people during its tour around the United States . . . .
In helping educate and, perhaps more importantly, inspire these young students, we
hope to raise the general awareness of the public at large as to what nanotechnology
is all about. This technology, and in general most technology, will have a very positive
impact on our lives. These challenges are to insure that the general public is informed
and cognizant of the potential. Mistakes made by failing to promote public understanding can lead to a wholesale rejection of the technology . . . nanotechnology as a field
evolved from engineering, but its extended roots will be found in fields including
physics, chemistry, and materials science. Its major new impact will clearly be in
the life sciences, presenting a challenge of organizing interdisciplinary groups that
can communicate and function effectively. This is no simple task as considerable
boundaries exist in language and culture. But, moreover, advances in the field will
be obscured if there is a failure to engage the general public and lay the foundation
for articulating how advances in the field will have a positive impact on their lives.[24]

The question of public discourse over whether and on what basis life science
applications of nanotechnology should proceed should be extended to how nanotechnology should be taught. The sooner in life students can be introduced to the
principles of nanotechnology (or any subject), the better grounded and more conver-
sant they will be. A primary challenge, however, is to tell primary students about
something they cannot see.

NANOTECHNOLOGY IN HIGH SCHOOL: A CASE STUDY

The town of Pulaski in south central Tennessee has the dubious distinction of being
the birthplace of the Ku Klux Klan.[28] It is also the home of one of the first high
school courses in nanotechnology in the nation. The following article appeared in
the Giles Today county newspaper in early 2006[29] and is reproduced as an
example of how public discourse might proceed.

When motivational speaker Ed Barlow came to Pulaski two months ago and addressed
a group of educators and business leaders, he generated excitement with one word . . .
nanotechnology. Since then Giles County Vocational Director Bill Davis has spent a lot
of time and energy generating interest and exploring ways to incorporate nanoscience
into the Giles County schools' curriculum. I mentioned it to Joe Fowlkes and Jan
McKeel, executive director of the South Central Tennessee Workforce Board
(SCTWB), after hearing Ed Barlow speak," Davis said. "From there it just got
bigger." Since January, Davis has talked with school board members, business and
industry leaders and educators throughout South Central Tennessee about the possibility of bringing nanotechnology to Giles County classrooms. Earlier this month, in
a thirst for knowledge about the subject, he and School Board Chairman Michael Gonzales traveled to Boston, Mass., to a nanotechnology conference. "We were the only
school system in the entire nation there looking at this," Gonzales told board

[28]The rise and fall of Jim Crow. Available at http://www.pbs.org/wnet/jimcrow/stories_events_kkk.
html.
[29]Giles educators, leaders examining nanotechnology. *Giles Today*, January 2006.

members at last Thursday's regular meeting. "We can prepare kids (for the future) by showing them what nanotechnology is about."

Instructor Thomas Smith has developed a course[30] that identifies areas of nanotechnology, researches its history and literature, and gives an overview of current research areas and safety concerns (e.g., since nanoparticles are so small conventional hazardous material masks are of little use; other protection must be adopted). Another aspect of the course deals with measurement techniques and material properties at the nanoscale, and introduces the mathematics necessary to analyze energy and forces in nanotechnology. Biological, hybrid, and biomimetic materials and tools used in their manufacture and application are discussed.

CONCLUSIONS

I remain hopeful that the development of nanotechnology can occur in a climate of reasoned public discussion. Recently, I asked noted biomedical ethicist Tom Beauchamp[31] how he thought the issue of stem cell research might proceed. He cited the precedent of gene therapy, in which an international committee of experts, the Recombinant DNA Advisory Committee (RAC) was convened 30 years ago to give international oversight to clinical studies. The early history of gene therapy was not very fruitful, primarily because little was known about the complexity of getting vectors to targets. Despite that lack of understanding and despite the death of one of the subjects in a trial due to lack of protocol oversight, RAC has provided a forum for gene therapy to proceed in a rational manner.

I have offered for several years a course in biomedical engineering ethics. At the first of that course I offer several goals for the students: to be able to perceive an ethical situation as it may arise, to become knowledgeable about details of the situation as it emerges, and perhaps most importantly, to be able to be conditional in one's approach, to see a situation from many different perspectives. Perhaps the science and technology of nanotechnology can proceed similarly.

[30]Smith, T. Nanotechnology standards. Unpublished course outline.
[31]Tom Beauchamp, personal communication.

CHAPTER 10

Technological Revolutions: Ethics and Policy in the Dark[1]

NICK BOSTROM

INTRODUCTION

We might define a technological revolution as a dramatic change brought about relatively quickly by the introduction of some new technology. As this definition is rather vague, it may be useful to complement it with a few candidate paradigm cases.

Some 11,000 years ago, in the neighborhood of Mesopotamia, some of our ancestors took up agriculture, beginning the end of the hunter–gatherer era. Improved food production led to population growth, causing average nutritional status and quality of life to decline below the hunter–gatherer level. Eventually, greater population densities led to vastly accelerated cultural and technological development. Standing armies became a possibility, allowing the ancient Sumerians to embark on unprecedented territorial expansion.

In 1448, Johan Gutenberg invented the movable-type printing process in Europe, enabling copies of the Bible to be mass produced. Gutenberg's invention became a major factor fueling the Renaissance, the Reformation, the scientific revolution, and helped give rise to mass literacy. A few hundred years later, *Mein Kampf* was mass produced using an improved version of the same technology.

Brilliant theoretical work in atomic physics and quantum mechanics in the first three decades of the twentieth century laid the foundation for the Manhattan project during World War II, which raced Hitler to the atomic bomb. Some believe that the subsequent buildup of enormous nuclear arsenals by both the United States and the former Union of Soviet Socialist Republics (USSR) created a balance of terror that prevented a third world war from being fought with conventional weapons, thereby saving

[1]I am grateful to Eric Drexler, Guy Kahane, Matthew Liao, and Rebecca Roache for helpful suggestions.

Nanoscale: Issues and Perspectives for the Nano Century. Edited by Nigel M. de S. Cameron and M. Ellen Mitchell

many tens of millions of lives. Others believe that it was only by luck that a nuclear Armageddon was avoided, which would have claimed the lives of many hundreds of millions, perhaps billions. These beliefs may both be true.

In 1957, Soviet scientists launched Sputnik 1. In the following year, the United States created the Defense Advanced Research Projects Agency (DARPA) to ensure that the United States would stay ahead of its rivals in military technology. This agency began developing a communication system that could survive nuclear bombardment by the USSR. The result, ARPANET, later became the Internet, which made available the World Wide Web, email, and other services. The long-term consequences remain to be seen.

It would appear that technological revolutions are among the most consequential things that happen to humanity, perhaps exceeded in their impact only by more gradual, nonrevolutionary technological developments. Technological change is in large part responsible for the evolution of such basic parameters of the human condition as the size of the world population, life expectancy, education levels, material standards of living, the nature of work, communication, healthcare, war, and the effects of human activities on the natural environment. Other aspects of society and our individual lives are also influenced by technology in many direct and indirect ways, including governance, entertainment, human relationships, and our views on morality, cosmology, and human nature. One does not have to embrace any strong form of technological determinism or be a historical materialist to acknowledge that technological capability—through its complex interactions with individuals, institutions, cultures, and the environment—is a key determinant of the ground rules within which the game of human civilization is played out at any given point in time.

In the course of a normal lifetime nowadays, we can all expect to be involved in one or more technological revolutions: if not as inventor, funder, investor, regulator, or opinion leader, then at least as voting citizen, worker, and consumer. Given that technological revolutions have such profound consequences, one might think that they should be the focus of intense ethical deliberation and feature centrally in public policy analysis. If so much is at stake, it would seem to behoove us to dedicate a corresponding amount of effort to ensuring that we make the right decisions. How is humanity measuring up to this challenge?

ETHICAL, LEGAL, AND SOCIETAL ISSUES RESEARCH, AND PUBLIC CONCERNS ABOUT SCIENCE AND TECHNOLOGY

One can perceive a slow trend since World War II of intensifying endeavors to connect science and technology (hereafter, S&T) policy to a broader discussion about desired social outcomes. The rise of environmentalism in the 1960s fostered this trend, reflecting public demand that more S&T resources be devoted to the betterment of water and air quality. The congressional Office of Technology Assessment was created to improve understanding of the societal implications of technological choices. Disease lobbies have formed, seeking, among other things, increased funding for medical research into a variety of conditions. Concerns about global warming has pushed greatly increased resources into climate science and led to calls for more funding for research into alternative energy sources, as well as more direct interventions to reduce greenhouse gas emissions.

Some 3% of the budget for the Human Genome Project was set aside for studying the ethical, legal, and societal issues (ELSI) connected to genetic information—not much in relative terms, but still enough for Art Caplan to describe the move as the "full employment act for bioethics." There is now a burgeoning of research in a number of technology-related fields of applied ethics such as computer ethics, neuroethics, and especially nanoethics. Research into the ethical, legal, and societal issues related to nanotechnology (NELSI) might over time outstrip that of the genetic ELSI program.

Today, there is also a large and diverse set of grassroots organizations, think tanks, and university centers that work on technology-related issues. In recent years, ethical issues related to human enhancement technology have particularly come to the fore. There is growing apprehension that anticipated technological developments—including nanotechnology, but also artificial intelligence, neurotechnology, and information technology—are likely in the twenty-first century to have transforming impacts on human society, and perhaps on human nature itself. Some speak of an "NBIC" convergence (referring to the integration of the neuro-bio-information- and cognitive sciences), and explicitly link this to the prospect of human enhancement.[2]

While the spectrum of opinion represented in these discussions is quite broad, there seem to be some points of consensus, at least within the Western "mainstream":

- Technological development will have major impacts on human society.
- These developments will create both problems and opportunities.
- "Turning back" is neither feasible nor desirable.
- There is a need for careful public examination of both the upsides and downsides of new technologies, and for exploration of possible ways of limiting potential harms (including technological, regulatory, intergovernmental, educational, and community-based responses).

In addition to disagreements about the *content* of S&T policy, there are also disagreements about the *process* whereby such policy should be determined, with challenges being raised to the "official" model of the appropriate relationship between science and society, which harks back to the Enlightenment. According to the Enlightenment model, "the only scientific citizens are the scientists themselves. For science to engage in the production of properly scientific knowledge it must live in a 'free state' and in a domain apart from the rest of society. Historically, science's grip on Truth is seen as having grown progressively stronger as society's

---

[2]See, for example, Roco and Bainbridge 2003; Bainbridge and Roco 2006.

grip on science has grown progressively weaker and ever more closely circum-scribed."[3] The Enlightenment model pictures science as the goose that lays the golden egg, but only when it is protected from external interference.

This model has come under increased scrutiny since the 1960s. In Europe, broad efforts are underway to change the "social contract" between science and society in order to create a larger role for public participation and deliberation in setting the priorities and limitations of science and technology. The notion that science is unproblematically associated with progress is no longer widely accepted, a more critical approach having been stimulated by developments, such as the nuclear arms race during the cold war, the Chernobyl disaster, and the increased salience of environmental concerns in later years. In Britain, the mismanagement of the out-break BSE (mad cow disease) eroded public confidence in Government science policy. The experiences from the BSE crisis later helped foment public opposition to the introduction of genetically modified crops.

Initiatives to build more opportunities for the public to become engaged in science and technology issues can be seen as an effort to rebuild public confidence and to secure science's "license to operate." But beyond such public relations goals, there are also many who argue that the S&T enterprise needs much more guidance from society in order to ensure that scientific and technological research is really directed to achieve socially beneficial outcomes. The aim is not necessarily to restrict research, or to contest any particular scientific theory, but to yoke the science and technology behemoth to ends chosen by the people after due delibera-tion and debate. If S&T is such an important shaper of the modern world, it should be brought under democratic control, the thinking goes, and its workings should become more transparent to the people who have to live with the consequences.[4]

This view is reflected in a recent paper by Michael Crow and Daniel Sarewitz:

> When resources are allocated for R&D [research and development] programs, the implications for complex societal transformation are not considered. The fundamental assumption underlying the allocation process is that all societal outcomes will be posi-tive, and that technological cause will lead directly to a desired societal effect. The lit-erature promoting the National Nanotechnology Initiative expresses this view.[5]

They continue:

> The fact that societal outcomes are not a serious part of the framework seems to derive from two beliefs: (1) that the science and technology enterprise has to be granted auton-omy to chose its own direction of advance and innovation; and (2) that because we cannot predict the future of science or technological innovation, we cannot prepare

---

[3]Elam and Bertilson 2002, p. 133.

[4]Part of this intellectual trend is the conglomeration of science and technology into "technoscience," the idea being (roughly) that science and technology are inextricably linked and that both are socially coded, historically situated, and sustained by actor networks consisting of both human and artifacts; see, for example, Latour 1987. In this essay, I for the most part do not sharply separate science and technology, but it seems to me that a more nuanced treatment would have to distinguish different components of the "science and technology enterprise" (to use the term favored by Crow and Sarewitz).

[5]Crow and Sarewitz 2001, p. 97.

for it in advance. These are oft-articulated arguments, not straw men. Yet the first is contradicted by reality, and the second is irrelevant. The direction of science and technology is in fact dictated by an enormous number of constraints (only one of which is the nature of nature itself). And preparation for the future obviously does not require accurate prediction; rather, it requires a foundation of knowledge upon which to base action, a capacity to learn from experience, close attention to what is going on in the present, and healthy and resilient institutions that can effectively respond or adapt to change in a timely manner.[6]

Let us look in a little more detail at these two issues, the autonomy of the S&T enterprise and unpredictability, starting with the latter.

UNPREDICTABILITY

Technological revolutions have far-reaching consequences that are difficult to predict. This poses a challenge for technology policy. The challenge, however, is not unique to technology policy. All major policy changes have far-reaching consequences that are difficult to predict. There does not exist an exact science that can tell us precisely what will happen in the long run when a government decides to abolish slavery, go to war, or give women the right to vote.

For more modest policy changes, such as a reduction of a sales tax or the introduction of stricter regulation on lead paint, expectations of the near-term consequences are more tightly constrained by economic and scientific models and by parallel experience in other countries. But social systems are complex, and even small interventions can have large unanticipated long-term consequences. Perhaps reduced lead levels will lead to increased intelligence in some children, and some of these might then grow up to become more successful scientists than they would otherwise have been. Some of these scientists might invent the future equivalent of the atomic bomb or antibiotics. Perhaps a reduced sales tax will increase profits in one sector of the economy, some of which might be used as campaign contributions that get a politician elected who will pass legislation that may, in turn, have wide-ranging and unpredictable ramifications.

Even the most trivial personal decisions can have monumental consequences that shape the fate of nations. Maybe one afternoon a thousand years ago in some Swiss village, a young woman decided to go for a stroll to the lake. There she met a lad, and later they married and had children. Thus she became the great-grandmother of Adolf Hitler. If she had gone to the forest instead of the lake, the Holocaust and perhaps World War II would not have happened.[7]

On the other hand, the unpredictability of the future should not be exaggerated. Crow and Sarewitz appear to concede that "we cannot predict the future of

[6]See footnote 5, p. 98.
[7]This example is borrowed from James Lenman (Lenman 2000). See also Bostrom 2006.

science or technological innovation," but in doing so they concede too much. As Eric Drexler notes:

> The future of technology is in some ways easy to predict. Computers will become faster, materials will become stronger, and medicine will cure more diseases. Nanotechnology, which works on the nanometer scale of molecules and atoms, will be a large part of this future, enabling great improvements in all these technologies.[8]

Predictability is a matter of degree, and the degree varies radically depending on what precisely it is that we are trying to predict.

Even such a seemingly platitudinous claim as "physics is better at prediction than social science" is on closer inspection quite problematic. Physics can predict some things and not others. We can use physics to predict with considerable accuracy where the planet Jupiter will be in the year 2020. But physics does not enable us to predict which particular atom will be closest to the center of gravity in the solar system in 2020. Social science can also predict some things: the prediction that there will exist some inequalities in social status and income in the United States in 2020 seems about as reliable as the prediction of where the planet Jupiter will be at that time.

If we want to make sense of the claim that physics is better at predicting than social science is, we have to work harder to explicate what it might mean. One possible way of explicating the claim is that when one says that physics is better at predicting than social science one might mean that experts in physics have a greater advantage over nonexperts in predicting interesting things in the domain of physics than experts in social science have over nonexperts in predicting interesting things in the domain of social science. This is still very imprecise since it relies on an undefined concept of "interesting things." Yet the explication does at least draw attention to one aspect of the idea of predictability that is relevant in the context of public policy, namely, the extent to which research and expertise can improve our ability to predict. The usefulness of ELSI-funded activities might depend not on the absolute obtainable degree of predictability of technological innovation and social outcomes but on how much improvement in predictive ability these activities will produce.

Hence, let us set aside the following unhelpful question: *Is the future of science or technological innovation predictable?* A better question would be *How predictable are various aspects of the future of science or technological innovation?*

But often, we will get more mileage out of asking: *How much more predictable can (a certain aspect of) the future of science or technological innovations become if we devote a certain amount of resources to study it?*

Or better still: *Which particular inquiries would do most to improve our ability to predict those aspects of the future of S&T that we most need to know about in advance?*

---

[8]Drexler 2003. Some technology impacts are equally predictable, for example, some new medicines will be used, will save lives, some of those people whose lives have been saved will draw state pensions, vote, and so on.

Pursuit of this question could lead us to explore many interesting avenues of research which might result in improved means of obtaining foresight about S&T developments and their policy consequences.[9]

Crow and Sarewitz, however, wishing to side-step the question about predictability, claim that it is "irrelevant":

> preparation for the future obviously does not require accurate prediction; rather, it requires a foundation of knowledge upon which to base action, a capacity to learn from experience, close attention to what is going on in the present, and healthy and resilient institutions that can effectively respond or adapt to change in a timely manner.

This answer is too quick. Each of the elements they mention as required for the preparation for the future relies in some way on accurate prediction. A capacity to learn from experience is not useful for preparing for the future unless we can correctly assume (predict) that the lessons we derive from the past will be applicable to future situations. Close attention to what is going on in the present is likewise futile unless we can assume that what is going on in the present will reveal stable trends or otherwise shed light on what is likely to happen next. It also requires prediction to figure out what kind of institutions will prove healthy, resilient, and effective in responding or adapting to future changes. Predicting the future quality and behavior of institutions that we create today is not an exact science.

It is possible, however, to reconstruct Crow and Sarewitz's argument in a way that makes more sense. Effective preparation for the future does require accurate prediction of at least certain aspects of the future. But some aspects are harder to predict than others. If we despair of predicting the future in detail, we may sensibly resort to courses of action that will do reasonably well independently of the details of how things turn out. One such course of action is to build institutional capacities that are able to respond effectively to future needs as they arise. Determining which institutional capacities will prove effective in the future does require prediction, but this is often a more feasible prediction task than predicting the details of the situations that they will have to respond to. The more the future is veiled in ignorance, the more it makes sense to focus on building general-purpose capabilities.

Recast in this way, the argument is more defensible as far as it goes. But it does not go very far. Its limitations become clear when we consider it within the context of S&T policy.

The tasks of S&T policy include setting priorities for the allocation of funding to research projects. It is hard to predict which lines of research will bear fruit and which will not. There are several possible ways of responding to this predicament. The first is to concentrate funding on those research avenues that we can be fairly certain will bear at least some fruit. The second is to diversify the research portfolio and fund a little bit of everything. A third approach is to bet on a few research avenues that seem especially promising and accept the risk of total failure.

---

[9]See, for example, Tetlock 2005.

Depending on the funders' attitude to risk, and other factors, some mixture of these approaches might be optimal.[10]

But the question of predictability does not go away. Of course, it is not possible to fund a little bit of (literally) *everything*, and spreading out funding as evenly as possible among all seekers seems unlikely to be the smartest way of going about things. So how much funding should be concentrated on a few promising fields, such as nanotechnology? How tightly focused should that funding be on particular approaches, methods, and research centers? There is no simple answer. The optimal strategy will depend on just how confident the funders are in their ability to pick winners, that is, predict future advances.

The situation becomes even more complicated if we consider that some research projects might not simply fail to come to fruition but might bear poisoned fruit— produce results that we would be better off without. Certain possible weapons technologies fall into this category, but some critics of technoscience would argue that it includes a great deal else beside. Presumably, the majority's feeling that humanity ought to pursue scientific and technological research rests on the assumption that the value of the consequences of such advances is likely to be, on balance, positive. But if this assumption is true, and if it is also granted that *some* technological advances will prove detrimental, then again the question becomes whether we can be confident enough in our ability to predict in advance which particular trees will produce poisoned fruit in order to be justified in cutting them down now, or whether we should instead them let all grow, in the name of our epistemic modesty. Universal cultivation seems to require that there be just the right amount of predictability: enough so that we can expect that on balance the orchard will be beneficial to humanity and that our cultivation of it will in fact promote its growth, but not so much that we would be better off by chopping down selected trees because their growth may in the long run cause harm.

The complexity of our prediction problem increases even further when we consider that the payoff of an individual research project is not independent of what happens with other research projects. Different advances may work synergistically (as in the case of NBIC technologies), or one might preempt another and make it obsolete. When such dependencies exist, the development of an optimal research portfolio becomes more difficult. If predictability is low, we might decide to ignore such dependencies; if it is higher, on the other hand, we would be remiss not to take them into account in deciding our research priorities. The question of dependencies between potential future advances might also have to be reflected in what institutional structures we should create for the process of S&T agenda setting and implementation; for example, whether to establish a separate committee for a particular subfield. Again, the question of the degree of predictability of various aspects of the

---

[10]The need to make these kinds of tradeoff, of course, is not confined to funding of the natural sciences and technology, but applies to funding of the social sciences and ELSI programs too.

future—far from being one that can be trivially answered or sidestepped—is in fact highly difficult and central to many S&T policy issues.

From this brief discussion of predictability, we can already draw several conclusions. First, while some scientific advances and technological innovations are hard to predict with accuracy far in advance, the problem is not unique to the science and technology context. Big policy decisions, small policy decisions, and trivial personal decisions all have important consequences that cannot be predicted in detail. Second, there are many aspects of scientific and technological developments that *can* be predicted. Third, all meaningful preparations for the future rely, explicitly or implicitly, on prediction. Fourth, the issue of the *relative* predictability of different aspects of the future, and how much the predictability can be *improved* by various kinds of investment, is important in thinking about how R&D programs should be structured.

A further lesson is that improvements in our ability to predict various potential S&T advances and their consequences could make a very valuable contribution to our capacity to make wise S&T policy decisions.

STRATEGIC CONSIDERATIONS IN S&T POLICY

Let us now turn to consider the source of an additional level of complexity in S&T governance: strategy and politics. These impose constraints on what can be done, and therefore on what it would make sense to attempt to do. As Ralph Waldo Emerson once wrote:

> Web to weave, and corn to grind;
> Things are in the saddle,
> And ride mankind.

In particular, one needs to question whether mankind is really riding science and technology, or whether it is the other way around.

One obvious sense in which mankind is not in the saddle is that the S&T policy decisions on this planet are not made by some one unified body of rational and beneficent representatives of humanity who are trying to get us to some particular destination. Instead, there are countless agents, pursuing different and often opposing objectives, influencing various aspects of our species' S&T activities—national and regional governments, corporations, private philanthropic foundations, special interest lobbies, journal editors, research councils, media organizations, university presidents, prize committees, consumers, voters, scientists, public intellectuals, and so forth. More specifically, we know two things: (1) there is no unified decision-making entity that has the power to direct or halt all research worldwide in any area, and (2) that many of the decision-making entities that influence S&T policies at various locations are themselves subject to influence from a variety of agents with diverging goals and agendas. Both of these facts have profound consequences for our thinking about S&T policy.

The first fact, the absence of global control of the world's S&T, makes it difficult or impossible to stop research and innovation in a particular direction even if it would be a good thing to do so. For example, even if some detailed study or public consultation concluded that a nanotech revolution would be detrimental to humanity, there is no clear path to preventing such a revolution from happening anyway. The government of one country might rescind public funding for research in certain areas thought likely to enable advances in nanotechnology; but research would continue (albeit perhaps at a slower pace) using funding from private sources. The government might then ban all such research, whether publicly or privately funded, but other nations would almost certainly continue to push forward, and if the technology is feasible, it will eventually see the day anyway. Global bans on technological developments are very difficult to negotiate and even harder to police. The difficulties are amplified in cases where significant incentives exist for some groups for moving forward, where development can be conducted with modest resources, where concealment is possible, and where there is no salient demarcation between the hypothetically proscribed activity and legitimate research. Nanotechnology satisfies many of these conditions, so the prospect of global relinquishment appears to be close to nil, at least in lieu of dramatic advances in both surveillance technology and global governance.[11]

The infeasibility of halting certain kinds of research is a point often repeated in S&T policy discussions: "If our country does not go forward with this, someone else will and we will fall behind," or "If nanotechnology is outlawed, only outlaws will have nanotechnology."[12] The appeal to national competitiveness seems to be one of the rhetorically most effective arguments both for increased spending on research, and against regulation that would slow development.

It is worth comparing this argument from economic competitiveness with another appealing argument for more research funding: that research is a global public good and should be supported out of love of humankind. The two arguments stand in some tension. The global public goods argument suggests that it might be in a nation's self-interest to free-ride on other nations' S&T investment (particularly foundational research, the benefits of which are especially difficult for the producer to monopolize). If this is the case then "national competitiveness" might actually suffer from the diversion of resources away from other sectors of society to S&T research. Yet, both arguments could be true. There might be high returns for a nation to its investments in R&D, and additional returns that cannot be captured domestically and instead become a positive externality benefiting other nations. In this case, both the love of humankind and the appeal to

---

[11]The latter proviso is not insignificant if we are thinking about longer time scales. One might also imagine that support for tough international action could increase dramatically following a big disaster, such as an act of nuclear terrorism. Of course, another way in which nanotechnology research could come to a halt is as a result of a civilization-destroying global catastrophe. This proviso is also not insignificant; see Bostrom 2002.

[12]For example, Vandermolen 2006.

national advantage would work in tandem as reasons for increasing R&D investment.[13]

As both of these arguments illustrate, there are important consequences for S&T policy from the fact that S&T policy is not perfectly globally coordinated. But the complexity of the strategic situation increases vastly when we take into account that even within a particular country, S&T decisions are not made by a single unified perfectly rational and perfectly beneficent agency. Policy recommendations directed to an imaginary ideal global or national decision-maker may form useful focal points for interim discussion, but ultimately they need to be transformed into recommendations addressed to some identifiable real agent. At that stage, recommendations must take into account the limitations of that agent's powers, understanding, attention, and interests. This transformation of "what should be done" in an abstract sense into sensible recommendations to an agent that can actually do things is far from straightforward.

For example, one argument that has been given for moving forward with nanotechnology research as rapid as possibly is as follows:[14]

1. The risks of advanced nanotech are great.
2. Reducing these risks will require a period of serious preparation.
3. Serious preparation will only begin once the prospect of advanced nanotech is taken seriously by broad sectors of society.
4. Broad sectors of society will only take the prospect of advanced nanotech seriously once there is a large research effort underway.
5. The earlier a serious research effort is initiated, the longer it will take to deliver advanced nanotech (because it starts from a lower level of pre-existing enabling technologies).
6. Therefore, the earlier a serious research effort is initiated, the longer the period during which serious preparation will take place, and the greater the reduction of the risks that will eventually have to be faced.
7. Therefore, a serious research effort should be initiated as soon as possible.

I present this argument not in order to evaluate it, but to illustrate the point about strategic complexity. What naively looks like a reason for going slowly or stopping (the risks of advanced nanotech being great) ends up, on this line of thinking, as a reason for the opposite conclusion. When one attempts to integrate such second

---

[13]Most studies of the economic returns to R&D have not focused on the international dimension. Domestically, it appears that the social returns of R&D, although they are difficult to measure, are very high and that optimal R&D investment substantially exceeds the actually level. See, for example, Jones and Williams 1998; Salter and Martin 2001.

[14]See Drexler 1992a, p. 242. Drexler (private communication) confirms that this reconstruction corresponds to the point he was making. Obviously, a number of implicit premises would have to be added if one wished to present the argument in the form of a deductively valid chain of reasoning. By "advanced nanotechnology" I here refer to a possible future form of radical nanotechnology, sometimes called molecular nanotechnology, or "machine-phase" nanotechnology; see also Drexler 1992b.

guessing of the responses of various actors in one's R&S policy recommendations, the result might differ radically from what one would get from a more simple-minded approach that ignores the strategic dimensions of the situation.

It is interesting to consider to whom it is that this kind of argument addresses itself. The "broad sectors of society," which will supposedly begin serious preparation only after a large research effort is already underway, are presumably not the intended recipients of the message. If they were capable of and willing to understand, agree with, and act on an argument like this, then they would not need to wait for a large research effort to get underway in order to take the need for preparation seriously. The argument appears to be esoteric. There are some people who are "in the know" about the prospects of advanced nanotechnology, and these people would have a reason (so the argument goes) to direct their efforts toward accelerating the implementation of a serious nanotech research effort even if they thought that the risks of advanced nanotech outweighed the benefits.

One way the cognoscenti could do this would be by publicizing another argument for the acceleration of nanotech research, such as the argument that if *we* do not move forward quickly, then *somebody else*, perhaps a hostile state, will get there first—and that would be the worst possible outcome of all. Note that even this second argument addresses itself not to everybody but to a select group—in this case our compatriots, or at least the citizens of "good" states. It would not be desirable that the citizens of "bad" states urge *their* compatriots and government officials to launch a crash program for the development of advanced nanotech so that *they* get there first.[15]

There are actually people (although perhaps not many) who think at this level of sophistication and attempt to take strategic considerations, such as those above into account in deciding what they ought to do. Some of these people are well meaning and honest and would not consent to putting forward an argument or an opinion that they did not sincerely hold to be true. Esoteric arguments do not require deception, or even active concealment, because to a significant extent audiences self-select which arguments they hear and absorb. The "nano-*cognoscenti*" might be the only ones who are receptive to the argument about a large research program being necessary to get broad sectors of society to take the risks seriously and start preparing. The citizens of "good" nations might be more likely to follow their compatriots' advice on the need to move forward with the research to avoid falling behind in a future arms race than the citizens of the "bad" nations who we fear might otherwise take the lead. (But how sure can we be that this is always the case?)

Predictability, or the lack thereof, again emerges as an important issue. Clearly, anticipating the responses of many different agents, how these responses will interact, and more generally how the ecology of ideas and opinions will be affected by the promulgation of one argument or another, is a daunting task—in many cases even more difficult than forecasting future developments in S&T.

---

[15]One may of course insist that the good states should develop only *defensive* nanotech capabilities. But offensive and defensive applications would require largely the same underlying technological advances.

LIMITING THE SCOPE OF OUR DELIBERATIONS?

It is temping to ignore all of these difficulties and focus on the simpler task of figuring out what we have most reason to do, subtle strategic considerations aside. At the individual level, a person might simply try to decide: Is nanotech likely to do more good than harm? If yes, then be in favor of its development; if no, then oppose it. Or alternatively: Is there some path involving nanotech development combined with certain regulations and/or public policies that would bring great benefits and only moderate risks? If yes, then promote that path.

We may note immediately that the choice here is not a dichotomous one, either to ignore all strategic considerations or else to take them all on board without limitation. Even in relatively uncomplicated deliberations, *some* strategic considerations might be admissible. For example, we might take into account the fact that other nations will proceed with nanotech development even if our nation does not, while setting aside all considerations having to do with the cognitive limitations of the *hoi polloi* or the way that special interests are likely to influence the implementation of any officially adopted policy directive.

At a collective level, too, we might decide to exclude certain kinds of reasons from various contexts of public deliberation. This is the idea, for example, behind John Rawls' concept of *public reason*.[16] In dealing with constitutional essentials and matters of basic justice, citizens abiding by the idea of public reason are, according to Rawls, entitled to justify the position they want adopted only on the basis of reasons that could reasonably be accepted by other citizens who do not necessarily share the same metaphysical, religious or cultural views.[17] This constraint has been criticized on grounds that it would require insincerity (not putting forward the real reasons for one's views) and lead to a shallowness in public discourse as participants are required to confine themselves to the lowest common denominator of shared background assumptions. Rawls believes that we should nevertheless abide by such a constraint:

> ... given our duty of civility to other citizens. After all, they share with us the same sense of its imperfection, though on different grounds, as they hold different comprehensive doctrines and believe different grounds are left out of account. But it is the only way, and by accepting that politics in a democratic society can never be guided by what we see as the whole truth, that we can realize the ideal expressed by the principle of legitimacy: to live politically with others in the light of reasons all might reasonably be expected to endorse.[18]

As a much more mundane example of discourse restriction, consider the convention against the use of *ad hominem* arguments in science and many other arenas of

---

[16]For a review of philosophical views on the idea of publicity, including those of Rawls, see Gosseries 2005.

[17]This is a simplification of Rawls' view, but the details are not essential for present purposes; see also Gosseries 2005.

[18]Rawls 1999, pp. 242–243.

disciplined discussion. The nominal justification for this rule is that the validity of a scientific claim is independent of the personal attributes of the person or the group who puts it forward. Construed as a narrow point about logic, this comment about ad hominem arguments is obviously correct. But it overlooks the epistemic significance of heuristics that rely on information about *how* something was said and *by whom* in order to evaluate the credibility of a statement. In reality, no scientist adopts or rejects scientific assertions solely on the basis of an independent examination of the primary evidence. Cumulative scientific progress is possible only because scientists take on trust statements made by other scientists—statements encountered in textbooks, journal articles, and informal conversations around the coffee machine. In deciding whether to trust such statements, an assessment has to be made of the reliability of the source. Clues about source reliability come in many forms—including information about factors, such as funding sources, peer esteem, academic affiliation, career incentives, and personal attributes, such as honesty, expertise, cognitive ability, and possible ideological biases. Taking that kind of information into account when evaluating the plausibility of a scientific hypothesis need involve no error of logic.

Why is it, then, that restrictions on the use of the ad hominem command such wide support? Why should arguments that highlight potentially relevant information be singled out for suspicion? I would suggest that this is because experience has demonstrated the potential for abuse. For reasons that may have to do with human psychology, discourses that tolerate the unrestricted use of *ad hominem* arguments manifest an enhanced tendency to degenerate into personal feuds in which the spirit of collaborative, reasoned inquiry is quickly extinguished. Ad hominem arguments bring out our inner Neanderthal.

The instrumental proscription of ad hominem and the more deeply normative Rawlsian idea of public reason both illustrate the general concept of constructing discourses partly by ruling out of court some types of consideration. There exist many such discourse boundary constraints, both in science and in other arenas, varying from context to context. It is perhaps not implausible to suppose that some strategic considerations, such as the ones introduced above, may also be appropriately excluded in some contexts of S&T policy deliberation.

How should we determine where the appropriate boundaries for a particular discourse should be placed? The complexities of this question protrude far beyond the scope of this essay, but it might be worth listing a few potentially relevant factors.[19] These might serve to hint at how which future investigations could explore this area further.

1. *Self-Deception and Bias.* We know that human cognition is susceptible to self-deception and biases of various kinds. Certain types of consideration might offer more foothold than others to irrationality and prejudice.

---

[19]These would be complementary to the "micro-level" maxims identified by Paul Grice in his work on conversational implicatures, Grice 1975.

2. *Manipulability*. Similarly, in public deliberation, certain types of consideration might be easier to manipulate for partisan purposes, suggesting that they should be given less weight or barred altogether.

3. *Unpredictability*. As argued already, the complexity of a certain level of strategic consideration is such that we might quickly lose ourselves in a fog of unknowability if we venture there.

4. *Direct versus Mediated Consequences*. This harks back both to unpredictability and foregoing discussion of strategic considerations. But one might also maintain, as a matter of basic ethical theory, that we are not responsible for (and should hence not take into account?) those consequences of our actions that are mediated by many external causal factors or other by people's choices, even if they should be partly predictable.

5. *Division of Labor*. In a public deliberation, it is not always necessary that all parties attempt to incorporate and digest all types of consideration. A sensible majority view might emerge even if some or all parties are biased and blinkered in what considerations they take into account.

6. *Accessibility to Stakeholders*. Some types of consideration might effectively exclude some stakeholders from participating meaningfully in a public deliberation. Does this sometimes make the appeal to such considerations inadmissible?

7. *Moral Side Constraints*. Honesty and truthfulness included, but perhaps there are also other moral constraints and desiderata that apply.

8. *The Potential for Cumulative Progress*. It might be that discussion of certain types of consideration produce results that are easier to preserve and carry over to future dialogs and new situations, enabling cumulative progress.

9. *The Potential for Connection to Other Domains*. Work that clarifies some types of consideration might more readily be integrated with results from other disciplines or cognitive domains. This might be an argument for paying more attention to such considerations.

Recall the two beliefs or "oft-articulated arguments" to which Crow and Sarewitz attribute responsibility for the lack of consideration given to the implications for complex societal transformation in the allocation of resources to R&D programs. One was the argument about unpredictability. The other is the argument that the science and technology enterprise has to be granted autonomy to choose its own direction of advance and innovation. This latter argument is, in Crow and Sarewitz's view,

> contradicted by reality . . . . The direction of science and technology is in fact dictated by an enormous number of constraints (only one of which is the nature of nature itself).

It is unclear that how this is supposed to be an objection to the view that the S&T enterprise should be granted autonomy. Those who hold this view might well agree with Crow and Sarewitz that the spirit of free inquiry is currently fettered by an enormous number of constraints while time insisting that it would be preferable if at least some of these shackles could be abolished, or at least relaxed, and that imposing additional constraints would make things worse. One needs to distinguish the normative question about how things ought to be from the positive question of how things are.[20]

Democracies often find it advisable to insulate certain functions from direct public influence. The U.S. constitution, for example, deliberately creates a judiciary that is substantially insulated from direct political influence. Central banks in many countries are similarly autonomous. In all democracies, the day-to-day operation of most government departments is significantly shielded from the direct impact of public opinion. In the universities, senior research staff enjoy a great deal of intellectual freedom and frequently have the opportunity to pursue research projects of their own choosing, even though their salaries might come from the taxpayer. In the private sector, corporations and individuals are free to pursue almost whatever direction of research they desire, provided they have the funding to do so. In addition, independent intellectuals and writers are accountable to nobody but themselves.

What advocates of increased public involvement and direct democracy in S&T are arguing is not that matters of scientific controversy ought to be settled by popular referendum, or that the public be brought in to micromanage the conduct of scientific research projects. Nor are they minded to restrict freedom or speech or any of the other intellectual freedoms currently enjoyed by individuals. What is being proposed is generally some more moderate position, for example, that the setting of the overall parameters and priorities of publicly funded S&T research ought to be made more transparent and more directly subject to democratic input than is currently the case. Efforts might be advocated to improve public understanding of science and to create more opportunities for genuine, two-way dialogue between scientists and the public. Often, the focus is on the implementation side of the S&T enterprise, urging greater direct democratic control over which new technologies are permitted, under what forms of regulation, and with what ancillary polices to modulate their societal impacts. Also, sometimes of concern are specific methods of scientific research, such as animal experimentation or the use human embryos.

The question of governance of S&T issues can probably not be separated from questions of governance in general. My focus here, however, is not on the structures of governance—with *who* should make the decisions—but rather on the terms of the discourse: what kind of considerations should be taken into account, and in what

---

[20]A distinction that is often overlooked in discussions about the future. There is a lamentable tendency to let prediction and evaluation blend into a confused blob of wishful (or fearful) thinking.

ways.[21] Of course, who makes decisions does, in reality, influence what considerations get taken into account. But we can still distinguish the normative question of what kinds of reasons ought to be guiding the S&T policy.

We have considered some possible grounds for various limitations of what reasons should be considered in the determination of S&T policy. Let us end the section with the words of Marie Curie:

> We must not forget that when radium was discovered no one knew that it would prove useful in hospitals. The work was one of pure science. And this is a proof that scientific work must not be considered from the point of view of the direct usefulness of it. It must be done for itself, for the beauty of science, and then there is always the chance that a scientific discovery may become like the radium a benefit for humanity.[22]

EXPANDING THE SCOPE OF OUR DELIBERATIONS?

The quote by Marie Curie expresses an extreme version of the view that we ought to narrow the considerations taken into account. According to Curie, scientific work must be done "for itself, for the beauty of science" and with no view to its direct usefulness.

One may contrast the innocence of Marie Curie's words with a well-known remark made some two and a half decades later by another distinguished physicist, Robert Oppenheimer, who had spearheaded the development of the nuclear bomb:

> In some sort of crude sense which no vulgarity, no humor, no overstatement can quite extinguish, the physicists have known sin; and this is a knowledge which they cannot lose.[23]

The explosion of the first nuclear weapon in the Trinity test, and the later use in Hiroshima and Nagasaki, are sometimes seen as emblems of the failure of scientists to concern themselves with the societal implications of their work. In fact, many of the scientists involved in the Manhattan project (and others, such as Linus Pauling, who declined to participate) were quite deeply concerned about societal implications. Among those agreeing to lend their skills to the project, a major motivation was the concern that Nazi Germany might otherwise get to the fission bomb first. This is an example of a strategic consideration mentioned earlier, national

---

[21]Another variable here would be what we may term *the format* of deliberation, for example, whether it should occur behind closed doors or in the full glare of publicity, and what the tone or spirit of a deliberation should be, for example, how much and what kind of emotion should be displayed. It is plausible that such format factors have a substantial effect on the nature of the deliberation. There is some direct empirical evidence for this. For example, one recent study of the deliberations of the Federal Reserve's Federal Open Market Committee found that transcript publication suppressed the expression of dissenting opinions and stifled debate over short-term interest rates. Meade, Stasavage, and London School of Economics and Political Science. Centre for Economic Performance 2004.

[22]Marie Curie (1867–1934), Lecture at Vassar College, May 14, 1921.

[23]Robert Oppenheimer (1904–1967), "Physics in the Contemporary World," lecture at M.I.T., November 25, 1947.

competitiveness. On the Marie Curie version of the Enlightenment model, neither this consideration nor scruples over how the scientific findings might later be used should be taken into account, at least not by the people doing the scientific work.

There are several general counterarguments against the view that the scope of our deliberations should be in some ways restricted.

First, ignoring considerations that are evidentially relevant to potential outcomes that we care about means ignoring relevant information. Ignoring relevant information might not be rational, and might impede our effectiveness in achieving our goals. The more relevant the information, and the more important the goals to which it is relevant, the greater the cost of such intentional ignorance.

Second, confining deliberations within a set of fixed constraints can yield power to those who determine what these constraints should be. This power can be misused. As rhetoricians and sophisticated technocrats are well aware, the framing of an issue, which implicitly determines what kinds of consideration will be seen as being appropriate and having a bearing, often effectively determines the conclusion a deliberation will reach. In political discourse, framings are often fiercely contested. To accept a set of discourse constraints might mean buying into someone else's agenda that tilts the deliberation in favor of some predetermined position.

Third, and related to the first two arguments, one fairly likely effect of adopting a narrow framework for our deliberations of the S&T enterprise is to create a bias in favor of a certain kind of "conservatism"—conservatism not in the sense of political ideology, but in the sense of a presumption in favor of business-as-usual. Scope-restrictions risk ruling out radical critiques, ones that challenge the fundamental assumptions behind the common way of thinking and doing things. In the context of S&T, the effect of this would not be to perpetuate the status quo, because change in science and technology is brought forth ineluctably by the intellectual advances generated by the enterprise itself. Rather, the effect could be to diminish the possibility of a *deliberate change of course*. Without recourse to radical critique, the locomotive will roll along its track, and the track might turn left or right; we might even be able to flick a switch here and there to select which branch of a bifurcation we take; but we exclude from our mental space a host of discontinuous possibilities, such as getting off the train and continuing our journey via another mode of transport.[24]

Radical critiques might challenge the metaphysical underpinnings of our world view. They might challenge our basic values or moral norms. They might undermine our confidence in the entire S&T project. Alternatively, they might suggest that our attempts to humanize the S&T enterprise will have the opposite effect from the one intended. They might argue that one particular anticipated technological breakthrough will have consequences overshadowing all the rest, and that by failing to act accordingly we are grossly misdirecting our attention and our resources. They might contend that increased public engagement and increased efforts to anticipate the societal implications will have obnoxious consequences.

---

[24]In the context of fundamental science, such a course change could be a Kuhnian paradigm shift.

They might argue for relinquishment of broad areas of technological research, or alternatively that everything should be put into accelerating some applications. They might suggest new ways of funding basic research that would sidestep expert panels and bureaucratic procedures. They might identify completely new ways of evaluating and measuring progress. The possibilities are myriad and impossible to specify in advance.[25]

Many radical critiques are utterly wrong, and some of them would be extremely dangerous if they became popular. Nevertheless, if we look back historically, and observe how many widely held conventional wisdoms of the past are revealed as blinkered or deeply flawed by our current lights, we must surely admit that, by induction, it is likely that many of our own central beliefs, too, are deeply flawed. We need to build into our processes of individual and collective deliberation some self-correcting mechanism that enables us to question and rectify even our most deep-seated assumptions.

One way in which the scope of our deliberations can be broadened is by taking into account the kind of meta-level reflections that I have tried to illustrate in this essay. More specifically, one could argue that more work should be done on the normative dimensions of the S&T enterprise.

One normative dimension is ethics, and this is to some extent already part of the official programs, for example, as the "E" in ELSI and NELSI, and as applied ethics more generally. One might have occasional misgivings about the quality, depth, or impact of this research, but at least there is some recognition of the significance and relevance of the questions it is supposed to address.

Another normative dimension that has been given rather less attention in these programs is that of applied (normative) epistemology. This encompasses a number of important problems. One such problem is to develop better higher order epistemic principles for the conduct of scientific research. As said by E.T. Jaynes:

> It appears to us that actual scientific practice is guided by instincts that have not yet been fully recognized, much less analyzed and justified. We must take into account not only the logic of science, but also the sociology of science (perhaps also its soteriology). But this is so complicated that we are not even sure whether the extremely skeptical conservatism with which new ideas are invariably received, is in the long run a beneficial stabilizing influence, or a harmful obstacle to progress.[26]

Yet the epistemological problems go beyond the challenge of how to maximize scientific and technological advancement. As we have seen, they also are central to our thinking about the ethical and policy issues prompted by the S&T enterprise. Applied epistemology also lies at the heart of the problem of how to evaluate radical critiques of this enterprise.

---

[25]Although not necessarily intended as "radical critiques," for a few recent examples see, for example, Joy 2000; Bostrom 2003b, 2003a, 2005; Hanson 2003.
[26]Jaynes and Bretthorst 2003, p. 525.

There are many interesting approaches to these matters in addition to philosophical reflection and theoretical analysis. Here is a sample:

- Study the heuristics and biases affecting human cognition, and figure our ways of applying the findings to improve our judgment.[27]
- Study the correlates of true opinion, among experts and the public, and use this information as clues to who is right in cases of disagreement and to how we might improve our own epistemic situation.[28]
- Information technologies. Develop our information infrastructure in ways that will facilitate the collection, integration, and evaluation of information.
- Cognitive enhancement. Improve individual reasoning ability (e.g., concentration, memory, and mental energy) by educational, pharmaceutical, and other means.
- Study how vested interests, the mass media, and other social realities shape and bias (or facilitate) the processes collective deliberation.
- Public deliberation. Develop procedures, formats, or rhetorical standards to improve the quality of public debate by confronting "plebiscitory" reason.[29]
- Subsidize and implement institutional innovations, such as information markets, which have been shown to outperform expert panels in many prediction tasks.[30]

The unifying theme is to explore how we could make ourselves smarter and wiser, both individually and as an epistemic community. The research could be slanted toward applications in S&T assessment, but it is likely to have important spill-over benefits in other areas. Such a program could be combined with more narrowly focused efforts to gather and analyze information in areas of particular concern, such as nanotechnology.

Finally, I want to call attention to one more "normative dimension," except that it is not really a dimension but rather the space spanned by all the other vectors. I am referring to the challenge of integrating all crucial considerations into some coherent unity that will let us determine what we have most reason to do all things considered. This might have to accommodate predictions about technology, social impacts, strategic considerations, value judgments, ethical constraints, and assorted meta-level thoughts about how all these things should fit together.

Such synthetic work is not in fashion at the present time. It is discouraged by the disciplinary structure of academia and by prevailing norms of academic publishing, and it does not appear to be strongly nurtured by the short-term grants funding available for interdisciplinary projects either. As a consequence, synthetic work is

---

[27]See, for example, Kahneman and Tversky 2000; Gilovich, et al. 2002. For an attempt to apply this kind of information to a technology-related issues, see Bostrom and Ord 2006.
[28]See, for example, Tetlock 2005.
[29]Chambers 2004.
[30]Hanson 1995; Wolfers and Zitzewitz 2004; Leigh and Wolfers 2006.

undertaken mainly by time-pressured journalists, popular science authors, retired scientists or senior scholars who are no longer willing or able to do serious work, crackpots, and miscellaneous eccentrics. The quality of the ensuing contributions is often poor, and even when something worthwhile comes out there is no guarantee that it will be used and built upon by others.

Let me try to indicate slightly more clearly what kind of synthetic task I have in mind. I am not referring to the erection of grand philosophical systems resting on foundations of indubitable first principles; nor do I refer to the painting of "visions" or the realization of comprehensive ideological "outlooks"; nor still am I pushing for some new kind of systematization of all knowledge, a universal taxonomy, or the creation of an encompassing database, or a library of commissioned studies. Rather, what I have in mind is the task of attempting to think through some of the big challenges for humanity in a way that does not leave out any *crucial* consideration—by which I mean a consideration such that if it were taken into account it would overturn the conclusions we would otherwise reach about how we should direct our efforts. For example, some of the strategic considerations related to the nanotechnology initiative that I mentioned earlier might be "crucial" in that they might plausibly, if they are sound and once taken into account, rationally deliver a practical conclusion pointing in the opposite direction from the one we might otherwise believe we ought to strive toward. But not only strategic considerations, but also other empirical, epistemological, axiological, and methodological considerations could be crucial in this sense.

Implicitly, we are confronting a challenge of integrating all crucial considerations every time we are attempting to make a reasoned decision about some matter that we think it is important to "get right." When our goal is very limited, we might at least sometimes succeed in meeting the challenge (albeit not usually by relying on reasoned deliberation alone). When the goal is more open-ended, such as if we are attempting to decide what we have most reason to do with our own life all things considered, or if we are seeking to form an opinion on a topic such as what public policy ought to be with regard to some anticipated technological revolution, then the complexity of the synthetic challenge grows enormously. It is not clear that we ever manage to meet it in any robust sense. Instead, what answer we end up espousing might depend mostly on contingent factors, such as the political inclinations of our parents, the idiosyncratic views of our thesis advisor, the current cultural climate in the place we happen to live, or the mere fact that we failed to think of some crucially relevant consideration that would have caused us to come to a very different conclusion.

What this seems to amount to is that we can have very little rational confidence that our efforts, insofar as they are aiming ultimately at important long-term goals for humanity, are not entirely wrongheaded. Our noblest and most carefully considered attempts to effect change in the world might well be pushing things further *away* from where they ought to be. Perhaps around the corner lurks some crucial consideration that we have ignored, such that if we thought of it and were able to accord it its due weight in our reasoning, it would convince us that our guiding beliefs and our struggles to date had been orthogonal or worse to the direction that would then come to appear to us as the right one.

Other than becoming generally more agnostic, what can we do about this? It is not clear how effectively the synthetic task could be broken into subtasks that can be delegated to a team of researchers, and then stitched back together. But it is also dubious that any single genius can accomplish the task alone. Perhaps room could be created for an open-ended enquiry by many minds to focus on the problems of integration. Considerable progress might be made over time, which would be interesting in its own right, whether or not it would succeed in creating a pragmatic synthesis of which one could be justifiably confident that it did not fail to recognize and do justice to any crucial consideration. A decisive breakthrough would mean that we would have the opportunity to make our choices in the full light of available facts and reasons, and with justified confidence that we are not pushing in entirely the wrong direction.

Whether things would be better that way, we do not know for sure. Nor do we know for sure what presently available actions would best encourage such an outcome, should we decide to strive for it, for this is of course one of the big open-ended goals to which the argument applies: Even if we think hard and honestly about this issue, we are apt to neglect at least one crucial consideration. Unless and until we achieve a dramatic enlightenment in our capacity for pragmatic synthesis, then on this and on other revolutionary prospects, we will continue to stake out our ethics and policy paths in the dark.

BIBLIOGRAPHY

Bainbridge, W. S. and M. C. Roco, eds. (2006), *Managing Nano-Bio-Info-Cogno Innovations*. Dordrecht, The Netherlands: Springer.

Bostrom, N. (2002), "Existential Risks: Analyzing Human Extinction Scenarios and Related Hazards," *Journal of Evolution and Technology* 9.

Bostrom, N. (2003a), "Are You Living in a Computer Simulation?" *Philosophical Quarterly* 53 (211):243–255.

Bostrom, N. (2003b), "Astronomical Waste: The Opportunity Cost of Delayed Technological Development," *Utilitas* 15 (3):308–314.

Bostrom, N. (2005), "The Fable of the Dragon-Tyrant," *Journal of Medical Ethics* 31 (5):273–277.

Bostrom, N. (2006), "Infinite Ethics," *Working Manuscript*. Available at http://www.nick-bostrom.com/ethics/infinite.pdf.

Bostrom N., and T. Ord (2006), "The Reversal Test: Eliminating Status Quo Bias in Bioethics," *Ethics* 116 (4):656–680.

Chambers, S. (2004), "Behind Closed Doors: Publicity, Secrecy, and the Quality of Deliberation," *Journal of Political Philosophy* 12 (4):389–410.

Crow, M. M. and D. Sarewitz (2001), "Nanotechnology and Societal Transformation," In A. H. Teich, S. D. Nelson, C. McEnaney, and S. J. Lita (eds.), *AAAS Science and*

Technology Policy Yearbook, Washington, DC: American Association for the Advancement of Science, pp. 89–101.

Drexler, Eric (2003), "Nanotechnology Essays: Revolutionizing the Future of Technology (Revised 2006)," *AAAS EurekAlert! InContext* April.

Drexler, K. E. (1992a), *Engines of Creation*. Oxford: Oxford University Press.

Drexler, K. E. (1992b), *Nanosystems: Molecular Machinery, Manufacturing, and Computation*. New York: John Wiley & Sons, Inc.

Elam, M. and M. Bertilson (2002), "Consuming, Engaging and Confronting Science: The Emerging Dimensions of Scientific Citizenship," In P. Healey (ed.), *STAGE (HPSE-CT2001-50003) Final Report*, 121–158. Available at http://www.stage-research.net/STAGE/documents/STAGE_Final_Report_final.pdf.

Gilovich, T., D. W. Griffin, and D. Kahneman (2002), *Heuristics and Biases: The Psychology of Intuitive Judgement*. Cambridge, U.K.: Cambridge University Press.

Gosseries, A. (2005), "Publicity," In E. N. Zalta (ed.), *The Stanford Encyclopedia of Philosophy*. Available at: http://plato.stanford.edu/entries/publicity/.

Grice, P. (1975), "Logic and Conversation," In P. Cole and J. L. Morgan (eds.), *Syntax and Semantics, Vol. 3, Speech Acts*, New York: Academic Press, pp. 41–58.

Hanson, R. (1995), "Could Gambling Save Science? Encouraging an Honest Consensus," *Social Epistemology* 9:(1):3–33.

Hanson, Robin *Shall We Vote on Values, But Bet on Beliefs?* 2003. Available at http://hanson.gmu.edu/futarchy.pdf.

Jaynes, E. T. and G. L. Bretthorst (2003), *Probability Theory: The Logic of Science*. Cambridge, UK: Cambridge University Press.

Jones, C. I. and J. C. Williams (1998), "Measuring the social return to R&D," *Quarterly Journal of Economics* 113 (4):1119–1135.

Joy, B. (2000), "Why the future doesn't need us," *Wired* 8.04.

Kahneman, D. and A. Tversky (2000), *Choices, Values, and Frames*. Cambridge, UK: Cambridge University Press.

Latour, B. (1987), *Science in Action: How to Follow Scientists and Engineers Through Society*. Cambridge, Mass.: Harvard University Press.

Leigh, A. and J. Wolfers (2006), "Competing Approaches to Forecasting Elections: Economic Models, Opinion Polling and Prediction Markets," *Economic Record* 82 (258):325–340.

Lenman, J. (2000), "Consequentialism and Cluelessness," *Philosophy and Public Affairs* 29 (4):342–370.

Meade, E. E., D. Stasavage, and London School of Economics and Political Science. Centre for Economic Performance. *Publicity of debate and the incentive to dissent evidence from the US Federal Reserve*.Centre for Economic Performance, London School of Economics and Political Science 2004. Available at http://cep.lse.ac.uk/pubs/download/dp0608.pdf.

Rawls, J. (1999), *The Law of Peoples*. Cambridge, Mass.: Harvard University Press.

Roco, M. C. and W. Sims Bainbridge (2003), *Converging Technologies for Improving Human Performance: Nanotechnology, Biotechnology, Information Technology, and Cognitive Science*. Dordrecht; Boston, Mass.: Kluwer Academic Publishers.

Salter, A. J. and B. R. Martin (2001), "The Economic Benefits of Publicly Funded Basic Research: A Critical Review," *Research Policy* 30 (3):509–532.

Tetlock, P. (2005), *Expert Political Judgment: How Good is it? How can we Know?* Princeton, N.J.: Princeton University Press.

Vandermolen, T. D. (2006), "Molecular Nanotechnology and National Security," *Air & Space Power Journal* 20 (3):96–106.

Wolfers, J. and E. Zitzewitz (2004), "Prediction Markets," *Journal of Economic Perspectives* 18 (2):107–126.

PART II

NANO LAW AND REGULATION

New technologies perennially present fresh challenges to lawmakers and regulators, and it is no surprise that in both the United States and Europe (as well as other jurisdictions) a spate of consultations, reviews, and controversies is presently focused on the adequacy and applicability of existing legal and regulatory regimes for a technology that is being presented as disruptive at every level. Moreover, a series of wide-ranging standards discussions is under way that cross-cuts with efforts at regulation. Any review of law and regulation risks being soon outdated, although one of the features of the current situation is the cautious (some would say slow) pace of response to the challenges of the technology and to those who are pressing for fresh regulatory frameworks.

For this section, we have invited experts to review aspects of the current situation. Sonia E. Miller, an attorney who founded the Converging Technologies Bar Association and has been closely involved with various policy reviews, sets the scene for U.S. regulation. Attorney George Kimbrell offers a perspective focused on the application of environmental law. Trudy Phelps, current chair of the European Commission's nanotechnology standards committee, reviews the European approach to regulation.

Two further chapters examine aspects of patent law. Julie Burger, Marianne Timm, and Lori B. Andrews broadly review the intellectual property landscape in which nanotechnology is being developed, and Jessica Fender reviews and assesses trends in nanotechnology patents.

Nanoscale: Issues and Perspectives for the Nano Century. Edited by Nigel M. de S. Cameron and M. Ellen Mitchell

CHAPTER 11

Regulating Nanotechnology: A Vicious Circle

SONIA E. MILLER

> We need to pursue economic and regulatory policies that are responsive to today's world.
>
> Henry M. Paulson, Jr., at his Swearing-in Ceremony as
> U.S. Treasury Secretary, 2006

INTRODUCTION

Nanotechnology is frequently the subject of "small" jokes. It is oftentimes even reduced to a four-letter word—"nano." And, it is casually attached as a prefix or suffix to non-nanoproducts or everyday idioms. Given the scale of its mutations, the issue of whether to regulate or not to regulate nanotechnology at this particular stage of its research and development generates ongoing discourse of immense concern, commonly driven by the imprint of spinmeister grassroots organizations.

Spin frames public discourse, and defines issues and kinds of information disguised as truth. Language activates frames, and shapes social policies and the institutions formed to carry out those policies.[1]

People tend to think in frames. For the truth to be accepted by the public, it must fit within their fixed frame of perception. The question is: What is the truth about nanotechnology? And, at the nanoscale, do the same regulations and laws apply—from international trade law, to treaties banning chemical and biological weapons, to regulations governing medicine and the environment?[2]

---

[1]Lakoff, G. *Don't Think of an Elephant! Know Your Values and Frame the Debate.* Vermont: Chelsea Green Publishing. 2004.
[2]Miller, S. E. *Regulating Nanotechnology: The EPA and FDA are Likely Watchdogs. New York Law Journal.* Volume 233-Number 64, April 5, 2005.

Nanoscale: Issues and Perspectives for the Nano Century. Edited by Nigel M. de S. Cameron and M. Ellen Mitchell

Regulatory discourse can readily advance or easily suffocate the progress of invention. Fears of the new can create phantom risks,[3] or real risks that provoke disproportionate public concern and arouse suspicion. Oftentimes, these fears tend to hover indefinitely in the background and never seem to crystallize.[4]

However, these dangling particles of disquietude wreak havoc on, create premature speculation in, and may strangle the very effectiveness of current regulatory evaluative procedures. Fears are the propelling influence behind exaggerated realities and the drafting of strained regulatory mandates. They create panic and a perceived comparable, irreversible threat of danger.

In *Breaking the Vicious Circle: Toward Effective Risk Regulation*,[5] Justice Stephen Breyer writes:

> The three elements of the vicious circle—public perception, Congressional reaction, and the uncertainties of the regulatory process—reinforce each other. Obviously, public perceptions influence Congress, Congress (through press reports of its activities in particular) helps to shape public perception, and both influence the response of agency administrators to the problems they consider important.[5]

Nanotechnology has been the reluctant recipient of such a vicious circle. Advocacy groups and non-governmental organizations (NGOs) are challenging the U.S. regulatory agencies, that maintain that the unique size and properties of nanoscale materials do not warrant new regulation.[2] As a result, no real regulatory policy has been formulated to deal with it.

Dissatisfied with the lack of immediate Congressional reaction, these self-appointed grassroots advocacy groups have gone so far as to call for a moratorium on the continued advance of nanotechnology until there is a 100% guarantee of its safety. The NGOs reason that simply tinkering with existing regulations will not address the broad social, health, environmental, and economic concerns of technologies converging at the nanoscale and propose the drafting of new regulations.[6]

Of course, unintended consequences ensure there are no guarantees, no matter how detailed the measured precautions or stringent the drafted regulations. Ultimately, only some, but not all, of the risks can be regulated. Before drafting new regulations, or recommending such, current laws ought to be reviewed and evaluated for their effectiveness once sufficient, valid, and reliable data are known about the risks of nanotechnology.

Data gaps in research initiatives and inconclusive toxicity studies underscore the speculative nature of the regulatory implications of the effects of commercialized nanotechnology on health and the environment. "Very little is known about the safety risks that engineered nanomaterials might pose, beyond limited research

---

[3]Foster, K. R. et al. *Phantom Risk: Scientific Inference and the Law*. Cambridge, MA: MIT Press, 1999.
[4]Huber, P. W. *Coping with Phantom Risks in the Courts*. Available at http://www.piercelaw.edu/risk/vol6/spring/huber.htm (last visited Dec. 29, 2003).
[5]Breyer, S. *Breaking the Vicious Circle: Toward Effective Risk Regulation*. Massachusetts: Harvard University Press, 1993.
[6]See footnote 2, p. 5.

data indicating they possess certain properties associated with safety hazards in traditional materials."[7]

Knowing this, which federal agencies may have jurisdiction over the responsible regulatory development of such a transformative, disruptive, emerging, converging, and enabling nanoscale technology? Should the relentless call for regulation and government controls be furthered in the face of scientific uncertainties, lack of data and guidance pertaining to the real risks of nanomaterials, and undeveloped hazard assessment frameworks? Or, should the process of nanotechnology innovation be allowed to continue on its evolutionary journey as a cautiously monitored and unfolding industry-in-the-making?

This chapter will provide a status update of the rhetoric and hyperbole confronting the invisible, an overview of its conflicting tension within the vicious circle, and attempt to answer the proverbial nanotechnology question: "Does size matter?"

THE REGULATORY SYSTEM

The regulatory system has two parts, a technical and a policy-oriented one. The technical part is called "risk assessment," and it is designed to measure risk associated with a substance. It can be divided into four activities: (1) identifying the potential hazard; (2) determining how the risk of harm varies with exposure; (3) estimating the amount of human exposure; and (4) categorizing the result of the hazard. Assessing risk is a matter of probability requiring complex judgments.

The policy-oriented component, or "risk management," determines the course of action the regulator should take regarding the risk revealed by the assessment—given the identified risks, the risks associated with the alternatives, the costs, practicalities, and the ultimate effect on benefits.[8]

Risks and benefits must both be considered, with all their consequences, to regulate a technology in a logically defensible manner. To consider its impact on individuals and society as a whole, a technology must be regulated in an ethically defensible way. The key lies in the procedures employed.[4]

In any society, governmental entities enact laws, make policies, and allocate resources.[9] Advocacy groups frequently attempt to influence the general public, as well as public policymakers, about the nature of problems—perceived or real, needed legislation to rectify those alleged problems, and the funding required to provide services or to conduct research.

Oftentimes though, unfamiliar, difficult to understand, invisible, involuntary, and/or potentially catastrophic perceived risks are overestimated, inspiring

[7]Approaches to Safe Nanotechnology: An Information Exchange with NIOSH. National Institute for Occupational Safety and Health. Available at http://www.cdc.gov/niosh/topics/nanotech/nano_exchange.html.
[8]Breyer, S. *Breaking the Vicious Circle: Toward Effective Risk Regulation*. Massachusetts: Harvard University Press, 1993, pp. 9, 10.
[9]Kilpatrick, D. G. *Definitions of Public Policy and the Law*. Available at www.vawprevention.org/policy/definition.shtml (last visited October 15, 2003).

uncertainty and fear verging on superstition. Contributing to this may be poor scientific literacy and unfamiliarity with the statistical aspects of those risks. However, activists regularly deluge the public and policymakers with misinformation, leaving behind a bewildered, overwhelmed, and conflicted constituency.[10]

To advance their funded agenda and attempt to raise public concerns at the interface of science and law, self-professed independent expert NGOs use language that is inflammatory but vague, with accompanied scare tactics, and scandalous accountings of, as yet, unproven risks associated with the wild unleashing of a new technology.

With nanotechnology, NGOs have already begun to penetrate all the elements of the vicious circle by stirring public fears of the unknown, evoking anxieties regarding change, coupled with the spinning of emotionally charged negative connotations.

Advocating for regulation of an innovation with yet uncertain and inconclusive demonstrated risks, without clearly understanding how the regulatory systems are designed to work, dilutes the very purpose for which they were intended while simultaneously impairing the progress of transformative, disruptive, emerging, converging, or enabling technologies.

How can one regulate something that cannot be seen, and that is not even here yet? Even more, why impose such an unnecessary restraint on its advance and choke the very opportunities nanotechnology may present? Ultimately, will promulgating regulation diminish the alleged risks, produce different risks, deprive the end users of expected benefits, or simply silence the political reaction of the nano-twisting activists?

Sadly, such rhetoric accompanies the small scale of such a potentially huge technology. It is already influencing public perception, the reaction of Congress, and the ability of the regulatory agencies to objectively assess its respective risks, if any.

THE RHETORIC BEHIND NANOTECHNOLOGY

U.S. regulators maintain that the unique size and properties of nanoscale materials do not warrant new regulation. In the absence of scientific evidence to the contrary, their position is that current safety and health regulations are adequate to address the risks associated with nanotechnology, and advocate more research and study.

In contrast to this argument are positions proffered by advocacy groups and NGOs, who allege that current regulations are not sufficiently elastic to address the unique and novel risks to people and the environment posed by nanoparticles.[2]

While these groups may recognize that there are a number of existing laws for reviewing and regulating nano materials, they purport that all these laws either

---

[10]Miller, H. I. *The Risky Business of Understanding Risk.* February 2, 2004. TCS: Tech Central Station—Where Free Markets Meet Technology. Available at http://www2.techcentralstation.com/1051/printer.jsp?CID=1051-020304B (last visited February 5, 2004).

suffer from major shortcomings of legal authority, or from a gross lack of resources, or both.[11] The conflicting tension behind the rhetoric of nanotechnology between U.S. regulators and advocacy groups and NGOs recently manifested when the International Center for Technology Assessment (ICTA), joined by a coalition of consumer, health, and environmental advocacy groups, mounted the first legal petition[12] on the risks of nanotechnology against the U.S. Food and Drug Administration (FDA). "Filed in conjunction with the release of a comprehensive report[13] by Friends of the Earth, one of the eight petitioning organizations, on the alleged dangers of sunscreens and cosmetics containing nanoparticles, the 80-page petition requests that the FDA take regulatory action regarding products composed of engineered nanoparticles and calls for the immediate recall of nano-sunscreens."[14]

Similarly, and in keeping with the Precautionary Principle, just 2 years ago, the ETC Group called for the removal from the shelves of all food, feed, and beverage products incorporating manufactured nanoparticles, with new ones prohibited from commercialization until such time as laboratory protocols and regulatory regimes were firmly established and the products shown to be safe.

Equally and for the same reasons, they called for the prohibition from environmental release of nanoscale formulations of agricultural input products, such as pesticides, fertilizers, and soil treatments, until the government could engage the public in an analysis of the broad health, environmental, safety, and socioeconomic implications of their distribution. In continued form and 4 years prior, the ETC Group had called for a moratorium on the release of manufactured nanoparticles until laboratory protocols were established to protect workers and regulations in place to protect consumers.

The ETC Group was not alone. In 2004, the Royal Society and Royal Academy of Engineering[15] recommended that the release of nanoparticles and nanotubes into the environment be avoided because there was little, if any, information about how they may behave in the air, water, or soil, or about their ability to accumulate in food chains. They further suggested that manufactured nanoparticles and nanotubes be treated as hazardous waste streams and be prohibited from use in the remediation of groundwater. While nanotechnology is called many things and projected to assault the public on its potential unsubstantiated risks, it remains a misunderstood regulatory anomaly.

---

[11]Davies, J. C. *Managing the Effects of Nanotechnology*. Woodrow Wilson International Center for Scholars, Project on Emerging Technologies. 2005.

[12]Petition Requesting FDA Amend Its Regulations for Products Composed of Engineered Nanoparticles Generally and Sunscreen Drug Products Composed of Engineered Nanoparticles Specifically. May 16, 2006. Available at http://www.icta.org/doc/Nano%20FDA%20petition%20final.pdf (visited May 17, 2006).

[13]*Nanomaterials, Sunscreens and Cosmetics: Small Ingredients, Big Risks Report*, May 2006. Friends of the Earth. Available at www.foe.org/doc/Nano%20PP%20final%20May%2016,%202006.pdf.

[14]Miller, S. E. *Sunscreen Risk? FDA Faces First Nanotechnology Legal Challenge. New York Law Journal*. Volume 235-Number 103, p. 5 (last visited May 30, 2006).

[15]*Nanoscience and Nanotechnologies: Opportunities and Uncertainties*. Royal Society and Royal Academy of Engineering. July 2004.

NANOTECHNOLOGY: IS IT LEGAL?

What is nanotechnology? Well, it depends on whom you ask. To date, there is not one universally accepted definition of nanotechnology.

Nanotechnology lacks a uniform nomenclature; an internationally valid standardization for the vast variety of nanotechnological substances, materials, products, or applications; a common language enabling shared global understanding; a universal assessment of its opportunities and impact on the risk landscape in the future, particularly as it relates to toxicology and exposure; and common terminology for a comparative analysis of the risks assessments of different institutions or countries.[16] As an emerging risk, it challenges the insurance industry due to uncertainty regarding nanotoxicity or nanopollution, its expected future ubiquitous presence, and the possibility of long latent, unforeseen claims.[17]

Suffice it to say that nanotechnology is a multidisciplinary field of discovery converging physics, chemistry, biology, engineering, information technology, with metrology and other significant areas of study.

However, a commonly used and significantly accepted definition of nanotechnology is the one proffered by the NSF, which provides: "Nanotechnology is the understanding and control of matter at dimensions of roughly 1–100 nm (one billionth of a meter), where unique phenomena enable novel applications. Encompassing nanoscale science, engineering and technology, nanotechnology involves imaging, measuring, modeling, and manipulating matter at this length scale".[18]

While nanotechnology may be capable of producing material 100 times stronger than steel at one-sixth the weight,[19] its true strength was seen on December 3, 2003, when the 21st Century Nanotechnology Research and Development Act (the Act)[20] was enacted into law.

Section 10(2) of the Act defined nanotechnology a little differently than the NSF. It interprets nanotechnology as the science and technology that will enable one to understand, measure, manipulate, and manufacture at the atomic, molecular, and supramolecular levels, aimed at creating materials, devices, and systems with fundamentally new molecular organization, properties, and functions.

The vast impact of this small, four-letter-abbreviated technology was glaringly seen in the financial investment and commitment of the U.S. Federal government.

---

[16]Miller, S. E. *A Matter of Torts: Why Nanotechnology Must Develop Processes of Risk Analysis. New York Law Journal.* October 5, 2004. Volume 232-Number 67, p. 5.
[17]*Nanotechnology: Small Matter, Many Unknowns, Few Considerations.* Swiss Re. May 10, 2004.
[18]National Science Foundation. Nanotech Facts. Available at http://www.nano.gov/html/facts/whatIsNano.html (last visited September 4, 2006).
[19]Miller, S. E. *The Convergence of N: On Nanotechnology, Nanobiotechnology and Nanomedicine. New York Law Journal.* Volume 230-Number 107 (last visited December 2, 2003).
[20]Public Law 108–153.

Authorizing nearly $3.7 billion from 2005–2008 for research and development programs across various Federal agencies, the Act ensured, under § 2(b)(10), that "ethical, legal, environmental, and other appropriate societal concerns ... are considered during the development of nanotechnology."[21] It also required, under § 5(c), that a one-time study be conducted to assess the need for standards, guidelines, or strategies for ensuring the responsible development of nanotechnology.[21]

To date, Federal funding for nanotechnology research and development has increased substantially since the formation of the National Nanotechnology Initiative (NNI) in 2001 under President William Jefferson Clinton—from approximately $464 million in 2001, to a 2007 budget request of $1.3 billion, an increase of 21% over the 2006 request, and nearly triple from that just 6 years ago.[22]

With programs and activities initiated across 25 federal agencies participating in the NNI today, a major increase from just six at its inception, funding for 2006 had been classified for the first time across seven component program areas with respective planned 2007 funding as follows: (1) fundamental nanoscale phenomena and processes ($401 million); (2) nanomaterials ($250 million); (3) nanoscale devices and systems ($263 million); (4) instrumentation research, metrology, and standards for nanotechnology ($77 million); (5) nanomanufacturing ($41 million); (6) major research facilities and instrumentation acquisitions ($164 million); and (7) societal dimensions ($82 million).[22]

The Act additionally firmly established the NNI, the multiagency U.S. government program aimed at accelerating the discovery, development, and deployment of nanoscale science, engineering, and technology. A supporting component of the President's American Competitiveness Initiative (ACI), the NNI is dedicated to four goals: (1) maintain a world-class research and development program; (2) facilitate technology transfer; (3) develop educational resources, a skilled workforce, and supporting research infrastructure and tools; and (4) support the responsible development of nanotechnology.[23]

Today, the NNI has been recognized for creating a highly successful, collaborative, and interdisciplinary nanotechnology community in the United States.[24] It has become the ultimate driving force behind interdisciplinary nanotechnology initiatives around the world, and the comparative model against which similar collaborative programs are launched and measured.

---

[21]U.S. Congress. Public Law 108–153. 2003. 21st Century Nanotechnology Research and Development Act. 15 U.S.C. § 7501. 108 Cong.

[22]The National Nanotechnology Initiative. *Research and Development Leading to a Revolution in Technology and Industry Supplement to the President's 2007 Budget.* July 2006. The 2006 figures can be found in the NNI Supplement to the President's FY 2006 Budget. Available at www.nano.gov/ NNI_06Budget.pdf (last visited September 4, 2006).

[23]2004 NNI Strategic Plan. Available at www.nano.gov/html/about/strategicplan2004.html (last visited September 4, 2006).

[24]Roco, M. C. National Nanotechnology Initiative—Past, Present, Future. *PREPRINT Handbook on Nanoscience, Engineering and Technology, 2nd ed.*, Taylor and Francis, February 20, 2006.

NANOTECHNOLOGY: A MARKET FORCE?

Historically, innovation is accompanied by the promise of new societal benefits (often exaggerated in time and expectation), but a hope, nonetheless. Current scientific advances in nanotechnology tout optimism for the future, with possibilities oftentimes presented in statements, such as "revolutionize the world," "the next frontier," "change our view of life," "pioneering the future," and so on, even though the scientists themselves are not always their own best self-advocates nor are their predictions considered the holy grail of the next generation.

While at the moment there is no nanotechnology market or industry, but simply a value chain, it is viewed as an enabler expected to impact all manufactured goods. According to Lux Research, Inc., in 2014, nanotechnology will be incorporated in products worth $2.9 trillion in revenue, with new, emerging nanotechnology accounting for 89%.[25]

Yet, in a cross-industry benchmark study conducted by the National Center for Manufacturing Sciences (NCMS)[26] that polled more than 6000 senior-level executives in leading U.S. organizations, the majority surveyed indicated that their organizations faced considerable difficulty in nanomanufacturing due to the following: (1) emergent technology issues; (2) raising capital for critical infrastructure investments; (3) attracting technical and business talent; (4) connecting with early end-users; (5) producing competitively to meet new market applications and volumes; (6) intellectual property issues and the sharing of knowledge; and (7) the lack of clear regulatory policy, which could impede industry and impact the public's reaction to future product developments.

While the nanomanufacturing industry is confronted with challenges at this stage, it recognizes similarities with previous technology waves, such as the Internet and biotechnology, offering several lessons learned for the formulation of sound anticipatory approaches. These include: (1) the continued education of the public and key policy makers at the state and federal levels, as well as government agencies and legislative bodies resulting in clearer product approval pathways; and (2) robust standards and responsible practices to ensure the continued dominance of the United States,[26] currently an acknowledged leader in nanotechnology research and development with an investment roughly one-quarter of the current government investments by all nations.[27]

Even though it is believed that nanotechnology poses differing issues from a regulatory perspective than previous innovation, the bottom line is that nanotechnology may not be all that unique.

---

[25]Lux Research, Inc. *Sizing Nanotechnology's Value Chain.* October 2004.
[26]2005 NCMS Survey of Nanotechnology in the U.S. Manufacturing Industry (Sponsored by NSF), March 6, 2006 Abstract. Available at www.ncms.org.
[27]The National Nanotechnology Initiative. *Research and Development Leading to a Revolution in Technology and Industry Supplement to the President's 2007 Budget.* July 2006.

PUBLIC PERCEPTION: THE VICIOUS CIRCLE—PART I

Science wields power in modern society because of its ability to create knowledge by discovering truth. However, when it comes to "crisis management"—the specialized area of public relations that helps clients fend off scandals and repair damaged reputations—the truth may not necessarily be a solid, but a liquid.[28]

What may be considered as true is not necessarily so when viewed from a different perspective. Truth depends on the critical message and position intended to be communicated. At times, it is a nonexistent creation comprised of manipulated perception-managed imagery, masquerading as science.

Message positioning pertaining to nanotechnology was vividly captured in a recent report entitled: "Informed Public Perceptions of Nanotechnology and Trust in Government,"[29] the result of a two-part exploration of citizen perceptions of nanotechnology.

Funded in 2004 by the NSF, and conducted as a national survey sampling 1250 people, the first part explored issue framing, trust in government to manage risks, and expectations of nanotechnology benefits versus risks.[29] It investigated the effect of science fiction films and novels such as Michael Crichton's book *Prey*,[30] and the resulting awareness and attitudes toward nanotechnology.

The second study used experimental issue groups in three different cities to investigate the reactions of 152 citizens who had been provided with background information on nanotechnology and potential scenarios of unpredictable possibilities.[30]

The report presents the results of a study conducted May–June of 2005 of 177 private citizens divided among 12 groups formed in Cleveland, Ohio; Dallas, Texas; and Spokane, Washington, detailing the public's perceptions of government, nanotechnology, and regulation. This was designed in response to the 2004 study that found low levels of public trust in government associated with the management of the potential risks surrounding nanotechnology.[31]

Intended to inform the stakeholders and leaders of the nanotechnology research and development communities of public wants and expectations, the report purports to examine in-depth the knowledge base and matters of importance for concerned American citizens about the use of nanotechnology and steps government and industry might undertake to improve trust.

Bearing in mind that 54% of the participants professed to know almost nothing about nanotechnology,[31] the report findings suggested the following public

---

[28]*Liquid Truth: Advice from the Spinmeisters.* Center for Media and Democracy. Available at http://www.prwatch.org/prissues/2000Q4/truth.html (last visited June 22, 2006).

[29]Macoubrie, J. *Informed Public Perceptions of Nanotechnology and Trust in Government.* Project on Emerging Nanotechnologies at the Woodrow Wilson International Center for Scholars created in partnership with The Pew Charitable Trusts. September 2005.

[30]Crichton, M. *Prey.* New York: (2004).

[31]*Attitudes Toward Nanotechnology and Federal Regulatory Agencies.* Conducted by Peter D. Hart Research Associates for The Woodrow Wilson International Center for Scholars Project on Emerging Nanotechnologies. September 19, 2006. Available at http://nanotechproject.org/78 (last visited June 22, 2006).

perceptions regarding nanotechnology: (1) 76% said "a ban is overreaching";[31] (2) 41% said the benefits would exceed the risks;[31] (3) true unknowns, regulation, and human health risks accounted for about 40% of the concerns—highlighting "politics getting into regulation," "who regulates the regulators," "that government can be manipulated to get the effect desired," and "whether it will be overregulated";[31] (4) 55% felt government control beyond voluntary industry standards is necessary, with 33% unsure whether voluntary standards would be sufficient;[31] and (5) increased consumer information was needed to enhance the public trust and to make better informed choices.[31]

Subsequently, the Project on Emerging Nanotechnologies at the Woodrow Wilson International Center for Scholars (WWIC) commissioned yet another national poll on public awareness of nanotechnology and engaged Peter D. Hart Research Associates, an independent research firm, to conduct the survey from August 23 to August 27, 2006. Research findings[31] of 1014 U.S. adults revealed that 42% of Americans had no awareness of nanotechnology with the elderly and women being the least informed.[31]

On balance, whether the 2004 studies and 2005 report, and subsequent 2006 poll reflect a valid and reliable source of results concerning the publics' perception of nanotechnology to date, or merely a contrived demographic selection of speculative rhetoric, is truly a question of framed perception, coupled with skeptical attitudes of trust. Just as equally, although involvement of the public is important to their under-standing of public policy, it is less useful for the formulation of policy.[32]

This issue of trust prompted several prominent scientists to recently form Scientists and Engineers for America,[33] a national organization dedicated exclusively to protecting the integrity of science and electing government leaders who envision a future of renewed respect for wise science and technology policy.

Calling for an open and transparent process of review, its Bill of Rights for Scientists and Engineers, demands that scientists and engineers enter the political debate when the nation's leaders systematically ignore scientific evidence and analysis, put ideological interest ahead of scientific truths, suppress valid scientific evidence, and harass and threaten scientists for speaking honestly about their research.[33]

Due to the dramatic expansion of the regulatory system, growing bodies of statutory and administrative laws, multiple regulatory agencies with disparate procedures and requirements, the expansion of liability for damages caused by defective products and toxic chemicals, coupled with the continued failure of the U.S. government to recognize the need to financially support meaningful public research on health and the environment, science has been placed under intense pressure.[34]

[32]Miller, H. I. *Public Opinion vs. Public Policy.* TCS: Tech Central Station—Where Free Markets Meet Technology. January 4, 2004. Available at http://www2.techcentralstation.com/1051/printer.jsp?CID= 1051-010504C (last visited February 5, 2004).

[33]Cornelia, D. *Scientists Form Group to Support Science-friendly Candidates. The New York Times,* September 28, 2006. See also www.sefora.org.

[34]Wagner, W. and R. Steinzor, eds. *Rescuing Science From Politics: Regulation and the Distortion of Scientific Research.* New York: Cambridge University Press. 2006.

A New Law for Nanotechnology?

On the heels of the release of the report findings on public perceptions, the call for a new law for nanotechnology, focusing on products and not on the environment, was advocated by J. Clarence Davies[11] in yet another report released by the WWIC, Project on Emerging Nanotechnologies.

While acknowledging that little is known about the possible "adverse" effects of nanotechnology due to a lack of data and the unpredictability of nanomaterials, a repetitive theme surrounding nanotechnology, Davies champions a new law to manage its potential unacceptable risks,[35] which he defines as no more or less stringent than "unreasonable risk," which he leaves undefined.

Davies proposes several, nonmutually exclusive approaches to defining "unacceptable risk." First, the manufacturer would have to anticipate likely product risks and be required to demonstrate preventative steps for occurrence, or nominal damage if unpreventable. Taking a rough cost–benefit approach, accept that the benefits of the product outweigh the risks. Or, a third comparative risk approach, accept a product's risk if it substitutes for a material having greater risks.

He describes options for government action to address the "adverse" effects of nanotechnology by requiring the establishment of new institutions or institutional mechanisms, together with the involvement of members of the general public. He allegedly supports his management plan with relevant evidence.

Still, Davies calls for a new law structured in such a way as to provide incentives for developing effects data and making it available, while placing the burden of product safety on the manufacturer. To him, a new focused law on nanotechnology would serve a twofold purpose: (1) it would avoid the pitfalls of previous regulatory laws; and (2) it could be tailored to the particular characteristics of nanotechnology while avoiding the need to pigeonhole it into existing law.[11] More so, he calls for the creation of new and better institutions for public participation[36] and the use of nanotechnology as a good opportunity to use and experiment with these new approaches.

The call for a new nanotechnology law is contrary to certain stakeholder interests. For some, "nanotechnology ... offers the greatest benefits for society if left to grow through modest regulation, civilian research, and an emphasis on self-regulation and responsible professional culture."[37] While:

> it is recognized that nanotechnology may need new regulatory approaches due to the implications of size, persistence in the environment, disposal and self-assembling nanosystems, for several countries, a first step towards researching the need for adapting existing legislation is a focus on developing appropriate monitoring and warning systems when current legislation provides insufficient.[38]

---

[35]See footnote 11, p. 3.
[36]See footnote 11, p. 30.
[37]Arrison, S. *New Regulations Not Needed Says Institute*. Pacific Research Institute Issues Statement on Nanotechnology. January 17, 2006.
[38]*Survey on Nanotechnology Governance: Volume A. The Role of Government*. International Risk Governance Council—Working Group on Nanotechnology. M. C. Roco, chair. Geneva. December 2005.

CONGRESSIONAL REACTION: THE VICIOUS CIRCLE—PART II

To justify the continued U.S. government financial investment in nanotechnology since the inception of the NNI 6 years ago, while simultaneously appeasing public concern of its potential risks as vocalized through NGOs, commissioned reports by Congress abound.

Four of the most recent and recognized reports include: (1) A Matter of Size: Triennial Review of the National Nanotechnology Initiative;[39] (2) The National Nanotechnology Initiative at Five Years: Assessment and Recommendation of the National Nanotechnology Advisory Panel;[40] (3) The National Nanotechnology Initiative Strategic Plan;[41] and (4) The National Nanotechnology Initiative Environmental, Health, and Safety Research Needs for Engineered Nanoscale Materials.[42]

To silence the noise makers, while meeting the requirements set forth in § 2(b)(10)(D) of the Act, the National Nanotechnology Coordinating Office (NNCO), as established in § 3, convened a Public Participation in Nanotechnology Workshop: An Initial Dialogue[43] from May 30–31, 2006, to glean approaches to engaging the public on nanotechnology related issues. Sponsored by the Nanoscale Science, Engineering and Technology Subcommittee (NSET) and supported by the U.S. Environmental Protection Agency (EPA), the International Association for Public Participation (IAP2), and the National Coalition for Dialogue and Deliberation (NCDD), the NNI sought input from citizens to aid in its decision-making processes.

1. *A Matter of Size: Triennial Review of the National Nanotechnology Initiative*

   Pursuant to the Act, the National Research Council created the Committee to Review the NNI (Committee) to conduct a 13-prong triennial external evaluation of the national nanotechnology program, together with a one-time assessment to determine the technical feasibility of molecular self-assembly, coupled with a study on the responsible development of nanotechnology.[21]

[39]*A Matter of Size: Triennial Review of the National Nanotechnology Initiative*. Prepublication Draft: September 25, 2006. Committee to Review the National Nanotechnology Initiative, National Materials Advisory Board, Division on Engineering and Physical Sciences—National Research Council of the National Academies.

[40]*The National Nanotechnology Initiative at Five Years: Assessment and Recommendations of the National Nanotechnology Advisory Panel*. Submitted by the President's Council of Advisors on Science and Technology. May 2005. Available at http://www.nano.gov/FINAL_PCAST_NANO_RE-PORT.pdf (last visited September 5, 2006).

[41]*The National Nanotechnology Initiative Strategic Plan*. Developed by the Nanoscale Science, Engineering, and Technology Subcommittee, Committee on Technology, National Science and Technology Council. December 2004. Available at http://www.nano.gov/NNI_Strategic_Plan_2004.pdf (visited September 5, 2006).

[42]*The National Nanotechnology Initiative Environmental, Health, and Safety Research Needs for Engineered Nanoscale Materials*. Nanoscale Science, Engineering, and Technology Subcommittee, Committee on Technology, National Science and Technology Council. September 2006.

[43]Available at https://nnco.nano.gov/p2/ (last visited September 11, 2006).

Oftentimes a precursor to regulation, the call for the responsible development of nanotechnology has become commonplace among stakeholders and government leaders. Yet, it remains loosely defined and open to vague and overly broad interpretation.

In *A Matter of Size*, the Committee defined responsible development as balancing the positive contributions of nanotechnology while minimizing its negative consequences.[44] To this end, the Committee focused on current environmental, health, and safety (EHS) research relevant to nanotechnology, addressing worker health and safety concerns, regulatory and standards-setting activities, in addition to government outreach efforts to include the public in discussions of ethical and social issues.[45] It thus analyzed an inventory of United States and international studies in the field.

Based on its examination, the Committee concluded that, at this stage of nanotechnology research and development, it was not yet possible to make a rigorous assessment of the level of risk posed by engineered nanomaterials, thereby recommending continued, increased, and expanded EHS research.[45] To effectively advance reproducible and statistically reliable data, the Committee suggested an integrated approach among scientists, engineers, social scientists, toxicologists, policymakers, and the public.[45]

2. *The National Nanotechnology Initiative at Five Years: Assessment and Recommendation of the National Nanotechnology Advisory Panel*

In response to direction in the President's Fiscal Year 2004 Budget, and pursuant to § 4 of the Act, a review was conducted of the multiagency NNI by the National Nanotechnology Advisory Panel (NNAP), designated to the President's Council of Advisors on Science and Technology (PCAST) by Executive Order in July 2004. This review represented the first periodic assessment of nanotechnology research efforts by the federal government and PCAST in its role as the NNAP, with a focus on U.S. competitiveness.

To strategically accomplish this feat, PCAST sought answers to four questions relative to the federal investment in nanotechnology research and development:

- Where do we stand?
- Is this money well spent and the program well managed?
- Are we addressing societal concerns and potential risks?
- How can we do better?

[44]National Nanotechnology Coordinating Office, *Public Participation in Nanotechnology Workshop: An Initial Dialogue* (2006). Available at https://nnco.nano.gov/p2/ (last visited September 11, 2006).
[45]*The National Nanotechnology Initiative Environmental, Health, and Safety Research Needs for Engineered Nanoscale Materials*. Nanoscale Science, Engineering, and Technology Subcommittee, Committee on Technology, National Science and Technology Council. September 2006.

In its review, the NNAP acknowledged that the United States currently holds a leadership position in nanotechnology and that continued coordinated interagency planning and programming provided an appropriate manner in which to organize and manage the NNI to maintain that leadership for the next 5–10 years.[46]

One of the seven duties of the NNAP, as stipulated in the Act, and subsequently addressed in question number three of the review, included advising the President on whether societal, ethical, legal, environmental, and workforce concerns were being adequately coordinated and communicated by the NNI— grouped together and coined as "societal implications"—a frequent interchangeable synonym to "responsible development."

While the review concluded that existing regulatory authorities appear to adequately protect the public and the environment at this stage of research and development, it recognized that nanotechnology products may require regulation and that such must be rational and based on science, not perceived fears. It encouraged the government regulatory bodies to work together to coordinate regulatory policies.

To this end, the NNAP drew special attention to the National Toxicology Program (NTP)—an interagency agenda within the U.S. Department of Health and Human Services (HHS)—to determine the toxicity of specific nanomaterials, as well as the National Institute for Occupational Safety and Health (NIOSH) to ensure worker safety. It also noted the formal establishment of the Nanotechnology Environmental and Health Implications (NEHI) Working Group (WG) created by the NSET Subcommittee of the Committee on Technology within the National Science and Technology Council (NSTC) for the purpose of describing the EHS research and information needed to identify, understand, and manage the potential risks of engineered nanoscale materials.

Still, the NNAP conceded that, at this point, "the state of knowledge with respect to the actual risks of nanotechnology is incomplete."[47] In its defense, the NNAP noted that many new technologies and products have associated risks that are successfully managed in order to gain their benefits, citing gasoline, electricity, and medical X-rays as examples of earlier innovation similarly situated.

3. *The National Nanotechnology Initiative Strategic Plan*

The NSTC, established by Executive Order in 1993, establishes national science and technology investment goals by preparing research and development strategies. It is the principal means through which the President coordinates science, space, and technology policies across the Federal government.

---

[46]*The National Nanotechnology Initiative at Five Years: Assessment and Recommendations of the National Nanotechnology Advisory Panel.* Submitted by the President's Council of Advisors on Science and Technology. May 2005. Available at http://www.nano.gov/FINAL_PCAST_NANO_REPORT.pdf.

[47]See footnote 46, p. 35.

Pursuant to § 2(c) of the Act, the NSTC is responsible for overseeing the planning, management, and coordination of the NNI.[21] As such, its mandate calls for the development of a strategic plan through which the establishment of goals and priorities, as well as program component areas, guide the activities and anticipated outcomes of participating agencies within the NNI for the next 5–10 years. This plan also addresses the societal dimensions of the development of new technologies, and their potential implications for health and the environment, as well as the importance of dialogue with the public.

While the vision of the NNI is to create a future where the understanding and control of matter at the nanoscale leads to a revolution in technology and industry, its goals are: (1) maintain a worldclass research and development program aimed at realizing the full potential of nanotechnology; (2) facilitate transfer of new technologies into products for economic growth, jobs, and other public benefit; (3) develop educational resources, a skilled workforce, and the supporting infrastructure and tools to advance nanotechnology; and (4) support the responsible development of nanotechnology.[48]

A recurring theme in anticipation of regulation, the NNI supports research to better understand the benefits and risks to human health and the environment, as well as the methods for nanotechnology risk assessment and management. Here, the responsible development of nanotechnology is divided into two categories: (1) EHS implications; and (2) ethical, legal, and all other societal issues. By establishing clear channels of communication to allow the public and the government to make well-informed decisions and build trust, the NNI expects to better identify and prioritize research needed to support regulatory decision making.

4. *The National Nanotechnology Initiative Environmental, Health, and Safety Research Needs for Engineered Nanoscale Materials*

The NEHI WG was formally chartered in 2005 to: (1) provide for exchange of information among agencies that support nanotechnology research and those responsible for regulation and guidelines related to nanomaterials; (2) facilitate the identification, prioritization, and implementation of research and other activities required for the responsible development, utilization, and oversight of nanotechnology, including research methods for life cycle analysis; and (3) promote communication of information related to research on environmental and health implications of nanotechnology to other government agencies and non-government parties.[49] Twenty-four Federal agencies and offices participate.

---

[48]*The National Nanotechnology Initiative Strategic Plan.* Developed by the Nanoscale Science, Engineering, and Technology Subcommittee, Committee on Technology, National Science and Technology Council. December 2004. At p. i.

[49]*The National Nanotechnology Initiative Environmental, Health, and Safety Research Needs for Engineered Nanoscale Materials.* Nanoscale Science, Engineering, and Technology Subcommittee, Committee on Technology, National Science and Technology Council. September 2006. p. 5.

In September 2006, 3 years after it was established, the NEHI WG identified EHS (defined as environmental health, human health, animal health, and safety) research needs necessary to glean a better understanding of potential risks associated with engineered nanoscale materials, or nanomaterials (those purposefully manufactured or synthesized with at least one dimension approximately of 1–100 nm, and exhibiting unique properties as a result of that size range), expected to be used in products, medical therapeutics, environmental applications, or in the workforce.

This document was released to inform, guide, and coordinate such nanoscale EHS research programs among the NNI agencies, and to communicate to NGOs and stakeholders effective approaches for obtaining the knowledge and understanding necessary to enable the sound risk assessment and management of nanomaterials. In so doing and to its credit, NEHI acknowledged the risk/benefit ratio concern as a phenomenon not solely unique to nanotechnology, but to any new technology. The insurance industry is well aware of the fact that risk accompanies innovation and change. For them, the introduction of nanotechnology signifies a paradigm shift—both in industrial applications and in the exposure mechanisms.[50] With nanoparticles, establishing a relationship between cause and effect for potential claims is difficult or almost impossible to assess.

Due to lack of data, the complexity of the materials, measurement difficulties, and undeveloped hazard assessment frameworks, there is little scientific guidance pertaining to the real risks of nanomaterials. Because the potential losses associated with nanoparticles may be either impossible or very difficult to evaluate because of their scale, location, and time of occurrence, the insurance industry remains vigilant, recognizing that, with the high level of uncertainty still attached to nanotechnology, it may be difficult to precisely determine the probability of a loss occurring or its possible extent, for a long time to come.

To date, understanding the interaction of engineered nanoscale materials with biological systems remains incomplete. Engineered nanoparticles, such as buckyballs and gold nanoshells, may constitute an entirely new class of particles. While a handful of toxicity studies indicate hazardous tendencies, they can also be engineered to be less, so by conjugating other chemicals to the surface of buckyballs and changing their chemical properties. So, the proper question for regulators and policy makers to ask of nanotechnology is not Is it safe? but instead How can we make nanotechnology safer?[51]

The NEHI WG endeavored to seek answers to several questions:

1. Are current toxicity testing methods appropriate for assessing the toxicity and potential biological effects of engineered nanoscale materials?
2. How do chemical–physical properties of those nanomaterials relate to their elicited biological responses?

[50]Swiss Re. *Nanotechnology: Small Matter, Many Unknowns*. Risk Perception. 2004. Available at www.swissre.com/INTERNET/pwsfilpr.nsf/vwFilebyIDKEYLu/ULUR-5YNGET/$FILE/Publ04_Nanotech_en.pdf.
[51]United Nations Educational, Scientific and Cultural Organization (UNESCO). *The Ethics and Politics of Nanotechnology*. September 2006. Available at http://www.unesco.org/shs/ethics.

3. What kinds of human and environmental exposures to nanomaterials can be anticipated and measured?

4. By which paths do nanomaterials move within the body?

5. Are there any special considerations for the measurements of nanomaterials?

Working with the Federal agencies, the NEHI WG drew from various reports before finalizing its inventory of research and scientific data available. These included the *Nanotechnology White Paper*[52] released by the EPA identifying research needs for environmental applications and implications of nanotechnology; and *Approaches to Safe Nanotechnology: An Information Exchange with NIOSH*,[53] a draft guidance document outlining current knowledge regarding the occupational health and safety implications and applications of engineered nanoscale materials.

Due to the early stage in development of nanomaterials and their applications, the NEHI WG established principles for identifying and prioritizing EHS research needs and investments to better coordinate and facilitate the NNI agencies' research programs. This was reflected in the recommendations found in the Triennial Review of the NNI, the NNAP assessment of nanotechnology research efforts, and the NNI Strategic Plan.

The Vicious Circle: Part II—Conclusions

While evident that Congress is responsive to public opinion and wishes to engage the public pursuant to statutory mandate, it lacks an active uniform approach to the responsible development of nanotechnology. Endless reports, journal articles, and conference proceedings mandated by legislation or commissioned and authored by self-appointed experts in their respective fields add to the already existing polarizing political debate. Investigator fear of not receiving or being awarded funding for future scientific research and development multiplies the already existing pressure creating skewed recommendations regarding risk assessment and management.

Who objectively decides the future safety of innovation, and what is in the best interests of society? Who is licensed to make and capable of making unbiased and nonprejudicial determinations warranting trust? How can today's Congress design the infrastructure for the successful and effective creation of new adaptive regulatory models and self-corrective proactive paradigms seamlessly capable of withstanding rapid and constant change from invasive, invisible, pervasive technologies and unimaginable next-generation convergent innovation within an ever-increasing, complex, internationally networked societal structure?

---

[52]U.S. Environmental Protection Agency (U.S. EPA), Nanotechnology Working Group, *Nanotechnology White Paper*, External Draft for Review. Available at http://www.epa.gov/osa/nanotech.htm (2005).

[53]National Institute for Occupational Safety and Health (NIOSH), *Approaches to Safe Nanotechnology: An Information Exchange with NIOSH*. Available at http://www.cdc.gov/niosh/topics/nanotech (2005).

REGULATORY UNCERTAINTIES: THE VICIOUS CIRCLE—PART III

Nano-products, materials, applications, and devices are governed today within the existing framework of statutes, laws, regulations, and policies. The still-unanswered question is whether current regulatory controls are adequate to meet the many concerns posed by the ability of nanotechnology to create products whose structures, devices, and systems have novel properties and structures because of their size. Yet, the most pressing issue may be not in the creation of new, but in the enforcement of old, regulations on the industries that create and process these new materials.

Several agencies with regulatory responsibilities participate in the NNI. These include the Environmental Protection Agency (EPA), Food and Drug Administration (FDA), Consumer Product Safety Commission (CPSC), Nuclear Regulatory Commission (NRC), Occupational Safety and Health Administration (OSHA) in the U.S. Department of Labor, and U.S. Department of Agriculture (USDA). A few have held public meetings to determine whether their regulatory systems in place are nano-friendly.

Still, are these Federal agencies in a position to institutionally combat the continued circular journey of regulatory hodgepodge and gridlock, and facilitate the effective and safe commercialization of nanoscale technologies when their practices and interpretations of science differ from one to the other? Federal regulatory agencies serve as a nexus for scientific fact finding and adjudicating controversies. Oftentimes, they are charged with technical incompetence, or with subordinating science to political ends.[54]

EPA

The EPA is one of the Federal watchdog agencies challenged by nanotechnology. Within its jurisdiction, nanotechnology crosses several core Federal environmental statutes: (1) the Toxic Substances Control Act (TSCA); (2) the Federal Insecticide, Fungicide, and Rodenticide Act of 1972 (FIFRA); (3) the Comprehensive Environmental Response, Compensation and Liability Act (Superfund) (CERCLA); (4) the Resource Conservation and Recover Act (RCRA); (5) the Clean Water Act (CWA); and (6) the Clean Air Act (CAA).

TSCA

Passed by Congress in 1976, the TSCA[55] gives the EPA the authority to prohibit or limit the manufacture of particular chemicals based on risk assessments, and the power to regulate and control new and existing chemical substances in commercial use with risk or potential risk to the environment.[2] It does this by ensuring the review of all new chemicals prior to their commercial manufacture by requiring a premanufacture notification (PMN).

[54]Jasanoff, S. *Procedural Choices in Regulatory Science.* Available at http://www.piercelaw.edu/risk/vol4/spring/jasanoff.htm (last visited December 29, 2003).
[55]15 U.S.C. § 2601 *et seq.*

The TSCA provides the EPA with the tools to respond where information comes to light that supports the finding that the manufacturing, processing, distribution, use, and/or disposal of a chemical substance will present "an unreasonable risk of injury to health or the environment."[56]

Because manmade structures less than 100 nm in size may exhibit unusual properties, preliminary studies speculate that nanoscale substances may be harmful to the environment. As nanoscale properties behave differently than macro-size substances of similar chemical compositions, still unresolved is whether nanoscale materials in the TSCA inventory are defined as new or existing chemicals.

If nanotechnology is a material in a chemical system, does it pose an unreasonable risk within the existing definition? Or, does the application of nanotechnology-based substances constitute a new use of an existing chemical under the TSCA? When classifying new substances, the TSCA does not address the differences between macro- and nanoscale behavior of substances.[57]

In 15 U.S.C. § 2602(2)(a), the TSCA defines a chemical substance as "any organic or inorganic substance of a particular molecular identity including any combination of such substance occurring, in whole or in part, as a result of chemical reaction or occurring in nature and any element or uncombined radical." In 15 U.S.C. § 2602(9), it defines a new chemical substance as "any chemical substance which is not included in the chemicals substance list compiled and published under § 2607(b) of the TSCA Chemical Substance Inventory."

While the EPA is attentive to the concerns associated with nanotechnology, it claims that its current regulatory structure is sufficiently adequate to ensure product safety and the potential risks posed by nanotechnology applications involving chemical substances.

According to the results of a comprehensive review of core Federal environmental statutes, the American Bar Association (ABA) Section of Environment, Energy, and Resources (SEER)[58] agrees, and concludes that: (1) nanomaterials include chemical substances and mixtures the EPA can regulate pursuant to the TSCA; (2) if a "new" chemical substance is manufactured at the nanoscale, it is subject to the same PMN review requirements under § 5(a)(1); and (3) the EPA may regulate nanomaterials as existing chemical substances pursuant to its § 5(a)(2) authority to promulgate significant new use rules (SNURs).[58]

FIFRA

The FIFRA[59] gives the EPA authority over genetically engineered crops that produce pesticides. It governs the commercial use and release of dangerous

---

[56]Bergeson, L. L. and B. Auerbach. *The Environmental Regulatory Implications of Nanotechnology.* Daily Environment Report, Bureau of National Affairs. ISSN 1060-2976. B-1, No. 71. April 14, 2004.

[57]Miller, S. E. *A Matter of Torts: Why Nanotechnology Must Develop Processes of Risk Analysis.* New York Law Journal. Volume 232-Number 67. October 5, 2004.

[58]Available at http://www.abanet.org/environ/.

[59]25 June 1947, Ch. 61 Stat. 163.

substances while in the research and development stage, as well as into the environment. Its primary focus is to provide Federal control over pesticide distribution.

Depending on its environmental risk assessment, the EPA can issue an experimental use permit (EUP) for research involving pesticides. If field testing proves there are no significant risks to the environment or human health, the EPA will grant a limited license. All pesticides used in the United States must be registered by the EPA, which ensures that they are properly labeled and will not cause unreasonable environmental harm.

The convergence of nanotechnology with biotechnology has allowed for the development of new pesticide products with enhanced effectiveness. Probable uses of nanotechnology in agriculture include agrochemical delivery, nanosensors, and new or modified active pesticidal ingredients.[60]

The EPA recently released a *Nanotechnology White Paper*[61] on whether the use of a nanoscale material will result in a change to a pesticide product registered under FIFRA. The FIFRA § 3 registration requirement prohibits, with limited exceptions, the distribution or sale of any unregistered pesticide and can require extensive information regarding a pesticide's risk/benefit assessments. Under FIFRA § 3(c)(5)(D), the EPA determines if a pesticide "will not generally cause unreasonable adverse effects on the environment" and the conditions under which a nanopesticide may be registered.

Genetically engineered microorganisms used as pesticides serve as an analogy for the EPA when it comes to effectively regulating nanopesticides. The EPA determined in 1986 that it could regulate pesticidal products of biotechnology through FIFRA, without the need for new legislative authority.[62] In 2001, without additional legislative authority, it promulgated regulations to address a particular class of bioengineered pesticides.[63]

Still, several regulatory challenges remain: (1) Does the use of a nanoscale material result in a change to a pesticide product registered under FIFRA?[64]; and (2) Are new registrations needed for nano versions of registered conventional pesticides?[65] Here again, the ABA concludes that the EPA is in a position to adequately regulate nanopesticides within its existing statutory authority.[65]

---

[60]"The Nanotechnology–Biology Interface: Exploring Models for Oversight, September 15, 2005, Workshop Report." Center for Science, Technology, and Public Policy, University of Minnesota. Available at http://www.hhh.umn.edu/img/assets/9685/nanotech_jan06.pdf (last visited September 5, 2006).
[61]EPA Science Policy Council, "Nanotechnology White Paper" (February 2006). Available at http://www.epa.gov/osa/pdfs/EPA_nanotechnology_white_paper_external_review_draft_12-02-2005.pdf.
[62]51 Fed. Reg. at 23313.
[63]66 Fed. Reg 37772 (July 19, 2001) (40 C.F.R. Part 174).
[64]EPA, Science Policy Council, "Nanotechnology White Paper" (external review draft) (December 2, 2005) at 26, 27. Available at http://www.epa.gov/osa/pdfs/EPA_nanotechnology_white_paper_external_review_draft_12-02-2005.pdf (last visited September 5, 2006).
[65]The Adequacy of FIFRA to Regulate Nanotechnology-Based Pesticides, American Bar Association—Section of Environment, Energy, and Resources. June 2006. Available at www.abanet.org/environ/ (last visited September 5, 2006).

CERCLA

Enacted by Congress on December 11, 1980, CERCLA allowed for a tax to be levied on the chemical and petroleum industries so the EPA could better respond to uncontrolled releases or threatened releases of hazardous substances endangering human health or the environment. The tax collected went into a trust fund, commonly known as the Superfund, for cleaning abandoned or uncontrolled hazardous waste sites.

Under CERCLA[66] § 102(a), the EPA has the authority to list in its National Priorities List (NPL) as "hazardous substances" those "which, when released into the environment may present substantial danger to the public health or welfare or the environment." This category also includes listed or characteristic "hazardous waste" as found under RCRA. Here, the parties are held jointly and severally liable and responsible for release of any amount of hazardous waste, even if late-emerging, with the EPA providing for cleanup when such parties cannot be identified.

Additionally, through CERCLA, the Federal government established requirements and prohibitions for closed and abandoned hazardous waste sites, as well as cleanup guidelines and procedures for accidents, spills, and other emergency releases of pollutants and contaminants through its National Contingency Plan (NCP).

Given the paucity of information, it is still uncertain whether nanomaterials pose adverse consequences for human health or the environment such that they may be classified hazardous substances within the definition of CERCLA. While it may be premature to apply CERCLA to nanomaterials at this time, its functional core elements of detection, production, use, or disposal of materials are sufficiently flexible and easily adaptable to prospectively regulate and determine liability should special considerations arise.

CERCLA is not bound by a statute of limitations, or the lawfulness of a particular action at the time committed and, therefore, may impose retroactive liability on a party for historic practices in the fate or transport of a hazardous substance. As a backward-looking statute, should it ultimately be determined that nanoscale materials are indeed a hazardous substance within the definition of CERCLA, the EPA has the authority to enforce its regulatory practices.

While it is often mentioned that nanoscale particles may pose a risk of harm to health or environment, at the forefront, it must also be remembered that these same materials may offer potential utility as remediation tools to mitigate known risks of conventional hazardous substances. As yet, it is too early to determine the risk/benefit ratio at this scale because the scientific and technical predicates for applying CERCLA to nanomaterials do not yet exist.

---

[66]The Comprehensive Environmental Response, Compensation and Liability Act, 42 U.S.C. § 9601 *et seq.*

RCRA

The Resource Conservation and Recovery Act (RCRA)[67] is the public law that creates the framework for the proper management and governance of the generation, transportation, treatment, storage, and disposal of hazardous and nonhazardous solid waste, and regulation of underground storage tanks containing these substances and petroleum products. In contrast to CERCLA, which manages abandoned and historical sites, RCRA focuses on active and future facilities.

A RCRA hazardous waste can be a liquid, solid, contained gas, or sludge and a byproduct of manufacturing processes or discarded commercial products, such as cleaning fluids, pesticides, or pharmaceutical products. It is a waste that appears in one of the EPA's hazardous wastes lists, including: (1) F-list, nonspecific source wastes; (2) K-list, source-specific wastes; and (3) P- and U-list, discarded commercial chemical products. If it does not fit into one of those three categories, it may still be considered a hazardous waste if it exhibits one of four characteristics: (1) ignitability—capable of creating a fire; (2) corrosivity—acids or bases capable of corroding metal containers; (3) reactivity—unstable under normal conditions; or (4) toxicity—harmful or fatal when ingested or absorbed.

While most innovation serves a dual purpose, the alleged potential toxicity of nanomaterials on the one hand, with the beneficial ramifications they may lend to the clean up of hazardous wastes and contamination via environmental detectors and sensors on the other, poses a dilemma of scale. While neither Federal nor state waste management programs have felt the need to develop specific regulatory protocols or guidance directives for nanoscale wastes, the RCRA mandate provides the EPA with extensive broad statutory powers and provisions to define and control waste at those dimensions, respond to novel characteristics or hazards, and to promulgate new regulations if needed.

Of course, it is easy to speculate the applicability and scope of current law when established determinations of nanoscale risks remain uncertain. Equally, and in the same instance, it is conjecture to call for the creation of new regulations. Neither a professed psychic with a crystal ball nor a tarot-reading astrologer could foresee the future positives or negatives of nanotechnology.

CWA

The Clean Water Act (CWA) gives authority to the EPA to regulate the discharge of "pollutants," defined as chemical wastes and "industrial, municipal, and agricultural waste discharged into navigable waters,"[68] or defined broadly as any materials added to a watercourse.[69] The term "toxic pollutant" includes:

> those pollutants, or combination of pollutants ... which after discharge and upon exposure, ingestion, inhalation or assimilation into any organism, either directly

[67]40 C.F.R. Parts 239–299.
[68]CWA § 502(6), 33 U.S.C. § 1362(6).
[69]CWA, § 302, 33 U.S.C. § 1312.

from the environment or indirectly by ingestion through food chairs, will, on the basis of information available to the Administrator, cause death, disease, behavioral abnormalities, cancer, genetic mutations, physiological malfunctions (including malfunctions in reproduction) or physical deformations, in such organisms or their offspring.[70]

The objective of CWA is to restore and maintain the chemical, physical, and biological integrity of the nation's waters. It finances municipal wastewater treatment facilities and manages polluted runoff. It sets wastewater standards for industry, and water quality standards for all contaminants in surface waters, requiring a permit to discharge any pollutant from a point source into navigable waters.

Given the broad interpretation of definitions assigned by Federal agencies to their regulatory mandates, it is therefore likely that the regulation of nanoparticles, if deemed a pollutant, or toxic pollutant, would fall under the jurisdiction of CWA, thereby giving the EPA the authority to establish guidelines and standards for their discharge.

Still, this is illustrative of the state of knowledge as it pertains to nanotechnology and premature speculation as to whether Federal agencies will need to alter their current methods of assessment, evaluation, monitoring, measurement, and management of substances that may be harmful to health, safety, or the environment.

CAA

Common air pollutants include nitrogen dioxide, carbon monoxide, particulate matter, sulfur dioxide, and lead. The Clean Air Act (CAA)[71] regulates air emissions from area, stationary, and mobile sources in addition to the problems of acid rain, ground-level ozone (smog), urban air pollution, stratospheric ozone depletion, and air toxics. It authorizes the EPA to establish air quality standards through the National Ambient Air Quality Standards (NAAQS) and set limits on hazardous air pollutants (HAPS). Through its clearinghouses, it provides scientific research, expert studies, and engineering designs, together with funds to support clean air programs. It contains a permit program for the categories of sources that may release the 189 chemicals listed by Congress in CAA into the air.

As an incentive, CAA provides economic credits for cleaning up pollution, protecting health, and preventing environmental and property damage. One of its requirements is for factories and businesses to develop plans to prevent accidental releases of highly toxic chemicals into the air. Through its Chemical Safety Board, CAA requires investigation of and reporting on accidental releases of hazardous air pollutants from industrial plants.

Some consumer products that fall under the regulation of CAA are hair sprays, paints, foam plastic products, and carburetor and choke sprays. These release ozone-destroying chemicals. Those products identified as containing less destructive ozone-destroying chemicals will require labels by 2015.

[70]CWA § 502(13), 33 U.S.C. § 1362(13).
[71]42 U.S.C. § 7401 *et seq.* (1970).

Richard P. Feynman, recipient of the Nobel Prize in physics, said in 1959:

I would like to describe a field in which little has been done, but in which an enormous amount can be done in principle .... What I want to talk about is the problem of manipulating and controlling things on a small scale .... Atoms on a small scale behave like nothing on a large scale .... At the atomic level, we have new kinds of forces and new kinds of possibilities, new kinds of effects.[72]

By operating at the same scale as biological processes, nanotechnology offers the capability to intervene in the blueprints of living and nonliving matter, and to recreate nature .... Yet, while some of the major applications for nanotechnology are not expected to be seen for five to 10 years, numerous products featuring the unique properties of nanoscale materials are already being used in electronic, biomedical, pharmaceutical, aerospace, cosmetic, energy, magnetic and optoelectronic, catalytic and materials applications.[73]

It is expected that engineered nanoparticles may fall under the jurisdiction of CAA and, in particular, its definition of particulate matter, should it be determined that they are a HAP. A new set of critical parameters due to its behavioral scale, shape, size, and composition may require that quantification be in the form of number, rather than the current mass limitations CAA uses.

Evaluating the risk resulting from different types of engineered nanoparticles is still not well defined. Because of this, it is still unknown whether they may pose a challenge, not just for EPA, but any regulatory agency and its conventional methods of identifying, monitoring, measuring, and controlling the emission of hazardous substances. As it is now, the EPA has the regulatory authority to oversee the emissions of engineered nanoparticles.

EPA Conclusion

In 2005, the EPA convened stakeholder meetings to promote and advocate for the establishment of voluntary guidelines to better address the release of nanotechnology-related products. Reinforcing its regulatory position, the recently released report by the ABA[74] concluded that the core environmental statutes and current regulations provided the EPA with sufficient legal authority to adequately address the potential challenges and risks associated with nanotechnology.

However, the difficulty in further assessing the adequacy of current laws and statutes as they relate to the nansocale is due to the yet-unknown concern—whether nanomaterials are hazardous and present an unreasonable risk to human health or the environment. Implementation of any regulation requires the ability to be able

---

[72]Miller, S. E. *A Matter of Scale: Nanotechnology's Novelty Poses Challenges to Patent Process.* *New York Law Journal*, Volume 232-Number 23, August 3, 2004, p. 5.

[73]*Ibid.* p. 5.

[74]Regulation of Nanoscale Materials under the Toxic Substances Control Act. American Bar Association—Section of Environment, Energy, and Resources. June 2006. Available at www.abanet.org/environ/ (last visited September 5, 2006).

to monitor, measure, and control nanoparticles. To date, the technology has not been sufficiently developed to allow for a definitive determination of its potential risks. Scientific uncertainty continues, and studies have been inconclusive in their results. In addition, epidemiological data are not reliable, and harmful links are difficult to establish.

Even so, the comprehensive framework of the EPA gives it ample statutory authority to promulgate new regulations and definitions, if needed, to deal with the "novel" characteristics of any new technology.

Still, the EPA, under TSCA, will convene a Public Meeting on Risk Management Practices for Nanoscale Materials under a possible voluntary Nanoscale Materials Stewardship Program (NMSP) in October 2006 for the purpose of exploring the encouragement of the responsible commercial development of nanoscale materials.[75] Through the NMSP, the EPA hopes to enhance stakeholder ability to assess the potential risks to human health and the environment from nanoscale materials, as well as identify effective risk management practices to reduce such risks. This certainly is in keeping with the goals and objectives of the NEHI WG and a proactive regulatory initiative by a Federal agency.

FDA

Operating under the HHS, the mission of the FDA is to ensure the safety and efficacy of drugs, drug delivery systems, cosmetics, medical devices, vaccines, and food products before reaching the marketplace.[2] It does not regulate processes or technology. Instead, this Federal agency regulates products based on their statutory classification—premarket approval, market clearance and postmarket review—with limited regulatory authority over certain categories of products.

Organized by product line, "the FDA cites that it has traditionally regulated many products with particulate materials in the nano-size range and that existing requirements may be sufficiently adequate for most nanotechnology products that it will regulate,"[76] with no safety concerns reported in the past because of particle size. Its six Centers specialize in regulating particular types of products: (1) the Center for Food Safety and Applied Nutrition (CFSAN); (2) the Center for Drug Evaluation and Research (CDER); (3) the Center for Biologics Evaluation and Research (CBER); (4) the Center for Devices and Radiological Health (CDRH); (5) the National Center for Toxicological Research (NCTR); and (6) the Center for Veterinary Medicine (CVM) within its two offices: (1) the Office of Regulatory Affairs (ORA); and (2) the Office of the Commissioner (OC).

Having learned from lessons past and to facilitate the regulation of nanotechnology products, the FDA proactively established a NanoTechnology Interest Group (NTIG) through which its Centers and Offices meet quarterly to share and coordinate nanoproduct concerns, solutions, and advances. It launched its nanotechnology website in January 2005. On August 9, 2006, it announced the formation of an

---

[75]Available at http://www2.ergweb.com/projects/conferences/nano (last visited October 4, 2006).
[76]See footnote 2, p. 5.

internal FDA Nanotechnology Task Force charged with determining regulatory approaches that encourage the continued advance of innovative, safe, and effective FDA-regulated products that use nanotechnology materials.[77] Chartered to identify and recommend ways in which to address knowledge or policy gaps, the Task Force is primed to:

1. Chair a public meeting to help further its understanding of nanotechnology developments.
2. Assess the current state of scientific knowledge.
3. Evaluate the effectiveness of its regulatory approaches.
4. Explore opportunities to better foster nanotechnology innovation.
5. Continue to strengthen its collaborative relationships with other Federal agencies.
6. Consider appropriate vehicles for communication with the public.
7. Submit its initial findings within 9 months of the public meeting.

On October 10, 2006, the FDA held its first Public Meeting on Nanotechnology Materials in FDA Regulated Products to learn about the kinds of new nanotechnology products currently under development in food and color additives, animal feeds, cosmetics, drugs and biologics, and medical devices, as well as new or emerging scientific issues of which the FDA should be apprised.

The area of cosmetics has already become a contentious issue for the FDA. Nanoscale titanium dioxide, zinc oxide, carbon fullerenes, and manganese nano oxides are already being used in cosmetics, sunscreens, and other personal care products. Yet, no worldwide regulations exist specifically addressing the manufacture and marketing of nanoparticles in the personal care industry.

In its May 2006 Friends of the Earth report[78] accompanying the first nano lawsuit against the FDA, the NGO calls for a moratorium on the further commercial release of personal care products containing engineered nanomaterials and the withdrawal of those products currently on the market. The report calls for the assessment of nanomaterials as new substances and the labeling of consumer products containing nanoparticle ingredients. Yet, the FDA does not have a premarket approval process for cosmetic products or ingredients. While the FDA does not comment on legal challenges, it has 6 months to respond to the petition and the first formal call to halt the advance of nanotechnology.[14]

Although cosmetic labeling is regulated by the FDA under its Food, Drug and Cosmetic Act (FDCA), enacted in 1938, and the Fair Packaging and Labeling Act (FPLA),[79] it remains the responsibility of the manufacturer, distributor, or packager

---

[77]"FDA Forms Internal Nanotechnology Task Force." August 9, 2006. Available at http://www.fda.gov/bbs/topics/NEWS/2006/NEW01426.html (last visited October 10, 2006).
[78]*Nanomaterials, Sunscreens and Cosmetics: Small Ingredients, Big Risks Report*, May 2006. Friends of the Earth. Available at www.foe.org/doc/Nano%20FDA%20petition%20final.pdf (last visited May 18, 2006).
[79]21 C.F.R. Parts 701 and 740.

to ensure proper labeling and safety of ingredients prior to marketing. No cosmetic may be labeled or advertised as FDA approved.

The FDA is expected to face extensive applications for novel therapies and technologies containing nanoscale particulates. To confront these next-generation scientific advances, it must maintain its expertise in cutting-edge technologies and provide for the effective training of its examiners.

Additionally, the FDA must independently prepare to effectively and expeditiously address technological advances without pull or push from a meddlesome Congress and the relentless imposition of its philosophical values and beliefs. Regulatory pathways must remain uncomplicated, uncompromised, unbiased, and impartial.

Congress is already considering legislation to reform the FDA process. Its Prescription Drug User Fee Act (PDUFE) expires in 2007. Given recent scandals, such as Vioxx, a reform bill is in draft aimed to improve public disclosure of clinical trials, while simultaneously reducing conflicts of interest on FDA advisory committees.

Receiving fierce opposition is S. 1956, The Access, Compassion, Care and Ethics for Seriously Ill Patients Act (The Access Act), which strives to expedite the receipt of drugs, not yet fully approved or tested by the FDA, to seriously ill or terminal patients urgently in need and willing to ingest them. It proposes to establish a tiered system for drug approvals corresponding to the three phases of clinical trials.

Clearly, concerns abound regarding the risk—benefit analysis, and the careful balancing act the FDA must undertake in its regulation of nanomedicine—promoting timely patient access and fostering innovation, while protecting public health by guarding against unsafe technologies.[80]

On balance, the FDA faces several nanotechnology regulatory issues, not necessarily through any fault of its own: (1) limited authority for potentially high risk nano-products; (2) lack of a commonly accepted nomenclature, definition of a nano-particle and an understanding of its properties; (3) limited basic public health research on nanomaterials; and (4) the possibility of new "tools" needed for new nano-materials.[81] One uncertainty is determining when the performance of a product is a function of the size of the particular material.[2] As nanomaterials and devices will be used to develop more advanced versions of existing products, the FDA will continue to confront many uncertainties and will be forced to make difficult decisions about the risks of new therapies and the data required for regulatory approval.

Several months prior to the FDA Public Meeting, the WWIC had already produced an online inventory of nanotechnology-based consumer products.[82] As of March 2006, the inventory contained 212 products or product-lines grouped into eight categories: health and fitness, electronics and computers, home and garden,

---

[80]Miller, J. C. et al. *The Handbook of Nanotechnology: Business, Policy, and Intellectual Property Law.* Hoboken, New Jersey: John Wiley & Sons, Inc. 2005.
[81]Available at http://www.fda.gov/nanotechnology.
[82]*A Nanotechnology Consumer Products Inventory.* Woodrow Wilson International Center for Scholars (2006). Available at http://www.nanotechproject.org/idex.php?id=44 (last visited October 11, 2006).

food and beverage, cross cutting, automotive, appliances, and goods for children. The largest category was health and fitness, with the United States having the most products, followed by Asia and Europe, respectively, and carbon—including fullerenes and nanotubes as the most common material—with silver, silica, titanium dioxide, zinc oxide, and cerium oxide following; calculating a total of 56 products applied directly to the skin. These, of course, were referenced by the Friends of the Earth report.

In short, since "preparedness" is a commonly used buzzword today, of course, the WWIC, Project on Emerging Nanotechnologies weighed in on FDA nano-readiness in strategy, expertise, and resources with the release of its October 2006 report: Regulating the Products of Nanotechnology: Does FDA Have the Tools it Needs?[83] To do so, it first analyzed the potential risks associated with nanotechnology products, adopting the presumption that engineered nanomaterials are "new for safety evaluation purposes."[84]

Its conclusions found gaps in three prime areas of the FDA's legal authority: (1) the lack of premarket oversight tools for cosmetics; (2) its ability to acquire information about nano-laden products sufficiently early in their development to proactively prepare for their regulation; and (3) inadequate authority for post-market adverse event reporting.[83]

Acknowledging the well-established fact that lack of resources impeded the ability of the FDA to provide effective oversight of nanotechnology products, the WWIC stressed the potential consequences of missing safety problems by latent discovery, and lagging in providing regulatory guidance and prompt regulatory reviews, jeopardizing public health and confidence, and hindering the advance of innovation as a result. Recommendations were divided into near-term actions, legal authority, and resource needs—requesting that Congress rebuild the FDA's capacity to meet the public's expectations[85]—whatever those might be.

CPSC

An independent regulatory agency created in 1973 under the Consumer Product Safety Act (CPSA), the CPSC assesses a product's potential chronic and acute health effects to consumers, once distributed in commerce, under its Federal Hazardous Substances Act (FHSA).[86] It is charged with protecting the public "against unreasonable risks of injuries associated with consumer products,"[86] and its jurisdiction includes more than 15,000 types of consumer products used in or around the home.

To be considered a "hazardous substance" under the FHSA, a product must satisfy a two-part definition: it must be toxic, and have the potential to cause "substantial personal injury or illness during or as a proximate result of any customary or

---

[83]Taylor, M. R. *Regulating the Products of Nanotechnology: Does FDA Have the Tools It Needs?* Woodrow Wilson International Center for Scholars, Project on Emerging Nanotechnologies. October 2006.

[84]See footnote 83, p. 17.

[85]See footnote 83, p. 58.

[86]15 U.S.C. § 1261.

reasonably foreseeable handling or use."[87] If a product meets that definition, the FHSA requires cautionary labeling regarding its safe use, handling, and storage. If such is inadequate, the CPSC has the authority to ban the product.

In contrast to the FDA, the FHSA does not provide for premarket registration or approval. Therefore, the responsibility rests with the manufacturers to ensure that their products are labeled in accordance with CPSC regulations.

Pursuant to its CPSC Nanomaterial Statement,[88] and in line with the EPA and FDA, the CPSC has determined that the potential safety and health risks of nanomaterials can be assessed under its existing statutes, regulations, and guidelines. Because some of these new nanomaterials are used in consumer products to improve their performance and/or durability, the CPSC cannot generalize, at this time, about the potential effects of exposure during consumer use and disposal.

NRC

An independent agency established by the Energy Reorganization Act of 1974 (ERA), the NRC regulates the civilian use of byproduct, source, and special nuclear materials. It ensures the protection of public health and safety, and promotes the common defense and security to better protect the environment. Its regulatory mission includes: (1) reactors (commercial and test); (2) materials; and (3) waste.

The ERA requires the licensure of civilian uses of nuclear materials and facilities. It empowers the NRC to establish by rule or order and to enforce standards deemed necessary to protect health and safety, and to minimize danger to life or property. It adheres to five Principles of Good Regulation: (1) independence; (2) openness; (3) efficiency; (4) clarity; and (5) reliability.

Given the uncertainty of nanotechnology, the safe disposal of its byproducts will more than likely fall under NRC jurisdiction.

OSHA

Almost everyone working in the United States falls under the jurisdiction of the Occupational Health and Safety Act,[89] whose aim it is to ensure safety and health in the workplace by establishing standards, providing training, outreach, and education for a better working environment. Under the administration of President George W. Bush, it focuses on three strategies: (1) strong, fair and effective enforcement; (2) outreach, education and compliance assistance; and (3) partnerships and cooperatives programs.

To establish standards for a workplace free of hazards to health and safety, the OSHA created NIOSH as the research institution for OSHA, its administrative arm.

The NIOSH, within the Centers for Disease Control and Prevention (CDC) of HHS, is at the forefront of conducting research and providing guidance on the

[87]15 U.S.C. § 1261 (f)(1)(A).
[88]Available at http://www.cpsc.gov.
[89]29 U.S.C. § 651 *et seq.* (1970).

occupational safety and health implications and applications of nanotechnology. This role stems from its mission as the Federal institute that conducts research on and makes recommendations for occupational health and safety.

To this end, its research focuses on worker exposure to nanoparticles during manufacturing or industrial use, plus their interaction with and effects on the body. NIOSH therein has identified 10 critical nanotechnology research topic areas: toxicity, risk assessment, epidemiology and surveillance, controls, measurement methods, exposure and dose, applications, safety, communication and education, and recommendations and guidance.

The ultimate commercialization of nanotechnology may directly impact the workforce. A potential benefit is the prevention, early detection, and treatment of occupational and environmental diseases, thereby lowering U.S. healthcare costs. However, because a thorough understanding of exposure to engineered nanoparticles and nanomaterials is unavailable and data are inconclusive, appropriate exposure monitoring and control strategies, as well as best practices, have not yet been put in place.

Still, as a member of NSET, NIOSH works with other Federal agencies involved with the NNI in supporting the responsible development and use of nanotechnology. With the FDA, it co-chairs the NEHI. Alongside other NNI agencies, NIOSH will sponsor the International Conference on Nanotechnology and Occupational Health: Research to Practice in December 2006.

To advance its research agenda in tandem with partnering government agencies, NIOSH has created a Nanotechnology Research Center, initiated a program under the National Occupational Research Agenda (NORA) to characterize the physical and chemical properties of nanomaterials, as well as and their effects and risks, established a nanotechnology web presence, and drafted for stakeholder review and public comment three documents: (1) Strategic Plan for NIOSH Nanotechnology Research: Filling the Knowledge Gaps; (2) Approaches to Safe Nanotechnology: An Information Exchange with NIOSH, which includes a chapter on "Guidelines for Working with Engineered Nanomaterials"; and (3) Evaluation of Health Hazard and Recommendations for Occupational Exposure to Titanium Dioxide—of particular interest to the ICTA in its suit against the FDA. Additionally, it established the Nanotechnology and Health & Safety Research Program, a 5-year study, to assess and evaluate the toxicity and health risks of occupational exposure to nanoparticles.

USDA

The vision of the USDA is to be recognized as a dynamic organization by providing leadership on food-related concerns based on "sound public policy, the best available science, and efficient management,"[90] and it has created a strategic plan to implement its mission. The FDA, EPA, and USDA are all subject to the National

---

[90]United States Department of Agriculture. Available at www.usda.gov.

Environmental Policy Act (NEPA),[91] requiring that all Federal agencies consider the consequences of their proposed actions on the environment prior to decision making[92] by submitting Environmental Assessments and Environmental Impact Statements.

The Cooperative State Research, Education, and Extension Service (CSREES), created by Congress through the 1994 Department Reorganization Act, is the component through which it participates as a USDA and agency representative on NSET responding to problems, such as: (1) improving agricultural productivity; (2) creating new products; (3) protecting animal and plant health; (4) promoting sound human nutrition and health; (5) strengthening children, youth, and families; and (6) revitalizing rural American communities.[93]

As a USDA agency partner in the NNI, the CSREES identifies opportunities and potentials to revolutionize agriculture and food systems through nanotechnology. In a national planning workshop, "Nanoscale Science and Engineering for Agriculture and Food Systems" in 2002, the USDA/CSREES developed a science roadmap, or strategic plan, with specific recommendations for implementing a new program in nanotechnologies.[94]

Proceedings from the workshop were drafted into a report highlighting priority research areas in nanotechnology complementary to and supportive of the goals of the CSREES. These include: (1) pathogen and contaminant detection; (2) identity preservation and tracking; (3) smart treatment delivery systems; (4) smart systems integration for agriculture and food processing; (5) nanodevices for molecular and cellular biology; (6) nanoscale materials science and engineering; (7) environmental issues and agricultural waste; and (8) education of the public and future workforce. The basic areas of nanotechnology having the potential to serve as an enabling technology for agriculture and food systems were concluded to be: (1) microfluidics; (2) BioMEMS; (3) nucleic acid biogineering; (4) smart delivery systems; (5) nanobioprocessing; (6) bioanalytical nanosensors; and (7) nanomaterials and bioselective surfaces.

In keeping with its mission, in October 2006, the CSREES supported the second annual "Nano4Food Conference"[95] for the purpose of (1) learning how nanotechnology could improve productivity and cost effectiveness; (2) better understanding current and future food safety concerns; (3) evaluating nanotechnology's market potential; and (4) acquiring the necessary knowledge to direct regulation.

---

[91]42 U.S.C. § 4321 *et seq.* (1969).

[92]Issues in the Regulation of Genetically Engineered Plants and Animals. Pew Initiative on Food and Biotechnology. April 2004.

[93]See footnote 92, at www.csrees.usda.gov/about/faqs.html.

[94]"Nanoscale Science and Engineering for Agriculture and Food Systems" submitted to CSREES/USDA, National Planning Workshop. November 18–19, 2002. Washington, DC, September 2003.

[95]Nano4Food Conference, Atlanta, GA, October 12–13, 2006. Available at http://www.csrees.usda.gov/nea/technology/events/nanotech_event_atlanta.html.

REGULATORY POLITICS BEHIND THE SCIENCE OF NANOTECHNOLOGY

As Justice Breyer concludes in *Breaking the Vicious Circle*:

> Since Congress created different safety regulatory programs at different times, under different circumstances, with differing statutory language, administered by different agencies with different institutional environments, employing different scientists from different disciplines, involving different publics with differing degrees of interest, why should we not expect to find inconsistent treatment of health and safety risks and inconsistent results (particularly when complex rule-creation and rule-review procedures tend to freeze old rules in place)? ... Given the uncertainties and regulatory methods, are the "tunnel vision" (political pressures for a stricter regulatory solution) results surprising?[96]

Certainly not! Statutes are laws made by legislatures, and agencies make law by writing procedures and regulations for the laws to be implemented. The Administrative Procedure Act (APA)[97] governs the processes of Federal administrative agencies. It offers interested stakeholders the right to petition an agency for the issuance, amendment, or repeal of a rule, as well as standards for the judicial review of agency actions. The APA incorporates the Freedom of Information Act (FOIA)[98] making government documents public (except those dealing with national security, works in progress, enforcement confidential information, and classified trade secrets) upon request. Through the Negotiated Rulemaking Act (NRA) and the Administrative Dispute Resolution Act (ADRA), the APA provides for alternative processes for resolving differences. It additionally includes the Regulatory Flexibility Act (RFA) requiring that the needs and concerns of small enterprises be considered when rulemaking, and the Congressional Review Act (CRA) mandating that all agency rules be submitted to Congress prior to effectuation.

Given such, the Federal government continues to place a bandage on the patchwork of current laws, statutes, and regulations in the hope that those earlier enacted will suffice and seamlessly apply today to nanotechnology or a next-generation innovation. The result is a mountain of unmanageable and contradictory interpretations of precedence controlled by multiple, disconnected jurisdictions within a bureaucracy of stagnant, inflexible, territorial, obsolete, dysfunctional, and nonadaptive models. Add to that the fact that federal and state courts apply different rules to the admission of scientific evidence, the determination of who is an "expert" in the field, and what constitutes "expert testimony."

> History is, in large measure, the study of change. Justice requires that the legal system operate on the basis of reality, not dysfunction justified by theory. The reality is that

---

[96]Breyer, Stephen. *Breaking the Vicious Circle: Toward Effective Risk Regulation*. Massachusetts: Harvard University Press, 1993, pp. 51.
[97]5 U.S.C. Chapters 5–8.
[98]5 U.S.C. § 552 (1966).

human experience has grown far more diverse, and knowledge far more complex and specialized, than even 25 years ago. The present system is the product of a particular era, with its own requirements. In many important ways our needs are different from theirs. And their reforms have given rise to our dysfunctions. The present civil justice system . . . is a product of the needs of the Industrial era, and unsuited to the 21st Century. The civil justice system, like every other institution, must adapt its workings to the forces which have converged upon us: computerization, globalization, ever-increasing complexity, and accelerating change.[99]

While technological change is encouraged and supported by the Federal government, fundamental progress in accelerating innovation is strategically stifled by bureaucratic controls, such as those found in export control laws, as well as in immigration laws, frequently hindering competitiveness, and impacting bottom-line economic prosperity.

Combining the known substance of nanotechnology, with regulatory procedural fairness, within a science-controlled and politically divisive administration, stirs conflict with inconsistent results. The ultimate outcome is a terse tension between those trying to create a better society through nanotechnology and next-generation innovation, and those imposing principles of constraint by crying "wolf." A balance must be brokered between protecting society's health and environment, and ensuring that the process of innovation is not unduly burdened by the anticipatory call for yet more government controls by unruly activists, disinformation experts, and intelligence manipulators who capitalize on an uncertain regulatory framework, while simultaneously ravaging and impairing collective socioeconomic opportunities and future successes.

In nanotechnology, it is not so much size, as scale, that is important. By manipulating matter on the atomic scale, optical, electrical, magnetic, and other characteristics of materials change.[100] Acknowledged as an international phenomenon that is not yet an industry, nanotechnology is a collection of tools and approaches that integrates with other technologies to provide new products, devices, systems, and applications[101] at that scale. It calls for an international system of cooperation and regulatory oversight, viewed from a full lifecycle perspective. The United States remains in danger of losing its worldwide leadership in nanotechnology due to the shortsighted, antiquated "vicious circle" framework of its regulatory landscape within this internationally diverse "flat world"[102] of quick sound bites, instant nanosecond messaging, and media convergent cyber juggling.

Still, it is important to maintain a proper perspective, move incrementally, and recognize that an aggressive call for regulation within the existing unsteady

---

[99]Katz, R. R. with Philip Gold. *Justice Matters: Rescuing the Legal System for the Twenty-first Century.* Washington: Discovery Institute. 1997.

[100]Miller, S. E. *A Matter of Scale: Nanotechnology's Novelty Poses Challenges to Patent Process. New York Law Journal.* Volume 232-Number 23, August 3, 2004.

[101]See footnote 100, p. 5.

[102]Friedman, T. L. *The World is Flat: A Brief History of the Twenty-First Century.* New York: Farrar, Straus and Giroux. 2005.

framework could aggravate and stifle the very development and promising opportunities of nanotechnology.[2] "Regulators must exercise as much care against unintended consequences as scientists because regulation leads to Frankensteinian results more often than does science."[103] And, "to the extent that science still plays an important role in most regulatory decisions, its role has become more suspect by those who find regulations burdensome or of questionable legitimacy."[34]

---

[103]Reynolds, G. H. Forward to the Future: Nanotechnology and Regulatory Policy. Pacific Research Institute, Executive Summary. November 2002.

CHAPTER 12

The European Approach to Nanoregulation

TRUDY A. PHELPS

INTRODUCTION

This chapter intends to provide a general overview of different aspects of regulation that are relevant to, or likely to affect, the European legislative framework being developed for materials or products made using nanotechnology. Included is a description of the current legislation, as well as how and why this may be amended in the future to accommodate the placing of nanoproducts on the market. Other topics that are relevant, such as the views of various stakeholder groups on their specific needs for regulation, standardization, the law concerning intellectual property (patents), and military uses of nanotechnology will also be explored. To a large extent, the process of determining suitable legislation is very much in its early phases of development.

European regulation is developed centrally, primarily in Brussels. Most of the laws that relate to regulation of products using nanotechnology (nanoproducts) are generated under one of the three types of the European Union (EU) law called "secondary legislation" and developed by the European Community. This legislation is binding on all Member States of the EU. The outcome of the legislative process is usually either a directive or a regulation. Directives have to be implemented by national legislation, each Member State writing its own national legislation based on the relevant directive. In practice, this means that the national implementations can be somewhat different between Member States, leaving them the freedom to achieve the intended result of the directive without specifying exactly how this is done. Regulations, in contrast, are given the immediate force of law in all Member States without change.

The process of determining suitable legislation is on going in Europe, both at the European level and at the Member State level. This chapter will focus more

Nanoscale: Issues and Perspectives for the Nano Century. Edited by Nigel M. de S. Cameron and M. Ellen Mitchell

specifically on the United Kingdom and the steps being taken in the United Kingdom to deal with nanoregulation, but these steps are being repeated in similar ways in other Member States within the EU.

THE LEGISLATIVE FOCUS

The government, European citizens, and society as a whole are rightly concerned about health, safety, and the environment. Regulations that will ensure health and safety for people, animals, and the environment are the top priority in Europe at the moment.

Treaty Articles 152[1] and 153[2] set out the policies concerning public health and consumer protection, and these require "a high level of human health protection," and that "consumer protection requirements ... be taken into account in defining and implementing other Community policies and activities."

But there are also other drivers for nanolegislation. The Lisbon Strategy (also known as the Lisbon Agenda or Lisbon Process) provided a strategy for economic reform to develop Europe into a knowledge-based society, declaring its goal of making the EU "the most competitive and dynamic knowledge-based economy in the world by 2010."[3] Nanotechnology, and other innovative technologies are seen to be key to fulfilling this ambition, and considerable effort and finance is being devoted to providing an optimum environment for nanocommerce to flourish. Therefore, the EU and its interested stakeholders (manufacturers, workers, financiers, insurers, consumers, and regulators) are concerned about the legal instruments that will be needed to make the wheels of knowledge-based commerce run smoothly, safely, and productively for the benefit of all the citizens.

Apart from legislation needed to ensure safety and promote commerce, there may be a requirement in the future for laws to address various ethical concerns that nanotechnology raises. For example, how can access to the promised societal benefits of nanotechnology be made available to all? There are concerns that the technology may be used in ways destructive to society (e.g., by the military or by terrorists). Some wonder whether nanotechnology will present society with entirely new dilemmas (e.g., can humans be somehow transformed into human machines through advances in technology, and if so, are legislative restrictions needed?).

In 2004, the European Commission's Directorate General for Health and Consumer Protection organized a workshop[4] to assess, among other things, the most suitable means of regulating nanotechnologies. Having considered a number of options

---

[1]Treaty establishing the European Community (Official Journal C325 of 24 Dec. 2002), Part Three—Community policies, Title XIII, Public Health, Article 152.

[2]Treaty establishing the European Community (Official Journal C325 of 24 Dec. 2002), Part Three—Community policies, Title XIII, Consumer Protection, Article 153.

[3]Available at http://www.euractiv.com/en/agenda2004/lisbon-agenda/article-117510.

[4]Nanotechnologies: A preliminary risk analysis on the basis of a workshop organized in Brussels on 1–2 March 2004 by the Health and Consumer Protection Directorate General of the European Commission, Available at http://ec.europa.eu/health/ph_risk/events_risk_en.htm.

from "do nothing" to "stop everything," the workshop concluded that the best option was to launch "an incremental process using existing legislative structures"—in other words, a legislative review of existing legislation to be followed by adjustments (when deemed to be prudent), and as evidence of need for change became available. This option included the need to issue recommendations, commission studies, promote risk assessment, encourage actions by existing institutions, and opt for a "minimalist, appropriate and proportionate regulatory intervention."

Following up on the initial workshop, the European Commission published its Action Plan[5] for nanosciences and nanotechnologies for Europe 2005–2009 (hereafter called the EU Action Plan), which is intended to provide a strategy to implement an integrated and responsible approach on nanotechnology at the EU level. The Action Plan has provisions to coordinate actions among the Member States to provide increased investment, the necessary infrastructures and human resources for essential research, and to ensure that the legal, ethical, and societal implications of nanotechnology are addressed.

To summarize, the main activities regarding nanoregulation in Europe are focused on providing for the health and safety of people, animals, and the environment, and on providing the necessary economic and legislative environment for safe and increasing global competitiveness. Although there is considerable talk and consultation concerning the legal, ethical, and societal implications of nanotechnologies, there are so far no concrete proposals from the European Commission on legislation in this area.

THE NEW APPROACH AND THE IMPORTANCE OF HARMONIZED STANDARDS

One type of legislative instrument that has been developed to remove the barriers to the free circulation of goods in Europe is the New Approach[6] to product regulation, which limits public intervention to what is essential and leaves business and industry the greatest possible choice on how to meet the public obligations.

A New Approach directive contains certain essential requirements, which set out the basic requirements for protection of health and safety that products must meet before they are placed on the market in Europe, but it does not say how these requirements are to be met. Whether or not they are met is assessed by an independent assessment body (a notified body). Approval by the notified body enables a manufacturer to affix a CE mark to its product, which is required before a product may be sold within Europe.

The European Committee for Standardization (CEN), the European Committee for Electrotechnical Standardization (CENELEC), and the European

---

[5]Nanosciences and nanotechnologies: An action plan for Europe 2005–2009, COM(2005) 243 final, Communication from the Commission to the Council, the European Parliament and the Economic and Social Committee.

[6]Guide to the implementation of directives based on the New Approach and the Global Approach, European Communities, 2000. Available at: http://europa.eu.int/comm./enterprise/newapproach/newapproach.htm.

Telecommunications Standards Institute (ETSI) have the task of drawing up technical specifications (standards) that describe how the essential requirements of the directives can be met. Such standards are referred to as "harmonized standards," and compliance with these is deemed to provide a presumption of conformity with the essential requirements contained in the New Approach directive.[7]

New Approach directives currently regulate a number of products, including medical devices, construction products, machinery, and explosives for civil uses; other legislation relies heavily on standards, such as the General Product Safety Directive. It has recently been suggested that the European Commission is considering amending the Cosmetics Regulation into a New Approach instrument. For this reason (among other reasons), the importance of development of harmonized and other European standards should not be underestimated.

WHO NEEDS WHAT?

Key to any discussion on regulation is to understand the regulatory needs of the various stakeholders. The Innovation Society[8] undertook a multistakeholder dialogue exercise in order to bring together a wide group of stakeholders to consult with them and discuss what is needed to provide for a successful, safe, and sustainable development and use of nanotechnology. The following information is taken from their report on Nano-Regulation:[9]

The summarized outcome provided the main points considered necessary for a structured and adjusted process toward a sustainable regulatory framework for nanotechnology. All stakeholders were agreed on the need for the following:

- Clear and consistent terminology (e.g., a definition for nano). Terminologies and standards are urgently needed, and CEN, along with the British Standards Institution (BSI) and the International Organization for Standardization (ISO), is working toward providing these. They are needed not only to make it possible for scientists and the public to know that they are talking about the same things, but also for legal reasons (e.g., patent applications).
- Risk data that will enable the regulatory measures needed—this needs to be evidence based and confirmed by research data.
- Review of existing legislations to identify regulatory gaps, particularly in the areas of occupational health and safety, product and consumer safety, and environmental safety.
- Adaptation of existing legislations, to fill the regulatory gaps, using a precautionary approach until such time as scientific evidence is available.

---

[7]Available at http://ec.europa.eu/enterprise/newapproach/standardization/harmstds/index_en.html.
[8]The Innovation Society is a Swiss research and consulting company focussing on business applications and economic impacts of nanotechnology. Available at www.innovationsociety.ch.
[9]Nano-Regulation—A multistakeholder-dialogue-approach toward a sustainable regulatory framework for nanotechnologies and nanosciences (March 2006), Christoph Meili, The Innovation Society, Ltd, St. Gallen, Switzerland.

- Development of *interim* guidance (e.g., on safe handling and safe production); evaluation of protective equipment (e.g., filters and fume cupboards); voluntary labeling of potentially hazardous materials and/or consumer products; compilation of inventories and databases of potentially hazardous nanoparticles[10] and applications; review of threshold values for declaration of presence of nanoparticles; development of standards for declaration/self-declaration; review of requirements for Material Safety Data Sheets; and life cycle analysis (LCA) studies of nanomaterials.
- Provision of authoritative information to enable producers and users of nanomaterials to take responsibility for the safety of their products.
- Cooperation among different stakeholders (e.g., provision of resources by the industry sector for needed research would be welcomed by governmental organizations and scientists). Mutual cooperation among stakeholders is seen as being crucial on the scientific, governmental, and economic levels, and needs to be strengthened at all levels, both regionally and internationally.
- Coordination and communication among the many national, regional, and international nanoprojects and initiatives, to avoid duplication and to make efficient use of time and money.
- Communication among experts and between experts and the public. Information provided by one group of stakeholders alone is seen as manipulative, so neutral communication channels need to be established.
- Upstream public engagement, to include communication about both risks and benefits and in order to prevent public distrust; this is difficult as public understanding and knowledge about nanotechnology is perceived as being limited. Accidents or unforeseen damages could influence regulatory processes in an unpredictable way.

Industry

A high safety standard in products and in processes is a key issue for industry, both to provide a high return on investment, and to prevent liability claims. Therefore, clear regulatory guidelines are crucial—ones that provide clear safety guidelines, while remaining as liberal as possible. "Overregulation" should be avoided in order not to dampen the nanotechnology innovation impetus. Legislation for nanoparticles is indispensable, requiring risk assessment data for handling, processing, and waste management. In addition, hazard classification of nanomaterials and substances should be provided. It is important that regulation is internationally harmonized. In terms of communication, industry sees the media as the most important aspect of framing the public attitude toward emerging technologies. One of the most feared risks of many industry representatives is "nano-bashing" by the media.

[10]Nanoparticles are tiny particles in the nano size range of 1–100 nanometers (nm), a nanometer being 1×10^{-9} m (1-billionth of a meter).

Government Agencies

These agencies see their main role as assessing and amending current regulations. To this end, harmonized testing procedures and risk assessment methodologies are key. Valid and standardized methodology for toxicity and ecotoxicity needs to be developed. There needs to be agreement on which are the important properties of nanoparticles to assess (e.g., size, surface reactivity, toxicity, exposure, and composition). It is important to these agencies that industry is integrated into the process of risk assessment and evaluation. International cooperation is also seen to be important.

Insurance Industry

As the insurance industry enables other stakeholders (e.g., industry) to take certain risks by guaranteeing payments in case of losses, accurate risk assessment and evaluation data are crucial. The insurance industry often lacks such data for emerging technologies. Therefore, risk assessment methodologies and hazard, exposure, and life cycle analysis data are required. A lack of technology awareness by the public is also perceived as a problem.

Retail Organizations

Retail organizations are similar to industry in that they are liable for the safety of the products they sell, so it is important that products containing nanomaterials are safe for consumers and the environment. Retail organizations often have to rely on external experts and knowledge. In addition, they require precise and clear regulations. The definition of the term "nano" and labeling is highly critical, as "nano" is often used for marketing purposes. Experience has shown that consumer acceptance often requires adequate information on a product, and sensitive products have to be appropriately labeled. A lack of information is a serious cause for distrust. The implications are that industry has an "information duty" to retailers, and should provide information not only to their retail partners, but also to consumers. It is also important that the consumer benefits are clearly communicated to the consumer. One key issue for retailers is the insurability of nanotechnology, as consumers perceive technologies that are not fully insurable as potentially dangerous.

Academia

Research scientists, on the whole, are well aware of the safety issues involved with research on nanoparticles. Nevertheless, research on risks is considered a high priority for academia, which should be done in collaboration with industry and according to a specific and coordinated research policy. In research and development, nanotechnology is regarded as a powerful tool for solving problems, obtaining new products, and improving processes. It is important that the risks are evaluated very carefully so that the risks can be weighed against the potential benefits.

Media

Although not specifically stated in the report, it would be safe to assume that the most important thing the media needs is accurate, nonbiased, scientifically based (where available), up-to-date information. All stakeholders have responsibility for providing this information.

STANDARDIZATION

Standardization is the process by which standards are developed. There are many different uses of the term standard or standards, but use in this chapter is restricted to those technical specifications, guides, and reports published by recognized standards development organizations (SDOs), such as British Standards (BSI), European Committee for Standardization (CEN), European Committee for Electro Technical Standardization (CENELEC), and European Telecommunication Standard Institute (ETSI).[11] Such standards are developed by consensus, with all relevant countries and stakeholders invited to take part. They indicate best practices at the time of development, are kept current through ongoing review, and are used by commercial and not-for-profit organizations (e.g., manufacturers and regulators) for competitiveness, safety, quality assurance, customer satisfaction and trade, and by government to support regulation. They are to be distinguished from standards that an individual company, organization, or national regulatory body might establish without broad consultation.

In Europe, standards developed by these SDOs have a status such that once a European standard is published all national standards in conflict with that standard must be withdrawn. This is to ensure that there are no unnecessary national barriers to trade within the EU due to differing standards being applied in different Member States.

The Need for Standards

Standards are of vital importance in order to ensure the quality and safety of products. In the field of nanotechnology, new standards are needed because nanotechnologies combine several other technologies in new ways. Not only are standards needed to provide the methods whereby safety and risk assessment can be carried out, but they are also needed to provide a foundation for the necessary research and resulting commercial applications, as well as to help create the public acceptance needed for the widespread adoption of these applications.

The starting point is the need for terminologies and definitions—people need to know that they are talking about the same thing. Exactly what is a "nanoproduct?" How is a nanoparticle measured? What are the important parameters of

[11]The international equivalent standardization organizations are ISO (International Standards Organization), IEC (International Electrotechnical Commision), and ITU (International Telecommunications Union).

nanoparticles (e.g., size, surface reactivity, chemical make up)? Methods of testing the toxicity of nanoparticles and assessing potential harm to the environment need to be developed. Standard procedures for safe handling, production, and distribution of nanomaterials need to be described.

Standardization is the foundation for any enterprise and nanotechnology is certainly no exception. Indeed, nanotechnologies introduce so many new and unique concepts that foundational standards are needed before many aspects of exploitation of the technology can progress.

European Standards Committee: CEN/TC 352 Nanotechnologies

In March 2006, the CEN Technical Board approved the establishment of a new committee CEN/TC 352 Nanotechnologies to be responsible for the development of standards specifically for nanotechnologies. It is envisaged that there will be strong cooperation in developing standards with the equivalent international standards committee ISO/TC 229 Nanotechnologies. BSI, the UK's National Standards Body, holds the secretariats and chairmanships for both of these committees. Because of the New Approach legislation in Europe and the need for harmonized standards to support this legislation, some standards may be developed in Europe that specifically address European legislative requirements. Specific tasks of CEN/TC 352 include: developing standards for classification, terminology and nomenclature; metrology and instrumentation; science-based health, safety and environmental practices; and nanotechnology products and processes.

CEN/TC 352 will provide European input into the work programme of ISO/TC 229, which held its first meeting in November 2005. The first work item is based on the BSI Publicly Available Specification PAS 71.[12] There is also a proposal to start work on a technical report on occupational safe practices regarding nanotechnologies.

The Role of Research in Standardization

Of course, it is not possible to standardize anything until the information on which the standard is based becomes available. Therefore, it is essential that the current investment in research to provide the basic information concerning methods of test, risk assessment methodology, toxicity tests, and so on, be translated into standards that are available for all interested stakeholders.

In implementing the EU Action Plan, the European Commission has provided extensive investment in research programmes relating to nanotechnology in its Framework Programmes for Research and Development. Research projects relevant to nanotechnology have been funded under both the fifth and sixth Framework Programmes (FP 5 and FP 6) and are due to receive substantially increased funding in the forthcoming FP 7 programme, which will provide funding for projects

---

[12]PAS 71: 2005 Vocabulary—Nanoparticles (ISBN 0-580-45925-X). Available at www.bsi-global.com/nano.

undertaken during 2007–2013. Indeed, it is part of the contract with researchers under FP 6 that contractors must inform the European standardization bodies about knowledge that may contribute to the preparation of European or other standards.

In relation to the FP 7 programme, CEN has recommended to the Commission that, since standards are effective in ensuring a wide-scale dissemination of the output of research, and are one key element for improved safety and increased competitiveness, the relevance for standardization should remain one of the criteria in the evaluation process of proposals for applied research.[13]

THE ROYAL SOCIETY REPORT AND THE GOVERNMENT RESPONSE

In June of 2003, the UK government commissioned the Royal Society and Royal Academy of Engineering to carry out an independent study into current and future developments in nanosciences and nanotechnologies and their impacts. Their report was published in July 2004,[14] and has become widely known as "the Royal Society report." Included in the remit was to identify what health and safety, environmental, ethical, and societal implications or uncertainties may arise from the use of nanotechnologies, both current and future. It was also to identify areas where additional regulation needs to be considered.

The Royal Society report concluded that almost all the concerns relate to deliberately manufactured (also called engineered) nanoparticles and nanotubes[15] that are free rather than fixed to or within a material. As a result, all the recommendations are specifically related to free, manufactured nanoparticles. The report concluded that there was no case for a moratorium on the laboratory or commercial production of manufactured nanomaterials. In addition, it found that the evidence suggests that present regulatory frameworks at EU and UK level are sufficiently broad and flexible to handle nanotechnologies at their current stage of development. However, some regulations will need to be modified, and a number of possible regulatory gaps were identified.

One of the recommendations was that all relevant regulatory bodies should consider whether existing regulations are appropriate to protect humans and the environment from potential hazards, and to publish their review and details of how they will address any regulatory gaps. In addition, a number of recommendations were made regarding specific legislation; these will be dealt with in the relevant sections below.

[13]CEN/STAR Recommendations for ERA (the European Research Area) and the 7th FWP (7th Framework Programme), Ref: SG CORR/11125, in letter to Mr. J. Potocnik, Commissioner for Research, European Commission dated December 6, 2004. Available at http://cordis.europa.eu/documents/documentlibrary/2483EN.pdf (accessed August 6, 2006).
[14]Nanoscience and nanotechnologies: opportunities and uncertainties. London: The Royal Society and The Royal Academy of Engineering, 2004. Available free of charge at www.royalsoc.ac.uk/policy and on www.raeng.org.uk.
[15]Nanotubes are a specific tube-like form of carbon nanoparticles.

The UK government published a response[16] (hereafter referred to as "the Government Response") to the recommendations in the Royal Society report in February 2005. The Government Response announced a new Nanotechnology Issues Dialogue Group (NIDG) to coordinate activities listed in the Government Response, one of the first of which was to carry out a cross-departmental review of current legislation with the Health and Safety Executive and the Medicines and Healthcare products Regulatory Agency. This review has already begun.

The Royal Society report made a number of specific recommendations for regulatory amendments to cover a number of aspects of health, safety, and the environment and in relation to different nanotechnology products. The Government Response answered the Royal Society report point by point and the following sections summarize these.

Most of the recommendations state that dialogue and cooperation with the European Commission will be necessary before coming to final conclusions, and so it can be assumed that these discussions will be broadly applicable to Europe as a whole.

INDUSTRIAL APPLICATIONS

A number of specific proposals were made in the Royal Society report that relate to different aspects of the industrial application of manufactured nanoparticles. In the Government Response, several "precautionary" recommendations were made as *interim* measures until there is sufficient scientific-based knowledge to establish whether nanoparticulate material is more toxic than the same material in the larger size range.

Health and Safety in the Workplace

Health and safety in the workplace in the United Kingdom is the responsibility of the Health and Safety Executive (HSE). Specifically, the HSE has responsibility for regulation of the health and safety hazards of industrial chemicals and the risks they pose in the work place. The UK regulations implement the several European directives, the main one being the Dangerous Substances Directive,[17] which contains classification and labeling rules for all substances as well as an obligation to notify certain substances.

Present requirements until June 1, 2008, are that "new" substances, that is, substances not recorded on the EINECS[18] database, should be notified. The question

---

[16]Government response to the Royal Society and Royal Academy of Engineering Report "Nanoscience and nanotechnologies: opportunities and uncertainties." London: DTI, HM Government (2005).

[17]Council Directive 67/548/EEC of 27 June 1967 on the approximation of laws, regulations, and administrative provisions relating to the classification, packaging and labeling of dangerous substances.

[18]EINECS (*E*uropean *I*nventory of *E*xisting *C*ommercial chemical *S*ubstances) database is a database of chemicals on the European Community market between January 1, 1971, and September 18, 1981. It is maintained by the European Chemicals Bureau. Any chemical marketed after September 18, 1981, is a "new" chemical.

arises whether manufactured nanoparticles are "new." All the evidence suggests that the behavior, including toxicity, of manufactured nanoparticles cannot be assumed to reflect that of the same material in larger sizes. However, there is no current distinction in the regulations based on size. The substance definition in the legal text does not distinguish between bulk substances and nanomaterials, as "substance means a chemical element and its compounds . . . ."[19] Particle size or intrinsic properties are not relevant for deciding whether the notification obligation applies. In light of the novel properties of manufactured nanoparticles and nanotubes, the Government Response recommended that they should be treated as new chemicals, and that industry should publish details of safety tests showing that the novel properties of nanoparticles have been taken into account. In addition, they recommended that nanoparticulate exposure (to workers) be minimized until the possible risks are better understood. They committed to undertake a review with the HSE to determine the adequacy of the current regulatory regimes to provide effective regulation of nanoparticles.

The HSE published the results of this regulatory review[20] in March 2006. The conclusion of this review was that the principles of the regulations and the interconnections between them are appropriate and applicable to nanomaterials. There was no need to fundamentally change the regulations themselves, or to introduce new regulations. However, there are some important areas in which there is insufficient information, notably in the toxicological and physicochemical hazards, the appropriate dose/exposure metric(s), the means of measuring exposures, the risks to health, and the effectiveness of control measures. The absence of such data means that all involved in the regulatory process will have great difficulty at present in confidently fulfilling their responsibilities within the various regulations.

Regarding notification of nanoparticles as "new" substances, the report suggests that, as a general rule, "top down" manufactured nanoparticles (those derived from breaking down a bulk material into nanosized particles, e.g., by grinding) are less likely to need to be notified; while "bottom up" nanoparticles (those constructed by building up individual atoms or molecules into a nanoparticle or nanostructure) are more likely to need to be notified. This is especially true for fullerenes and their derivatives (e.g., carbon nanotubes). Fullerenes are a particular type of carbon structure, but they are not included in the existing materials registry.[21]

A recent workshop[22] concluded that the current system for classification and labeling of substances and preparations is generally considered to be adequate, because different entries for the same chemical compound in different forms

---

[19]Dangerous Substances Directive (67/548/EEC).

[20]Review of the adequacy of current regulatory regimes to secure effective regulation of nanoparticles created by nanotechnology, HSE. Available at www.hse.gov.uk/horizons/nanotech, published March 2006.

[21]European Inventory of Existing Chemical Substances (EINECS).

[22]The 12th International Workshop on Quantitative Structure–Activity Relationships in Environmental Toxicology, 2006 (Poster) (accessed August 11, 2006). Available at http://ecb.jrc.it/home.php? CONTENU+/DOCUMENTS/QSAR/INFORMATION_SOURCES/PRESENTATIONS/Bassan_Lyon_ 0605_poster.pdf.

(e.g., metal block and metal powder) are possible. In addition, it was stated that the European Chemicals Bureau is planning to develop central databases for the EU on the chemical characteristics, potential hazards and environmental and human-health effects of nanomaterials.

However, when new substances are notified, information is to be provided on its identity and properties, hazards associated with its use, an assessment of potential exposure to the substance, and risk management; information is lacking on all of these for nanoparticles, and it is likely to be different for each type of nanoparticle.

Much of the current EU legislation will be superseded by new European legislation, referred to as Registration, Evaluation, Authorisation and Restriction of Chemicals (REACH), but until its registration provisions start applying (as of June 1, 2008), the notification scheme under Directive 67/548/EEC[23] will apply for new substances and notified substances with significantly new uses. REACH specifies a minimum trigger level for registration of substances at production or importation levels more than 1 ton/annum and per manufacturer or importer. The Government Response considered this to be too high a trigger level for nanoparticles, which may be produced or imported in a commercial setting at levels below this threshold. It was therefore suggested that sector specific regulations could be produced should the need arise.

The REACH regulation[24] was adopted on December 18, 2006, without stipulating a reduction of the minimum trigger level for nanoparticles, and without requiring nanoparticles to be considered "new" materials, as long as the material they are derived from is listed in EINICS. In addition, it is envisaged that a new Globally Harmonised Scheme (GHS) for classification and labeling of substances and preparations will be introduced in the EU in the next few years. For these reasons, it was suggested in the HSE regulatory review that Member States should concentrate their efforts on issues surrounding the treatment of nanomaterials in REACH (and GHS) rather than amending existing legislation, despite the fact that the Action Plan suggested that Member States should modify their national legislation (now) to take into account the specificities of nanosciences and nanotechnology applications and use.

REACH will enter into force gradually, starting on June 1, 2007, and will take a number of years to implement.

Explosive Hazard

One of the issues of relevance for nanoparticles, specifically nanopowders,[25] is the inherent explosibility of powders. Most organic materials, many metals, and even

---

[23]Council Directive 67/548/EEC of 27 June 1967 on the approximation of laws, regulations and administrative provisions relating to the classification, packaging and labeling of dangerous substances.
[24]Regulation (EC) No. 1907/2006.
[25]Powders composed of particles on the nanoscale range of approximately 1–100 nm. Other terms for nanopowders are "nanoparticles," "nanomaterials," and "ultrafine particles."

some nonmetallic inorganic materials will explode if finely divided and dispersed in air and if ignited by a strong enough ignition source. Explosive hazard is regulated by Directives 99/92/EC[26] and 94/9/EC.[27]

The demand for nanopowders is growing as the potential applications of nanotechnology increase and products begin to come on the market. Although a great deal of research is being done on the toxicological effects of nanoparticles, the potential hazard of explosibility has received little attention. The HSE commissioned a literature search to explore the use of nanopowders in industry and the potential explosion hazards.

The literature review was published in 2004.[28] It was reported that the upper size limit for the formation of an explosive dust cloud is the order of 500 μm. The general trend is for the violence of the dust explosion and the ease of ignition to increase as the particle size decreases, though for many dusts the trend begins to plateau at particle sizes of the order of tens of microns (μm). No lower particle size limit has been established below which dust explosions cannot occur. The main findings were that there is an increasing range of materials capable of producing explosive dust clouds being produced as nanopowders. At the same time, new uses of nanopowders are adding to the demand. The production of these is likely to increase in the future.

The literature search revealed no data for nanopowders. It was considered that the extrapolation of data for larger particles to the nanosize range cannot be carried out with confidence, due to the change in chemical and physical properties of particles below sizes of approximately 100 nm.[29] The report of the literature search recommended that the explosion characteristics of a representative range of nanopowders be determined using the standard apparatus and procedures already used for assessing dust explosion hazards, and that these be compared with data for micron-scale powders of the same materials. This should allow knowledge of particle size effects to be extended into the nanosize range, although care will have to be taken to ensure that the nanoparticles do not clump together (agglomerate) in the test vessel, thus producing larger than normal particles.

Environment

The Royal Society report recommended that manufactured nanoparticles and nanotubes should be treated as if they were hazardous and that the use of these in

[26]Directive 1999/92/EC of the European Parliament and of the Council of December 16, 1999 on minimum requirements for improving the safety and health protection of workers potentially at risk from explosive atmospheres.
[27]Directive 94/9/EC of the European Parliament and the Council of March 23, 1994 on the approximation of the laws of the Member States concerning equipment and protective systems intended for use in potentially explosive atmospheres.
[28]Literature review—explosion hazards associated with nanopowders (HSL/2004/12). Author(s): D K Pritchard, BSc, PhD, CChem, MRSC. Available at http://www.hse.gov.uk/research/hsl_pdf/2004/hsl04-12.pdf#search = 'HSE%20explosion%20hazards%20associated%20with%20nanopowders%20%28HSL%2F2004%2F12%29' (accessed October 18, 2006).
[29]The nanosize range is usually thought of as being between 1 and 100 nm.

environmental applications (e.g., remediation)[30] is prohibited until appropriate research has been undertaken and it can be demonstrated that the potential benefits outweigh the potential risks. The Government Response supported this and agreed that nanoparticulate releases to the environment from the workplace be minimized until the possible risks are better understood. It was also suggested that there should be liaison between researchers monitoring airborne manufactured nanoparticulates and those monitoring pollutant nanoparticles from vehicle emissions.

The government committed to undertake a detailed and ongoing review of the manufacture and uses of the products of nanotechnologies in order to ensure that there is clear information identifying any inputs to the environment. This review is described below under Voluntary Reporting Scheme. Further work has been commissioned by HSE to identify suitable test protocols.

End of Life and the Waste Stream

The Royal Society report recommended that manufacturers of products covered by extended producer responsibilities regimes, such as end-of-life regulation, should be required to publish procedures outlining how manufactured nanoparticles will be managed to minimize human and environmental exposure. This was suggested because it was thought that the most likely time of release of any nanoparticles from the materials to which they have been fixed would be greatest during disposal, destruction, or recycling. The Government Response to this proposal was to point out that existing EU Directives covering extended producer responsibilities, such as the End of Life Vehicles Directive,[31] already deal with the treatment of materials including those presenting special hazards at the end of life of products. Incorporation of the Royal Society's recommendation would require extensive consultation to agree which materials to include and the format for published procedures.

However, as a precautionary measure, the government has asked industry to reduce or remove waste stream discharges containing manufactured nanoparticles and nanotubes; the government undertook to work in partnership with industry to help implement this request.

Voluntary Reporting Scheme

In the Government Response to the Royal Society report, the government committed to review the manufacture and uses of nanomaterials, particularly free manufactured nanoparticles, in order to ensure that there is clear information identifying any inputs to the environment. This was assigned to the Department for Environment, Food and Rural Affairs (Defra). A research study was undertaken to identify what materials were being manufactured or imported into the United Kingdom. In addition, Defra devised a Voluntary Reporting Scheme (VRS), the goal of which is to

---

[30]"Remediation" is the cleanup of an environmentally contaminated site. Nanofilters are able to remove biological toxins and toxic metals from water, and nanoremediation is a promising application of nanotechnology to environmental problems.
[31]Directive 2000/53/EC.

provide an evidence-based overview of the current UK production and use the life cycle of nanoparticles, with a view to eventually developing appropriate controls where necessary. A consultation took place whereby UK companies, universities, associations, and interested organizations (e.g., Greenpeace and Friends of the Earth) were asked to comment on a proposed set of questions. A standardized data submission sheet has been prepared and the United Kingdom participating companies and organizations will be asked (voluntarily) to provide data as on-going information to Defra over a period of 2 years, starting September 2006.[32] Defra has decided to include manufacturers, users, importers, researchers, and waste managers in the scheme and is encouraging them to assist in this activity as full partners in the regulatory process.

The EU Action Plan called upon Member States to "make inventories of use and exposure of nanosciences and nanotechnology applications, in particular, manufactured nanoscale entities." The information, collected by Defra, will be fed into the European regulatory process as it is collected and analyzed.

Consumer Products

In the Government Response, the UK government agreed with the Royal Society recommendation that ingredients in the form of manufactured, free nanoparticles should undergo a thorough safety assessment by the relevant scientific advisory body[33] before they are used in consumer products. The means by which this would take place is to be discussed and agreed at the European level. It was also suggested that the testing methodologies used by industry to assess the safety of free, manufactured nanoparticles used in consumer products should be disclosed to regulators.

The government also thought it would be useful to consider whether the presence of nanoparticles needs to be included in the list of ingredients on the labels of consumer products, because, without this, the consumer would be unable to make a fully informed choice.

Food and Food Packaging

As part of the regulatory review in the United Kingdom, the Food Standards Agency (FSA) reported its initial results in May 2006.[34] The goal of the draft report was to identify potential gaps in regulation or risk assessment relating to the use of nanotechnologies and the potential deliberate or adventitious (accidental) presence of

---

[32]Defra consultation on a Voluntary Reporting Scheme for engineered nanoscale materials—Summary of findings and government's response, August 2006. Available at http://www.defra.gov.uk/corporate/consult/nanotech-vrs/nanotechvrs-consultfindings.pdf.

[33]Each UK regulatory authority has an independent scientific advisory body to give advice relevant to the topic concerned (e.g., there are advisory committees for carcinogenicity, mutagenicity, and toxicity of chemicals in food, consumer products, and the environment).

[34]Draft report of FSA regulatory review of the use of nanotechnologies in relation to food, March 2006. Available at http://www.foodstandards.gov.uk/Consultations/ukwideconsults/2006/nanotech.

manufactured nanomaterials in food. A public consultation on the draft report ended in July 2006. The FSA concluded that on the basis of current information, most potential uses of nanotechnologies that could affect the food area would come under some form of approval process before being permitted for us. For example, the Novel Foods Regulation (EC) 258/97 establishes a mandatory premarket approval system for all novel foods and processes. In addition, all permitted food additives have to be assessed for safety by the independent Scientific Committees that advise the European Commission. However, there are currently no specific criteria to consider particle size under the EC food legislation. The FSA has called for research into potential applications of nanotechnology for food additives, to help identify how near to market any developments are.

The FSA review, therefore, did not identify any major gaps in regulations, but points out there are major gaps in information for hazard identification. The knowledge gaps, as in other areas, are to be discussed and outcomes decided at the EU level. The European Food Safety Authority (EFSA) is the European authority responsible for regulation of food, and any new nanomaterials would need to undergo safety assessments by the EFSA before they were included on the relevant positive list of approved food additives.

The Food and Drink Federation (FDF) published their response[35] to the FSA review. Their opinion was that food products should be evaluated for their safety as they are presented to the consumer, not for the process by which they were made. They, therefore, would not support specific labeling of products of nanotechnologies (as was suggested in the Government Response to the Royal Society report). The FDF pointed out that many nanosized particles in food are naturally occurring, therefore definitions in describing what is meant by "nanoparticle" for the purposes of regulation will be key. In addition, any reference to particle size as a requirement for specific review or assessment should relate only to those nanoparticles that are intentionally manipulated for a specific purpose. In relation to the admitted knowledge gaps, they believed that concerted action is needed at the EU level to identify safety assessment and data requirements to maintain confidence in the system. They pointed out that the current Regulation[36] relating to the possible migration of "nanocomponents" into food does not currently make any obligation on the suppliers of food contact materials to inform their customers of the nature and amount of such migrations. A number of European regulations relating to food are currently under review.

Cosmetics and Sunscreens

Cosmetics and sunscreens are regulated under the Cosmetics Directive 76/768/EEC.[37] Under the cosmetics regulations in the EU, ingredients (including those in

[35]FDF response to FSA draft report of regulatory review of the use of nanotechnologies in relation to food. Available at http://www.fdf.org.uk/responses/fdf_response_nano.pdf.
[36]Regulation (EC) 1935/2004 of the European Parliament and of the Council of 27 October 2004 on materials and articles intended to come into contact with food.
[37]Council Directive 76/768/EEC of 27 July 1976 on the approximation of the laws of the Member States relating to cosmetic products.

the form of nanoparticles) can be used for most purposes without prior approval, as long as they are not on the list of banned or restricted-use chemicals and that manufacturers declare the final products to be safe. It was for this reason that the Royal Society report recommended that free nanoparticles should undergo a safety assessment by the relevant scientific advisory body before they are used in consumer products. By far, the most commonly used nanoparticle in cosmetics and sun creams is titanium dioxide. The nanoparticulate form of titanium dioxide was given a favorable opinion [as a chemical used in ultraviolet (UV) filters] by the SCCNFP,[38,39] but apparently insufficient information had been provided for zinc oxide. The Government Response states that the European Cosmetic, Toiletry and Perfumery Association (Colipa) is compiling the additional dossier on microfine zinc oxide, after which it will be submitted to the SCCNFP. Scientific data on the risk of absorption of nanoparticles through the skin (dermal route) is still being generated and debated.

Medicinal Products (Medicines) and Medical Devices

In Europe, medicines and medical devices are regulated separately. At present, therefore, nanomedicinal products and nanodevices (medicinal products and medical devices produced using nanotechnology) are regulated according to whether they are deemed to be medicinal products or medical devices. There is currently no distinction made on the basis of the type of technology used to produce them.

The regulations for both medicinal products and devices require the manufacturer to carry out an analysis of the risks associated with them, to eliminate or reduce the risks to as low as reasonably possible, and to assess the balance of risks versus benefits. Particular attention must be paid to the chemical, physical, and biological properties of the materials used with regard to toxicity and biocompatibility with tissues, cells, and body fluids. Products are monitored through post market surveillance and other vigilance activities.

Medicinal products are regulated by Directive 2001/83/EC,[40] which defines a medicinal product as follows:

Medicinal Product

Any substance or combination of substances presented for treating or preventing disease in human beings.

Any substance or combination of substances which may be administered to human beings with a view to making a medical diagnosis or to restoring, correcting or modifying physiological functions in human beings is likewise considered a medical product.

[38]Scientific Committee on Cosmetic Products and Non-Food Products.
[39]Opinion concerning Titanium Dioxide, Colipa n S75 adopted by the SCCNFP during the 14th plenary meeting of October 24, 2000. Available at http://ec.europa.eu/health/ph_risk/committees/sccp/docshtml/sccp_out135_en.htm.
[40]Medicinal Products Directive 2001/83/EC, Council Directive of 6 November, 2001. Available from: http://ec.europa.eu/enterprise/pharmaccuticals/eudralex/vol-1/consol_2004/human_code.pdf.

Medical devices are regulated by three different New Approach directives,[41-43] depending on the type of device. A medical device is defined as follows:

Medical Device

instrument, apparatus, appliance, material or other article, whether used alone or in combination, including software necessary for its proper application intended by the manufacturer to be used for human beings for the purpose of:
-diagnosis, prevention, monitoring, treatment or alleviation of disease,
-diagnosis, monitoring, treatment, alleviation of or compensation for an injury or handicap,
-investigation, replacement or modification of the anatomy or of a physiological process,
-control of conception,
and which does not achieve its principle intended action in or on the human body by pharmacological, immunological or metabolic means, but which may be assisted in its function by such means.

In practice, there is often a blurred dividing line between medicinal products and medical devices, and there are products that combine aspects of both (combination products). The dividing line is drawn on the basis of the intended action, with a medical device being one that does not achieve its principle intended action by "pharmacological, immunological, or metabolic means; for example, a wound dressing (medical device) with a silver nanoparticle component with an antimicrobial action (medicinal product).

With the advent of a plethora of innovative medical technologies on the horizon, including tissue engineering,[44] the regulatory regime is becoming more and more complicated. Many of the structures used in tissue engineering in the future are likely to use nanotechnology, since nanostructuring of surfaces has a profound effect on the way cells grow onto the scaffolds. A new draft regulation for advanced therapy medicinal products is being developed (the Advanced Therapy Medicinal Products, or ATMP regulation), which aims to establish harmonized rules for marketing human tissue engineered products (hTEPs), gene therapy products, and somatic cell therapy products. The current thinking is that nanomedicinal products and nanodevices would only be regulated under the ATMP regulation if they were also tissue engineered, gene therapy, or somatic cell therapy products. In other words, there is to be no stand-alone legislation addressing nanomedicinal products, nanodevices, and nanotissue-engineered products.

There is currently a regulatory review underway within a number of individual Member States, as well as within the European Commission, to determine if there

---

[41]Council Directive 93/42/EEC of June 14, 1993 concerning medical devices.
[42]Council Directive 90/385/EEC of June 20, 1990 on the approximation of the laws of the Member States relating to active implantable medical devices.
[43]Council Directive 98/79/EC of October 27, 1998 on *in vitro* diagnostic medical devices.
[44]Tissue engineering is the regeneration of biological tissue through the use of cells, with the aid of supporting structures (that may be mechanical scaffolds) and/or biomolecules. The cells used may be human or animal derived.

are any changes needed to existing legislation to accommodate nanotechnology. There is no doubt that a watching brief will be kept on the uses of nanotechnology in medicine and healthcare. The European Medicines Agency (EMEA) has circulated a paper[45] in which it is acknowledged that many novel applications of nanotechnology will span the regulatory boundaries between medicinal products and medical devices and that additional specialized expertise may be required for the evaluation of the quality, safety, efficacy, and risk management of such nano products. The Agency has created a new Innovation Task Force[46] to ensure EMEA-wide coordination of scientific and regulatory competence in the field of emerging therapies and technologies, including nanotechnologies, and to provide a forum for early dialogue with applicants on regulatory, scientific, or other issues that may arise from the development.

The view expressed to date is that existing medicines and medical device legislation is adequate to safeguard the public, however, there is probably a need for recommendations on how the legislation should be implemented and interpreted in relation to nanotechnology. There has been a suggestion that nanomedical devices containing free nanoparticles should be placed in the Class III (the highest risk) device category. There will also probably be a need to amend some of the harmonized standards relating to medical devices. For example, the Directive 93/42/EEC requires that there must be compatibility between the materials used in a medical device and biological tissues (biocompatibility), but it is not specified how that is to be demonstrated. There is a series of "harmonized" standards that set out how biocompatibility can be assessed. A manufacturer can therefore claim conformity with the essential requirement for biocompatibility by making a medical device according to these standards. Amending these standards to include new methods of test for nanotoxicity, therefore, would address concerns about nanotoxicity of nanodevices, without the need to amend the Directive itself. Similar amendments could also be introduced to the harmonized standards for medical device labeling, risk assessment, quality systems, and so on.

ETHICAL, LEGAL, AND SOCIETAL ISSUES

The Action Plan commits to ensuring that European Community funding of research and development in nanoscience and nanotechnologies continues to be carried out in a responsible manner, for example, via the use of ethical reviews, and lists possible ethical issues as nontherapeutic human enhancement by nanotechnology and invasion of privacy due to invisible sensors. The Action Plan also commits to ask the European Group on Ethics in Science and New Technologies to carry out an ethical analysis of nanomedicine to identify the primary ethical concerns and enable future ethical reviews of proposed research projects to be carried out appropriately.

[45]Reflection paper on nanotechnology-based medicinal products for human use, EMEA/CHMP/79769/2006, June 29, 2006.
[46]Announcement of formation of new task force available at: http://www.emea.eu.int/htms/human/itf/itfintro.htm.

Public Dialogue

Both the (UK) Royal Society report and the (EU) Action Plan make proposals for extensive public dialogue to discuss and debate both the benefits and risks of nanotechnologies, and to consider social and ethical issues expected to arise from the development of some nanotechnologies.

The Government Response highlights a 10-year Science and Innovation Investment Framework[47] for science and innovation that is expected to fund such dialogue. Within the principles established in the Government Response for this dialogue, it is to "Feed into public policy—with commitment and buy-in from policy actors" and to "Take place within a culture of openness, transparency, and participation." At the same time, however, it is clearly stated in the Scope that the dialogue is (only) to be focused on informing, rather than determining policy and decisions.

The Action Plan promises to create the conditions for and pursue a true dialogue with the stakeholders. It also plans to produce multilingual information material including films, brochures, and other internet-based material to raise awareness of the topic for different age groups.

Intellectual Property

The concerns of intellectual property are gradually coming into focus, raising questions as to whether and when nanoproducts are patentable. The use of nanosized entities in the natural world is among the most amazing and profound imaginable. So questions arise as to when a nanosomething is a natural object, and thus not patentable, and when is it a constructed object? The protection of intellectual property rights is essential for innovation both in terms of attracting initial investment and for ensuring future revenue. However, it is claimed that entrepreneurs are striving to claim patents over as many key nanotechnologies a possible, and that nano patents are likely to overlap and come into conflict, resulting in legal battles that benefit no one.[48] The concern was expressed that one fundamental nanopatent might dominate developments in many industrial sectors, and even enable the ownership of nature.

The Commission proposes to support the establishment of a nanosciences and nanotechnology Patent Monitoring System, for example, by the European Patent Office (EPO), as well as to support the harmonization of practices in the processing of nanosciences and nanotechnology patent applications between patent offices, such as the EPO, United States Patent and Trademark Office (USPTO), and Japan Patent Office (JPO).

---

[47]Science and innovation investment framework 2004–2014, Part 7 "Science and Society." Available at http://www.hm-treasury.gov.uk/spending_review/spend_sr04/associated_documents/spending_sr04_science.cfm.
[48]4th Nanoforum Report: Benefits, Risks, Ethical, Legal and Social Aspects of Nanotechnology. Part 7: the need for and rise of new legislation and regulation caused by the emergency of Nanotechnology, 2nd Edition—October 2005.

MILITARY USES OF NANOTECHNOLOGY AND OTHER SECURITY RISKS

Nanotechnologies have many potential uses in the military context; however, it is very difficult to see how these might be regulated. A preliminary risk analysis took place in an EU workshop[49] in 2004 on the risks to security of nanotechnologies. It was concluded that there was no immediate risk (to society); however, in the future there are potential dangerous applications of nanotechnologies that should be preventitively limited. These include nano-bioterrorism, the misuse of unmanned devices incorporating nanotechnologies and the possible social effects regarding implants, either for performance enhancement or as state monitors of humans.

A paper presented during the workshop by Jürgen Altmann highlighted the fact that nanotechnologies could be abused to invade privacy, damage property, injure, or kill people, or to harm the environment. Those responsible for abuse could be criminals, terrorists, enterprises, government agencies, and the armed forces. It was suggested that the military might proceed into research and applications before a broad debate in civilian society had determined which lines should be drawn and where. Altmann suggested that in the future the risks on the international level will need legislation, investigation and criminal prosecution on a similar level as within societies.

CONCLUSIONS

Considerable attention is being given to assessing the present regulatory frameworks in individual Member States and at EU level. The current thinking is that these are sufficiently broad and flexible to handle nanotechnologies at their current stage of development. However, some regulations will need to be modified and a number of possible regulatory gaps are being identified. It is not envisaged to enact specific legislation relating to nanotechnologies, though this may change as legislation is reviewed and amended.

The current focus is on making sure that health and safety in the workplace, safety of nanoproducts, and safety to the environment is maintained. There are many gaps in the methodologies that support legislation that is meant to assure such safety. The use of standardization as a means of publishing these terminologies, guides and recommendations is very important, and work has begun in Europe and internationally to develop these.

At this stage, much talk and some resources have gone into ensuring full stakeholder dialogue, including with the public, on ethical, legal, and societal concerns that may be raised by nanotechnology; however, at present, these are not being translated into any concrete suggestions for legislative change.

---

[49]Nanotechnologies: A preliminary risk analysis on the basis of a workshop organized in Brussels on 1–2 March 2004 by the Health and Consumer Protection Directorate General of the European Commission.

There is a widespread determination to ensure that revised regulations are coordinated at an international level and based on scientifically obtained evidence, where possible. In the meantime, *interim* measures will be implemented as a precaution, until a sufficient body of evidence and/or experience is obtained to determine what changes are needed.

CHAPTER 13

The Potential Environmental Hazards of Nanotechnology and the Applicability of Existing Law

GEORGE A. KIMBRELL[1]

INTRODUCTION

A woman shopping at a well-known department store's cosmetics counter purchases a high-end cosmetic product, a face cream. This particular face cream is one of several cosmetics on the market that happens to contain a new form of manufactured material made using nanotechnology. Like most of the public, the woman is unaware of nanotechnology, or that nanomaterials are being manufactured and inserted into consumer products. The woman applies her new face cream as directed, and it washes off in the shower. From her shower drain, the cosmetic product and its ingredients enter the household's waste stream, traversing the sewers beneath the woman's town and eventually make its way out into the tributaries and waterways surrounding the town. Once in the natural environment and water cycle, the nanomaterial separates and interacts with different elements in the aquatic environment, working its way up the food chain. Now imagine that this particular nanomaterial—known as carbon fullerenes (C_{60}) or more commonly as "buckyballs"—was found to cause brain damage to fish and be toxic to other aquatic life, as well as be toxic to human liver cells in low levels. Would she feel safe placing this material on her body, or comfortable having it wash off, knowing it would enter the waste stream and the environment? Would she wonder about the applicability of existing laws to address any possible risks to her health or the environment from this material?

Indeed, the small segment of the public that is aware of nanotechnology seems to focus on the promised future applications of nanoscience, such as cancer-curing drug

[1]Staff Attorney, The International Center for Technology Assessment, Washington, DC.

Nanoscale: Issues and Perspectives for the Nano Century. Edited by Nigel M. de S. Cameron and M. Ellen Mitchell
Copyright © 2007 John Wiley & Sons, Inc.

vector devices. Similarly, when picturing nanotechnology's risks, minds immediately conjure images of nanotechnology pioneer Eric K. Drexler's now-infamous "Grey Goo" scenario,[2] or the predatory nano-swarms of fiction writer Michael Crichton.[3] Perhaps it is human nature to focus on those visions that fuel the imagination. Nanotechnology's current reality, however, is equally compelling, but a bit more down to earth: the commercialization of the first nanomaterial-laced consumer products, from manufactured nanoparticles of zinc oxide and titanium dioxide used in sunscreens and cosmetics, to carbon nanotube-reinforced tennis rackets, to stain-resistant, nanomaterial-coated pants. These manufactured consumer products containing nanomaterials make up the first commercialization wave of nanotechnology in many industries. But just because there are no nanoswarms or grey goo does not mean that there are not serious concerns: environmental impacts are part and parcel of this nanomaterial manufacturing and commercialization process, and present difficult and unique challenges to our existing framework of environmental laws.

This chapter explores the potential environmental impacts of nanotechnology and analyzes the existing framework of environmental laws as applied to the regulation of nanomaterials. The first section recounts the current state of nanotechnology's commercialization, describes the foreseeable means by which nanomaterials are entering the natural environment, and explains why manufactured nanomaterials entering the environment represent a new class of manufactured pollutants. The second section analyzes the potentially harmful environmental impacts of nanomaterials. The third section surveys the landscape of our current environmental laws as applied to nanomaterials. Finally, the fourth section offers some general conclusions on the adequacy and problems of our existing legal framework as applied to nanotechnology and nanomaterials.

NANOTECHNOLOGY AND NANOMATERIALS: THE FUTURE IS NOW

Signs of nanotechnology's continuing maturation abound. Most pundits focus on the continuing surge in nanotechology[4] research and development

---

[2]Eric Drexler, *Engines of Creation* (1986).
[3]Michael Crichton, *Prey* (2003).
[4]The National Nanotechnology Initiative (NNI) defines "nanotechnology" as:

> the understanding and control of matter at dimensions of roughly 1 to 100 nanometers, where unique phenomena enable novel applications. Encompassing nanoscale science, engineering and technology, nanotechnology involves imaging, measuring, modeling, and manipulating matter at this length scale.

National Nanotechnology Initiative, Factsheet: What Is Nanotechnology? Available at http://www.nano. gov/html/facts/WhatIsNano.html. This chapter also uses the following definitions for other relevant nanoterminology:

> Engineered/Manufactured Nanoparticle: A particle of less than 100-nm engineered or manufactured by humans on the nanoscale with specific physicochemical composition and structure to exploit properties and functions associated with its dimensions and exhibits new or enhanced size-dependent properties compared with larger particles of the same material.

(R&D): federally funded research and investments coordinated through the National Nanotechnology Initiative (NNI) were approximately $1 billion in 2005,[5] with around $2 billion in yearly R&D investment spent by non-federal sectors e.g., states, academia, and private industry.[6] Global nanotech R&D is estimated at approximately $9 billion, with U.S. global spending estimated to reach $1 trillion by 2015.[7] But so much more is happening than merely R&D. In the media and in society, the buzzword "nano" is rapidly approaching ubiquitous status.[8] The nanopatent "gold rush"[9] continues, with a total of 3966 U.S. patents issued to date.[10] More than 1600 nanotechnology companies are currently operating in the United States and an estimated 20,000 people deal with nanomaterials in the workplace.[11]

Perhaps most surprising is the fact that nanotechnology commercialization is moving forward at lightening speed. Thousands of tons of nanomaterials are already being produced each year,[12] and consumer products containing nanomaterials are entering the market at a steady pace: more than $32 billion in products incorporating nanotechnology were sold last year, more than double the previous year.[13] Already, at least 300 nanoproducts are currently on U.S. markets, including cosmetics, sunscreens, wound dressings, fuel additives, sports gear, paints, foods, cleansers, stain-resistant clothing, cigarette filters, computer hardware, pesticides, and antimicrobal coatings for refrigerators, washing machines, and other appliances.[14] By 2014, $2.6

---

[5]See, e.g., External Review Draft *Nanotechnology White Paper* (hereafter EPA White Paper), at p. 11, Prepared for the U.S. Environmental Protection Agency by members of the Nanotechnology Workgroup, a group of EPA's Science Policy Council, Science Policy Council, U.S. Environmental Protection Agency, Washington, DC. (December 2, 2005). Available at http://www.epa.gov/osa/nanotech.htm.

[6]See footnote 5, p. 12.

[7]See footnote 5, p. 11.

[8]Genetic Engineering News, *Nanotechnology in $32 Billion Worth of Products; Global Funding for Nanotech R&D Reaches $9.6 Billion; Lux Research Releases The Nanotech Report, 4th Edition, the Indispensable Reference Guide to Nanotechnology*, May 8, 2006. Available at http://www.genengnews.com/news/bnitem.aspx?name = 1255070 (finding the use of "nanotech" rising 40% in 2005 media articles, to more than 18,000 citations).

[9]See, e.g., Charles Choi, *NanoWorld: Nano Patents in Conflict*, Wash. Times, April 25, 2005. Available at http://washingtontimes.com/upi-breaking/20050422-011739-1902r.htm; Nanodot.org, *Nanotechnology Gold Rush Yields Crowded, Entangled Patents*. Available at http://nanodot.org/article.pl?sid=05/04/22/181229.

[10]Genetic Engineering News, See footnote 7.

[11]Evan Michelson, Presentation, Woodrow Wilson Center's Project on Emerging Nanotechnologies, *Falling through the Cracks: Issues with Nanotechnology Oversight*, The Nanotechnology-Biology Interface: Exploring Models for Oversight, University of Minnesota, September 15, 2005.

[12]See, e.g., The Royal Society and the Royal Academy of Engineering, *Nanoscience and Nanotechnologies: Opportunities and Uncertainties*, London, July 2004, pp. 26–27 and Table 4.1. Available at http://www.nanotec.org.uk/finalReport.htm (hereafter Royal Society Report).

[13]Genetic Engineering News, See footnote 7.

[14]Howard Wolinsky, *Nanoregulation*, 7 European Molecular Biology Organization Reports 858, 859 (2006); see generally, Woodrow Wilson Center for International Scholars, Project on Emerging Nanotechnologies, *Nanotechnology Consumer Product Database*. Available at http://www.nanotechproject.org/44.

trillion in manufactured products will be nanoproducts, making up 15% of total global manufacturing.[15]

On the Loose: Manufactured Nanomaterials in Nature

The nanomaterials now being manufactured, marketed, and purchased are inevitably finding their way into the natural environment. Entry can occur accidentally or intentionally over the course of a nanomaterial's lifecycle, during manufacturing,[16] transportation, use, recycle, or disposal. The current wave of nanoproducts includes an inordinate number of sunscreens, cosmetics, and other personal care products, as the personal care industry is a leading sector in the manufacturing and marketing of nanoproducts.[17] These nanoproducts enter the environment via the household waste stream, as they are washed off in showers, or are directly transported from human skin into oceans, rivers, lakes, ponds, community, and private pools.[18] Other nanomaterials, such as those used in electronics, fuel cells, and tires, will be worn off or leak out over a period of use or after product disposal. Still other nanomaterials will reach the environment through landfills or other methods of disposal (e.g., residual sunscreens or cosmetics in containers). Finally, some nanomaterials may be introduced deliberately into the natural environment for environmental remediation purposes. For example, studies have indicated that iron nanoparticles could be used to clean up contaminated soil by neutralizing contaminants (e.g., DDT and dioxin).[19] As the many industries involved in nanotechnology expand, and increase the number and variety of nano-enhanced products available; both industrial and domestic nanowaste or nanopollution will logically increase in quantity, as well.

A New Class of Nonbiodegradable Pollutants

Once loose in nature, nanomaterials constitute a completely new class of manufactured nonbiodegradable pollutants.[20] "Nano" does not merely mean tiny, a billionth

---

[15]Genetic Engineering News, See footnote 7.

[16]Nanomaterial manufacturing can be done by either the "top-down" or the "bottom-up" method. Top-down manufacturing is the grinding or breaking down of a substance to the nanoscale, while bottom-up involves building materials through chemical synthesis including self-assembly. See generally Royal Society Report, See footnote 11, at 25 and Table 4.1.

[17]See generally, Friends of the Earth, Report, *Nanomaterials, Sunscreens and Cosmetics: Small Ingredients, Big Risks*, May 2006.

[18]Royal Society Report, see footnote 11, at 46 ("Any widespread use of nanoparticles in products such, as medicines (if the particles are excreted from the body rather than biodegraded) and *cosmetics* (that are washed off) will present a diffuse source of nanoparticles to the environment, for example through the sewage system. Whether this presents a risk to the environment will depend on the toxicity of nanoparticles to organisms, *about which almost nothing is known*, and the quantities that are discharged") (emphasis added).

[19]Ernie Hood, *Nanotechnology: Looking as We Leap*, 112 Envtl. Health Persp. A741, A744 (2004); see e.g., Zhang, W., *Nanoscale Iron Particles for Environmental Remediation: An Overview*, 5 J. Nanoparticle Res. 323–332 (2003).

[20]Humans and animals have been encountering naturally occurring nanomaterials for millions of years. Nature produces some nanoparticles, like salt nanocrystals found in ocean air or carbon nanoparticles emitted from fire. Thus, one could conclude that there is no danger in the nanoscale *per se*. However,

of a meter in scale; rather, "nano" is best understood to also mean *fundamentally different*. It is well known that materials engineered or manufactured to the nanoscale can exhibit different fundamental physical, biological, and chemical properties from bulk materials of the same substance.[21] One reason for these fundamentally different properties is that a different realm of physics, quantum physics, comes into play at the nanoscale.[22] Another is that the reduction in size to the nanoscale results in an enormous increase of the surface/volume ratio, giving nanoparticles a much greater surface area per unit mass compared to larger particles.[23] For example, a gram of nanoparticles has a surface area of $1000 \, m^2$. Because growth and catalytic chemical reactions occur at the particle surface, a given mass of nanoparticles will have an increased potential for biological interaction and be much more reactive than the same mass made up of larger particles, thus enhancing intrinsic toxicity.[24] Aluminum is an oft-cited example: inert in the bulk form used to make soda cans, it is highly explosive when manufactured on the nanoscale.

Just as the size and physics properties of engineered nanoparticles can give them exciting properties, those same new properties—tiny size, high surface area/volume ratio, high reactivity—can also create unique and unpredictable human health and environmental risks.[25] Swiss insurance giant Swiss Re noted that:

it is only recently that scientists have developed the techniques for synthesizing and characterizing many new materials with at least one dimension on the nanoscale. The concern is that nanomaterials now in development are different than anything that exists in nature. In fact, the very reason that nanotechnology is hyped so much is because it allows people to create products that do things that natural particles cannot. These new manufactured and engineered nanoparticles are patented for their novelty. Accordingly, the assessment of environmental and human health risks associated with nanomaterials is largely regarding the new materials that are being so formed and generated, the increased exposure levels from engineered nanostructures now being manufactured and marketed in greater and greater quantities, and the new routes/scenerios by which human and environmental exposure can occur with the current and anticipated nanomaterial applications.

[21] National Nanotechnology Initiative, *What Is Nanotechnology?* Available at http://www.nano.gov/ html/facts/whatIsNano.html. These properties include electrical, optical, magnetic, toxicity, chemical or photoreactive, persistence, bio-accumulation, and explosiveness, to name a few. Hood, see footnote 18.

[22] Nanotechnology Now, *Nanotechnology Basics.* Available at http://www.nanotech-now.com/basics.htm.

[23] See, e.g., Andre Nel et al., *Toxic Potential of Materials at the Nanolevel*, 311 Science 622–627, 622, 623 Fig. 1 (2006) (showing the inverse relationship between particle size and the number of surface expressed molecules). "In the size range < 100 nm, the number of surface molecules (expressed as a % of the molecules in the particle) is inversely related to particle size. For instance, in a particle of 30-nm size, about 10% of its molecules are expressed on the surface, whereas at 10 and 3 nm size the ratios increase to 20% and 50%, respectively. Because the number of atoms or molecules on the surface of the particle may determine the material reactivity, this is key to defining the chemical and biological properties of nanoparticles." See footnote 22, p. 623 (Fig. 1).

[24] See footnote 23; see, e.g., European Commission's Scientific Committee on Emerging and Newly Identified Health Risks (SCENIHR), Opinion on the Appropriateness of Existing Methodologies to Assess the Potential Risks Associated with Engineered and Adventitious Products of Nanotechnologies, p. 22 (adopted September 28–29, 2005) (hereafter SCENIHR opinion on existing methodologies); Warheit, D.D., *Nanoparticles: Health Impacts?*, 7 Materials Today 32–35 (2004).

[25] See, e.g., Nel, see footnote 22, p. 622, 623 Fig. 1; see generally Florini et al., *Nanotechnology: Getting It Right the First Time*, 3 Nanotechnology L. & Bus. 38, 41–43 (2006) (giving an overview of risks stemming from nanoparticles' inherent characteristics and the results of existing health and safety studies); *id.* at 622 ("[T]heir properties differ substantially from those bulk materials of the same composition, allowing them to perform exceptional feats of conductivity, reactivity, and optical sensitivity. Possible

Never before have the risks and opportunities of a new technology been as closely linked as they are in nanotechnology. It is precisely those characteristics which make nanoparticles so valuable that give rise to concern regarding hazards to human beings and the environment alike.[26]

As a result, nanomaterials present novel health and environmental risks that cannot be predicted from conventional materials. The European Commission's *Scientific Committee on Emerging and Newly Identified Health Risks* (SCENIHR): "Experts are of the *unanimous* opinion that the adverse effects of nanoparticles cannot be predicted (or derived) from the known toxicity of material of macroscopic size, which obey the laws of classical physics."[27] Thus, the "safety evaulations of nanoparticles and nanostructures cannot rely on toxicological and ecotoxicological profile of bulk material that has been historically determined."[28] Similarly, the *Institute of Occupational Medicine*:

> [b]ecause of their size and the ways they are used, they [engineered nanomaterials] have specific physical-chemical properties and therefore may behave differently from their parent materials when released and interact differently with living systems. It is accepted, therefore, that it is not possible to infer the safety of nanomaterials by using information derived from the bulk parent material.[29]

These new properties create numerous human health risks. For starters, due to their size, nanoparticles have unprecedented mobility: they are more easily taken up by the human body and can cross biological membranes, cells, tissues, and organs more efficiently than larger particles.[30] Once in the blood stream, nanomaterials can circulate throughout the body and can be taken up by the organs and tissues, including the brain, liver, heart, kidneys, spleen, bone marrow, and nervous system.[31] When inhaled, they reach all regions of the respiratory tract, and can move out of it via different pathways and mechanisms.[32] When in contact with

---

undesirable results of these capabilities are harmful interactions with biological systems and the environment, with the potential to generate toxicity").

[26]Swiss Re, *Nanotechnology: Small Matter, Many Unknowns*, (2004), p. 17. Available at http://www.swissre.com/INTERNET/pwswpspr.nsf/fmBookMarkFrameSet?ReadForm&BM=../vwAllbyID-KeyLu/ulur-5yaffs?OpenDocument.

[27]SCENIHR opinion on existing methodologies, see footnote 23, p. 6 (emphasis added).

[28]See footnote 22, p. 34.

[29]Tran et al., *A Scoping Study to Identify Hazard Data Needs for Addressing the Risks Presented by Nanoparticles and Nanotubes, Institute of Occupational Medicine Research Report* (December 2005), p. 34.

[30]See, e.g., Holsapple et al., *Research Strategies for Safety Evaluation of Nanomaterials, Part II: Toxicological and Safety Evaluation of Nanomaterials, Current Challenges and Data Needs*, 88 Toxicological Sciences 12 (2005).

[31]See, e.g., Oberdörster et al., *Nanotoxicology: An Emerging Discipline from Studies of Ultrafine Particles*, 113 Environmental Health Perspectives 823–839 (2005).

[32]See footnote 31, p. 837.

the skin, there is evidence of penetration of the dermis and subsequent translocation via the lymph nodes.[33] When ingested, systematic uptake can occur.[34]

Second, the change in the physicochemical and structural properties of engineered nanoparticles can also be responsible for a number of material interactions that could lead to toxicological effects. Once inside cells, they can interfere with cell signaling, cause structural damage, and cause harmful damage to DNA.[35] There is a dependent relationship between size and surface area and nanoparticle toxicity; as particles are engineered smaller on the nanolevel, they are more likely to be toxic.[36] Many relatively inert and stable chemicals (e.g., carbon) pose toxic risk in their nanoscale form.[37]

THE POTENTIAL ENVIRONMENTAL IMPACTS OF NANOMATERIALS

As with potential direct impacts on human health, various potentially damaging environmental impacts stem from the novel nature of manufactured nanomaterials. However, despite moving at light speed with nanomaterial applications and commercialization, only a few studies on the environmental impacts of engineered nanoparticles exist or are available in the public domain, leaving many potential risks dangerously untested. This is mainly due to the paucity of federal funding for environmental, health, and safety (EHS) research as compared to federal funding for nanotechnology's commercial applications: even according to government calculations, only a paltry 4% of the NNI's FY06 $1-billion budget was earmarked for EHS studies.[38] Other estimates put the EHS funding number as actually closer to 1%.[39] Organizations from diverse sectors have called on Congress to increase this number substantially.[40] Despite this lack of funding, the existing studies have raised some red flags for scientists, indicating the potential for

[33]See footnote 32.

[34]See footnote 32.

[35]See footnote 32; Oberdörster et al., *Principles for Characterizing the Potential Human Health Effects from Exposure to Nanomaterials: Elements of a Screening Strategy*, 2 Particle and Fibre Toxicology 8, at 1.0 (2005).

[36]See generally Tran, see footnote 28, p. 21.

[37]See, e.g., Nel, see footnote 22, p. 622 ("Thus, as particle size shrinks, there is a tendency for pulmonary toxicity to increase, even if the same material is relatively inert in bulkier form (e.g., carbon black and TiO_2").

[38]See, e.g., International Center for Technology Assessment, Congressional Letter on NNI 2006 Budget. Available at http://www.icta.org/doc/nano%20approp%20letter_Feb_2006.pdf.

[39]Woodrow Wilson International Center for Scholars, Project on Emerging Nanotechnologies, Press Release, *Nanotechnology Development Suffers from Lack of Risk Research Plan, Inadequate Funding & Leadership*, September 21, 2006. Available at http://www.wilsoncenter.org/index.cfm? topic_id = 166192&fuseaction = topics.item&news_id = 201894.

[40]See, e.g., Letter to Senate and House Appropriations Committees Urging Significant Increases in Nano EHS Appropriations, (February 14, 2006). Available at http://www.environmentaldefense.org/documents/5067_nano-appropsLetter.pdf (signed by a number of groups including, *inter alia*, the NanoBusiness Alliance and the Natural Resources Defense Council).

significant environmental impacts. In addition, nanomaterials' unique chemical and physical characteristics create foreseeable, yet unexplored, risks. These potential hazards further underscoring the need for dramatically increasing the federal EHS nanotechnology funding.

Mobility/Absorption and Transportation of Pollutants

Generally, nanoparticles have the ability to reach places that larger particles cannot.[41] Because of their tiny size, nanoparticles move with great speed through aquifers and soils, and settle more slowly than larger particles. Because of their large surface area, nanoparticles provide a large and active surface for absorbing smaller contaminants, such as cadmium and organics. In the soil, nanoparticles could bond with pollutants and transport them, causing the pollutants to be absorbed by soil layers in larger quantities and at a faster rate. The enhanced bonding and mobility characteristics of nanoparticles create a means by which ordinarily less mobile pollutants like fertilizers or pesticides could "hitch a ride" over long distances.[42] Further, because nanoparticles tend to be more reactive than larger particles, interactions with substances present in the soil could lead to new and possibly toxic compounds. Similarly, nanoparticles originating from industrial processes, consumer products, or other sources could easily be transported by runoff or rain to water bodies. Also, nanoparticles could provide a means for long-range transport of pollution in underground water, similar to colloids.[43]

The case study of engineered nanoparticles of iron, investigated as part of possible environmental remediation technology, illustrates these environmental impact concerns. Field tests have shown that the engineered nanoparticles remain active in soil and water for several weeks, and that they can travel in groundwater as far as 20 m. However, the impact that the high surface reactivity of engineered nanoparticles used for remediation might have on plants, animals, microorganisms, and ecosystem processes is unknown, as testing to determine the safety of these nanoparticles to environmentally relevant species has not yet been done. The basis of many food chains depends on the soil flora and fauna, which could be seriously impacted by injected manufactured nanomaterials. As a consequence, the UK Royal Society has recommended that the release of free manufactured nanoparticles into the environment for remediation be prohibited until more research is completed.[44] More generally, the Royal Society recommended that:

---

[41]For example, EPA White Paper, see footnote 4, p. 36–37 ("There are limited data on the fate and transport of nanoparticles, but existing data show that their behavior can be very different from much larger particles of the same material. Nanoparticles generally will be retained in the water column due to diffusion and dispersion.").

[42]EPA White Paper, see footnote 4, pp. 37, 40.

[43]Colvin, *Responsible Nanotechnology: Looking Beyond the Good News*, EurekAlert!: Nanotechnology in Content: Nov 2002. Available at http://www.eurekalert.org/context.php?context = nano&show = essays&essaydate = 1102. Naturally occurring colloids are particles that remain mobile in liquids because they do not form conglomerations and are not deposited.

[44]Royal Society Report, see footnote 11, p. 80.

until more is known about their environmental impact we are keen that the release of nanoparticles and nanotubes in the environment be avoided as far as possible. Specifically we recommend as a precautionary measure that factories and research laboratories treat manufactured nanoparticles and nanotubes as hazardous, and seek to reduce or remove them from waste streams.[45]

Durability/Bioaccumulation of Nanomaterials

Even if nanoparticles are "fixed" inside a product matrix,[46] nanomaterials are "highly durable" and will remain in nature long after the disposal of their host products.[47] The longevity of nanomaterials theoretically could create accumulation that could upset ecological balances, even if that particular nanomaterial is harmless to humans. One such example is manufactured nanoparticles of silver. Nanoparticles of silver are currently being used in a plethora of products, including washing machines and food packaging.[48] The same property that makes these nanoparticles attractive to manufacturers—their highly enhanced antimicrobial properties—can be highly destructive to ecosystems.[49]

Studies have also suggested that some nanomaterials will bioaccumulate in microorganisms and plants. For example, scientists have found that engineered nanoparticles of aluminum oxide slowed the growth of roots in at least five species of plants: corn, cucumbers, cabbage, carrots, and soybeans.[50] Seedlings can interact with the nanoparticles and stunt their growth.[51] Such nanoparticles are commonly used in coatings and sunscreens.[52]

Detection/Removal

Even simply detecting and measuring engineered nanomaterials in the environment is a new challenge created by their unique physical and chemical characteristics.[53] The methods and protocols needed are just beginning to be developed.[54] Most particle measurement technology was designed to function at the micron particle size level, many hundreds to thousands times larger than nanoparticles, and are not

---

[45]See footnote 44, p. 46.

[46]"Fixed" nanomaterials are immobilized in a solid matrix (for example, tennis rackets reinforced with carbon nanotubes). On the other hand, "free" nanoparticles are suspended in the liquid or cream. Free particles are more easily dispersed and more quickly spread around. Free particles are also a form more conducive to being absorbed by organisms. These types of particles make up the largest percentage of the known nanomaterials in the consumer product market, including cosmetics and sunscreens.

[47]Andrew Maynard, *Nanotechnology: A Research Strategy for Addressing Risk*, Woodrow Wilson International Center for Scholars, Project on Emerging Nanotechnologies, at 12 (July 2006).

[48]See nano FIFRA discussion, see footnote 49, pp. 31–33 and accompanying footnotes.

[49]See footnote 48.

[50]*Study Shows Nanoparticles Could Damage Plant Life*, Science Daily (November 22, 2005). Available at http://www.sciencedaily.com/releases/2005/11/051122210910.htm.

[51]See footnote 50.

[52]See footnote 13.

[53]Royal Society Report, see footnote 11, at 42; EPA Nanotechnology White Paper, see footnote 4, p. 42.

[54]See Section III.

effective at the nanometer scale.[55] Once detected, to remove them from water or air requires new filtering techniques. Nanoparticles pass through most available filters, such as those used to filter drinking water.[56] Extraction could pose a challenge because of the nanoparticles' strong adsorption properties.

The Case of Carbon Fullerenes

Of the scant existing research on nanomaterials' environmental impacts, one of the most well-known examples is the study of the effects of carbon$_{60}$ fullerenes (or buckyballs).[57] As noted in the example at the beginning of this chapter, fullerenes are found in commercial nanomaterial consumer products, including several face creams and moisturizers.[58] Significant lipid peroxidation was found in the brains of fish (largemouth bass) after exposure, demonstrating the toxic effects of these engineered nanoparticles on aquatic and possibly other organisms, given that the fish species is seen as a model for defining ecotoxicological effects.[59] Fullerenes also have been found to cause death in water fleas and water to become clear, due to nanoparticle interference with bacterial growth.[60]

Another study on fullerenes showed that they clump together in water to form soluble nanoparticles and persist up to 15 weeks, raising concerns of water as a vector for nanoparticle movement through the environment.[61] That 2005 study also found that, even in very low concentrations, fullerenes are toxic to soil bacteria, raising concerns about how they interact with natural ecosystems.[62] Unpublished studies by the same group of researchers at Rice University's Center for Biological and Environmental Nanotechnology (CBEN), showed that the nanoparticles could "easily be absorbed by earthworms, possibly allowing them to move up the food chain and reach humans."[63] Finally, low levels of exposure to fullerenes have been shown to be toxic to human liver cells.[64]

---

[55]EPA Nanotechnology White Paper, see footnote 4, p. 42.

[56]NRDC et al., Comments to EPA, Re: EPA Proposal to *Regulate Nanomaterials through a Voluntary Pilot Program*, Docket ID: OPPT-2004-0122, July 5, 2005, p. 7.

[57]Oberdörster, Manufactured Nanomaterials (Fullerenes, C60) Induce Oxidative Stress in the Brain of Juvenile Largemouth Bass, 112 Environmental Health Perspectives 10 (2004). Fullerenes are a relatively recently discovered new form of carbon, a hollow cluster of 60 carbon atoms shaped in spherical, ellipsoid, or tubular form. A carbon nanotube is a spherical fullerene.

[58]Bethany Halford, *Fullerene for the Face: Cosmetics Containing C60 Nanoparticles are Entering the Market, Even if Their Safety is Unclear*, Chemical and Engineering News, March 27, 2006. Available at http://pubs.acs.org/cen/science/84/8413sci3.html.

[59]Oberdörster, see footnote 54.

[60]Rick Weiss, Nanoparticles Toxic in Aquatic Habitat, Study Says, Wash. Post (March 29, 2004) at A2.

[61]Press Release Rice University's Center for Biological and Environmental Nanotechnology, *CBEN: Buckyball Aggregates are Soluble, Antibacterial*, (June 22, 2005). Available at http://www.eurekalert.org/pub_releases/2005-06/ru-cba062205.php.

[62]See footnote 61; see J. D. Fortner et al., *C60 in Water: Nanocrystal Formulation and Microbial Response*, 39 Envtl. Sci. Tech. 4307, 4307–4316 (2005).

[63]Brumfiel G., *A Little Knowledge . . .*, 424 Nature 246 (July 17, 2003) (citing Vicki Colvin, Rice University's CBEN director).

[64]Sayes C. et al., *The Differential Cytotoxicity of Water-Soluble Fullerenes*, 4 Nanotechnology Letters 1881–1887 (2004).

APPLYING EXISTING ENVIRONMENTAL LAWS TO NANOMATERIALS

The unique chemical and physical properties of nanomaterials present immensely difficult challenges for existing environmental statutes. No current U.S. laws or regulations are specifically designed to regulate nanotechnology. Nonetheless, the situation is not wholly unprecedented, as the history of law is said to be "a history of borrowing of legal materials from other legal systems and of assimilation of materials from outside the law."[65] A number of laws provide some basis for regulatory oversight of some aspects of nanotechnology's effects on the environment and human health. The Environmental Protection Agency (EPA) has varied regulatory authority over nanomaterials' environmental impacts pursuant to the Clean Air Act (CAA), Clean Water Act (CWA), the Toxic Substances Control Act (TSCA), and the Federal Insecticide, Fungicide, and Rodenticide Act (FIFRA). The Federal Food, Drug, and Cosmetic Act (FDCA) grants the Food and Drug Administration (FDA) purview of the impacts of many nanomaterial products, including drugs, food, medical devices, and cosmetics. The Occupational Safety and Health Administration (OSHA) has authority over workplace health and safety issues, including the manufacturing of nanomaterials and nanoproducts. Finally, the National Environmental Policy Act (NEPA) applies to all federal agencies, requiring them to address the environmental impacts of their actions.

The overall adequacy of these laws as applied to nanomaterials is being debated. The American Bar Association's (ABA) Section of Environment, Energy, and Resources completed a review of the main existing environmental laws in summer 2006, concluding that, in general, existing laws provide EPA with sufficient legal authority to deal with the risks of nanotechnology.[66] However, the ABA review also found the statutes problematic in application to nanomaterials in some cases, absent regulatory amendments or adjustments, as well as raising questions on how the statutes should be applied.[67] Of the existing environmental laws, TSCA is generally seen as the best-suited law for creating the primary regulatory framework for nanomaterials, and EPA is in the process of initiating a voluntary pilot program for nanomaterials pursuant to TSCA.[68]

On the other hand, in another 2006 report, J. Clarence Davies, a former EPA and Council on Environmental Quality official who, among other things, drafted the original version of TSCA, concluded that TSCA and the other environmental laws were fundamentally flawed when applied to nanomaterials, and provided, at best,

---

[65]A. Watson, *Legal Transplants: An Approach to Comparative Law* 22 (2nd ed. 1993) (quoting Roscoe Pound).
[66]American Bar Association, Section of Environment, Energy, and Resources, Nanotechnology Project, at http://www.abanet.org/environ/nanotech; Colin Finan, *Bar Association Says EPA Has Adequate Power to Regulate Nanotech*, Inside EPA (July 31, 2006).
[67]Finan, see footnote 63.
[68]See footnote 69 pp. 21–22 and accompanying footnotes.

a short-term solution to nanomaterial regulation.[69] Davies called for new legislation to create a nanospecific regulatory framework.[70] The following survey of the existing legal landscape as applied to nanomaterials is merely a simplified overview of these complicated and lengthy statutes, highlighting topics of particular interest, outstanding questions, and regulatory challenges.

Toxic Substances Control Act

Toxic Substances Control Act (TSCA) coverage is broad and flexible, encompassing all "chemical substances and mixtures," organic or inorganic.[71] Unlike other environmental laws it is not limited to one subset of nature (e.g., water, air). The EPA has a wide array of regulatory tools available under TSCA, including in some cases premanufacture review[72] and information-gathering powers.[73] The EPA can also promulgate regulations, *inter alia*: prohibiting or limiting a chemical's manufacturing or distribution, in general or for a particular use; requiring labeling; requiring testing and/or monitoring; and regulating the chemical's disposal.[74] Further, TSCA can arguably adjust to the fact that nanomaterials have fundamentally different characteristics that can create novel or unique regulatory challenges, as seen by EPA's regulation of genetically modified microorganisms pursuant to TSCA.[75] For these reasons, TSCA is seen by many commentators as the best candidate of existing statutes for the regulation of nanomaterials.[76]

On the other hand, TSCA has some fundamental structural problems. In general, it is "a weak regulatory instrument."[77] First, TSCA assumes that no information equals no risk. If EPA does not have enough information to evaluate the health and environmental effects of a chemical, it can prohibit or limit its manufacture only if the agency can show that the chemical may present an unreasonable risk.[78] This places onerous data and risk burdens of production on the agency.

---

[69]J. Clarence Davies, Report, Woodrow Wilson International Center for Scholars, Project on Emerging Nanotechnologies, Managing the Effects of Nanotechnology (March 2006). Available at http://www.wilsoncenter.org/events/docs/Effectsnanotechfinal.pdf.

[70]See footnote 69, pp. 18–23.

[71]See 15 U.S.C. § 2602(2)(A) (defining "chemical substance" as "any organic or inorganic substance of a particular molecular identity . . ."). TSCA was meant to fill the gaps in the existing regulation of chemicals, covering chemicals other than those used as drugs, food additives, cosmetics, fuel additives, and pesticides.

[72]15 U.S.C. § 2604.

[73]15 U.S.C. § 2607 (providing the power to impose on chemical manufacturers recordkeeping and reporting requirements, health and safety study submission, and notice of new risks).

[74]15 U.S.C. § 2605(a).

[75]See 59 Fed. Reg. 45526 (Sept. 1, 1994) (proposed rule including microorganisms produced by biotechnology under TSCA jurisdiction).

[76]Davies, see footnote 69, p. 10.

[77]See footnote 69, p. 12.

[78]15 U.S.C. § 2603(a) (A-B) (TSCA Section 4) (explaining that EPA may only require testing if it first determines that a chemical substance presents an unreasonable risk to human health and the environment or that the chemical is produced in substantial quantities, and that there may be substantial human or environmental exposure, and that there is insufficient data available to provide the data, and testing is

Furthermore, it creates a disincentive for manufacturers to generate information on the possible risks of a chemical. This structural deficit is particularly problematic in the context of nanotechnology, where the opposite—the urgent generation of more health and safety data is called for.[79]

Second, even if the agency has that information available to it, or can produce it internally (both big ifs when discussing nanomaterials), the TSCA has further barriers to regulation. Although EPA has various regulatory tools, adopting any new standards requires the agency to undertake lengthy, data-intensive rulemaking processes supporting the agency finding of unreasonable risk, including a requirement that EPA impose the least economically burdensome regulations to manage the risk.[80]

Third, any rulemaking must satisfy a high technical standard of judicial review if challenged in court.[81] The TSCA standard—"supported by substantial evidence in the rulemaking record"—is a more rigorous standard than the "arbitrary and capricious" standard of review imposed upon agency action under the CWA, CAA, and other environmental statutes,[82] and affords courts a "considerably more generous judicial review."[83] The EPA's inability to ban asbestos under TSCA illustrates this burden nicely.[84] The EPA did *10 years* of analysis in support of its proposed asbestos rule, and a federal Court of Appeals still struck down the rule, finding the agency's analysis insufficient.[85]

Finally, in addition to these general structural weaknesses, questions exist about TSCA's adaption to nanomaterials. The EPA's current TSCA chemical notification requirements exempt several categories of chemicals, including a low volume exemption for chemicals produced in volumes of 10,000 kg or less a year (or less than 11 tons a year) and a "low release/low exposure" exemption.[86] Applying such exemptions could be dangerous because, as discussed above, nanomaterials can exhibit dramatically higher levels of activity per mass unit than conventional materials.[87]

---

necessary to do so); 15 U.S.C. § 2605(a-b) (TSCA Section 6) (authorizing EPA to regulate existing chemicals or order more information from a manufacturer but only when it has a reasonable basis to conclude that the substance "presents or will present an unreasonable risk to health or the environment").

[79]See, e.g., Rick Weiss, *Nanotechnology Risks Unknown; Insufficient Attention Paid to Potential Dangers, Report Says*, Wash. Post, p. A12 (September 26, 2006); see also, footnotes 35–37, and accompanying text.

[80]See generally, 15 U.S.C. § 2605; Linda K. Breggin, ELI, *Securing the Promise of Nanotechnology: Is U.S. Environmental Law Up to the Job?* (2005). Available at http://www.elistore.org/reports_detail.asp?ID = 11116.

[81]15 U.S.C. § 2618(c)(B)(i) (creating the technical standard of judicial review for the act as "supported by substantial evidence in the rulemaking record").

[82]*Environmental Defense Fund v. EPA*, 636 F.2d 1267, 1277 (D.C. Cir. 1980).

[83]*Abbott Laboratories v. Gardner*, 387 U.S. 136, 143 (1967) overruled on other grounds *Califano v. Sanders*, 430 U.S. 99 (1977).

[84]*Corrosion Proof Fittings v. EPA*, 947 F.2d 1201 (5th Cir. 1991).

[85]See footnote 89 p. 1215 (concluding that EPA had failed to carry its burden because it did not consider all evidence and did not give enough weight to the statutory directive to promulgate the least burdensome regulation required to adequately protect the environment).

[86]40 C.F.R. § 423.5.

[87]This conclusion is consistent with the recommendation of the Royal Society on the production thresholds that trigger testing for new chemicals. See Royal Society Report, see footnote 11, at Recommendation 10, p. 6.

TSCA's Section 5: New Chemicals versus Existing Chemicals

The most discussed provision of TSCA is Section 5, which provides EPA with the power to assess the risks of both "new" chemical substances and "existing" chemical substances and impose restrictions on their manufacture and use.[88] "New" chemical substances are substances not already listed in EPA's TSCA inventory,[89] and they require a premanufacture notice (PMN) and review by EPA.[90] That premarket review in theory allows EPA to review and assess the potential risks of a new chemical before its commercialization (and if necessary limit or prohibit its release).

There is currently a debate about whether nanomaterials, as a class or individually, should be considered "new" or "existing" chemical substances by EPA.[91] The TSCA defines the term by reference to its "particular molecular identity."[92] Thus, because carbon in bulk form (e.g., graphite) and carbon at the nanoscale (carbon nanotubes or carbon fullerenes) have the same molecular structure, under a strict reading of the statute, one could argue that a nanomaterial substance is not "new" for purposes of TSCA regulation, if a bulk material counterpart is already listed.[93] However, this assumption ignores the obvious fact that nanomaterials can have radically different physical and chemical characteristics and properties than those generally assumed of their existing and already-listed bulk material counterparts, properties that can give them a very different risk profile. By definition,[94] engineered nanoparticles are developed (and patented) for their novel properties that differ significantly from those of the conventional material. Thus, in accord with the statute's purposefully broad reach and flexible wording,[95] one can also argue that nanomaterials should be considered new for purposes of TSCA regulation, even if bulk forms of the same material are already listed. Along this line, the UK Royal Society and Royal Academy of Engineering

---

[88]See 15 U.S.C. § 2604(a)(1) & (a)(2). For a detailed discussion, see ABA Section of Environment, Energy, and Resources, Regulation of Nanoscale Materials under the Toxic Substances Control Act, pp. 5–17 (June 2006) (hereafter ABA TSCA Nano Paper).

[89]15 U.S.C. § 2602(9) (definition of new chemical substance); 15 U.S.C. § 2607(b)(1) (TSCA Inventory) (requiring EPA to "compile, keep current, and publish a list of each chemical substance which is manufactured or processed in the U.S.").

[90]15 U.S.C. § 2604.

[91]Finan, see footnote 63; Colin Finan, *Environmentalist Urge EPA to Draft Nanotech Rule To Refute Industry Position*, Inside EPA, June 1, 2006.

[92]15 U.S.C. § 2602(2)(A).

[93]Indeed, without any guidance from EPA on this issue, at least one major nanomaterial manufacturer and supplier has classified its carbon nanotubes as registered under TSCA as the same as synthetic graphite in its material data safety sheet, despite the emerging toxicity data and related risks unique to nanotubes. CNI (2003). Buckytube MSDS. Carbon Nanotechnologies Incorporated. Available at http://www.cnanotech.com/download_files/MSDS%20for%20CNI%20SWNT.pdf.

[94]See footnote 3.

[95]See, e.g., H.R. Rep. No. 1341, 94th Cong., 2nd Sess. 10 (1976), reprinted in H. Comm. On Interstate and Foreign Commerce, Legislative History of the Toxic Substances Control Act (1976) at 418 (stating that the definition of chemical substances is "necessarily" broad); ABA TSCA Nano Paper, see footnote 85, at 10 (noting that "EPA has occasionally been inconsistent in including different physical forms of the same particular identity on the inventory").

recommended that nanomaterials be treated as new substances for regulatory purposes.[96]

Even if EPA concludes that a nanomaterial is not a "new" chemical substance, it can still regulate the "existing" substance as a "significant new use" under TSCA if it was put to a use that might change its effects (as a nanomaterial form would in some cases).[97] Once EPA issues a significant new use rule (SNUR), manufacturers intending to manufacture a chemical for a significant new use must then submit a premanufacture notice to EPA.[98] However, issuing such a rule requires the agency to complete notice-and-comment rulemaking in accordance with the Administrative Procedure Act (APA)[99] and promulgate a rule for the particular nanomaterial[100] or possibly a category of nanomaterials with similar physical, chemical, or biological properties.[101] This is a huge difference in burden of production and agency resources. For a new substance, the burden is on the manufacturer to provide EPA with health and safety data on the substance for its review. The EPA can also require further data if it chooses, and can restrict or prohibit the manufacture of the chemical if there is inadequate data upon which to make an evaluation of health and environmental effects.[102] On the other hand, in order to promulgate a new SNUR for a nanomaterial, EPA must develop the data supporting the rule's necessity and undertake a detailed rulemaking process, a process not realistically feasible if needed for each nanomaterial chemical.[103]

EPA's TSCA Voluntary Pilot Program

In the summer of 2005, EPA's Office of Pollution Prevention and Toxics (OPPT) began discussions regarding a potential voluntary pilot program for nanomaterials under TSCA,[104] and proposed a voluntary program that fall.[105] An interim Ad Hoc Working Group held meetings to further discuss issues pertaining to the voluntary pilot program.[106] The program would request that manufacturers of nanomaterials submit to EPA basic materials data like material characterization, hazard information, use and exposure potential, and risk

---

[96]Royal Society Report, see footnote 11, at Recommendation 10.

[97]15 U.S.C. § 2604(a)(B).

[98]15 U.S.C. § 2604(a)(1)(B).

[99]5 U.S.C. § 553.

[100]See 40 C.F.R. § 720.22.

[101]15 U.S.C. § 2625(c)(2)(A).

[102]15 U.S.C. § 2604(e).

[103]See Davies, see footnote 69, p. 10 ("The new use provision would not be a feasible method of regulating [nanomaterials] if each particular nanomaterial had to be subject to a SNUR [significant new use rule] because that approach would require an unrealistically large amount of time and resources").

[104]Nanoscale Materials; Notice of a Public Meeting, 70 Fed. Reg. 24574 (May 10, 2005).

[105] EPA White Paper, see footnote 4, p. 14; EPA, Interim Ad Hoc Work Group on Nanoscale Materials, National Pollution Prevention and Toxics Advisory Committee (NPPTAC), Overview of Issues for Consideration by NPPTAC, October 8, 2005.

[106]See footnote 105.

management procedures.[107] The voluntary program's goal would be, in part, to assist EPA in developing a permanent regulatory program for nanomaterials. One major disadvantage of such a voluntary program is, of course, the absence of any incentive for "bad actors," or those with risky products are not likely to volunteer to do health and safety testing and submit to EPA any information indicating risk.[108] Given the already advancing state of nanomaterial development and commercialization, the EPA voluntary program was sharply criticized by a coalition of consumer and environmental advocacy groups as "inadequate and inappropriate" for the regulation of nanomaterials.[109] The commenting groups recommended, inter alia, that: EPA use its TSCA authority to prevent the release of nanomaterials into the environment until more is known; declare all nanomaterials to be "new" chemical substances under TSCA that require premanufacture EPA notice and review; require toxicity testing for nanomaterials intended for commercial use; and develop a nanomaterial inventory and tracking system.[110] The EPA is not expected to finalize the TSCA voluntary pilot program until 2007.

Clean Air Act

The Clean Air Act (CAA) provides the framework for EPA's regulation of pollutants into the ambient air.[111] The purpose of the CAA is, inter alia, to "protect and enhance the quality of the air in order to promote the public and welfare .... "[112] This process includes identifying types of pollutants, characterizing risk of exposure, controlling the release of pollutants to the degree necessary to protect human health and the environment, and monitoring regulated entities compliance with preventing or mitigating pollutant release.

While EPA has the authority to regulate nanoparticulate air emissions under the CAA,[113] such application is problematic because of the fundamental differences between nanomaterials and larger particles.[114] Existing air pollution monitoring, modeling, sampling, analysis, and control methods were designed to identify, measure by mass, capture, and control larger particles that behave in predictable chemical and physical ways.[115] The statute's triggers rely heavily on monitoring, which will require new or adjusted protocols for nanomaterials. In addition, EPA

[107]See footnote, p. 5.
[108]NRDC et al. Comments on EPA White Paper, see footnote 53.
[109]NRDC et al., Comments to EPA, Re: EPA Proposal to Regulate Nanomaterials through a Voluntary Pilot Program, Docket ID: OPPT-2004-0122, July 5, 2005, p. 11.
[110]See footnote, p. 12–13.
[111]42 U.S.C. Chapter 85, §§ 7401–7671 et seq.
[112]42 U.S.C. § 7401(b)(1) (congressional findings and declaration of purpose).
[113]EPA White Paper, see footnote 4, p. 28.
[114]American Bar Association, Section on Environment, Energy, and Resources, *ABA SEER CAA Nanotechnology Briefing Paper*, at p. 12 (June 2006) (hereafter ABA CAA Nano Paper) ("[U]ntil measurement and modeling methods are developed for nanoparticles that take into account the unique nature of these pollutants, nanoparticulate emissions cannot be reliably measured, and their fate and transport in the atmosphere cannot be predicted.").
[115]ABA CAA Nano Paper, see footnote 111, p. 8; Finan, see footnote 63.

must be able to distinguish between types of nanoparticles in the air (including engineered nanoparticles, naturally occurring nanoparticles, or nanoparticles resulting from combustion sources). Even after such protocols are developed, the sophisticated monitoring and control technologies necessary may not be affordable. The same goes for pollution control and removal technologies; they too will require development of nano-specific methods, as the unique chemical and physical characteristics of nanoparticles make existing conventional control devices and techniques ineffective.[116] Adapting CAA's regulatory programs like the National Ambient Air Quality Standards (NAAQS)[117] and the Hazardous Air Pollutants (HAPs) listing,[118] to include and regulate nanoparticles would require a fundamental paradigm shift from reliance on mass/volume measurement and limitations.[119] This relationship is not sufficient for nanomaterials, where other measures of toxicity—surface area, chemical composition, and shape—come into play.

Finally, one area of particular interest is CAA Section 211, which provides EPA with the authority to regulate fuels or fuel additives and require certain information from manufacturers, including health-effects data.[120] Nanotechnology-derived or nanofuel additives are currently coming to market, like Oxonica's Envirox, which is composed of nanoparticles of cerium oxide only 10 nm in size, creating a large catalyst surface.[121] Cerium oxide is a lung irritant and may be a greater irritant in nanoparticle form based on the greater surface area and enhanced reactivity;[122] however, Oxonica has stated that its product "does not present a substantial risk to human health or the environment."[123] The EPA is reviewing Oxonica's fuel additive application, and that process will inform future EPAs regulatory policy and guidelines for analyzing other dispersive nanomaterials.[124]

Clean Water Act

The Clean Water Act (CWA) definition of pollution is extremely broad[125] and nanomaterials discharged into navigable water would very likely be subject to

---

[116]ABA CAA Nano Paper, see footnote 111, p. 14; Finan, see footnote 63.

[117]42 U.S.C. § 7408 (NAAQS).

[118]42 U.S.C. § 7409 (HAPS) No nanomaterials are not currently listed as HAPs (189 HAPS currently listed), but EPA could add a nanomaterial if it finds that it is known to cause or may be reasonably anticipated to cause adverse health or environmental effects.

[119]CAA Nano Paper, see footnote 111, pp. 17–18; Finan, see footnote 63.

[120]EPA White Paper, see footnote 4, at 29; see 40 C.F.R. Part 79 (health effects testing requirements for fuels and fuel additives).

[121]Azonano, Efficiency Trials for Oxonica's Nano Fuel Additive, Envirox. Available at http://www.azo-nano.com/details.asp?ArticleID = 31.

[122]Available at http://ptcl.chem.ox.ac.uk/MSDS/CE/cerium_IV_oxide.html; APA CAA Nano Paper, see footnote 111, p. 19.

[123]Colin Finan, Fuel Additive Could Offer First-Time Air Act Test for Nanotechnology, Inside EPA (July 11, 2006).

[124]See footnote 123.

[125]33 U.S.C. §§ 1362(6) (pollutant), (12) (discharge of a pollutant), (13) (toxic pollutant), (19) (pollution). Note that the definition of toxic pollutant is broad enough to include materials found to be harmful to aquatic life even if not humans.

regulation.[126] However, any CWA regulation of nanomaterials will entail both further scientific evidence and new technological advancements. First, it will be necessary for EPA to demonstrate that specific nanomaterials or a class of nanomaterials discharged has a potential adverse effect on human health or the environment.[127] Currently, determinations as to nanomaterials' toxicity, persistence, and degradability are all difficult to make do to limited data and resources; extensive research and data development is needed. The studies showing adverse impacts of carbon fullerenes on aquatic species are one such example.[128] Second, similar to nanomaterial air pollution regulation, any regulation under the CWA will necessitate the development of new technologies for accurate monitoring, measurement, and control of nanomaterials.[129] Thus, while CWS permits[130] apply to nanomaterials, to be regulated, the nanomaterial must be detectable, measurable, and conducive to treatment by technology-based limitations, which are all current problems.[131]

One possible alternative source of authority CWA Section 401, which requires that applicants for a federal license or permit for activity resulting in a discharge into the navigable waters obtain a certification from the State that the discharge complies with the state's water quality standards. Such standards normally prohibit the degradation of water quality and the impairment of beneficial uses. Depending on the particular state's water quality standards, states could assert that nanoparticle discharge violated the water quality standards—based upon existing evidence of impairment or perhaps the uncertainty of impacts due to the paucity of study—and should be prohibited.[132]

Resource Conservation and Recovery Act

Resource Conservation and Recovery Act (RCRA) provides EPA the authority to regulate the generation, transportation, treatment, storage, and disposal of solid or

---

[126]EPA White Paper, see footnote 4, p. 30; see also, American Bar Association, Section of Environment, Energy, and Resources, *Nanotechnology Briefing Paper Clean Water Act*, (June 2006) (hereafter ABA CWA Nano Paper), p. 4 ("[I]t can be assumed that all provisions of the Act dealing with the creation of and implementation of water quality standards will apply to the discharge of any form of nanoparticles to any water of the United States covered by the Act."). Potential CWA authorities include: effluent limitations for point sources, 33 U.S.C. § 1311(b)(1)(A) & (b)(2); national pollutant discharge and elimination system permits (NPDES), 33 U.S.C. § 1342(a)(1); and toxic and pretreatment effluent standards, 33 U.S.C. § 1317(a).

[127]In order to create water quality standards, EPA will need to create a database of effects of nanomaterials on water bodies, including, inter alia, toxicology studies, chemical effects, transportation data, uptake and bioaccumulation information, and other studies necessary to evaluate the possible adverse impacts of specific nanomaterials on the environment and people. Based on information showing harm, EPA could issue regulations for nanoparticles under 40 C.F.R. Part 129.

[128]See footnote 54, see footnote 127 and accompanying text.

[129]ABA CWA Nano Paper, see footnote 123, p. 4. "[U]ntil reasonable and effective monitoring technology is developed for nanoparticles, EPA may be limited to obtaining [] data . . . ." See footnote 129, p. 7.

[130]33 U.S.C. §§ 1311(a), 1342(a) (NPDES permits).

[131]ABA CWA Nano Paper, see footnote 123, p. 10. NPDES permits could provide some regulatory oversight in the interim by requiring source-specific special conditions, including the collection of discharge effects data and the undertaking of effects studies by dischargers. See footnote 130, p. 11.

[132]ABA CWA Nano Paper, see footnote 123, p. 9.

hazardous waste.[133] No current EPA RCRA regulations address nanomaterials, but EPA has noted that "[n]anomaterials that meet the definition of RCRA hazardous wastes would be subject to these regulations."[134] For RCRA to apply, nanomaterial wastes must have one of four listed hazardous waste "characteristics" or meet a listed waste description.[135] Most nanomaterial wastes will likely not qualify under these existing categories,[136] and as with the other environmental statutes already discussed, some of RCRA's assumptions will have to be adjusted to account for nanomaterial differences. For example, EPA's current assumptions for toxicity of waste may be problematic as applied because of built-in assumptions about the way in which waste is disposed (e.g., it may not fully assess how toxic wastes with nanomaterials might affect groundwater).[137] Similarly, RCRA's regulation of waste generators varies based on amounts generated annually.[138] Smaller generators have fewer requirements and can store waste on-site for longer periods of time. But nanomaterials with fundamentally different characteristics in relatively small quantities will likely change the risk analysis involved with equivalent volumes of bulk material waste, and EPA will likely have to adjust its regulatory standards to reflect that.

Further, some of RCRA's exemptions from its definitions of "waste" will likely pose problems if applied to nanomaterials. The RCRA's definition of hazardous waste exempts household hazardous waste;[139] yet, as noted earlier, a broad spectrum of household consumer goods composed of nanomaterials are already in the marketplace and entering the waste stream.[140] Thus, one avenue for the "uncontrolled release of nanomaterials into the environment will be the discarding of consumer goods that qualify as household hazardous wastes."[141]

Comprehensive Environmental Response, Compensation, and Liability Act

Comprehensive Environmental Response, Compensation, and Liability Act (CERCLA or the Superfund Law), provides EPA with the authority to address the releases or threatened releases of hazardous substances,[142] and nanomaterials that meet CERCLA's broad criteria would qualify.[143] The EPA can list any material as

---

[133]See generally, 40 C.F.R. Parts 260–279 (RCRA regulations for hazardous waste management).
[134]EPA White Paper, see footnote 4, p. 25.
[135]There are four characteristic criteria used to designate "hazardous waste": toxicity, corrosivity, reactivity, and ignitability.
[136]Florini, see footnote 24, p. 47 n. 42 (Feb./Mar. 2006).
[137]American Bar Association, Section of Environment, Energy, and Resources, RCRA Regulation of Wastes from the Production, Use, and Disposal of Nanomaterials, p. 8 n. 19.
[138]40 C.F.R. Part 262.
[139]40 C.F.R. § 261.4(b)(1).
[140]See Section I pp. 4–6, see footnote 139.
[141]ABA RCRA Nano Paper, see footnote 134, p. 11.
[142]42 U.S.C. § 9601 *et seq.*
[143]EPA White Paper, see footnote 4, p. 31. CERCLA defines "hazardous substances" in the broadest possible manner, including waste deemed hazardous pursuant to RCRA and other statutes. 42 U.S.C. § 9601(4).

a new hazardous substance if it concludes that it presents "substantial danger to the public health or welfare or the environment."[144] However, while the authority exists for the regulation of nanomaterials under CERCLA,[145] given the limited scientific study of the environmental and human health impacts of nanomaterials, it is unlikely any nanomaterials could be listed currently, and therefore be subject to CERCLA.[146] Further, CERCLA also labors under the same fundamental misconception as the previously discussed statutes; namely, it assumes larger quantities pose greater risk, which may not be valid with nanomaterials that cause toxic effects at low volumes.[147] Finally, while there is a similar lack of the necessary scientific and technical prerequisites for CERCLA to apply to nanomaterials as with the previous statutes discussed, this knowledge gap is not as worrisome in the CERCLA context. The EPA could classify nanomaterials as hazardous at some future time, and CERCLA would apply retroactively. Indeed, CERCLA was expressly drafted to deal with the adverse impacts of unanticipated previous activities, creating liability for past actions, which makes it a good safety net for the unanticipated consequences.[148] Then again, if nanomaterials now entering the environment cause unprepared for harm, it may be too late to take remedial measures.

Federal Insecticide, Fungicide, and Rodenticide Act

The Federal Insecticide, Fungicide, and Rodenticide Act (FIFRA) is the federal regulatory scheme for the manufacture, labeling, sale, and application of pesticides. The EPA has premarket approval and testing authority over pesticides under FIFRA: A pesticide must be registered with the EPA before it can be distributed or sold.[149] If a substance is found to have "unreasonably adverse effects on the environment," it cannot be registered and brought to market; approval and registration is conditioned upon use in a manner designed to prevent unreasonable adverse effects.[150] A pesticide is defined broadly as any substance or mixture of substances intended for preventing, destroying, repelling, or mitigating any pest,[151] and EPA has said that it believes pesticide products containing nanomaterials will come under FIFRA review and registration.[152] This registration requirement is EPA's strongest tool for controlling the potential risks of nanopesticides, permitting EPA to, among other things, require that manufacturers test pesticides and submit the risk

---

[144]42 U.S.C. § 9602(a).
[145]EPA White Paper, see footnote 4, p. 31.
[146]American Bar Association, Section of Environment, Energy, and Resources, *CERCLA Nanotechnology Issues*, at 4.
[147]See footnote 146, p. 6.
[148]Finan, see footnote 63. Given the problematic application of environmental statutes that focus on current activities to nanomaterials, unless regulatory adjustments are made, CERCLA will likely be needed to clean up nanomaterial wastes.
[149]7 U.S.C. § 136a(a).
[150]*No Spray Coalition, Inc. v. City of New York*, 351 F.3d 602, 604–605 (2d. Cir. 2003) (citing 7 U.S.C. § 136a(C)(5)(D)).
[151]7 U.S.C. § 136(u)(1).
[152]EPA White Paper, see footnote 4, p. 27.

data.[153] Compared to TSCA, EPA's pesticide authority is much stronger than that provided for existing chemicals and more akin to EPA's authority over new chemicals.[154] The registration requirement is also supported by strong enforcement powers over unregistered pesticides,[155] and post-registration requirements, such as testing, reregistration, notification of post-registration adverse effects, and future cancellation or suspension of registration.[156]

As an alternative to the application of existing FIFRA regulation, EPA could, in some cases, create a special nanopesticide category to directly address the unique characteristics of nanopesticides, similar to the regulations for genetically modified microbial pesticides[157] and a particular class of bioengineered pesticides: plant-incorporated protectants.[158] Further, in some places, EPA's current FIFRA regulations need amending in order to account for the unique character of nanomaterials.[159] Similar to other statutes, regulatory volume thresholds and low volume exemptions may not be apropos with regard to nanopesticides.[160]

One question is whether EPA will consider a nanopesticide an unregistered pesticide if there is an already-registered pesticide that contains the same substance in bulk or conventional form. This is a similar query to the TSCA 'new' versus existing chemical question.[161] Recall that, under TSCA, whether a chemical was 'new' depended in the main on whether the chemical in question had the same 'molecular identity' as an existing listed chemical.[162] In contrast, under FIFRA, a pesticide registration depends on whether "when used in accordance with widespread and commonly recognized practice [a pesticide] will not generally cause unreasonable adverse effects on the environment."[163] Thus, at the heart of EPA's decision as to whether a pesticide is already registered is a risk–benefit analysis. Because of their fundamentally different chemical and physical properties—put another way, their very "nano-ness"—the nanoingredients in nanopesticides create a different risk–benefit balance than those of a pesticide composed of a bulk material counterpart. Thus, nanopesticides should require a new or amended registration.[164] Furthermore, it is unlawful to sell a pesticide as registered if it makes claims substantially different from the registered pesticide, or differs in

---

[153] 7 U.S.C. § 136a.

[154] American Bar Association, Section of Environment, Energy and Resources, *The Adequacy of FIFRA to Regulate Nanotechnology-based Pesticides*, p. 5.

[155] See 7 U.S.C. §§ 136j, 136k, 136l, and 136q.

[156] See 7 U.S.C. §§ 136a(g), 136a(c)(2)(B), 136a-1, 136d(a)(2), & 136d(b)–(c).

[157] See 40 C.F.R. Part 172, Subpart C (experimental use permits), 40 C.F.R. §§ 158.690, 158.740 (data registration requirements).

[158] 40 C.F.R. Part 174.

[159] ABA FIFRA Nano Paper, see footnote 151, p. 12 (citing EPA's data requirements at 40 C.F.R. § 158.190 as not including crucial characteristics of nanomaterials).

[160] See footnote 159, pp. 12–13.

[161] See Section III pp. 20–23, see footnote 159, see also, Finan, see footnote 63.

[162] 15 U.S.C. §§ 2504(a)(1), 2602(2)(A).

[163] 7 U.S.C. § 136a(c)(5)(D).

[164] ABA FIFRA Nano Paper, see footnote 151, pp. 10, 11. (The unique characteristics of a nanopesticide will most likely result in different risks and benefits than its macro version.).

composition.[165] Nanopesticides will undoubtedly make new claims based on the "nanoness" of their ingredients; similarly, given the unique nature of nanomaterials, those pesticides will be composed of different substances. Thus manufacturers with registered pesticides likely cannot distribute nanopesticides based on their earlier registrations for pesticides composed of conventional materials.

Recently, various consumer products containing (or purporting to contain) silver nanoparticles have come to market,[166] leading to questions about their environmental impacts once released in the waste stream, and calls for their regulation as pesticides.[167] Silver can be highly toxic to aquatic organisms, such as plankton, and has the potential to bioaccumulate in some aquatic species.[168] The EPA has said it is currently studying the issue in order to develop appropriate policies, but, as of May 2006, "[did] not know when it will make a decision."[169]

National Environmental Policy Act

Any survey of environmental laws would be incomplete without a discussion of the National Environmental Policy Act (NEPA), which is the "basic national charter for protection for the environment."[170] The NEPA is intended to "promote efforts which will prevent or eliminate damage to the environment and biosphere and stimulate the health and welfare of man."[171] Recognizing the effects of new technologies on the environment, Congress explicitly states in NEPA that "new and expanding technological advances" are activities that could threaten the environment.[172] Thus, in order to understand and control the effects of new technologies such as nanotechnology, Congress requires federal agencies to consider the environmental effects of a new technology by complying with the mandates of NEPA.

[165]7 U.S.C. § 136j(a)(1)(B)–(C).

[166]R. L. Rundle, *This War Against Germs Has a Silver Lining*, Wall St. J., June 6, 2006, at D1; ABA Nano FIFRA Paper, see footnote 151, p. 3.

[167]In early 2006, the National Association of Clean Water Agencies (NACWA) and Tri-TAC, a technical advisory group for Publicly Owned Treatment Works in California, both requested that household products, particularly washing machines containing "silver ions," be classified and regulated by EPA as pesticides. Letter from Ken Kirk, Executive Director, National Association of Clean Water Agencies, to Stephen Johnson, Administrator, Environmental Protection Agency (February 14, 2006) (on file with author); Letter from Chuck Weir, Chair, Tri-TAC, to James Jones, Director, Office of Pesticide Programs, Environmental Protection Agency (January 27, 2006) (on file with author). Pat Phibbs, *Pesticides: Examining Use of Nanoscale Silver in Washing Machines as Possible Pesticide*, Daily Environment Report, May 15, 2006, at A-5–A-6 (quoting Phil Bobel, who works with Tri-TAC).

[168]See footnote 167.

[169]Pat Phibbs, *Pesticides: Firms Making Nanoengineered Pesticides Urged to Meet with EPA Staff on Data Needs*, Daily Environment Report, May 15, 2006, at A-6; Pat Phibbs, *Pesticides: Examining Use of Nanoscale Silver in Washing Machines as Possible Pesticide*, Daily Environment Report, May 15, 2006, pp. A-5–A-6 (quoting EPA spokeswoman Enesta Jones).

[170]40 C.F.R. § 1500.1.

[171]42 U.S.C. § 4321.

[172]42 U.S.C. § 4331(a). In NEPA's legislative history, Congress expressed its concern with "[a] growing technological power … far outstripping man's capacity to understand and ability to control its impact on the environment." *Found on Economic Trends v. Heckler*, 756 F.2d 143, 147 (D.C. Cir. 1985) [quoting S. Rep. No. 91–296 (1969)].

To accomplish NEPA's purposes, all federal agencies are required to prepare a "detailed statement"—known as an Environmental Impact Statement (EIS)—regarding all "major federal actions significantly affecting the quality of the human environment . . . ."[173] To determine whether an EIS is required, federal agencies must prepare an Environmental Assessment (EA), that provides sufficient evidence and analysis to support the agency's determination on whether a proposed action will significantly affect the environment.[174] In addition to environmental concerns, the proposed action's possible direct, indirect, and cumulative impacts on public health must be reviewed if they are linked to its environmental impacts.[175] Nanotechnology-related research, projects, programs,[176] and activities that are funded or carried out by the federal government can be considered "major federal actions significantly affecting the quality of the human environment" for NEPA-purposes, and such activities arguably are, thereby, subject to NEPAs environmental impact assessment procedures and requirements.[177]

The Federal Food, Drug, and Cosmetic Act

Many nanoconsumer products currently on market fall under the broad regulatory umbrella of Food and Drug Administration (FDA), which is charged with regulating the safety and effectiveness of most food, drugs, and cosmetics, as well as other substances, such as medical devices, radiation-emitting products, animal feed, and combination products.[178] In addition to the growing number of consumer products, more than 100 nanomaterial drugs and medical devices are undergoing animal or clinical trials.[179] The FDA regulates "products, not technology."[180] It is aware of "several FDA regulated products [that] employ nanotechnology," including "cosmetic

---

[173]42 U.S.C. § 4332(C). The EIS must describe: (1) the "environmental impact of the proposed action"; (2) any "adverse environmental effects which cannot be avoided should the proposal be implemented;" (3) "alternatives to the proposed action;" (4) "the relationship between local short-term uses of man's environment and the maintenance and enhancement of long-term productivity;" and (5) any "irreversible or irretrievable commitment of resources which would be involved in the proposed action should it be implemented." See footnote 172.

[174]40 C.F.R. §§ 1501.4(b), 1508.9.

[175]40 C.F.R. § 1508.8(b); *Baltimore Gas & Elec. Co. v. NRDC*, 462 U.S. 87, 106 (1983) (explaining that "NEPA requires an EIS to disclose the significant health, socioeconomic, and cumulative consequences of the environmental impact of a proposed action").

[176]Beyond just assessing the impacts of particular project-related actions, agencies are also required to assess the broader impacts of its programmatic actions and to consider alternative program approaches. A programmatic EIS (PEIS) is called for under NEPA regulations, which define a "Federal action" broadly to include, in pertinent part, when there is: "Adoption of programs, such as a group of concerted actions to implement a specific policy or plan; systematic or connected agency decisions allocating agency resources to implement a specific statutory program or executive directive." 40 C.F.R. § 1508.18(b)(3) (defining "Federal action").

[177]Bergeson, L. and Auerbach, B. *Reading the Small Print*, Environmental Law Institute, Environmental Forum (Mar./Apr. 2004), p. 40.

[178]See generally, The Federal Food, Drug, and Cosmetic Act (FFDCA), 21 U.S.C. Chapter 9, *et seq.*

[179]Carrie Dahlberg, *Nanotech's Tiny Revolution Raises Caution*, Sacramento Bee, (August 19, 2006).

[180]FDA, *FDA Regulation of Nanotechnology Products*. Available at http://www.fda.gov/nanotechnology/regulation.html.

products claim[ing] to contain nanoparticles to increase the stability or modify the release of ingredients" and "nanotechnology-related claims made for certain sunscreens."[181]

The FDA's jurisdiction can be divided into sections. Drugs, biologics, and medical devices require premarket approval from FDA.[182] Such approvals are rigorous, and the burden of proof is on the manufacturer.[183] The FDA must ensure that drugs are "safe and effective."[184] Sunscreens, including nanosunscreens, are classified as human drugs because they make health claims.[185] In addition, food, drugs, and cosmetics cannot be adulterated or misbranded.[186] However, in contrast to other substances, FDA has relatively limited authority over cosmetics, including potentially high-risk nanocosmetics, that does not include premarket approval.[187] Finally, regarding its regulation and safety testing of nanomaterials generally, FDA believes that its existing battery of testing methods is "probably adequate for most nanotechnology products that [FDA] regulate[s]."[188]

In May of 2006, the International Center for Technology Assessment (CTA) and a coalition of consumer, health, and environmental groups filed a formal legal petition with the FDA, calling on the agency to address the human health and environmental risks of nanomaterials in consumer products.[189] The petition is the first U.S.

---

[181]FDA, Nanotechnology Products, Frequently Asked Questions. Available at http://www.fda.gov/nanotechnology/faqs.html.

[182]See, e.g., 21 U.S.C. §§ 321(p) (new drug), 355(a) (requiring new drug application before manufacture), 360c(a)(1)(C).

[183]See footnote 181; FDA's drug regulation is more rigorous than its regulations for most other consumer products. Drugs must be pre-approved by FDA, during which time their safety and efficacy need to be established; drugs and drug manufacturing facilities must be registered with FDA; product-related injuries must be reported to FDA; and current Good Manufacturing Procedures (GMPs) must be followed during drug manufacture. 21 C.F.R. Parts 200 through 499 (FDA's drug regulations).

[184]21 U.S.C. § 393(b)(2)(B).

[185]See, e.g., 58 Fed. Reg. 28195. Drugs are defined in relevant part as "(B) articles intended for use in the diagnosis, cure, mitigation, treatment, or prevention of disease in man or other animals, (C) articles (other than food) intended to affect the structure or function of the body of man, or (D) articles intended for use as a component of any articles specified in (A), (B), or (C) above." 21 U.S.C. § 321(g)(1).

[186]See, e.g., 21 U.S.C. §§ 331 (a) (prohibiting the introduction into commerce of any food, drug, device, or cosmetic that is misbranded), 343(a) (Foods are misbranded if their labeling is "false or misleading in any particular"), 352(a) (Drugs and Devices are misbranded if their labeling is "false or misleading in any particular"), 362(a) (Cosmetics are misbranded if their labeling is "false or misleading in any particular").

[187]FDA's regulation of cosmetics and cosmetic ingredients does not include premarket approval, besides the addition of color additives. FDA, Center for Food Safety and Applied Nutrition, *FDA Authority Over Cosmetics* (2006). Available at http://www.cfsan.fda.gov/~dms/cos-206.html. FDA protects the public's health and safety by prohibiting the adulteration or misbranding of cosmetics and has the authority to require warning labels. 21 C.F.R. pts. 361–363, § 740.10(a). FDA can also pursue enforcement actions against cosmetics manufacturers in violation of the law and request product recalls. 21 C.F.R. §§ 7.40–7.59.

[188]FDA, Regulation of Nanotechnology Products. Available at http://www.fda.gov/nanotechnology/regulation.html.

[189]CTA FDA Nano Petition. Available at http://www.icta.org/doc/Nano%20FDA%20petition%20final.pdf.

legal action filed on the potential human health and environmental risks of nanotechnology.[190]

The Occupational Safety and Health Act

Finally, the Occupational Safety and Health Administration (OSHA) promulgates occupational health and safety workplace standards, standards broad enough to cover nanomaterials.[191] The National Institute for Occupational Safety and Health (NIOSH) is the federal agency that conducts scientific research in the field of occupational safety and health, and makes recommendations for preventing work-related injuries, illnesses, and deaths. NIOSH is researching and developing "best practices" guidelines for nanomaterial manufacturing. In the meantime, NIOSH has some interim recommendations,[192] including recommending against inhaling nanoparticles in the workplace and advising workers to wear gloves and respirators, although the efficacy of these protective methods is unknown presently.[193] Workplace health and safety difficulties are also similar to those with environmental statutes to the extent that nanoparticle detection requires expensive and advanced equipment, and as of yet, it is still uncertain what factors are crucial for measuring nanomaterial toxicity.[194]

CONCLUSIONS

Various legal authorities that grant federal agencies the power to oversee the environmental impacts of nanomaterials exist. If implemented in a coordinated manner, these multiple statutes provide the legal and regulatory underpinnings for adequate regulation of some aspects of nanotechnology, in the short term. However, an examination of these laws shows that their application in existing form is, at best, problematic. At a minimum, many of the statutes require regulatory adjustments. Moreover, it appears that the gaps in existing statutory authority are most obvious with respect to two of the most common current uses of nanomaterials: cosmetics and consumer products. This analysis then also offers some insight into what a statutory framework should include for the adequate oversight of the environmental impacts of nanomaterials.

[190]See, e.g., Keay Davidson, *FDA Urged to Limit Nanoparticle Use in Cosmetics and Sunscreens*, San Francisco Chronicle, (May 17, 2006). For a more detailed breakdown of FDA's regulatory authority and CTAs legal challenge, see George A. Kimbrell, *Nanomaterial Consumer Products and FDA Regulation: Regulatory Challenges and Necessary Amendments*, 3:3 Nanotechnology, Law & Business (Fall 2006).

[191]OSHA § 3(8) ("A standard which requires conditions, or the adoption or use of one or more practices, means, methods, operations, or processes, reasonably necessary or appropriate to provide safe or healthful employment and places of employment.").

[192]NIOSH, *Approaches to Safe Nanotechnology: An Information Exchange with NIOSH.* Available at http://www.cdc.gov/niosh/topics/nanotech/safenano.

[193]Carrie Dahlberg, *Nanotech's Tiny Revolution Raises Caution*, Sacramento Bee, (August 19, 2006); Rick Weiss, *Nanotech Raises Worker Safety Questions*, Wash. Post, p. A1 (April 8, 2006).

[194]Davies, see footnote 69, p. 13.

First, while nanoparticles may consist of molecules that are regulated under existing statutes in larger forms, nanomaterials can behave very differently due to their "nanoness" (small size, negligible mass, and higher reactivity). Accordingly, applying conventional methods alone to identify, monitor, measure, and control nanoparticles is inappropriate and insufficient. New technologies and adapted protocols for measuring, monitoring, and controlling nanomaterials are required.

Second, environmental policy and regulation currently relies on well-known and understood chemical and physical properties, including solubility, reactivity, toxicity, and mass. Almost all environmental release restrictions and risk assessment measurements in existing environmental law are premised on a direct relationship between volume or mass and exposure. But adequate regulatory oversight of nanomaterials necessitates an entirely new set of analyses factors, including, but not limited to, particle size, surface area, shape, composition, conductivity, and reactivity. Urgent study is needed to further flesh out which parameters are crucial for nanotoxicology measurements.

Third and relatedly, there is a dangerous lack of data on potential environmental and human health risks, which compounds the challenge of analyzing the adequacy of existing laws to nanomaterials and attempting to judge their limitations and gaps. This dearth of data is explained, in large part, by the inadequate federal funding provided for EHS study of nanomaterials to date.[195] Further study of human health and environmental impacts is urgently needed to protect public heath and the environment and to provide the bases of adequate regulatory oversight of nanomaterials.

Fourth, because of their ubiquitous nature, nanomaterials have the potential to affect every area of environmental concern. Environmental impacts can occur at any stage of a nanomaterials' lifecycle—R&D, manufacturing, transportation, product use, recycling, disposal, or some time after disposal—and a nanomaterial lifecycle frame work helps assess how various statutory regimes apply and where regulatory gaps exist.[196] To adequately address all possible exposures and environmental impacts, a nanomaterial's complete lifecycle must be considered.

Fifth, if voluntary measures are going to be useful, they can only be a stopgap, short-term solution to better inform and formulate pending mandatory regulation.[197] Voluntary programs neglect the entities that most need to be regulated. They also lack transparency and accountability, failings that do not give the public confidence that the government is protecting its interests. For public and

[195]Weiss, see footnote 76 (describing both a September 2006 National Academies' National Research Council Report concluding that the U.S. government is "not paying enough attention to the environmental, health and safety risks posed by nanoscale products," as well as a September 21, 2006, U.S. House of Representatives Science Committee hearing at which "Republicans and Democrats alike took the Bush administration to task over the lack of a plan to learn more about nanotech's risks.").

[196]A lifecycle assessment is the "systematic analysis of the resources usages (e.g., energy, water, raw materials) and the emissions over the complete supply chain from the cradle of primary resources to the grave of recycling or disposal." Royal Society Report, see footnote 11, p. 32.

[197]Only 11% in a recent public opinion poll felt voluntary regulation was adequate. Jane Macourbrie, Woodrow Wilson International Center for Scholars, Project on Emerging Nanotechnologies, Report/Poll, *Informed Public Perceptions of Nanotechnology and Trust in Government* (2005). Available at http://www.wilsoncenter.org/news/docs/macoubriereport.pdf.

environmental safety, mandatory regulation and a comprehensive regulatory scheme that has a place for public participation and review, is necessary.

Sixth, in the long term, a new nanotechnology, nanomaterial-specific law, incorporating a lifecycle analysis, nanospecific testing and regulation, will be necessary. Given the lack of effects data, the burden of safety should be on the manufacturer to show that a product is safe before it is introduced into the market. This is one manifestation of the precautionary principle,[198] the approach being enacted regarding chemicals generally (including nanomaterials) by the EU under its new REACH (Registration, Evaluation, and Authorisation of Chemicals) Regulation, scheduled to take effect in June 2007.[199] REACH shifts the burden of proof to the manufacturer or industry to provide information, assess risk, and provide reasonable assurances of safety prior to marketing and use, rather than placing the burden on regulators to prove harm.[200] The precautionary principle is arguably even more important when dealing with new technological advances, such as those stemming from nanotechnologies, where long-term health and environmental impacts have not been adequately studied and are unpredictable. Accordingly, a moratorium on the production and marketing of nanomaterials should be in place until a legal and regulatory framework is in place that adequately protects human health and the environment.

Finally, in contemplating the adequate regulatory oversight of nanomaterials and what that framework should look like, perspective on the predictions and hype of nanotechnology is also helpful. Today's nanomaterial products are categorized as the "first phase" or stage of nanotechnology, known as the "passive stage" because the nanostructures developed are passive parts of existing products (e.g., zinc oxide nanoparticles added to sunscreens or carbon nanotubes added to electronics).[201] The so-called "second stage," beginning after 2005, focuses on "active" nanostructures that change their size, shape, or other properties during use (e.g., drug delivery devices).[202] Further "phases" of development predicted include systems of nanostructures, including guided assembly (circa 2010) and molecular nanosystems (circa 2015).[203] In fact, the hype and promise (and its always difficult to separate the two) promise:

> nothing less than complete control over the physical structure of matter—the same kind of control over molecular and structural makeup of physical objects that a word processor provides over the form and content of a text.[204]

---

[198]Simply stated, the precautionary principle stands for the idea that inaction is preferable to action in circumstances where taking action could result in serious or irreversible harm. See generally, Ronnie Harding and Elizabeth Fisher, eds., Perspectives on the Precautionary Principle 2–3 (1999).

[199]Cliff Betton, Presentation, *Reach*, Product Safety Assessment Ltd., Health and Beauty America Regulatory Summit, September 14, 2006, New York, NY.

[200]See footnote 199.

[201]See generally, M. C. Roco, National Science Foundation and National Nanotechnology Initiative, *Governance of Nanotechnology for Human Development*, Presentation, Science and Technology for Human Future, Apr. 28, 2006; M. C. Roco, Nanotechnology's Future, Scientific American (July 24, 2006).

[202]See footnote 201.

[203]See footnote 201.

[204]Reynolds, G. *Nanotechnology and Regulatory Policy: Three Futures*, 17 Harv. J. L. and Tech. 179, 185 (2003).

Thus, even if only a small portion of nanotechnology's predicted promise comes to pass, as a long-term solution, it is obvious that current laws are not equipped to regulate such fundamentally different products and processes. Traditional regulatory frameworks, benchmarks, and distinctions will be less—not more—useful as applied to nanotechnology's processes and applications over time. A new nano-specific law will be needed; it is only a matter of when.

It is worth reiterating that this is not the first "wonder" material or technology that the world has seen. History is strewn with once-thought miraculous substances that turned out to be deadly or harmful to the environment. Asbestos was once considered an ideal material for clothing, buildings, and other goods; today, it kills 10,000 people annually. Similarly, for more than 50 years, chlorofluorocarbons (CFCs) were thought to be a miracle substance, used in innumerable household appliances and consumer products; scientists today know that CFCs are a catalytic agent in ozone destruction, leading to less protection from the sun's UVB rays, increasing the risk of skin cancer, and eventually leading to international and national bans on their release. As illustrated by asbestos, CFCs, DDT, leaded gasoline, PCBs, mercury, and numerous other former "wonder" substances and technologies, some nanomaterials will undoubtedly have significant and unintended negative consequences on human health and the environment; whether our policymakers and regulators wait until that occurs or adapt pre-emptively in an attempt to avoid such an accident remains to be seen.

CHAPTER 14

Nanotechnology and the Intellectual Property Landscape[1]

JULIE A. BURGER, MARIANNE R. TIMM, and LORI B. ANDREWS

INTRODUCTION

Advocates of nanotechnology offer many promises for the future—cleaner, more efficient energy sources, drugs that will fight cancer, computer chips that can be implanted in the brain to help the blind see and the disabled walk, and devices that can detect biowarfare agents and fight terrorism. The federal government has invested heavily in this promise of nanotechnology, with spending reaching an estimated $1.3 billion in 2006.[2] The National Cancer Institute alone has implemented a 5-year, $144.3 million program to use nanotechnology to improve options for the prevention, diagnosis and treatment of cancer.[3] Numerous individual states have enacted statutes designed to promote nanotechnology through direct funding, tax incentives, educational grants, or otherwise encouraging nanotechnology research and development.[4]

The National Nanotechnology Initiative (NNI), a U.S. multi-agency endeavor that coordinates nanotechnology research and development, defines nanotechnology

[1]This material is based upon work supported by the National Science Foundation (NSF) under grant SES-0508321 and the Office of Science, U.S. Department of Energy (DOE) under Award Number DE-FG02-06ER64276.
[2]National Nanotechnology Initiative. Available at http://www.nano.gov/html/about/funding.html (last visited October 8, 2006).
[3]M. Sherman, "Exploring the World of Nano Medical Devices," Medical Device and Diagnostic Industry, May 2006. Available at http://www.devicelink.com/mddi/archive/06/05/008.html (last visited October 8, 2006).
[4]For example, Arkansas, ARK. CODE ANN. § 15-4-2104 (West 2006); California, CAL. EDUC. CODE § 88500 (Deering 2006); Connecticut, CONN. GEN. STAT. § 4-124hh (West Supp. 2006); Illinois, 2005 ILL. LAWS 094-079; Indiana, IND. CODE ANN. § 5-28-10-1 et seq. (West 2005); Michigan, MICH. COMP. LAWS ANN. §§ 125.2088 et seq. (West 2005) and 206.30 (West 2006).

as: "[T]he understanding and control of matter at dimensions of roughly 1 to 100 nanometers, where unique phenomena enable novel applications. Encompassing nanoscale science, engineering and technology, nanotechnology involves imaging, measuring, modeling, and manipulating matter at this length scale."[5] The NNI recognizes that the physical, chemical, and biological differences in properties of materials at the nanoscale have the potential to be harnessed in valuable applications.[5] Other definitions exist, however. The State of Michigan, in a statute designed to promote nanotechnology, defines nanotechnology as "materials, devices, or systems at the atomic, molecular, or macromolecular level, with a scale measured in nanometers."[6] This definition does not require any new property or function for a material or product to be labeled as nanotechnology.

Development of nanotechnology, and the impact it has on our health, economy, environment, security, and society will be influenced extensively by the application of the U.S. intellectual property laws. The intellectual property system is designed to provide incentives for innovation—a concept that is important in an emerging field, such as nanotechnology. When people know that their innovation will be rewarded, they have more of an incentive to invent. Yet, when intellectual property laws are improperly applied, patents may be granted that are overbroad, stifling innovation, or patents may be granted that are overly narrow, and do not provide sufficient incentive to continue to invent.

The differences in definitions for nanotechnology in the marketplace, combined with the existing mechanisms of review of nanotechnology patents at the United States Patent and Trademark Office (USPTO), make it problematic to determine exactly how prevalent patents covering nanotechnology are in the United States. Electronic word-based searches designed to locate patents claiming nanotechnologies can be performed on patents that have been issued by the USPTO since 1976, leading some researchers to attempt to analyze the number of nanotechnology patents that have been granted in the past three decades. But applicants can claim that their technology is "nano," even if it does not meet any of several definitions for nanotechnology. Depending on search terms used, such patents might be counted as referring to nanotechnology, though they really do not meet a definition such as that used by the NNI. For similar reasons, it is difficult to pinpoint the exact number of nanotechnology patent applications are currently pending. Not all patent applications are published. The application will remain confidential if the applicant promises not to file for patent protection outside the United States. With respect to published applications, they are only searchable since 2001, when the American Inventors Protection Act[7] went into effect, and are generally published 18 months after their filing date. Therefore, it is also difficult to estimate how many nanopatents applications might be pending.

A brief review of studies of the prevalence of nanotechnology patents illustrates the problems with estimating the incidence of nanopatents. One study undertaking a

[5]National Nanotechnology Initiative, "What is Nanotechnology." Available at http://www.nano.gov/html/facts/whatIsNano.html (last visited October 8, 2006).
[6]MICH. COMP. LAWS ANN. §§ 125.2088A(2) (West 2005) and 206.30(1)(bb)(i)(B) (West 2006).
[7]American Inventors Protection Act of 1999, Pub. L. No. 106-113, § 1000(a)(9), 113 Stat. 1536 (1999).

search of all patent descriptions (the portion of the patent describing the new invention) using the prefix "nano" identified more than 96,000 patents.[8] However, this is an overinclusive search because it includes patents that merely reference "nanoseconds," but do not relate to nanotechnology. Search results may also vary depending on whether just the title of the patent is searched, whether portions of the patent are searched, or whether the entire patent is searched. For studies that conducted searches of titles and claims (the portion that sets forth the legal boundaries of the patent) to include derivations of terms, such as "quantum dot" or "self-assembling," the number of patents found was in excess of 11,000.[9] Other studies reported the number of nanopatents drops to approximately 2000 when only the title is searched for the limited term "nano."[10]

Whatever search terms are used, however, there is evidence that nanotechnology patents are on the rise and that there are more nanotechnology patents than there were biotechnology patents at this stage of the latter technology's development.[11] An understanding of the patent system can help predict future stumbling blocks to innovation and potential costly litigation.

This chapter highlights the importance of the intellectual property system in the development of the new field of nanotechnology, briefly explains laws and principles governing intellectual property, examines how the application of patent law will shape nanotechnology research, development, and progress, and then discusses unique issues that may be raised when intellectual property laws are applied in the nanotechnology field.

PATENT LAW AND NANOINVENTION

The patent system is designed to provide incentive to inventors—people who use their ingenuity to create something truly new. In exchange for disclosing the details of their invention, inventors are given exclusive rights to their invention for a period of 20 years. Patent rights can spur development of technologies. For example, imagine a university develops a nanoparticle-containing coating that can be used as an antimicrobial material for use in hospitals. It receives a patent on this new material it has developed. One day researchers at the National Aeronautics and Space Administration (NASA) read about this material and, after obtaining a license from the university, experiment with the material, and develop heat-shielding tiles for spacecraft and patent that use. Two industries have benefited from this technology.

---

[8]T. K. Tullis, Comment, "Application of the Government License Defense to Federally Funded Nanotechnology Research: The Case for a Limited Patent Compulsory Licensing Regime," 53 UCLA Law Review 279, 282, 282 n.11 (2005).

[9]Z. Huang et al., "International Nanotechnology Development in 2003: Country, Institution and Technology Field Analysis Based on USPTO Patent Database," 6 Journal of Nanoparticle Research 325–354, 327 (2004).

[10] T. K. Tullis, Comment, "Application of the Government License Defense to Federally Funded Nanotechnology Research: The Case for a Limited Patent Compulsory Licensing Regime," 53 UCLA Law Review 279, 282 (2005).

[11]See Chapter 15.

Yet, inappropriate patent policies can have detrimental effects on the development of new technologies, such as nanotechnology, preventing society from reaping the benefits of the new field.

What if the nanoparticle coating was found unpatentable as "obvious" merely because someone had patented an antibacterial paint, with none of the key, unique properties of the nanoparticles? Or what if the first university were to patent all uses of the nanoparticle coating? There would be no incentive for other entities, like NASA, to develop other uses.

The USPTO is the agency charged with granting or refusing to grant patent rights on inventions it examines. If USPTO examiners are too stringent in their analysis of the first patent applications in a new or emerging field of study, they will improperly reject valid patent applications. The lack of adequate property rights and protection will deter future research and investment, delay knowledge of advances to other researchers, create more overlapping and unnecessary research, and consume judicial resources, time, and money. Improper rejections could be extremely costly, either through inventors making use of the appeals process within the USPTO or litigating against the USPTO in the courts.[12] This increase in cost, time, and energy to obtain patent protection could deter some companies from investing in nanotechnology for fear that their investment will not be protected or recouped.[13]

As an example, consider a company that spends 2 years developing a new method of making Buckminster-fullerenes (buckyballs), the famous soccerball-shaped carbon configuration, C_{60}. The method is completely new, unanticipated by other work in the area. When the company applies for a patent on the method, the examiner, who is unfamiliar with nanotechnology, rejects it, which allows other companies to use the new method without compensating the company who invented it. Now the company has lost its incentive to invest in research and development in the future because it may not be able to recoup its costs or profit from the investment. Or, the company may continue to research new methods, but might choose to keep its results a trade secret. Then the technology is not available for sale or license to others who might go on to develop new products or technologies from it.

On the other hand, if patents are granted that are too broad, developments in nanotechnology might be stifled. Overly broad patents could prevent other researchers and developers of technology from working in that area.[14] If examiners issue overly broad patents, then conflicting property rights will be created, which will

[12]S. J. Ainsworth, "Nanotech IP: As Nanometer-Scale Materials Start Making Money, Intellectual Property Issues are Heating Up," 82 Chemical and Engineering News 17–22 (April 12, 2004).

[13]A. Regalado, "Nanotechnology Patents Surge as Companies Vie to Stake Claim," Wall Street Journal, June 18, 2004, at A1; S. J. Ainsworth, "Nanotech IP: As Nanometer-Scale Materials Start Making Money, Intellectual Property Issues are Heating Up," 82 Chemical and Engineering News 17–22 (April 12, 2004); R. A. Bleeker et al., "Patenting Nanotechnology," Materials Today, 44–48, 46 (February 2004).

[14]R. A. Bleeker et al., "Patenting Nanotechnology," Materials Today 44–48, 47 (February 2004). For example, in the field of gene therapy, W. French Anderson and his collaborators at the National Institutes of Health were granted a patent on all human gene therapy that involved the removal, alteration, and reinjection of a patient's cells. That broad patent, covering an entire field, was later criticized as potentially thwarting innovation. Lori Andrews and Dorothy Nelkin, Body Bazaar: The Market for Human Tissue in the Biotechnology Age, 62–63 (Crown Publishers: New York 2001).

also waste public resources through the cost and time of litigation. In addition, research will be discouraged, end products for consumers will increase in cost,[15] and a high-tech bubble could form and burst. Improperly granted claims may have a chilling effect on other researchers' use of the technology because they may not realize that the patent is legally deficient or may not be able to afford to challenge the patent in court.[16]

The Constitutional and Statutory Foundation of the U.S. Patent System

The patent system is designed to provide an incentive for inventors to create and disclose new products and inventions—discoveries that will be beneficial to the public. Nanotechnology offers many promises that would be beneficial, but the success of nanotechnology might depend on how the laws of intellectual property are applied. The U.S. Constitution grants Congress the power "[t]o promote the progress of Science and the useful Arts, by securing for limited Times to Authors and Inventors the exclusive Right to their respective Writings and Discoveries."[17] Congress executed its power by enacting the Patent Act.[18] Under the Patent Act, inventors are essentially granted a monopoly—the exclusive rights for 20 years to make, use, sell, and import their invention.[19] If anyone makes, uses, sells, or imports the patented invention without the patent owner's permission, that individual has infringed the patent owner's rights[20] and is liable for damages.[21] The patent holder may also seek an injunction in federal court against the infringer and stop him or her from using the invention.[22] But there is a check to this system—not all inventions and discoveries may be patented.

An inventor may receive a patent on "any new and useful process, machine, manufacture, or composition of matter, or any new and useful improvement on these things."[23] The applicant must demonstrate that the invention is novel,

[15]A. Regalado "Nanotechnology Patents Surge as Companies Vie to Stake Claim," Wall Street Journal, June 18, 2004, at A1.

[16]The litigation process to challenge a patent's validity has been estimated to cost between $650,000 and $4.5 million. American Intellectual Property Law Association, Report of the Economic Survey 102 (2005).

[17]U.S. Constitution, Article I, Section 8, Clause 8.

[18]35 U.S.C.A. § 101 *et seq.* (2001 and West Supp. 2006).

[19]35 U.S.C. § 154 (2001 and West Supp. 2006).

[20]35 U.S.C.A. § 271 (2001 and West Supp. 2006).

[21]35 U.S.C. § 284 (2000).

[22]35 U.S.C. § 283 (2000). Courts, following Federal Circuit precedent, routinely awarded injunctions as a matter of course when a patent holder demonstrated the existence of a valid patent and infringement. In 2006, the U.S. Supreme Court ruled that in patent cases, courts must consider the traditional four factors to determine if an injunction should issue. *eBay Inc. v. MercExchange, L.L.C*, No. 05-130, 126 S. Ct. 1837, 547 U.S. _, 2006 U.S. Lexis 3872 at *2–3 (2006). One factor the courts must consider is whether the public interest weighs in favor of enjoining the infringer from future infringement. In nanotechnology, where it is probable that developed technologies will have an impact on the public interest (such as life-saving nanodrugs or devices necessary for national security), the public interest may weigh against granting a permanent, or even preliminary, injunction.

[23]35 U.S.C. § 101 (2000).

nonobvious, and useful.[24] The inventor must also provide a written description of the invention sufficient to "enable" someone skilled in that field to make and use the invention.[25] In exchange for revealing and describing the invention, the inventor receives the exclusive rights described above. The system is designed to benefit the public and to provide incentives to the inventor.

The patent system has become a three-way give-and-take among Congress, the USPTO, and the courts. All three have active roles in ensuring that the goals of the patent system are met and that the monopoly granted is not too broad. Most often, this means that the courts and Congress winnow back patents granted by the USPTO. When Samuel Morse convinced the USPTO to grant him a patent on the use of electromagnetic waves to write at a distance, the Supreme Court ruled the patent was overbroad; Morse could only patent his invention—the telegraph.[26] In another example, in the mid-1990s surgeons began to patent their surgical methods in larger numbers. The American Medical Association took its case to Congress and said this practice was not good for medicine or for research. Congress amended the law and now, while surgical methods can be patented, under the medical use exemption doctors can use patented medical procedures and cannot be compelled to pay a royalty.[27]

Nanotechnology could raise concerns similar to both of these situations—where overbroad patents harm business and innovation, and where patents create risks to the public health. Nanomedicine holds great promise in the detection and treatment of disease and the improvement of the human condition. However, as with the patenting of surgical methods, improper patent policies can impede the advance of medical research and the availability of technologies to patients.

The incentive a patent promises has driven the development of products in other fields (e.g., as pharmaceuticals) for years. Yet, just as there may be problems in the scientific development of a technology, there may be problems in the legal system's response to that technology. Some of the problems at the intersection of nanotechnology and intellectual property may be analogous to those encountered by any new technology for which patent protection is sought. Other problems, however, are likely to be unique because of the extraordinary characteristics of nanotechnology.

[24]35 U.S.C.A. §§ 101–103 (2001 and West Supp. 2006).

[25]35 U.S.C. § 112 (2000). The disclosure provisions require that an applicant satisfy four basic requirements in patent specification: written description, enablement, best mode, and definiteness. 35 U.S.C. § 112 (2000). Specifically, the law requires that the patent application "contain a written description of the invention, and of the manner and process of making and using it, in such full, clear, concise, and exact terms as to enable any person skilled in the art to which it pertains, or with which it is most nearly connected, to make and use the same, and shall set forth the best mode contemplated by the inventor of carrying out his invention." 35 U.S.C. § 112 (2000). Written description relates to whether the invention as claimed has been sufficiently disclosed in the specification. Definiteness relates to the way the claim is written; the claim must "particularly point out and distinctly claim[] the subject matter which the applicant regards as his invention." 35 U.S.C. § 112 (2000).

[26]*O'Reilly v. Morse*, 56 U.S. 62, 113 (1853).

[27]35 U.S.C. 287(c) (2000).

Patents May Only Be Granted on Eligible Subject Matter

The area of patentable subject matter has certain boundaries, outside of which no patents should be granted. For more than 150 years, the U.S. Supreme Court has held that patents are not allowed on laws and products of nature.[26] Basic laws of science are not patentable. The Supreme Court has emphasized:

> The laws of nature, physical phenomena, and abstract ideas have been held not paten-
> table. Thus, a new mineral discovered in the earth or a new plant found in the wild is not
> patentable subject matter. Likewise, Einstein could not patent his celebrated law that
> $E = mc^2$; nor could Newton have patented the law of gravity. Such discoveries are
> "manifestations of . . . nature, free to all men and reserved exclusively to none."[28]

If this were not so, future innovations could not be based on basic scientific ideas.

One way that overbroad nanopatents could be granted would be if nanopatents were granted on laws or products of nature. An example of overly broad patents may be in the area of nanotube technology. Nanotubes are cylinders made up of a layer of carbon atoms, either a single tube (single-wall carbon nanotubes) or multiple tubes within each other (multiwall carbon nanotubes). Credit for the discovery of nanotubes was asserted as early as 1991.[29] Two years after the discovery of the carbon nanotube was reported, IBM filed a patent application that included a claim for "[a] hollow carbon fiber having a wall consisting essentially of a single layer of carbon atoms."[30] This language is broad enough that it could encompass a single-wall carbon nanotube.[31] Obviously, the timing of the patent application raises questions as to whether it was truly novel when such a compound was discussed in scientific literature several years earlier. But just as importantly, nanotubes exist in nature.[32] Carbon occurs naturally in this form, and thus, as a product of nature, might be appropriately considered to be unpatentable subject matter.

Even methods of producing naturally occurring nanocompounds might mimic naturally occurring processes, resulting in overbroad patents if the processes are allowed to be patented. There are numerous patents for methods of producing bucky-balls, for example. Yet, as with nanotubes, buckyballs are found in exhaust from vehicles, soot,[33] and even after lighting strikes sand. The heating of a substance to increase the presence of C_{60} is a fundamental principle of chemistry and a process that occurs naturally in nature.

Thus, in the nanotechnology sphere, questions arise as to whether certain nano-processes are actually fundamental principles of biology, chemistry, and physics,

---

[28]*Diamond v. Chakrabarty*, 447 U.S. 303, 309 (1980) (quoting *Funk Bros. Seed Co. v. Kalo Inoculant Co.*, 333 U.S. 127, 130 (1948)).

[29]Sumio Iijima, "Helical Microtubules of Graphitic Carbon," 354 Nature 56–58 (Nov. 7, 1991).

[30]Carbon fibers and method for their production, U.S. Patent No. 5,424,054 cl. 3 (filed May 21, 1993).

[31]J. C. Miller et al., The Handbook of Nanotechnology, 70 (John Wiley & Sons, Inc.: New Jersey 2004).

[32]S. Iijima, "Helical Microtubules of Graphitic Carbon," 354 Nature 56–58 (Nov. 7, 1991).

[33]L. E. Murr et al., "Carbon Nanotubes, Nanocrystal Forms, and Complex Nanoparticle Aggregates in Common Fuel-gas Combustion Sources and Ambient Air," 6 Journal of Nanoparticle Research 241–251 (2004).

and should not be patented. If the first patent on a nanoinvention improperly includes a claim stating a law of nature or improperly encompasses a product of nature, it may impede future and better inventions. It could also result in time-consuming and costly patent litigation. For a startup company, these costs could be oppressive, leading it to avoid entering the market. Or money may be diverted into litigation or into licensing fees that could be better spent on research and development.

The Supreme Court has consistently invalidated patents that claimed laws of nature. However, the Federal Circuit (which reviews all patent cases that are appealed after the trial court's decision and before appeal to the Supreme Court) has taken the contrary position that a product of nature or law of nature may be patented if it produces a useful and tangible result or has a real-world function.[34] Yet, laws and products of nature inherently produce useful results and have real functions. A carbon nanotube filled with a metal may act as a semiconductor, but might also exist on its own in nature. While "anything under the sun that is made by man" may be patentable,[35] laws of nature are basic facts and processes that pre-existed human intervention. It is likely that, if the Federal Circuit upheld a patent on a nanotechnology that was a mere product of nature or a bare application of a law of nature were patented, and upheld by the U.S. Supreme Court would reverse it. In a case dealing with similar issues that was dismissed on jurisdictional grounds, Justice Breyer stated that regardless of whether research is difficult or costly, laws of nature should not be patented because sometimes "*too much* patent protection can impede rather than 'promote the Progress of Science and useful Arts.'"[36]

Novelty and Nonobviousness

Under the patent statute, inventions must be both novel and nonobvious.[37] But it appears that the U.S. Food and Drug Administration (FDA) is making findings that certain nanoproducts are not novel, which could impede their patentability if the USPTO were to follow the FDA ruling. Patents are to be truly new innovations (novel). But they must be more than just new. They cannot be inventions that can be easily created by combining existing technologies and they must represent an advance over earlier technology, that is, they must be "nonobvious." Existing technology and information in a field that are examined to determine whether the invention is novel and nonobvious are called "prior art."

Nanotechnology takes advantage of the fact that smaller size alone may give substances unique properties. Yet, for the past several years, the FDA has determined that it will treat nanotechnology products the same as any other product falling under its regulation.[38] Its policy has been that a nanotechnology product that has an

---

[34]*State Street Bank & Trust Co. v. Signature Financial Group, Inc.*, 149 F.3d 1368, 1373 (Fed. Cir. 1998).
[35]*Diamond v. Chakrabarty*, 447 U.S. 303, 309 (1980) (quotation omitted).
[36]*Laboratory Corporation of America v. Metabolite Laboratories, Inc.*, No. 04-607, 548 U.S. _, 126 S. Ct. 2921, 2006 U.S. Lexis 4893 at *4 (2006) (Breyer J. dissent) (quoting U.S. Const., Art. I, § 8, cl. 8).
[37]35 U.S.C.A. §§ 102–103 (West 2006).
[38]J. Miller, Note, "Beyond Biotechnology: FDA Regulation of Nanomedicine," 4 Columbia Science & Technology Law Review 1 at *9 (2002–2003). While the FDA has recently set up a task force to make recommendations about its nanotechnology policies, the results and the FDA's ultimate decision

identical composition to its larger common version would be considered to be equivalent under the FDA approval process. Accordingly, these smaller products might require no new premarket approval testing or might be eligible for an abbreviated approval process.[39]

The FDA has approved numerous drugs and devices that employ nanotechnology, such as particles for imaging, wound dressings, bone implants, drugs, makeup and cosmetics, dental implants, and sunscreens. It has determined that these drugs and devices are substantially equivalent to products that do not use nanotechnology. Several years ago, for example, the FDA determined that nanosized particles of titanium dioxide and zinc oxide, ingredients commonly found in sunscreen, are to be regulated the same as their larger sized counterparts.[40] The small size of the nanosized particles gives the sunscreen what seems to be a novel property—better absorbability that reduces the white skin appearance that otherwise results from these compounds. The small size might also allow the particles to cross the blood–brain barrier, thus raising health concerns.[41]

The product NanOss™ is another such device that capitalizes on nanotechnology and has benefited from expedited approval by the FDA. NanOss™ is a bone implant that uses nanocrystals of calcium phosphate created from a patented precipitation process.[42] The manufacturer claims that the nanocrystals, which will be reabsorbed by living bone, are very strong and resist cracking as compared to larger particles.[43] It advertises the nanocrystals as duplicating "the microstructure, composition and performance of human bone."[44] The FDA determined NanOss™ is "substantially equivalent" to other resorbable calcium phosphate bone void filler devices because its intended use, design, and functional characteristics are substantially the same as previously approved devices, each of which was intended to fill gaps in bone, was not intended to be load-bearing, and consisted of calcium compounds.[45] Yet, the USPTO has also granted a patent on NanOss™ covering both the nanocrystals and the method of producing them.[46] If NanOss™ is substantially equivalent to larger versions of bone implants, as the FDA has found, it might not be novel and nonobvious, as the USPTO has determined.

will not be known for many months, if not longer. Regardless, the FDA's determination in the past that products utilizing nanotechnology will be treated the same as their larger counterparts is an interesting comparison to the USPTO's nanopolicies.

[39]R. Monastersky, "The Dark Side of Small," Chronicle of Higher Education, September 10, 2004.

[40]Sunscreen Drug Products for Over-the-Counter Human Use; Final Monograph, 64 Fed. Reg. 27666 (May 21, 1999).

[41]A. Nel et al., "Toxic Potential of Materials at the Nanolevel," 311 Science 622-627 (February 3, 2006).

[42]Nanocrystalline Apatites and Composites, Prostheses Incorporating Them, and Method for Their Production, U.S. Patent No. 6,013,591 (filed January 16, 1998).

[43]A. Baluch, "Angstom Medica: Securing FDA Approval and Commercializing a Nanomedical Device," 2 Nanotechnology Law and Business 168, 169 (2005). This article can be found on the "Press Releases" page of Angstrom Medica's website. Available at http://www.angstrommedica.com/images/Nanotech%20L&B.htm (last visited October 8, 2006).

[44]Angstrom Medica, "Technology." Available at http://www.angstrommedica.com/technology/default.htm (last visited October 8, 2006).

[45]501(k) Summary for Angstrom Medica NanOss™ Bone Void Filler, K050025, February 3, 2005.

[46]Nanocrystalline Apatites and Composites, Prostheses Incorporating Them, and Method for Their Production, U.S. Patent No. 6,013,591 (filed January 16, 1998).

Antimicrobial silver wound dressings are another device that employ nanotechnology, have FDA approval, and have been patented. NUCRYST Pharmaceuticals uses a patented process to isolate silver-containing nanoparticles, which are then placed on a substrate of polyethylene mesh as atomically disordered nanocrystals. The substrate is used in wound and burn care devices. The FDA determined that the dressings are "substantially equivalent" to prior silver coated dressings that release silver ions into wound sites to provide an antimicrobial effect.[47] Yet the company holds numerous patents that cover the manufacturing process, as well as compositions of matter (including coatings, powders, and flakes) and uses that incorporate the technology.

The USPTO thus treats various products as "novel" and "nonobvious," while another U.S. government agency determines that the same product is "substantially equivalent" to already-existing technologies. This indicates either that the USPTO may have granted patents on nanotechnologies that do not meet the statutory requirements[48] or that the FDA is allowing products to be put on the market whose novel properties have not been adequately investigated.

THE USPTO'S RESPONSE TO NANOTECHNOLOGY

Ensuring that patents are properly granted pursuant to the Patent Act requires, in part, looking at patents that have been granted in the past and at technology and literature in the field. Yet in the field of nanotechnology, it may be difficult to find prior technology and literature. If the invention contains the same claims as another invention that was patented or described in a printed publication more than one-year prior to the U.S. filing date of the patent, the examiner should deny the patent.[49] But, as there is no single, universally accepted definition of nanotechnology, encompassing either the field or its products, materials, and applications, examiners may be unable to perform a proper search.[50] The lack of a uniform nomenclature, as well as the patent applicant's prerogative to act as his or her own lexographer and define terms as he or she chooses makes searches to determine

---

[47]501(k) Summary for Westaim Technologies, Inc.'s Acticoat™ Silver Coated Dressing, K955453, May 31, 1996.

[48]35 U.S.C. § 102 (2000 and Supp. 2003); 35 U.S.C.A. § 103 (2001 and West Supp. 2006).

[49]35 U.S.C. § 102(b) (2000 and Supp. 2003) ("A person shall be entitled to a patent unless . . . the invention was patented or described in a printed publication in this or a foreign country . . . more than one year prior to the date of the application for patent in the United States . . . "). Note that the patent or publication can come from any country.

[50]Some states, for example, provide funding or other incentives for nanotechnology research and development which incorporate a requirement that the nanoscale research involve structures with novel properties; others do not. Compare the definition Arkansas employs ("the materials and systems whose structures and components exhibit novel and significantly improved physical, chemical, and biological properties, phenomena, and processes due to their nanoscale size," ARK. CODE ANN. § 15-4-2103(5) (West 2006), with Michigan, which defines nanotechnology as "materials, devices, or systems at the atomic, molecular, or macromolecular level, with a scale measured in nanometers," MICH. COMP. LAWS § 206.30 (West 2006).

whether the invention has already been invented, or whether it was obvious in light of prior inventions, more difficult and could create or exacerbate future patent disputes.[51]

Additionally, although nano is currently a popular marketing term,[52] its widespread use does not help to define a technology that already covers a wide array of scientific and engineering disciplines. However, it does provide an incentive for patent applicants to use the prefix "nano." More than 1200 American nanotechnology startups are basing their existence solely on the promise of nanotechnology.[53] Claiming they have nanopatent portfolios would make them seem more attractive to potential investors or licencees. However, some of the technologies described as nanotechnologies may, in fact, be similar to inventions that were previously patented in the field of molecular biology. A search of the prior nanotechnology patents would not reveal this prior art and could lead to patents being improperly granted.

The rush to patent nanotechnologies or even technologies that label themselves in some way as nano raise issues for the USPTO that have important implications for future research, development, and innovation. With nanotechnology's substantial funding and hypothesized potential, the USPTO has already seen an influx in nanotechnology patents.[15] However, as discussed, because of the lack of standardization in the use of terminology, the number of patents issued covering inventions in the field of nanotechnology is difficult to estimate. The USPTO's response to nanotechnology has only begun to be quantified.

The USPTO examines patent applications within technology centers, which are comprised of examiners responsible for related technologies and disclosures. It organizes patent applications by describing them with a class number that identifies similar prior art. This system is designed to facilitate searches for related technologies and disclosures.

In 2004, the USPTO created a class (Class 977) in which nanotechnology related prior art should be catalogued. After a subsequent amendment, the class, which is used to index the technology and not used to assign patents to examiners for review, now encompasses a collection of prior disclosures and technologies related to "nanostructures." The USPTO defines a nanostructure as "an atomic, molecular, or macromolecular structure that: (a) has at least one physical dimension of approximately 1–100 nm; and (b) possesses a special property, provides a special function, or produces a special effect that is uniquely attributable to the structure's

---

[51]For a survey of past studies that have attempted to quantify the number of nanopatents that have been issued, see Chapter 15.

[52]Products capitalizing on the "nano" craze include the Apple "iPod nano," the Whisper Light Nano-Ionic Conditioning Hair Dryer by BioIonic iDry, the Samsung Silver Nano Health System washing machine, and the GM Hummer H3, popularly referred to as the "Nano Hummer."

[53]R. Bailey, "The Smaller the Better: The Limitless Promise of Nanotechnology—and the Growing Peril of a Moratorium," Reasonline, (December 2003). Available at http://reason.com/0312/fe.rb.the.html. (Last visited October 8, 2006.) Currently, more than 200 consumer products in the United States utilize nanotechnology. Project on Emerging Nanotechnologies, "A Nanotechnology Consumer Products Inventory." Available at http://www.nanotechproject.org/index.php?id=44 (last visited October 8, 2006).

nanoscale physical size."[54] The USPTO has reviewed previously issued patents to determine retroactively which should be classified under 977. However, the 977 classification has many exceptions. For example, enzyme and protein complexes are generally excluded from 977. Similarly, viruses utilized for viral functions are categorized in separate classes, rather than 977. But, a virus utilized to form a nanostructure is included in 977 classifications.[54]

To determine how the USPTO is using this class, we analyzed all patents issued between January 1976 and July 1, 2006, that contain "quantum dot" or its synonym "nanocrystal" in their title.[55] "Quantum dot" refers to semiconducting crystals created on the nanoscale.[56] It might seem that disclosures and technologies related to quantum dots would be catalogued in Class 977. But of the 280 patents found by our search, the USPTO placed only 48 patents examined in the study (17.1%) into Class 977. This may indicate that not all patents claiming nanotechnology are being put into a class that will be useful for future searches, or it may indicate that the numerous exceptions may keep patents claiming nanotechnology out of Class 977. This makes it possible that patents that do not meet the statutory requirements, or that are overlapping with already issued patents, will be granted.

In addition to overlapping patents, there is evidence that overly broad nanotechnology patents have been issued.[57] This may be problematic later when the patent holders step in to assert broad rights. Patent holders sometimes allow researchers and institutions to use a patented technology in research without alleging infringement. This allows the patent holder to create a demand for its technology with the potential of benefiting later. Once a commercial application is derived, patent holders will typically assert patent rights to the subsequent researcher's invention based on their previous patent.[58]

Returning to carbon nanotube technology, Japan's NEC Corporation declares it holds patents on the basic building blocks of nanotube technology, and within the last several years began asserting that any company wishing to work with that material must obtain licenses from it.[58] One of the patents NEC holds is U.S. Pat. No. 5,457,343.[59] The first claim is:

> A carbon tubule of a nanometer size in diameter which comprises: a plurality of tubular monoatomic graphite sheets coaxially arranged; and a foreign material enclosed in a

[54]Class 977 Definition. Available at http://www.uspto.gov/go/classification/uspc977/defs977.htm. The class functions as a cross-reference collection of art and is not a primary classification.

[55]The search was conducted on the USPTO's online issued patents database. Available at http://patft.uspto.gov/netahtml/PTO/search-adv.htm using the search terms: ttl/(nanocrystal$ or nano-crystal$ or "nano crystal$") or ttl/(quantumdot$ or quantum-dot$ or "quantum dot$"). The first patent to meet these criteria was issued in 1990.

[56]P. Weiss, "Quantum-Dot Leap: Tapping Tiny Crystals' Inexplicable Light-Harvesting Talent," 169 Science News 344 (June 3, 2006).

[57]A. Regalado, "Nanotechnology Patents Surge as Companies Vie to Stake Claim," Wall Street Journal, June 18, 2004, at A1; Susan J. Ainsworth, "Nanotech IP: As Nanometer-Scale Materials Start Making Money, Intellectual Property Issues are Heating Up," 82 Chemical and Engineering News 17–22 (April 12, 2004).

[58]S. J. Ainsworth, "Nanotech IP: As Nanometer-Scale Materials Start Making Money, Intellectual Property Issues are Heating Up," 82 Chemical and Engineering News 17–22 (April 12, 2004).

[59]Carbon Nanotubule Enclosing a Foreign Material, U.S. Patent No. 5,457,343 (filed December 21, 1993).

center hollow space which is defined by an internal surface of the most inner tubular monoatomic graphite sheet, said foreign material being a metal selected from the group consisting of lead, tin, copper, indium, mercury, and alkali metals.[59]

NEC, therefore, claims patent rights to concentric tubes of single layers of carbon atoms with a metal filling. This claim appears to be overbroad. It could be read to include any multiwalled carbon nanotube with any quantity of a listed material in it. A researcher filling a nanotube with certain metals to experiment with their conducting capabilities would infringe the patent. The patenting of so basic a building block for nanotechnology could run counter to the creation of incentives for innovation, the very foundation of patent law.

When NEC began enforcing its carbon nanotube patents,[60] several companies, such as Houston-based Carbon Nanotechnologies, Inc. (CNI), decided to pay NEC's royalty request rather than pursue costly litigation even though CNI believes that NEC's patents are most likely invalid.[61] In 2006, nanotube manufacturer SouthWest NanoTechnologies licensed NEC patents to facilitate production and distribution of the tubes.[62] As more nanotechnology-based products are brought to market, litigation of patents thought to be overbroad will be inevitable as companies like NEC enforce patents that many believe to be invalid.

Practical Review Issues Faced by the USPTO

In addition to the legal issues raised by nanotechnologies, there are practical review issues faced by the patent office. Some of the issues facing the USPTO are analogous to those it encounters when inventors seek patent protection for any dramatically new technology. The USPTO reports being underfunded and understaffed, and generally underequipped to deal with the number of patent applications filed annually.[63] Other issues the USPTO faces, however, are unique because of the extraordinary characteristics of nanotechnology.

Nanotechnology crosses several scientific fields and the potential benefits of nanoscale research "reach into electronics, biotechnology, medicine, transportation, agriculture, environment, national security, and other fields."[64] It is likely nanopatents will cross several areas, but the USPTO is not organized for analyzing

[60]S. J. Ainsworth, "Nanotech IP: As Nanometer-Scale Materials Start Making Money, Intellectual Property Issues are Heating Up," 82 Chemical and Engineering News, 17–22 (April 12, 2004).

[61]S. J. Ainsworth, "Nanotech IP: As Nanometer-Scale Materials Start Making Money, Intellectual Property Issues are Heating Up," 82 Chemical and Engineering News, 17–22 (April 12, 2004). (As quoted from Bob Gower, president and chief executive officer of CNI: "We have acted as if some claims are valid because we don't want to fight about it. One could argue that single-wall nanotubes were discovered much earlier than NEC claims, but that really isn't the issue we think is important at this stage.")

[62]S. Shankland, "Nanotube Manufacturer Licenses NEC Patents," CNET News.com (August 3, 2006). Available at http://news.com.com/2061-11204_3-6101848.html (last visited October 8, 2006).

[63]V. Koppikar et al., "Current Trends in Nanotech Patents: A View From Inside the Patent Office," 1 Nanotechnology Law & Business 24, 24 (2004).

[64]U.S. Department of Energy, "Nanoscale Science, Engineering, and Technology in the Department of Energy," at 4. Available at http://www.sc.doe.gov/bes/brochures/files/NSRC_brochure.pdf (last visited October 8, 2006).

multidisciplinary patents. Instead it is divided into eight specific technology centers: biotechnology and organic chemistry; chemical and materials engineering; computer architecture, software and information security; communications; semiconductors, electrical and optical systems and components; designs; transportation, construction, electronic commerce, agriculture, national security, and license and review; and mechanical engineering, manufacturing, and products. When an inventor submits a patent application, the USPTO routes it to the technology center with expertise in the particular discipline covered by the patent application for examination. Each technology center is responsible for reviewing patent applications that fall within its particular area of expertise. However, nanotechnology has the possibility of falling within several areas simultaneously, and the USPTO does not have a technology center devoted to nanotechnology.

Nanotechnology inventions create problems for the USPTO during the examination process because true nanotechnology inventions possess unique properties that require a different type of expertise (e.g., knowledge of quantum physics) than that typically found in many USPTO technology centers.[65] Examiners may not be gaining sufficient expertise in dealing with nanopatents.

To analyze the way in which nanopatents are assigned, we analyzed all the patents issued between January 1976 and July 1, 2006, that contain "quantum dot" or its synonym "nanocrystal" in their title.[66] Quantum dots have wide-ranging applications in highly diverse fields, such as healthcare and medical procedures, cosmetics, environmental remediation, and national security.[67] The survey revealed: 45.4% were assigned to the chemical and materials engineering technology center (center number 1700); 41.4% were assigned to the semiconductor, electrical and optical systems technology center (center number 2800); and the remainder were scattered among biotechnology and organic chemistry, transportation, and mechanical engineering.[68] The spread of patent reviews across centers may not facilitate the necessary build up of expertise.

In addition, examiners do not seem to be developing specialties in nanotechnology. These 280 patents we identified that dealt with "quantum dots" or "nanocrystals" were examined by 147 different USPTO examiners. Sixty-six percent of these examiners examined only one quantum dot patent. Almost 80% of the examiners examined only one or two quantum dot patents. Only 8.2% of these examiners looked at five or more quantum dot patents. This data is consistent with the

[65]See, e.g., T. K. Tullis, Comment, "Application of the Government License Defense to Federally Funded Nanotechnology Research: The Case for a Limited Patent Compulsory Licensing Regime," 53 UCLA Law Review 279, 291–293 (2005).

[66]The search was conducted on the USPTO's online issued patents database. Available at http://patft.uspto.gov/netahtml/PTO/search-adv.htm using the search terms: *ttl/(nanocrystal$ or nano-crystal$ or "nano crystal$") or ttl/(quantumdot$ or quantum-dot$ or "quantum dot$")*. The first patent to meet these criteria was issued in 1990.

[67]Lux Research Inc., "Statement of Findings: The Nanotech IP Landscape," (2005). Available at http://www.foley.com/files/tbl_s31Publications/FileUpload137/2655/SOF_NTS-R-05-002.pdf (last visited October 8, 2006).

[68]In a very small percentage of patents, it was not possible to discern to which technology center the nanopatent application had been assigned.

concern that nanotechnology patents are too broadly distributed across the patent office, possibly to examiners lacking expertise in the field.[69]

If examiners are inexperienced within a specific technology, or never see more than a few nanopatents, it might be expected that patents that are overbroad and overlapping will be granted. Our initial results returned overlapping patents. For example, patent numbers 6,444,143[70] and 7,060,252[71] both claim quantum dots 1.2–15 nm, that are water soluble (or do not require insolubility), that are coated by an organic outer layer, and that emit light or fluorescence. These two patents appear to claim the same or a very similar technology. They were examined by different examiners. Overlapping patents create conflicting intellectual property rights, inhibit research, and could result in costly litigation.

Patent Infringement and the Strict Liability Standard

As a new technology, nanotechnology raises issues similar to those of other new technologies with respect to the need for trained examiners and the proper application of legal standards. But the intellectual property issues go far beyond either. Some of the unique properties of nanoproducts that make them so exciting to use also create problems in enforcement.

On the one hand, the small size of nanotechnologies may make infringing uses difficult to discover and lead to less protection of patent holders than may be optimal. On the other hand, the potential for nanoproducts to spread in unintended ways could lead to an even more problematic scenario where people unwittingly infringe and are inappropriately found to owe royalties.

The patent holder can demand royalties from anyone who "uses" the invention. While usually it is fairly easy for an individual to avoid infringing on a patent, the unique characteristics of nanotechnology make it possible that an individual could "use" a nanotechnology without meaning to do so. Researchers are currently working on nanosized machines that will be inserted into the blood stream to clear cholesterol from clogged arteries.[72,73] Nanotechnology eventually may be used to help fight a person's cold or flu by the insertion of nanosized machines or particles into a person's blood stream or airway that could hunt and destroy viruses.[73] Depending on the nature of these devices, a person may only need to share fluids, mix blood, or sneeze to pass on his or her nanotechnology device to another.

[69]See also B. N. Sampat, "Examining Patent Examination: an Analysis of Examiner and Applicant Generated Prior Art," NBER Summer Institute, Working Paper, 1–62, 25 (2004). Available at http://faculty.haas.berkeley.edu/wakeman/ba297tspring05/Sampat.pdf (finding similar results with respect to patent examiners).

[70]Water-soluble Fluorescent Nanocrystals, U.S. Patent No. 6,444,143 (filed May 29, 2001).

[71]Functionalized Encapsulated Fluorescent Nanocrystals, U.S. Patent No. 7,060,252 (filed June 1, 2004).

[72]B. Behkam and M. Sitti, "Design Methodology for Biomimetic Propulsion of Miniature Swimming Robots," 128 Journal of Dynamic Systems, Measurement, and Control 36–43 (March 2006).

[73]L. Rubinstein, "A Practicle NanoRobot for Treatment of Various Medical Problems," The Foresight Nanotech Institute, Eighth Foresight Conference on Molecular Nanotechnology, Nov. 3–5, 2000. Available at http://www.foresight.org/conference/MNT8/Papers/Rubinstein/index.html (last visited October 8, 2006).

Imagine a person going to visit her brother who is a recent recipient of an injection of artery cleaning nanobots. While talking, the brother sneezes, exhaling some nanobots that are immediately and unwittingly inhaled by the sister. Now the nanobots begin coursing through the sister's arteries, clearing them of plaque. For purposes of the Patent Act, the sister is "using" the nanobots, even though she did not intend to use them and did not take any action to start using the technology, apart from breathing. Under the patent statute, the sister is liable for infringement of the patent.

This scenario is possible because patent infringement is judged by a strict liability standard.[74] Under strict liability, one will be held liable for infringement even if the infringing activity was unintentional, inadvertent, or unknowingly committed.[75] Intent is irrelevant to infringement,[76] and damages can be awarded regardless of the infringer's state of mind.[77]

The purpose behind the strict liability standard is to enhance social welfare by minimizing the social costs of wrongdoing through encouraging careful conduct and deterring wrongdoing.[78,79] It provides an incentive for companies and individuals to take preventative measures to avoid liability.[79] In patent law, this harsh standard strongly encourages potential infringers to take all safeguards possible against infringement. It is designed to prevent companies and inventors from avoiding liability by claiming they were unaware of another inventor's patent. Because their awareness is irrelevant, they will be held accountable for infringement whether or not they had knowledge of a patent. Therefore, prudent companies and inventors will take precautions to determine if they risk infringing another inventor's patent prior to creating an invention. The precautions should include making certain no one else has any rights in the invention. Companies and inventors that do not perform this search may face costly patent infringement litigation, damages for infringing on the patent, an order enjoining it from the

---

[74]35 U.S.C.A. § 271 (2001 and West Supp. 2006); R. D. Blair and T. F. Cotter, "Strict Liability and its Alternatives in Patent Law," 17 Berkeley Technology Law Journal 799, 821 (2002).

[75]R. D. Blair and T. F. Cotter, "Strict Liability and its Alternatives in Patent Law," 17 Berkeley Technology Law Journal 799, 821 (2002).

[76]35 U.S.C.A. § 271(a) (2001 and West Supp. 2006). In *Hilton Davis Chemical Co. v. Warner-Jenkinson Co.*, the Federal Circuit reiterated, "intent is not an element of direct infringement, whether literal or by equivalents. Neither Graver Tank nor any other authority supports the proposition that preventing 'fraud on a patent'... turns on the subjective awareness or intent of the accused infringer .... Infringement is, and should remain, a strict liability offense." 62 F.3d 1512, 1527 (Fed. Cir. 1995) (citing *Graver Tank & Mfg. Co. v. Linde Air Products Co.*, 339 U.S. 609, 610 (1950)), *overruled on other grounds. See also Eye-Ticket Corp. v. Unisys Corp.*, 155 F. Supp. 2d 527, 544 (E.D. Va. 2001).

[77]35 U.S.C. § 284 (2000). *Jurgens v. CBK, Ltd., Inc.*, 80 F.3d 1566, 1570 n.2 (Fed. Cir. 1996) (stating that infringement is a strict liability offense and damages must be awarded regardless of "the intent, culpability or motivation of the infringer").

[78]A. Hamdani and A. Klement, "The Class Defense," 93 California Law Review 685, 708 (May 2005).

[79]J. Arlen and R. Kraakman, "Controlling Corporate Misconduct: An Analysis of Corporate Liability Regimes," 72 New York University Law Review 687, 692 (October 1997).

infringing activity, and, if found liable of willful infringement, risks treble damages and attorney's fees.[80]

Moreover, the harshness behind the strict liability standard also encourages individuals and companies to obtain patents. They are willing to fulfill their statutory obligation to disclose their inventions to the public because the intellectual property rights conferred upon them are strictly enforced.

To infringe a patent, a person or company has to make, use, sell, or import the invention. In most other areas that means individuals or companies would have to take some sort of action. Before taking the action, they could attempt to ascertain whether the action would cause them to infringe on a patent, rendering them liable. With nanotechnology, however, people engaging in no action or choice could be held liable for infringement. The sister who inhaled the nanobots did not take any action beyond breathing, which is required for her survival and that probably could not be reasonably anticipated to result in an infringing activity. The actions that would allow her to avoid infringement (namely, not breathing or not visiting her brother) are not reasonable precautions.

The result of applying the strict liability standard to nanotechnology is that rather than being able to avoid liability by taking reasonable precautions, the individual might have to take some sort of affirmative action to avoid liability for infringement. In a Canadian genetically modified (GM) crop patent case in which a farmer was found liable for infringement where he had saved seed from patented GM plants, which blew on to his land, the court suggested that a truly innocent bystander who did not intend to use patented GM seed might be able to avoid liability by acting to arrange for the seed's removal.[81] The court, therefore, left open the possibility that a farmer had at least a minimal affirmative duty to ensure that if the patent holder's property (its patented GM seed) enters his land and contaminates his crops, he must take some action to remove the patented material. This could also leave open the possibility that the farmer has some affirmative duty to determine whether his crops have been contaminated with GM pollen, and if so, act accordingly.

Applying this reasoning to the sister who inhaled the nanobots, to avoid infringement she might need to determine whether she had inhaled nanobots and then take affirmative steps to have them removed. Even if she could make the company pay for the removal of the nanobots, she would still have to undergo a medical procedure to have them removed. Now, her bodily integrity has been violated twice, first by the nanobots entering her body without her consent, and next, by being required to

---

[80]The American Intellectual Property Law Association reported in 1997 a median cost of $2,510,000 per party for a patent infringement suit totaling over $5 million for the entire lawsuit. M. A. Lemley, "Rational Ignorance at the Patent Office," 95 Northwestern University Law Review 1495, 1502 (Summer 2001) *citing* AIPLA Report of Economic Survey (of U.S. IP Practitioners) (1997). By 2005, the reported cost for litigation (depending on at what stage of the litigation the case was resolved) ranged from $650,000 and $4.5 million. American Intellectual Property Law Association, Report of the Economic Survey 102 (2005).

[81]*Monsanto Canada Inc. v. Schmeiser*, [2004] 1 S.C.R. 902, ¶ 86 (Can.).

undergo a medical procedure to avoid liability for patent infringement. Clearly, this is not a viable alternative.

The integration of these nanobots into innocent bystanders could also have damaging effects far beyond infringement liability. The particles and machines could affect the bystanders' health. For example, the sister might not need any cholesterol removed, so this function could negatively affect her health. A person whose health is damaged might still be liable for infringement and damages. In addition, a patent holder could intentionally or negligently infect bystanders with his or her invention. Although the patent holder is the cause of the infringement and might be held liable for battery, the bystander is enmeshed in litigation, and still liable for infringement, rewarding the patent holder for his or her improper conduct.

The purpose of the strict liability standard is to promote diligence and encourage precautions, and more specifically, in patent law, to encourage research into the patent rights of others before action. But, as the previous scenario illustrates, these infringers engaged in no action. If the infringer did not choose to use or make anything, he or she cannot take the precautions the patent system is designed to promote. Nanotechnology patent holders can also abuse their patent rights by using the unique properties of nanotechnology to intentionally cause another to infringe. Therefore, regardless of research and even sometimes regardless of choice, the bystander will be an infringer. Nanotechnology extends patent rights and the strict liability standard beyond their intended scopes. These problems could be exacerbated if patents are granted on discoveries and inventions that do not meet the statutory requirements.

The strict liability standard creates even more mischief when viewed in the context of reproduction. Nanotechnology may be used someday to modify a person's DNA to cure a genetic or another type of disease. Children receive one-half of their DNA from each of their parents. Therefore, a mother who has purchased a nanotechnology that modifies her DNA to cure a disease could pass on her altered DNA to her child. But just because a patented technology can replicate itself, it does not necessarily mean the purchaser of a patent has the right to use replicated copies of the technology.[82] If companies allow people to pass on genetic cures through reproduction, they will not have a future market for their products. A company might attempt to hold a parent liable for inducement and/or contributory infringement for the child's inheritance of the modified genes because the parent would be inducing the child to infringe the patent.[83] Consequently, a child who inherits the replicated DNA might be subject to a claim for infringement, and liable for damages or an injunction.[82] A company might attempt to require

---

[82]*Monsanto Co. v. Scruggs*, 459 F.3d 1328, 1336 (Fed. Cir. 2006).

[83]"Whoever actively induces infringement of a patent shall be liable as an infringer." 35 U.S.C.A. § 271(b) (2001 and West Supp. 2006); "Whoever offers to sell or sells within the United States or imports into the United States a component of a patented machine, manufacture, combination or composition, or a material or apparatus for use in practicing a patented process, constituting a material part of the invention, knowing the same to be especially made or especially adapted for use in an infringement of such patent, and not a staple article or commodity of commerce suitable for substantial noninfringing use, shall be liable as a contributory infringer." 35 U.S.C.A. § 271(c) (2001 and West Supp. 2006).

parents to pay additional fees if they intended to have children, but this scheme would infringe on the parents' freedom to make reproductive decisions.

The application of intellectual property law and policies to nanotechnology can also inhibit people's right to travel. For example, if an overly broad patent has been granted on a building block of nanotechnology, it may increase research and development costs, and eventually will increase consumer costs. An end product, such as nanobots, may cost more in the United States than in countries that have not been granting overbroad patents. A person may thus seek to have a nanomedical device implanted in another country, where costs are lower. When that person returns to the United States, he or she could be sued by the patent holder for infringing on the patent by using and importing it. When a technology is inside a person, however, issues of bodily integrity conflict with corporate interests in the enforcement of nanotechnology patents.

CONCLUSIONS

Appropriate legal regulation will be critical to the development of nanotechnology. As indicated by M. C. Roco, "Nanotechnology success is determined by an architecture of factors such as creativity of individual researchers, training of students in nanoscale science and engineering, connections between organizations, patent regulations, physical infrastructure, legal aspects, state and federal policies, and the international context."[84]

An analysis of the current intellectual property landscape and comparisons with other technologies suggests that we can expect some contentious debates and court cases arising from the development, patenting, and commercialization of nano-technologies, added to, in the near future, by the unfamiliarity of the USPTO and the courts with this new science. These debates and cases have the potential to impede research and stifle innovation. The patent system's role of encouraging innovation must apply to nanotechnology.

Patents that are overly broad and overlapping may inhibit research, prevent new inventions, and waste judicial resources through patent disputes. The USPTO should not be granting, and courts should not be upholding, patents on laws and products of nature. Eligible subject matter for patents must be more than a discovery of a basic scientific fact even if it has a useful or tangible result—there must be human invention that produces a result beyond what the law or product of nature produces itself. Patents should add to the public store of knowledge, not remove knowledge from public domain. Patents should also be invalidated if it is shown that they impede people's rights, including freedom of speech, of reproduction, to travel, and to research.

[84]M.C. Roco, "Broader Societal Issues of Nanotechnology," 5 Journal of Nanoparticle Research 181–189, 181 (2003).

CHAPTER 15

Patenting Trends in Nanotechnology[1]

JESSICA K. FENDER

INTRODUCTION

The United States is experiencing tremendous growth in nanotechnology research. The National Science Foundation (NSF) has predicted that nanotechnology-related goods and services will reach $1 trillion by 2015, exceeding the combined economic impact of the telecommunications and information technology industries during the technology boom of the 1990s.[2,3] One of the ways to measure this growth is to track nanotechnology patenting activity. As nanotechnology inventions move from theory to commercialization, the number of nanotechnology patents granted by the U.S. Patent and Trademark Office (USPTO) is expected to increase dramatically. Today, nanotechnology patents account for approximately 0.7% of all patents issued by the USPTO.[4] There are more nanotechnology patents now than there were biotechnology patents at a similar stage in the biotechnology field's development.[5] The number of nanotechnology-related scientific and technological articles is also significantly higher than the number of biotechnology-related articles at a comparable time, indicating nanotechnology patents may soon overtake biotechnology patents in number.[6]

[1]This material is based upon work supported by the National Science Foundation (NSF) under grant SES-0508321 and the Office of Science, U.S. Department of Energy (DOE) under Award Number DE-FG02-06ER64276.
[2]R. Bawa, "Nanotechnology Patenting in the US," 1 *Nanotechnology Law & Business* 31–50, 37 (2004).
[3]M. C. Roco, "The US National Nanotechnology Initiative After 3 Years (2001–2003)," 6 *Journal of Nanoparticle Research* 1–10, 4 (2004).
[4]L. G. Zucker and M. R. Darby, "Socio-Economic Impact of Nanoscale Science: Initial Results and Nanobank," *NBER Working Paper Series* 11181, 1–31, 9 (2005).
[5]L. G. Zucker and M. R. Darby, "Socio-Economic Impact of Nanoscale Science: Initial Results and Nanobank," *NBER Working Paper Series* 11181, 1–31, 9 (2005).
[6]M. A. Lemley, "Patenting Nanotechnology," 58 *Stanford Law Review* 601–630, 605 (2005).

This chapter addresses the issues and trends in nanotechnology patents in the United States by comparing studies that identified a set of nanotechnology patents and analyzed them.[7] The studies illustrate how nanotechnology patenting has changed and grown over the last three decades. They provide key information, such as which entities are obtaining the most patents and which technological fields are experiencing the largest impact from nanotechnology research. The studies report results from as early as 1976, with some data being collected as recently as 2005. The researchers range from corporate research groups to law firms and academics. Methodologies vary widely; for example, some researchers identified nanotechnology patents by searching for nanotechnology-related terms in the patent title, whereas others searched for a given term throughout the entire body of the patent. The data reported in each study varies as a result of these methodological differences, and the methodologies are, therefore, presented in conjunction with the results to put them into context. The studies are referred to by the primary author's last name, or the group name when applicable.

RESULTS

Nanotechnology Patents on the Rise

Each study reported on the number of nanotechnology patents issued by the USPTO. The USPTO website allows one to search published patent applications from March 2001, whereas complete data on issued patents is available from 1976.[8] The number of patents reported by each study varies depending on three factors: which date range the author searched, which search terms the author used, and which part of the patent the author analyzed.[9] These parameters provide an explanation for most observed variances between the reported results.

In general, the study authors compiled their results by searching issued patents or published patent applications that contained terms relating to nanotechnology. The two studies that searched the entirety of the issued patent language (e.g., as

---

[7]There are nine studies discussed in this chapter. The studies were chosen for comparison because the researchers disclosed something about their search methodologies in addition to reporting on the number of nanotechnology patents they identified. Methodology is key to meaningful interpretation of each study's data.

[8]Pre-grant patent applications are available as of 2001, when the American Inventors Protection Act (AIPA) went into effect. *American Inventors Protection Act of 1999, Pub. L. No. 106–113, § 1000(a)(9), 113 Stat. 1536 (1999)*. Patent applications were not routinely published prior to 2001, but now applications are generally published 18 months after their filing date. Issued patents are available in full text after 1976, whereas patents issued prior to 1976 are searchable only by patent number, issue date, and current U.S. classifications. USPTO, "Patent Full-Text and Full-Page Image Databases." Available at http://www.uspto.gov/patft/ index.html, last viewed June 15, 2006. The applicant can request the application remain secret but only if the applicant has not filed for the same invention in a foreign country subject to the Patent Cooperation Treaty.

[9]For those unfamiliar with patents, a patent consists of various distinct parts. These include: a title; an abstract, which is a brief summary of the invention; the specification, which provides background information on the invention, as well as a detailed description of the new invention; and the claims, which define the boundaries of the patent applicant's legal right to exclude others from practicing the invention.

opposed to only patent titles or abstracts) reported the highest number of nanotechnology patents.

The Tullis study reported the highest numbers overall, with 96,312 patents that purportedly relate to nanotechnology. These results are the product of two factors. First, Tullis did not provide a range of dates, but referenced the USPTO website, indicating that he likely searched issued patents from 1976 until the day he ran the search on September 2, 2005.[10] Second, he did not filter his results, and he used the broad prefix "nano" as his only search term. Most of the other studies either chose specific search terms (e.g., quantum dot) or excluded measurement terms when they searched for the term "nano." Measurement terms may appear in patents that do not address nanotechnology; for example, the term "nanometer" could appear in any patent referencing the electromagnetic spectrum, such as a patent for an incandescent light bulb.[11]

This issue comes into focus upon repeating Tullis' search. One patent that comes up as a "hit" is a patent for a cleaning agent that uses the chemical compound sodium nitrate ($NaNO_3$, which the computer search engine recognized as containing the term nano).[12] The patent makes no claim to any nanotechnology-related invention. On the other hand, excluding measurement terms may also lead to artificially low numbers, as patents claiming nanotechnology more than likely include a term, such as "nanometer." U.S. Patent No. 7,005,669 provides just such an example—the patent claims quantum dots, nanocomposite materials of quantum dots, and devices using quantum dots, and therefore clearly contains nanotechnology-related inventions. However, the patent also contains the term nanometer, and, therefore, would have been excluded if all measurement terms were culled from the search results

The Huang et al.,[13] study largely avoided these problems by constructing a specific set of search terms to identify nanotechnology-related patents. The Huang group performed two studies. The first study, published in 2003, examined nanotechnology patents from January 1976 to December 2002, while the second study updated and refined those data and collected new data from 2003.[14,15] It is

---

[10]T. K. Tullis, Comment, "Application of the Government License Defense to Federally Funded Nanotechnology Research: The Case for a Limited Patent Compulsory Licensing Regime," 53 *UCLA Law Review* 279, 282, 282 n.11 (2005).

[11]See, e.g., *Improving Incandescent Bulb Efficiency*, U.S. Patent No. 4,196,368 (filed Sept. 7, 1977).

[12]*Cleaning Composition*, U.S. Patent No. 3,948,819 (filed June 18, 1973).

[13]Huang searched nanotechnology patents terms specifically related to nanotechnology, such as biomotor, molecular device, quantum dot*, and nano*. Patents that contained only the term "nanosecond" or "nanoliter" were excluded, and patents that contained more than one search term were represented just once in the study's accounting. Z. Huang et al. "International Nanotechnology Development in 2003: Country, Institution and Technology Field Analysis Based on USPTO Patent Database," 6 *Journal of Nanoparticle Research* 325–354, 326 (2004).

[14]Z. Huang et al. "Longitudinal Patent Analysis for Nanoscale Science and Engineering: Country, Institution and Technology Field," 5 *Journal of Nanoparticle Research* 333–363 (2003).

[15]Z. Huang et al. "International Nanotechnology Development in 2003: Country, Institution and Technology Field Analysis Based on USPTO Patent Database," 6 *Journal of Nanoparticle Research* 325–354 (2004).

not clear whether Huang et al., searched only issued patents, or whether they searched published patent applications, as well. Given that published patent applications are only available from 2001, however, it is likely that only issued patents were examined. Upon completing the search, Huang et al. reported 70,039 patents, the second largest number of nanotechnology patents. When they analyzed just the patent titles and claims, however, the number of reported patents dropped to 11,206.

Other studies reported numbers ranging from almost 4000 to fewer than 2000. These were largely studies that limited their search to just one part of the patent, such as the title or abstract. For those studies searching only the title of issued patents, Tullis found 2042 patents and Huang et al., found 1538 patents. For those studies searching only the claims of issued patents, Sampat reported that 3748 nanotechnology-related patents were issued between 2001 and 2003.[16] Lemley, repeating Sampat's method for 2004, identified another 1929 patents, for a total of 5677 patents.[17] Two research groups did not state which portion of the patent they searched. These are Lux Research, Inc., which identified 3,818 patents issued between 1985 and 2005,[18] and the Glänzel et al., study, which identified 3969 patents issued between 1992 and 2001.[19]

The studies reporting the lowest numbers were those that analyzed patents relating to a specific nanomaterial or application. A nanomaterial is any material created on the nanoscale. For example, a nanowire resembles a wire in that electrons are confined to move in one dimension, but the "wire" can be made from individual atoms and has a diameter on the order of nanometers. Lux Research, Inc., analyzed the abstracts and claims for patents relating to five different

[16]Sampat used the search terms provided in Huang's study. Huang, see footnote 13. B. N. Sampat, "Examining Patent Examination: An Analysis of Examiner and Applicant Generated Prior Art," NBER Summer Institute, Working Paper, 1–62, 24 (2004).

[17]Lemley criticized the Sampat study (and by extension, the Huang et al., study) by suggesting that Sampat's definition of nanotechnology was overly conservative, thereby underestimating the total number of nanotechnology patents. In addition, Lemley suggested that Sampat may have missed relevant nanotechnology patents when the search terms were located in the specifications, instead of the claims. Regardless, Lemley used the Huang et al. search terms for his general nanotechnology patent search. Huang, see footnote 13. M. A. Lemley, "Patenting Nanotechnology," 58 *Stanford Law Review* 601–630, 604 (2005).

[18]Lux Research, Inc., searched patent abstracts and claims for dendrimers, quantum dots, carbon nanotubes, fullerenes, and nanowires. Lux used these terms in conjunction with synonyms (e.g., "carbon fibril" or "carbonaceous cylinder" for carbon nanotubes). Lux did not provide an exhaustive list of the terms. Based on the patents Lux initially located, Lux identified key inventors and assignees and used those names to locate additional nanomaterial patents. Lux Research, Inc., "Statement of Findings: The Nanotech IP Landscape," Available at http://www.foley.com/files/tbl_s31Publications/File Upload137/2655/SOF_NTS-R-05-002.pdf (last visited June 14, 2006).

[19]Glänzel et al., did not provide search terms that would be recognized by the USPTO search engine. However, Glänzel et al., also performed a search of publications, using search terms similar to those employed by Huang. All microsystem-related technologies were purposely excluded. W. Glänzel et al. *Steunpunt O&O Statistieken*, "Nanotechnology: Analysis of an Emerging Domain of Scientific and Technological Endeavor," 1–73, 43 (2003). Available at http://www.steunpuntoos.be/nanotech_domain_study.pdf (last visited June 6, 2006).

nanomaterials, including dendrimers, quantum dots, carbon nanotubes, fullerenes, and nanowires. There were 1084 patents issued for these five nanomaterials through the beginning of March 2004. The ETC Group searched the Delphion patent database for issued patent abstracts to locate four types of nanomaterial patents.[20] It reported that 735 patents relating to some type of scanning probe microscopy (e.g., scanning tunneling microscopy or atomic force microscopy) were issued between 1982 and 2004. It also reported that 272 dendrimer-related patents and 146 quantum dot-related patents were issued between 1999 and 2004. The ETC's last case study focused on nanotubes, where it found 257 patents were issued between 1999 and 2004. This is in accordance with the Featherstone study, which identified nanotube patents by searching issued patent claims, and located just 206 in 2004.[21]

Only the ETC Group reported numbers for both published patent applications and issued patents combined. It searched the USPTO database for both published patent application and issued patent abstracts between 1999 and 2004, and identified 7004 such documents.[22]

Finally, of those studies that reported data for published patent applications, the numbers ranged from 42,293 applications, when Tullis searched the entire application description between 2001 and September of 2005, to 1235 applications when he searched just the title. Falling within that range were Lemley, listing 9184 patent applications between 2001 and 2004, and Lux Research, Inc., listing 1777 outstanding patent applications from 2001 until the date it compiled the report. The methodologies for the patent application searches are otherwise identical to those listed for the studies' issued patent search methodologies previously provided.

The Zucker and Darby study did not provide a specific number of nanotechnology patents, only relativistic data (i.e., the proportion of nanotechnology patents relative to nanotechnology articles), but the authors intend to make their results available on their online database, NanoBank, in the near future.[23]

---

[20]The ETC Group reported information for patents claiming quantum dots, dendrimers, and nanotubes, but they did not list the search terms used. For its analysis of atomic force or scanning tunneling microscope patents, ETC Group used specific terms, such as "atomic AND force AND microscope" and "scanning AND tunneling AND microscope." The ETC Group, "Nanotech's 'Second Nature' Patents: Implications for the Global South," 1–36, 21–29 (2005). Available at http://www.etcgroup. org/documents/Com8788SpecialPNanoMar-Jun05ENG.pdf (last visited June 6, 2006).

[21]Featherstone and Specht searched for issued nanotube patents by searching for the word "nanotube" in the claims. No alternate terms or nomenclatures, such as "nanocylinder," were used. D. J. Featherstone and M. D. Specht, "SKGF Nanotube Patent Study 2004," 1–22, 2 (2005). Available at http:// www.skgf.com/media/news/news.165.pdf (last visited June 8, 2006).

[22]The ETC Group searched for the term nano anywhere in the patent or application abstract without any other terms or filtering. The ETC Group, "Nanotech's 'Second Nature' Patents: Implications for the Global South," 1–36, 7 n.11 (2005). Available at http://www.etcgroup.org/documents/ Com8788SpecialPNanoMar-Jun05ENG.pdf (last visited June 6, 2006).

[23]Zucker and Darby searched nanotechnology patent titles and abstracts using two overlapping text searches, one with the string "nano" and the other using 475 nanoscale-specific terms, which were not provided in the study. They excluded all measurement terms, such as "nanometer" or "nanoliter" from their results. L. G. Zucker and M. R. Darby, "Socio-Economic Impact of Nanoscale Science: Initial Results and Nanobank," *NBER Working Paper Series* 11181, 1–31 (2005).

TABLE 15.1. Study Methodology and Results

Study	Dates of Examination	Type of Document Analyzed	Database Searched	Search Terms	Part of Document Searched	Number of Nanopatents
Tullis	1976–Sept 2005	Issued Patents Only	USPTO	"nano"	Description	96,312
Huang	Jan 1976–Dec 2002	Unclear	USPTO	selfassembl*, self-assembl*, atomic force microscop*, scanning tunneling microscop*, atomistic simulation, biomotor, molecular device, molecular electronics, molecular modeling, molecular motor, molecular sensor, molecular simulation, quantum computing, quantum dot*, quantum effect*, and nano* (excluding nanosecond and nanoliter).	Entire Patent	61,409 (number updated in 2003 study)
Tullis	Mar 2001–Sept 2005	Published Patent Applications	USPTO	"nano"	Description	42,293
Huang	Jan 1976–Dec 2002	Unclear	USPTO	same as in Huang (above)	Title + Claims	9,562
Lemley	Jan 2001–Dec 2004	Published Patent Applications	USPTO	same as in Huang (above)	Claims Only	9,184
Huang	Jan 2003–Dec 2003	Unclear	USPTO	same as in Huang (above)	Entire Patent	8,630
ETC Group	1999–2004	Published Patent Applications and Issued Patents	USPTO	"nano"	Abstract Only	7,004
Glänzel	Through 2001	Issued Patents Only	USPTO	Unclear	Unclear	3,969
Lux Research	1985–Mar 2005	Issued Patents Only	USPTO	Unclear	Unclear	3,818
Sampat	Jan 2001–Dec 2003	Issued Patents Only	USPTO	same as in Huang (above)	Claims only	3,748
Tullis	1976–Sept 2005	Issued Patents Only	USPTO	"nano"	Title Only	2,042
Lemley	Jan 2004–Dec 2004	Issued Patents Only	USPTO	same as in Huang (above)	Claims Only	1,929

Source	Date Range	Patent Type	Database	Search Terms	Fields Searched	Count
Lux Research	2001–Mar 2005	Outstanding Published Patent Applications	USPTO	Unclear	Unclear	1,777
Huang	Jan 2003–Dec 2003	Unclear	USPTO	same as in Huang (above)	Title + Claims	1,644
Tullis	Mar 2001–Sept 2005	Published Patent Applications	USPTO	"nano"	Title Only	1,235
Huang	Jan 1976–Dec 2002	Unclear	USPTO	same as in Huang (above)	Title Only	1,196
Lux Research	1985–Mar 2004	Unclear	USPTO	Dendrimer, quantum dot, carbon nanotube, fullerene, nanowire; synonyms such as carbon fibril and carbonaceous cylinder; common assignees and inventors	Abstract + Claims	1,084
ETC Group	1982–2004	Issued Patents	Delphion	atomic AND force AND microscope; scanning AND tunneling AND microscope; scanning AND tunneling AND microscope; scanning AND probe AND microscope	Abstract Only	735
Lemley	Jan 2001–Dec 2004	Issued Patents that are assigned to a University	USPTO	same as in Huang (above)	Claims Only	664
Huang	Jan 2003–Dec 2003	Unclear	USPTO	same as in Huang (above)	Title Only	342
ETC Group	1999–2004	Issued Patents	USPTO	Unclear; but relating to dendrimers	Unclear	272
ETC Group	1999–2004	Issued Patents	USPTO	Unclear; but relating to nanotubes "nanotube"	Unclear	257
Featherstone	Jan 2004–Dec 2004	Issued Patents Only	USPTO	Unclear	Claims Only	206
ETC Group	1999–2004	Issued Patents	USPTO	Unclear; but relating to quantum dots	Unclear	146
Zucker	Jan 1986–Dec 2003	Unclear	USPTO	475 "nano"-specific search terms, and the string "nano" (list unavailable; check NanoBank)	Title + Abstract	not reported

Table 15.1 provides information about each study's methodology and reported number of nanotechnology patents. Each study's data is reported separately.[24] Recall that important factors influencing each study's results include which portion of the patent was searched and whether the patent was an issued patent or only a patent application. The other key factor is the date of each study. A timeline (Fig. 15.1) has been created to put each study into chronological perspective. Figure 15.1 is not to scale, and is intended only to provide a rough guide for comparison purposes. The studies that searched the most restrictive field, the title, identified the fewest number of patents. Those searching the patent abstracts identified roughly as many patents as those who limited their search to the patent claims. In addition, most of the studies examining particular nanotechnology-related materials or products, such as fullerenes or dendrimers, reported between 200 and 300 patents. Although Lux Research, Inc., found 1084 patents relating to specific nanomaterials (e.g., quantum dots), this number was based on combining the total number of patents for the five different nanomaterials examined in the study.

Beyond the Numbers: Emerging Trends in Nanotechnology Patenting

The studies described above focused largely on the total number of nanotechnology patents. In addition, however, they provide other data that illustrates how nanotechnology is affecting our society. For example, commonly addressed topics include whether the bulk of patents are obtained by companies, universities, or individuals, and whether more patents were obtained by U.S. or foreign entities. A few of the authors discussed in detail the common perception that nanotechnology patents are of poor quality and claim overlapping subject matter, and proposed solutions for remedying that situation. Each of these areas will be explored in more detail below. When weighing the information provided, recall that the studies vary in their methodologies and that no study addresses every issue.

One of the ways to predict how nanotechnology will shape the future is to determine which fields are experiencing the most patenting activity. Nanotechnology is unique in that any innovation in the field can potentially impact a variety of diverse technologies, such as electrical engineering and pharmaceuticals.[25] The USPTO divides all of the patent applications it receives into different technology centers based upon the primary technology used by the invention. Nanotechnology patents are found across all the technology centers, but are generally concentrated in just four. These include the "Biotechnology and Organic Chemistry" technology

---

[24]For example, when Lemley repeated Sampat's 2001–2003 search for the year 2004, the combined total was not recorded on the table. Similarly, Huang et al. repeated their team's 1976–2002 search for 2003, and the 2003 data are recorded separately. This both represents each researcher's work with greater accuracy and allows a more detailed look at the observable trends.

[25]See, e.g., *Method of Precise Laser Nanomachining with UV Ultrafast Laser Pulses*, U.S. Patent No. 7,057,135 (filed Mar. 4, 2004), which claims a method of making a microstructure. The microstructure, which has at least one feature that measures more than 200 nm, may be used as a coupled quantum dot device, a micro-electrical-mechanical system, a micro-surface-acoustic-wave device, a biochip for

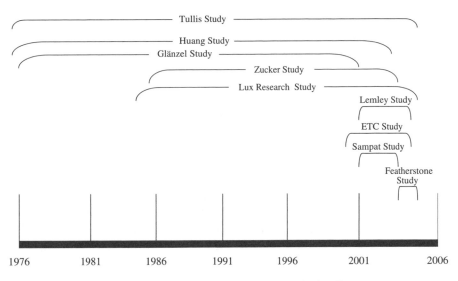

Figure 15.1. Nanotechnology patent study timeline.

center, which contains about 30% of the total nanotechnology patents; the "Chemical and Materials Engineering" center, which accounts for about 25% of the patents; the "Semiconductor, Electrical, Optical Systems" center, which contains 23% of the nanotechnology patents; and the "Mechanical Engineering, Manufacturing, and Products" center, which contains around 14% of the patents.[26,27] Together, these technology centers contain approximately 92% of all nanotechnology patents.

After the patent application is assigned to a technology center, it is assigned to a patent examiner. The examiners analyze the application to see if the USPTO should issue a patent. To do this, the examiners must first determine whether the application's invention has already been patented or described by another. Examiners assign a classification number to each application based upon the specific type of invention described therein. This allows examiners of new applications to easily identify the relevant patents and applications that they should search first as they determine whether the invention is patentable. These classification numbers provide a more detailed look at where the majority of nanotechnology patenting is taking place.

Three studies specifically determined which classes the examiners assign nanotechnology patents to most often. Glänzel et al., determined the top 10 technological

detection of hazardous chemical and biological agents, or a high-throughput drug screening and selection microsystem.

[26]B. Kisliuk, "Nanotechnology-Related Issues at the United States Patent and Trademark Office," presented at NC Nanotech 2006, March 9, 2006. Available at http://www.ncnano.com/ (last visited June 6, 2006).

[27]D. J. Robeson, "Nanotechnology and the USPTO," *The Disclosure*, May 2006. Available at http://www.napp.org/disclosure/ (last visited June 6, 2006).

classes between 1992 and 1999. They divided the patents into two data sets, one from 1992 to 1994, and the other from 1997 to 1999, to determine which areas were experiencing the largest growth. The study did not provide the precise USPTO classification, but listed the following ten areas in decreasing order of patent activity: semiconductors; surface technology and coating; analysis, measurement, and control technology; electrical machine and apparatus or electrical energy; optics; materials and metallurgy; macromolecular and polymer chemistry; chemical engineering; pharmaccuticals and cosmetics; and audiovisual technology. Although they found that each of the listed areas experienced absolute growth in later years, some areas experienced a decrease in their percentage of total nanotechnology patents. Glänzel et al., reported that the areas related to instrumentation, including analysis, measurement and control technology; electrical machinery and apparatus or electrical energy; optics; materials and metallurgy, pharmaceuticals and cosmetics, and audiovisual technology all decreased in share. For example, optics-related patents decreased slightly from 10.31 to 8.74%. On the other hand, semi-conductors, surface technology and coating, macromolecular chemistry, and chemical engineering all gained a small percentage increase.

Huang et al., found that the nanotechnology patents they identified covered 423 out of a possible 462 patent classes. They also analyzed which classes were experiencing the largest growth in patent activity by comparing the resulting patent class dispersion for the year 2003 against the results from 1976 to 2002. They listed the top 20 patent classes and observed that the fastest growth for the year 2003 (as compared to the most prominent classes between 1976 and 2002) occurred in molecular biology and microbiology, bioaffecting and body treating drugs, processes for manufacturing semiconductor devices, and a certain class of organic compounds. Between 2002 and 2003, however, the class for active solid-state devices grew the fastest overall.[28]

The Featherstone and Specht study is consistent with these results. They found that the most common patent classes were for semiconductor device manufacturing, inorganic compound chemistry, electric lamp and discharge devices, active solid-state devices, and radiant energy. All but the inorganic compound class appear in the top 20 lists for 1976–2002 and 2003 by Huang et al. See Table 15.2 for a comparison of all three studies' results.[29]

Next, five studies reported data regarding which entities receive the largest number of nanotechnology patents. All of the reports agree that U.S.-based patentees and assignees receive the bulk of USPTO patents, and that companies and corporations are receiving the largest percentage of those patents. For example, Glänzel et al., reported the United States accounted for 46% of the nanotechnology patents issued by the USPTO between 1992 and 2001. In addition, they found that companies received almost 80% of the patents, followed by universities and

---

[28]The method of measuring the growth of each class used by Huang et al., may be skewed, as they compare overall patent classifications for 1976–2002 (a 26-year total) against the classifications used for the year 2003.
[29]The Glänzel et al., study did not provide a USPTO classification number.

TABLE 15.2. USPTO Subclasses with the Largest Number of Nanotechnology Patents

Rank	Glänzel	Huang (1976–2002)	Huang (2003)	Featherstone
1	Semiconductors	435—Chemistry: molecular biology and microbiology	435—Chemistry: molecular biology and microbiology	438—Semiconductor device manufacturing: process
2	Surface technology, coating	514—Drug: bioaffecting and body treating compositions	257—Active solid-state devices	423—Chemistry of inorganic compounds
3	Analysis, measurement and control technology	424—Drug: bio affecting and body treating compositions	438—Semiconductor device manufacturing: process	313—Electric lamp and discharge devices
4	Electrical machine and apparatus, electrical energy	428—Stock material or miscellaneous articles	514—Drug: bio-affecting and body treating compositions	257—Active solid-state devices
5	Optics	250—Radiant energy	424—Drug: bio-affecting and body treating compositions	250—Radiant energy
6	Materials, metallurgy	530—Chemistry: natural resins or derivatives; peptides or proteins; lignins or reaction	428—Stock material or miscellaneous articles	—
7	Macromolecular chemistry, polymers	536—Organic compounds —part of the class 532 – 570 series	436—Chemistry: analytical and immunological testing	—
8	Chemical engineering	438—Semiconductor device manufacturing; process	427—Coating processes	—
9	Pharmaceuticals, cosmetics	257—Active solid-state devices	530—Chemistry: natural resins or derivatives; peptides or proteins; lignins or reaction	—
10	Audio-visual technology	427—Coating processes	250—Radiant energy	—

(Continued)

TABLE 15.2. *Continued.*

Rank	Glänzel	Huang (1976–2002)	Huang (2003)	Featherstone
11	—	436—Chemistry: analytical and immunological testing	430—Radiation imagery chemistry process, composition, or product thereof	—
12	—	430—Radiation imagery chemistry process, composition, or product thereof	356—Optics: measuring and testing	—
13	—	359—Optics: systems (including communication) and elements	359—Optics: systems (including communication) and elements	—
14	—	356—Optics: measuring and testing	436—Chemistry: analytical and immunological testing	—
15	—	422—Chemical apparatus and process disinfecting, deodorizing, preserving, or sterilizing	385—Optical waveguides	—
16	—	204—Chemistry: electrical and wave energy	422—Chemical apparatus and process disinfecting, deodorizing, preserving, or sterilizing	—
17	—	252—Compositions	524—Synthetic resins or natural rubbers - part of the class 520 series	—
18	—	524—Synthetic resins or natural rubbers - part of the class 520 series	313—Electric lamp and discharge devices	—
19	—	546—Organic compounds - part of the class 532–570 series	204—Chemistry: electrical and wave energy	—
20	—	210—Liquid purification or separation	252—Compositions	—

other higher educational institutions at approximately 11%, and administrative or public institutions at almost 8%. Only 1.4% of the nanotechnology patents were either owned by individuals or not otherwise reported. Featherstone and Specht found that 51% of the 2004 nanotube patents were assigned to a U.S. entity, and that all but one of the eight most common assignees were companies or corporations. Huang et al., reported that 60% of USPTO nanotechnology patents were filed by U.S. entities between 1976 and 2003. The ETC Group reported 48% of the patents in 2003 were assigned to U.S. companies. Top 10 or top 20 lists of entities receiving nanotechnology patents are given in each of the studies listed above (i.e., Featherstone and Specht, Huang et al., ETC Group, Glänzel et al., and Sampat). Though the particulars vary, certain entities commonly appear on these lists. Tables 15.3 and 15.4 compare each study's findings as they relate to the top nano-technology patent assignees. Table 15.3 lists the top nanotechnology patent owners or assignees reported by each study that addressed the question. Many of the entities appear in more than one list. Each of the entities is listed in Table 3 according to its rank. For example, if three companies tied for position four, then those companies would all be listed as fourth and the next available position would be position seven. If more than one entity received the same ranking, every other entity is ita-licized so that the full name of the each entity is clear. Table 15.4 sets forth the number of times each company was listed as a top assignee by each study.

A surprising trend is reported in three of the studies. Lemley found that, on average, 12% of inventors receiving a patent assigned their nanotechnology patents to a university between 2001 and 2004.[30] Across all fields, however, only about 1% of patentees assign their patents to a university, so the observed increase is significant.[31] The ETC Group and Glänzel et al., studies confirmed this result, finding that 11–12% of the analyzed nanotechnology patents were assigned to U.S.-based universities. The University of California accounted for a large portion of the patents, appearing on three of the top patent assignee lists (see Table 15.4).

Lemley also identified 10 of what he termed "building-block patents," or patents on foundational technology, such as carbon nanotubes, semiconducting nano-crystals, and self-assembling nanolayers. Of these patents, universities held seven. This result could explain why universities own between 11 and 12% of nanotechno-logy patents overall. If universities are obtaining more foundational patents, and nanotechnology is new enough that the bulk of the research is still taking place at the foundational level, then it follows that universities would have an increasing proportion of nanotechnology patents. One would expect this disparity to resolve itself over time, as patenting activity moves away from foundational technologies

---

[30]To determine the number of patents obtained by universities, Lemley searched for the terms "university," "college," "trustee," or "foundation" in the patent assignee field. However, he notes that the search may be both over- and under-inclusive. The search terms are over-inclusive because a "foundation" could refer to a private foundation as opposed to a university nonprofit organization; the terms are under-inclusive because a university-controlled patent may be held under a different name than that of the university (e.g., Com-petitive Technologies, Inc. acting as the patent owner for the University of Colorado). Mark A. Lemley, "*Patenting Nanotechnology*," 58 *Stanford Law Review* 601-630, 615 n.69 (2005).

[31]M. A. Lemley, "Patenting Nanotechnology," 58 *Stanford Law Review* 601–630, 616 (2005).

TABLE 15.3. Top Nanotechnology Patent Assignees or Owners[a]

Rank	Featherstone	Huang (1976–2002)	Huang (2003)	ETC Group	Glänzel	Sampat (rankings not explicit)
1	Samsung	IBM	IBM	Canon K.K.	IBM	IBM
2	Rice University	Xerox Corp.	Micron Technology, Inc.	IBM	Eastman Kodak	Micron Technology
3	Daiken Chemical Co.	Minnesota Mining and Manufacturing Co. (3M)	Advanced Micro Devices, Inc.	Silverbrook Research	Matsushita	Advanced Micro Devices
4	Hitachi, Ltd.	Eastman Kodak Co.	Intel Corp.	USA *Hitachi, Ltd.* Seagate Technology	Hitachi Ltd.	Xerox Corp.
5	Nakayama Yoshikazu	Motorola, Inc.	University of California	—	University of California	L'Oreal
6	Industrial Tech Res. Inst.	University of California	3M	—	Fuji Co.	University of California
7	Nantero	NEC Corp.	Motorola, Inc.	Micron Technology, Inc.	Toshiba K.K.	Motorola
8	Advanced Micro Devices, Inc.	Micron Technology, Inc.	Hitachi, Ltd. *Xerox Corp.*	Eastman Kodak Co.	NEC Co. *Sony Co.*	Eastman Kodak
9	—	Canon K.K.	—	Olympus Optical Co., Ltd.	Canon K.K.	General Electric
10	—	E.I. Du Pont de Nemours and Co.	Canon K.K. *Eastman Kodak Co.*	University of California *Rohm and Haas Co.* Polaroid Corp.	BASF	3M

11	—	General Electric Co.	—	—	—	U.S. Secretary of the Navy *Sumitomo*	—
12	—	Texas Instruments, Inc.	—	NEC Corp.	—	Phillips Corp.	—
13	—	Hitachi Ltd.	—	Corning Inc.	Sony Corp. *Molecular Imaging Corp.*	Nippon Co.	—
14	—	U.S. Secretary of the Navy	—	Applied Materials, Inc.	—	L'Oreal	—
15	—	Dow Chemical Co.	—	Fuji Photo Film Co., Ltd.	—	Agency of Industrial Science & Technology	—
16	—	Toshiba K.K.	—	Matsushita Electric Industrial Co., Ltd.	—	AMD *Lucent Technologies, Inc.* Texas Instruments, Inc.	—
17	—	Abbott Laboratories	—	Lucent Technologies Inc. *Texas Instruments, inc.*	—	Xerox Co.	—
18	—	Advanced Micro Devices, Inc.	—		—	Olympus Optical Co.	—
19	—	Massachusetts Institute of Technology	—	Genentech, Inc. *Toshiba K.K.* Massachusetts Institute of Technology	—	Dow Corp. *Fujitsu Ltd.*	—
20	—	Merck & Co., Inc.	—	—	—	Exxon Co.	—

*a*Rankings with more than one member are italicized to distinguish each member.

TABLE 15.4. Most Commonly Appearing Assignees on Table 15.3

Number of Lists on which Company Appears	Company Name
5	International Business Machines (IBM)
	Eastman Kodak
	Hitachi Ltd.
	University of California
4	Advanced Micro Devices, Inc.
	Xerox Corp.
	Micron Technology Inc.
	Canon K.K.
3	Minnesota Mining and Manufacturing Co. (3M)
	Motorola, Inc.
	NEC Corp.
	Toshiba K.K.
	Texas Instruments, Inc.
2	General Electric
	Olympus Optical Co.
	U.S. Secretary of the Navy
	Fuji Co.
	L'Oreal
	Lucent Technologies, Inc.
	Dow Chemical Co.
	Massachusetts Institute of Technology
	Matsushita Co.
1	Samsung
	Rice University
	Daiken Chemical Co.
	Nakayama Yoshikazu
	Industrial Technology Research Institute
	Nantero
	E.I. Du Pont de Nemours and Co.
	Abbott Laboratories
	Merck & Co.
	Intel Corp.
	Corning Inc.
	Applied Materials, Inc.
	Genentech, Inc.
	Silverbrook Research
	Seagate Technology
	Rohm and Haas Co.
	Polaroid Corp.
	Sony Corp.
	Molecular Imaging Corp.
	BASF
	Sumitomo
	Phillips Corp.
	Nippon Co.
	Agency of Industrial Science & Technology
	AMD
	Exxon Co.

and toward more advanced applications and marketable products. Although it is arguably beneficial that universities are on the cutting edge of this new technology, the fact that universities hold such a high percentage of nanotechnology patents may actually be cause for some concern.

Lemley reports that 60% of the publicly reported licensing deals in 2003 involved a university. All of these licenses were exclusive; in fact, the ETC Group reported that between 2003 and 2005 universities announced 20 nanotechnology patent licenses, and at least 19 were exclusive. In general, Lemley found that for the year 2003, universities granted exclusive nanotechnology patent licenses 89–100% of the time, whereas corporations or companies issued exclusive licenses around 50–67% of the time.[32] Nanotechnology-related patenting in universities may, therefore, be problematic, as it could restrict or limit future downstream research and development. There are two reasons why this may occur. First, exclusive licensing agreements inherently restrict innovation by limiting the number of researchers working on a given technology. Only the minds that happen to be employed by the particular company that obtained an exclusive license will be thinking about and analyzing the licensed technology. Second, universities own a relatively large portion of the foundational nanotechnology patents. These patented "building blocks" are applicable across many different technologies, not just the particular area in which the university researcher works. If the universities own a large number of nanotechnology patents and refuse to grant broad licensing rights to the foundational technology, fewer researchers will enter the affected fields, including those completely unrelated to the university researcher's work, and development in these unrelated fields will slow.

Another oft-voiced concern is that a "patent thicket" is forming. A patent thicket is created when overlapping patents are issued within a variety of diverse industries, making it almost impossible to identify and obtain the requisite licenses. Lemley offers several examples of nanoscale technologies that have overlapping patents, including carbon nanotubes, semiconducting nanocrystals, and drug delivery nanoparticles. For example, Lemley referenced another study that found 306 nanotube patents, including 10 patents claiming the nanotube itself and 20 patents on nanotube production methods.[33] In addition, he argued that older patents claiming a submicron scale invention could technically claim nanoscale inventions, even when as a practical matter the two would behave very differently. This would be the case even when the first inventor was unaware of the special physical properties stemming from reduction of the invention scale to below 100 nm. If submicron patent claims apply to nanotechnology, the patent thicket will become that much more difficult to navigate.

In addition to patents overlapping each other, there is concern that a single patent may contain claims that are too broad. For example, the ETC Group points to patents

---

[32]Nor is it only universities that have a propensity for granting exclusive licenses. Tullis reported that in 2003, 12 of the 15 publicly announced licensing deals for nanotechnology-related patents were exclusive; in 2004, 17 of 20 were exclusive.

[33]J. C. Miller et al. *The Handbook of Nanotechnology: Business, Policy, and Intellectual Property Law*, 68–71, 224 (John Wiley & Sons: Hoboken, New Jersey 2005).

that claim one basic nanotechnology application, but cover large portions of the periodic table. In one patent, the applicant claimed a metal oxide nanorod made with any one of 33 different chemical elements.[34] Similarly, another quantum dot patent claimed semiconducting nanocrystals made from any of the elements contained within groups III–V of the periodic table.[35] The claims included semiconducting nanocrystals that resulted both from the *combination* of any of the claimed elements as well as the elements alone.

Some of the authors argue that overly broad claims in nanotechnology patents are caused by examiners' lack of experience with nanotechnology. Many pointed out that a number of different examiners analyzed the patents at issue, indicating that few examiners had experience with more than one or two nanotechnology patent applications.[36] Between 2001 and 2003, for example, Sampat found that 794 different primary patent examiners examined the nanotechnology patents he identified in his study. This represented almost *one-fourth* of the primary patent examiners employed by the USPTO during that time period. Similarly, Featherstone and Specht reported that 142 examiners analyzed the 206 nanotube patents issued in 2004. Only 33 of these examiners examined more than one nanotube patent. The ETC Group reported that of the 726 patents they examined, more than 290 different patent examiners had been assigned to examine the patents.

Sampat noted that the wide dispersal of nanotechnology patents across many examiners raises concerns that the examiners do not have the needed expertise to properly analyze nanotechnology patents. Patent examiners are usually assigned to specific technology centers and art units so that they can gain proficiency in examining certain types of patents. If nanotechnology patents are spread out over many examiners, none of whom have the opportunity to examine many nanotechnology patents, the concern is that they will not gain experience in this emerging technology and will allow patents of poor quality to issue. Certainly the studies' results described herein provide support for this concern.

A few of the studies made suggestions to avoid or mitigate the effects of a patent thicket. Tullis argues that a "government license defense" should be created. Under the Bayh-Dole Act, a federal funding agency reserves the right to obtain a royalty-free government license for any patented technology funded by the government. Tullis argues that the language of the Bayh-Dole Act, which states in part, "the Federal agency shall have a ... license to practice or have practiced for or on behalf of the United States any subject invention throughout the world," could be extended to include any contractor being funded by the government. The contractor would then be "practicing for or on behalf of" the United States. Of course, this solution is only useful for those patents in which the federal government has an interest, and would not affect the bulk of nanotechnology-related patents.

---

[34]*Metal Oxide Nanorods*, U.S. Patent No. 5,897,945 (filed Feb. 26, 1996).

[35]*Preparation of III-V Semiconductor Nanocrystals*, U.S. Patent No. 5,505,928 (filed Apr. 21, 1994).

[36]For example, B. H. Sampat, "Examining Patent Examination: An Analysis of Examiner and Applicant Generated Prior Art," NBER Summer Institute, Working Paper, 1-62, 24 (2004); M. H. Heines, "Nano-Aerobics and the Patent System," 2 *Nanotechnology Law and Business* 335, 338 (2005).

Alternatively, Lemley suggests that the USPTO could implement the same strict utility requirement that already exists for chemistry and biotechnology patents to all patents, including nanotechnology-related patents.[37] In addition, the government could use the Bayh-Dole Act to require broad licensing of foundational nanotechnology patents. The Bayh-Dole Act allows the government to "march in" and "require the contractor, an assignee or exclusive licensee of a subject invention to grant a nonexclusive, partially exclusive, or exclusive license in any field of use to a responsible applicant" when, for example, the patent owner or licensee has not reasonably satisfied the public use requirements as provided by the federal government.[38] This solution, however, only applies to those patents supported by government funding.

The benefit to these methods is that each would shift activity away from foundational patents and towards "downstream implementations," though Lemley does not advise implementing such restrictions quite yet. He argues that because nanotechnology is still in a nascent stage, the incentive of obtaining broad patents may be needed to encourage researchers to invest their time and energy in developing the field. Instead, he recommends a few changes to key USPTO examination rules. First, he argues that the USPTO should not permit an unlimited number of continuation and continuation-in-part applications;[39] second, that the USPTO should publish all applications except those subject to secrecy orders;[40] and third, that treble damages should not be assessed against an inventor who independently invents an infringing product. Finally, Lemley believes that when a patent covers only a small component of a larger invention, an injunction should not be allowed to block the entire product.

CONCLUSIONS

Some of the trends observed in these studies are perhaps expected. As far as the number of nanotechnology patents identified, the researchers that searched a greater portion of patent language tended to identify more patents. Researchers

---

[37]J. D. Forman, Comment, "A Timing Perspective on the Utility Requirement in Biotechnology Patent Applications," 12 *Albany Law Journal of Science & Technology* 647, 655 (2002). Forman discusses the USPTO utility examination regulations for biotechnology patents. These regulations were issued in large part as a USPTO response to the influx of patent applications for DNA sequences that did not have a known functionality at the time the application was filed.

[38]35 U.S.C. § 203(a)(3) (2000).

[39]An inventor files a continuation once his or her initial patent application is rejected by an examiner. This filing keeps the patent application pending, and as the applicant can file an unlimited number of continuations, he or she can theoretically keep his or her application pending for a very long time. An inventor files a continuation-in-part when he or she improves upon his or her initial invention and adds subject matter that was not disclosed in the initial application. In this situation, the applicant gets a new filing date for the added subject matter, but keeps the old filing date for the original disclosure.

[40]Inventors can "opt out" of U.S. publication if they promise not to file their applications in any foreign country which would publish the applications. An inventor might opt out so that if he or she ultimately decides not to obtain a patent, he or she has not widely disclosed the invention to others. Patent applications that claim inventions that are, for example, important to national security concerns may be designated "secret" by the USPTO and will not be published to protect national interests.

that searched a large span of time or that searched the USPTO database more recently also tended to report higher numbers; in addition, researchers that used broader search terms, such as "nano," generally reported higher numbers. Based on these trends, it becomes clear why Tullis' study identified more than 95,000 patents. He searched the entire issued patent description for the term nano in the USPTO database from 1976 to September of 2005.[41] In the same vein, the ETC Group reported the smallest number of patents when it searched patents issued over a 5-year period for terms specifically relating to quantum dots.[42]

Beyond numerical data, however, some preliminary conclusions may be drawn about how nanotechnology will shape our world. As can be seen in Table 15.2, certain technological fields have a disproportionate number of nanotechnology patents. The majority of top classifications fall into a few broad categories, such as chemistry, pharmaceuticals, optics, and electronics. Semiconductors appear as a main classification in each of the four studies listed in Table 15.2, and nanotechnology-related advances in this field are already reaching the market in products like the 90 nm microchip, which is currently being incorporated into computers and cellular phones.

In addition, most nanotechnology patents are assigned to U.S. businesses. Certain companies appearing consistently on lists of top assignees, like IBM and Eastman Kodak, are well positioned to take advantage of the nanotechnology boom and are expected to reap significant economic benefits from future product development. Although universities obtain a disproportionate number of nanopatents relative to the number of patents they receive in other fields, only two universities appear among the top assignees. Further, if universities are obtaining these patents as a function of doing foundational research, the percentage of patents issued to universities should decrease in the coming years as the basic tools of nanotechnology become well established, and a shift to creating marketable products occurs.

In contrast to most scientific or technological fields, where innovation in a particular field affects that field alone, innovations in nanotechnology have the potential to affect many different fields. This cross-field applicability provides a unique opportunity to determine whether early patenting activity can accurately predict the impact of a new technology on our future. Although some early trends have emerged, only time will tell whether we are truly experiencing another industrial revolution.

[41]T. K. Tullis, Comment, "Application of the Government License Defense to Federally Funded Nanotechnology Research: The Case for a Limited Patent Compulsory Licensing Regime," 53 *UCLA Law Review* 279, 282 (2005).
[42]The ETC Group, "Nanotech's 'Second Nature' Patents: Implications for the Global South," 1–36, 30 (2005). Available at http://www.etcgroup.org/documents/Com8788SpecialPNanoMar-Jun05ENG.pdf (last visited June 6, 2006).

PART III

NANOMEDICINE, ETHICS, AND THE HUMAN CONDITION

Many of the most exciting predictions being made for nanotechnology applications lie in the field of medicine. The prospect of imminent cures for cancer and other chronic and terminal disease has led to charges of hype and irresponsibility, though there would seem to be well-founded excitement in the research community that dramatic breakthroughs lie within reach.

Ethical issues arise especially in the context of the application of nanotechnology to medicine. It is the development of the discipline of "bioethics" that has been seen by many as offering the basis for a "nanoethics" that will enable the human community to grapple with ethical challenges that go well beyond nanomedicine.

Such issues are explored in this section. Nigel Cameron explores current American bioethics and questions its fitness to be the basis for a nanoethics unless it is reframed in a manner that leads it to draw on the substantive moral vision of our various communities and traditions, with their shared Enlightenment heritage. Debra Bennett-Woods reviews the broader implications of nanotechnology in healthcare from the perspective of a scholar and teacher with experience in healthcare administration. William Cheshire, a neurologist at the Mayo Clinic, reviews the prospects for nanomedicine, and assesses competing utopian and dystopian visions. Christopher Hook, Director of Ethics Education and a hematologist at Mayo, embarks on an assessment of nanomedicine with special reference to the goals of medicine and the treatment–enhancement distinction.

Nanoscale: Issues and Perspectives for the Nano Century. Edited by Nigel M. de S. Cameron and M. Ellen Mitchell

CHAPTER 16

Toward Nanoethics?[1]

NIGEL M. DE S. CAMERON

ETHICS, POLICY, AND NEW TECHNOLOGIES

Discussion of the values that should frame nanotechnology policy is at an early stage in the United States. The Administration has, however, repeatedly expressed its commitment to the ethical development of new technologies. At the same time, the President's Council on Bioethics has released an extensive report on the principles at stake in one of the central issues raised by nanotechnology, that of human enhancement.[2] The 2003 Act of Congress that articulates the shape and funding of the National Nanotechnology Initiative (NNI)[3] emphasizes the need for ethical and broader societal implications of nanotechnology to be addressed, with particular focus on the development of superhuman intelligence and the use of nanotechnology to enhance the intelligence of human beings, themselves. While special funding has not been set aside for research in these areas, the lead federal agency, the National Science Foundation (NSF), has begun to fund such projects. A recent congressional mandate requires reporting on efforts to address nanotechnology's ethical and societal implications, and recommends that 3% of appropriated funds be set aside for it.[4] As a result, rapid development of projects exploring the ethical and societal agenda is anticipated.

---

[1]This chapter includes material presented by the author in testimony to the European Commission's European Group on Ethics (EGE) hearing on nanomedicine, Brussels, Belgium, March 21, 2006.
[2]The President's Council on Bioethics. (2003). *Beyond Therapy: Biotechnology and the Pursuit of Happiness.* Available at http://www.bioethics.gov/reports/beyondtherapy/beyond_therapy_final_web corrected.pdf (retrieved on October 19, 2006).
[3]21st Century Nanotechnology Research and Development Act, 15 U.S.C. § 7510 (2004). Available at http://www.nano-and-society.org/NELSI/documents/21stcenturynanor&dact.pdf (retrieved October 14, 2006).
[4]U.S. House of Representatives, Conference Report on H.R. 2862, Science, State, Justice, Commerce, and Related Agencies Appropriations Act, 2006, Title III—Science at p. H9797 (Nov. 7, 2006).

Nanoscale: Issues and Perspectives for the Nano Century. Edited by Nigel M. de S. Cameron and M. Ellen Mitchell

While he has yet to address directly the questions raised by research and development on the nanoscale, President George W. Bush noted the importance of nanotechnology in his 2006 State of the Union address,[5] and, in this same address, he stressed:

> A hopeful society has institutions of science and medicine that do not cut ethical corners, and that recognize the matchless value of every life. Tonight I ask you to pass legislation to prohibit the most egregious abuses of medical research: human cloning in all its forms, creating or implanting embryos for experiments, creating human-animal hybrids, and buying, selling, or patenting human embryos. Human life is a gift from our Creator— and that gift should never be discarded, devalued or put up for sale.[5]

In addition, he has made programmatic speeches on science and technology policy. One speech focused on a liberalization of U.S. funding policy of embryonic stem-cell research, and one called for a prohibition on human cloning (for research or reproductive purposes). The first, in 2001, included this declaration:

> As the discoveries of modern science create tremendous hope, they also lay vast ethical mine fields. As the genius of science extends the horizons of what we can do, we increasingly confront complex questions about what we should do. We have arrived at that brave new world that seemed so distant in 1932, when Aldous Huxley wrote about human beings created in test tubes in what he called a "hatchery".... The most noble ends do not justify any means.[6]

The 2002 cloning speech went further:

> Our age may be known to history as the age of genetic medicine, a time when many of the most feared illnesses were overcome. Our age must also be defined by the care and restraint and responsibility with which we take up these new scientific powers.
>
> Advances in biomedical technology must never come at the expense of human conscience. As we seek what is possible, we must always ask what is right, and we must not forget that even the most noble ends do not justify any means. Science has set before us decisions of immense consequence. We can pursue medical research with a clear sense of moral purpose or we can travel without an ethical compass into a world we could live to regret.[7]

---

[5]In this address, President Bush also stated:
> First, I propose to double the federal commitment to the most critical basic research programs in the physical sciences over the next 10 years. This funding will support the work of America's most creative minds as they explore promising areas such as nanotechnology, supercomputing, and alternative energy sources.

2006 State of the Union Address by the President. Available at http://www.whitehouse.gov/stateoftheunion/2006/index.html (retrieved on October 19, 2006).

[6]Bush, G. W. (2001). President Discusses Stem Cell Research. Available at http://www.whitehouse.gov/news/releases/2001/08/20010809-2.html (retrieved October 19, 2006).

[7]Bush, G. W. (2002). President Bush Calls on Senate to Back Human Cloning Ban: Remarks by the President on Human Cloning Legislation. Available at www.whitehouse.gov/news/releases/2002/04/20020410-4.html (retrieved on October 19, 2006).

In a speech setting the tone for the NSF's first conference on Converging Technologies (CTs),[8] former Undersecretary Phillip Bond, who lead the Office of Technology Administration in the Department of Commerce (DOC), laid out two key principles for the ethical development of nanotechnology: (1) "to achieve the human potential of everybody;" and (2) "to avoid offending the human condition."[8]

The goal of achieving everyone's "human potential" stresses the human dimension and raises the question, "potential for *what*?" The countervailing goal is that of avoiding "offense" to the "human condition." His charge offers a mirror image of the classic statement of medical values, the Hippocratic Oath, which sets forth the goals of doing good, and not doing harm; the principles of beneficence and nonmalificence.

The assumption of our engagement in the discussion of ethics and technology policy is that policy is necessary. Technical innovation and economic benefit will not necessarily prove congruent with the human good. That is to say, the choices of individuals as to where they see the fulfillment of their potential may have the effect of militating against the "human condition." Developments in what has been called "cosmetic neuropharmacology," as well as the prospect of "gene-doping" that could deliver potentially undetectable long-term "steroid" effects for athletes, are illustrative of the capacity of new technologies to rephrase old questions.[9] In such cases, few would challenge the need for regulatory controls.

The Human Genome Project in the United States spawned the acronym "ELSI" for the "ethical, legal and social issues" raised by that technology, and funded ELSI research (at a rate of first 3, then 5%, of total project expenditures) with two purposes: (1) to ensure early awareness of potential problems; and (2) in the process, to aid in the cultural acceptance of the technology, and thereby reduce its risk profile. As in every respect the stakes are higher for nanotechnology, the development of an effective nano ELSI (NELSI) process is central to a strategy for its success.

Following suit, the 21st Century Nanotechnology Research and Development Act of 2003 (the Act) seeks to ensure the vitality of NELSI in the unfolding of the technology and to provide commensurate resources.[2] These are key policy priorities. The Act repeatedly addresses the NELSI agenda.[2] Similarly, the NNI's strategic plan identifies several key research areas (economic, legal, ethical, cultural, science and education, quality of life, and national security) and divides the "responsible development of nanotechnology" into two categories: (1) environment, health and safety implications (EHS); and (2) nano technology's NELSI.[2]

The Act calls for a program that provides "public input and outreach to be integrated into the Program by the convening of regular and ongoing public discussions, through mechanisms such as citizens' panels, consensus conferences, and educational events, as appropriate."[2] In an effort to encourage effective public

[8]Roco, M. and Bainbridge, W. S., eds. (2002). *Converging Technologies for Improving Human Performance.* Available at http://wtec.org/ConvergingTechnologies (retrieved October 17, 2006).
[9]Kass, L., et al. (2003) *Proceedings of the President's Council on Bioethics Session 6: Neuropsychopharmacology and Public Policy Council Discussion.* Available at http://www.bioethics.gov/transcripts/jan03/session6.html (retrieved on October 19, 2006); see also *Session 4: Enhancement 4: Happiness and Sadness: Depression and the Pharmacological Elevation of Mood.* Available at http://(bioethicsprint.bioethics.gov/transcripts/sep02/session4.html (Sept. 12, 2002).

engagement, the NNI has begun to support forums for dialogue with the public and other stakeholders (at museums and science centers, and through other agency outreach mechanisms), to disseminate information to the public, and to evaluate public perceptions of nanotechnology. In particular, the NNI intends to encourage interdisciplinary dialogue and to incorporate research on societal implications at university-based nanotechnology centers. The NNI anticipates funding for research involving the identification and assessment of barriers to the adoption of nanotechnology in commerce, healthcare, education, and environmental protection. Moreover, the Act singles out for special consideration "the potential use of nanotechnology in enhancing human intelligence and in developing artificial intelligence which exceeds human capacity."[10]

The Emerging Ethical Agenda

The transformative significance of nanoscale research and development across the technology, economy, and broader culture of the twenty-first century has been widely asserted, although there has yet to emerge a commensurate public debate about its implications.[11] The lack of public awareness of what many see as the most dramatic driver of change in the next generation has had the short-term benefit of exempting nanoscale research funding from political and press scrutiny, yet at the price of uncertainty and concomitant risk. How the public hears about nano, and the frame of reference with which initial awareness develops, may prove crucial to sustaining confidence in the future of the technology—both in political terms and in the establishment of markets. The scale of European disenchantment with genetically modified foods provides a sobering example of the immense problems that can be engendered by public unease with a new technology. As the application of nanoscale research in medicine, seen by many researchers and investors as the most exciting of all fields of nanoendeavor, would bring the technology closer to individuals in a more intimate context than any other, the ethical issues raised by nanobiotechnology lie at the heart of the risk profile of nano as the putative transformative driver of the economy and a fundamental change agent for human society itself.

The word "nanoethics" has begun to be employed to focus the ethical implications of nanoscale research and development, though the term has yet to find clear meaning. In some cases it seems to be used to refer to the entire range of nontechnical issues raised by the technology, essentially subsuming distinct legal and societal issues, as well as their implications for policy. This mirrors, in some degree, the loose manner in which "bioethics" is used, as it spans both a category of applied ethics and wider discussions of law and policy. A more focused basis is found in the cognate field of "engineering ethics." It remains to be seen how the semantic range of "nanoethics" will unfold. Its prime relevance, at least in the

---

[10]See footnote 3, see also 21st Century Nanotechnology Research and Development Act, S. 189, 108th Cong. § 9 (b)(10) (2004).

[11]This paradox has been helpfully discussed in: Keiper, A. "The Nanotechnology Revolution," *The New Atlantis*, Number 2, Summer 2003, pp. 17–34.

shorter term, would seem to be to issues of nanomedicine, though in that context the older term bioethics might be thought to serve equally well.

The Context of Bioethics

The last quarter of the twentieth century witnessed the emergence of "bioethics,"[12] a term coined around 1970, as a frame of reference for discussion and potential resolution of questions raised at the intersection of bioscience, medicine, law, policy, and ethics. Supplanting the older term "medical ethics," especially in the United States, both to illustrate the wider range of questions raised by the development of new technologies and to distinguish the discussion from one that had traditionally been focused chiefly within the medical profession, "bioethics" has sought to reframe the discussion of ethics outside of the traditional moral tradition of Western thought, which was grounded in an amalgam of its classical roots (especially the pervasive influence of the Hippocratic tradition from the Greece of late antiquity) and the Judeo–Christian legacy—from which, through the Enlightenment, the moral and legal assumptions of Western societies, and the instruments of international law, have largely been drawn.[13] As an interdisciplinary field with its strongest underpinning in philosophy, bioethics has sought to set the tone of public debate on new developments in medicine and bioscience, although, so far, it has shown little interest in emerging technologies, of which nanotechnology is the most prominent.

While various approaches have been proposed, the general trend of bioethics has been to move away from deontological models of ethics (in which there is "right" and "wrong," however it is established) toward utilitarian models (in which an assessment is made of degrees of benefit and detriment to the end that the "greatest good of the greatest number" may be established). While bioethicists have generally pulled back from approaches that would allow the good of some to override entirely the good of others (an approach that could readily be used, e.g., to justify not only the eugenics movement that swept much of the Western world in the first part of the twentieth century, and the barbaric human experimentation to which it helped lead the way in the 1940s in Nazi Germany and imperial Japan), there is no question that the utilitarian weighing of the good of some against the good of others offers a potent threat to the more traditional focus on individual human rights. As the potency and cost of interventions in human health and well-being grow, these considerations could achieve decisive significance in shaping the human future.

The second fundamental trend in contemporary bioethics is to some degree at odds with its focus on utilitarianism. It is common for bioethics writers to view

---

[12]For a survey of "bioethics," see Jonson, AR. The Birth of Bioethics. NY: Oxford University Press 1998; see also, Stevens, M. Bioethics in America: Origins and Cultural Politics. Baltimore, MD: Johns Hopkins University Press. 2000.

[13]See the United Nations Educational, Scienific, and Cultural Organization (UNESCO). Universal Declaration on Bioethics and Human Rights (Paris 2005). This Convention draws on the *Universal Declaration of Human Rights* (1948). See also, Council of Europe. *Convention for the protection of human rights and dignity of the human being with regard to the application of biology and medicine: Convention on Human Rights and Biomedicine* (Oviedo, 1997) CETS No.: 164.

the mainstream Western medical-ethical tradition in terms that are referred to as "paternalist," in which the physician essentially dictated treatment decisions to patients who were passive in the decision-making process. This seemingly monolithic model has been overturned in several ways—through court decisions in which individuals (or their proxies) have been accorded final responsibility for treatment (or nontreatment) decisions; through fresh approaches to medical education and practice that have stressed the patient's role; through a legal focus, especially in the United States, on "informed consent" procedures—exemplified in the Patient Self-Determination Act of 1991,[14] which required that all hospitals receiving federal funding ask patients on admittance if they possessed, or wished to write, an advanced directive for their care should they cease to be competent; and, above all, through the theoretical development of fresh understandings of the nature of ethics in medical practice, in which patient "autonomy" is the central organizing principle. The fit of this theoretical approach with the emphasis on autonomy in U.S. culture and legal practice has ensured its wide adoption.

The most common placement of "autonomy" is in a fourfold package of principles, together with beneficence, nonmaleficence, and justice. This approach adds considerations with broader societal implications to two keystones of the Hippocratic tradition (beneficence and nonmaleficence).

Within the Hippocratic tradition, the focus of medical ethics is on the individual physician–patient relationship. Despite accusations of "paternalism," and the acknowledged fact that individual physicians have abused the privilege of the relationship, the Hippocratic tradition sought to address the unique character of professional decision-making within the bounds of the physician–patient relationship. Indeed, it is widely credited with setting out not simply a model for medical consulting and decision making, but also with laying the foundations for medicine as a profession, and even for the notion of a profession. In an often-quoted statement, Margaret Mead, doyenne of anthropologists and one of the most influential social thinkers of the twentieth century, paid tribute to the vast influence of Hippocratism in these terms: "With the Greeks, the distinction was made clear. One profession ... were to be dedicated completely to life under all circumstances, regardless of rank, age, or intellect—the life of a slave, the life of the Emperor, the life of a foreign man, the life of a defective child ...."[15]

That is to say, far from serving as a charter for unbridled paternalism, the Hippocratic model constrains the freedom of physician and patient alike, with limits borne of clinical experience and unique pressures on all parties in the face of the ultimate life-and-death issues.[16] Conduct was constrained both in general terms (the Oath's stress on serving the good of the patient, and the famous maxim "do no harm," though it appears in the Hippocratic *Epidemics* and not the Oath) and in specific

---

[14]Patient Self-Determination Act. 42 U.S.C. §§1395cc, 1396a (1994).

[15]Levin M., *Psychiatry and Ethics* (New York: Braziller, 1972), citing a personal communication.

[16]See Cameron N. *The New Medicine: Life and Death after Hippocrates* (repr. Chicago: Bioethics Press, 2001). For the most influential account of Hippocratic origins, see Edelstein L. *The Hippocratic Oath* (Baltimore: Johns Hopkins University Press, 1943).

terms (the Oath prohibits both abortion and physician-assisted suicide, requires referral to experts for surgery, and stresses the privileged nature of the consulting relationship—no sex with patients or even their slaves, and utmost confidentiality). Yet, the context of these constraints is highly significant. Not only is the Oath a vow, sworn before the conventional Greek gods and goddesses, but it also sets out the social–professional framework of a community that is mutually accountable. This aspect of Hippocratism has been little noted, though it serves as the foundation for the professional model that in medicine and other fields survives to this day.[17] Its best illustration lies in the Oath's mandate that physicians teach medicine without charge to the sons of their colleagues, for medical teachers, in turn, to be supported as if they were fathers, and the requirement that, before any aspiring medical student could be taught clinical practice, he would first need to swear the Oath (i.e., in the language of the university, as it is not conceived as a graduation oath, but one taken upon matriculation).

The summarizing of the Hippocratic principles as "beneficence" and "nonmaleficence" is not improper from the patient perspective, but it neglects the mutually accountable professional community of medicine, and sunders the connection between medical ethics and the privileged particulars of the physician–patient relationship. The addition of autonomy to these two, while underlining the dignity and rights of the patient, is problematic, because it raises complex questions of competence and informed consent. By suggesting that, in place of the substantive values of the Hippocratic tradition, or the Western moral tradition in general, individual patients are expected to shoulder the moral burden for their own treatment is to place a heavy responsibility on people who are typically unable to bear it. That does not mean that their wishes should be ignored, but it draws attention to the limitations of information or competence to consent on the part of a typical patient. These limitations compromise the capacity of patients to provide consent that derives from what we might call "strong" autonomy. A model of "weak" or modified autonomy requires that the physician, the medical community, or wider society, accept shared responsibility for the norms according to which treatment is undertaken. It could be argued that traditional Hippocratic assumptions about the physician–patient relationship evince essentially such a "weak" autonomy model (e.g., the patient sought out the Hippocratic physician whose values were publicly known, in the pluralist medical context of late antique Greece). By the same token, medical practice in the early years of the twenty-first century is driven by a similar model, whatever its theoreticians may declare; most patients look to physicians for advice on treatment options, and will often ask outright, "What would you do, doctor?"

Moreover, the adoption of "justice" as a key principle of bioethics is also problematic in strong–weak terms. Like the stress on autonomy, it tends to "prove too much." Is the reference intended to suggest that the individual physician should weigh the claims of a particular patient based on time and resources as against those of other patients—in the same hospital, the same healthcare system, the

[17]See Freidson E. *Profession of Medicine: a Study in the Sociology of Applied Knowledge* (New York: Harper and Row, 1970).

same nation, or around the world? Or is it intended to guide medical managers and policymakers in their allocation of resources? The latter fits better a model of decision-making, and removes from the individual physician a responsibility that might be seen to undermine entirely his or her commitment to a particular patient—in the interests of other patients, real or theoretical. Of course, it raises acutely the fundamental ethical problem of how any such comparison may validly be made. While ideas of distributive justice play a key part in policy processes, both within and to a lesser degree between nations, the idea that an individual physician should weigh such considerations in treatment decisions for his or her patient is problematic, and could be held to be deeply unethical. It represents a radical departure from the Hippocratic norm, with its focus on the privileged nature of the individual consulting relationship, and opens the door to utilitarian ideas of value. Taken to its logical conclusion, it also has the ironic effect of effacing much of the significance of patient "autonomy" by assigning serious (basically cost) constraints on its exercise. The unresolved and problematic character of contemporary bioethics offers a salutary context for potentially more complex and demanding issues that will be raised by nanotechnology—in medicine, as well as in wider applications.

AN ETHICAL AGENDA FOR NANOTECHNOLOGY

It will be seen that these questions from the discussion of medical ethics and bioethics have radical implications for the prospect of nanomedicine, because, in essence, applications of nano to human medicine will offer more dramatic choices, more compelling prospects of cure, and (surely) options that involve a greater element of cost. They will also spill over into the question of the integrity of the human condition, breaching the divide between applications that are unambiguously therapeutic in character and those that focus on the "enhancement" of human nature. Such issues have been presaged in existing practices, chiefly those of "cosmetic" plastic surgery and the use of steroids and other performance-enhancing drugs.[18] We return to this issue later in the chapter.

From one perspective, the ethical questions raised by nanotechnology represent the familiar issues that all technologies entail. Yet, the hopes and expectations that have been raised for the applications of nanotechnology for human well-being are so great that the ethical implications are potentially of a different order of magnitude. Indeed, they have the effect of transforming discussion of the particular applications of a particular technology at the nanoscale into a point of focus for our consideration of the place of technology in relation to human nature and human society as such. The emerging discussion on nanotechnology and human values, therefore, takes on a symbolic role, as a surrogate for our entire social conversation about what we do with the work of our hands and brains, and who we are. These following questions offer an ethical

---

[18]The U.S. President's Council on Bioethics' report *Beyond Therapy: Biotechnology and the Pursuit of Happiness* (Washington, DC: President's Council on Bioethics, 2003) offers the most useful review of the therapy/enhancement distinction and its problematic though vital character.

lens through which to assess the claims and proposals of nanotechnology in its many applications, in medicine and beyond.

1. The question of the use of hype to raise expectations and make predictions that are not properly grounded.

This is an issue of ethics, yet it could prove to be a significant factor in determining whether public confidence in this technology—and indeed in technology as a whole—will be maintained. This problem is already being discussed outside the professional literature. A 2006 issue of *The New Yorker* magazine quoted Nobel prize winner Paul Nurse's sharp criticism of the claims of Andrew Von Eschenbach, then head of the U.S. National Cancer Institute, that thanks to advanced in nanomedicine we will "eliminate suffering and death due to cancer by 2015."[19] Nurse commented: These statements "cannot be justified even as a statement of aspiration . . . because when we fail to deliver, as we surely will . . . we will lose the confidence of both the politicians and the public."[20] As public debate in the United States and Europe on embryonic stem-cell research has demonstrated, the public is simultaneously susceptible to hyped claims for the outcomes of science and technology programs, and suspicious of them.

2. The need for the public to be fully informed.

This extends the particular problems raised by hype onto a much broader canvas. Within the context of democratic accountability, there are always anxieties on the part of the scientific community that people will misunderstand their work, or will underestimate its significance, or will be unduly fearful of its outcomes. This tends to result in reticence and resistance on the part of scientists that their efforts should be transparent—and it also results in their own uncertainties about the outcomes of their work. The need for honesty in the relationship between researchers and the public, which often funds such work, is great.

3. The question of hazard: What risks are appropriate?

While issues of safety are always also issues of ethics, the ethical dimension of nanotechnology risk is in proportion to the potential dangers of the technology. The cautious approach taken in the 2004 Swiss Re report[21] suggests that while some of the detractors of nanotechnology may overstate its risks to health and the environment, and while the likelihood of unintended harm may be low, the scale of damage that could result from misjudgment could prove vast.

---

[19]Miller M. (May 16, 2003). 2015: A Target Date for Eliminating Suffering and Death Due to Cancer. Benchmarks vol. 3 issue 2.
[20]Specter M. (Mar. 13, 2006). Editorial. *New Yorker*, vol. 82 issue 4, pp. 58, 68.
[21]Hett A. Nanotechnology: Small Matter, Many Unknowns (Swiss Re, 2004).

4. The clinically-related ethical questions that focus on the application of technology to individuals.

Approaches to recombinant DNA therapy—focused in the United States in the listing of key ethical "points to consider"[22]—have framed the ethical application of human genetics. This has special relevance, as *mutatis mutandis*, such principles apply to clinical applications at the nano–bio interface.

5. Broader challenges that these new technologies present for the social order and the wider human community include threats to confidentiality.

Such prospects as large-scale diffusion of radio frequency identifier (RFID) chips, retinal scanning, face identification technologies, and so-far-undeveloped options may render privacy a costly (or even nonexistent) commodity. The preservation of medical confidentiality has already been rendered enormously more difficult by the development of electronic databases.

6. Military applications raise special questions.

A substantial portion of public funding for nanotechnology is focused on defense budgets. Plainly, some of this work will be classified, although, as with intelligence data, it should be reported to politicians with appropriate security clearances, and, by the same token, to ethicists outside of the chain of command. The extent to which military exceptions to general civilian concerns (e.g., in the area of "enhancements") should be permitted is a question for democratic accountability.

7. Issues of equity, which have been termed the "nanodivide."

Despite the hopes of some that technology at the nanoscale will prove ultimately very cheap, it is a reasonable assumption that its applications to medicine will result in very costly treatments. Thus, the suggestion that "all cancer" will be curable by 2015 is unlikely to include the cancer of all persons afflicted with the disease, but rather the cancers of the wealthy and those with access to medical insurance, private or social.

8. The fundamental question is that of the human condition itself; the hardest to address, and yet the most significant.

At an intuitive level, this question is both clear and, for many, of first importance. Yet, it is questionable whether we can articulate with sufficient clarity, for policy purposes, a description of the "human condition." A major theme of the U.S. President's Council on Bioethics report on enhancement technologies, *Beyond Therapy*,[23] is the difficulty we face in drawing such lines. But there is no more important question, as the fundamental challenge of this technology is to our anthropology and the assumptions we make about human being and what is proper for ourselves.

---

[22]"Recombinant DNA Research: Actions under the Guidelines," *Federal Register* 60 (81): 20731–20737; 27 April, 1995; repr. various places including LeRoy Walters and Julie Gage Palmer, *The Ethics of Human Gene Therapy* (New York: Oxford University Press, 1997), pp. 171–185.

[23]The President's Council on Bioethics (Oct. 2003). *Beyond Therapy: Biotechnology and the Pursuit of Happiness* (Washington, DC).

THE PROSPECT OF "ENHANCEMENT"

One recent writer on nanotechnology ethics has put the matter thus:

> Among the applications of nanotechnology that some researchers consider "science fiction," while others are actively attempting to implement, are enhancements to human memory, physical strength, and other characteristics. Though usually framed as attempts to monitor or repair ailments or disabilities such as Parkinson's disease or genetic abnormalities, some of these technologies can simultaneously be used to control or enhance particular human characteristics in "normal" humans as well.[24]

In his notorious essay, "Why the Future Doesn't Need Us," technology guru Bill Joy proffered alternative scenarios of doom: either unintended disaster or intentional enhancement will ensure the end of human nature as we know it.[25]

Recent discussion of "converging technologies" as the context for nanotechnology draws attention to the interconnected challenges they present, above all to human nature. Leon Kass, later chairman of the U.S. President's Council on Bioethics, remarked on the interrelations of these technologies, such that advances in genetics "cannot be treated in isolation," but must be correlated with "other advances in reproductive and developmental biology, in neurobiology, and in the genetics of behavior—indeed, with all the techniques now and soon to be marshaled to intervene ever more directly and precisely into the bodies and minds of human beings."[26]

In that respect, the recent response to some U.S. approaches to "converging technologies" [nanotechnology, biotechnology, information technology, and cognitive science (NBIC)] from the European Commission's High Level Expert Group (HLEG) offers a valuable counterweight. The appointment of the HLEG was sparked by the appearance of the 2002 NSF report, *Converging Technologies for Improving Human Performance* (the NBIC report), which gave credence to "transhumanist" aspirations that see nanotechnology as a route to the transformation of human nature into some "posthuman" form—via progressively "enhanced" human being to machine intelligence that could supplant corporeal *Homo sapiens* altogether. The prominent inclusion of these perspectives in a document published by the U.S. government had the effect of giving them greater significance than was warranted. The NBIC report, of course, simply reflects the proceedings of a particular conference. While some readers have enthused about the visionary character of the NBIC report, others see parts of it as both bizarre and naïve. It does, however, usefully provide a basis for discussion of claims that work on the nanoscale will lead to "enhancement" applications, which transhumanists view as paving the way to their vision of becoming something beyond human. The extent to which

---

[24]Lewenstein B. (2005). What Counts as a 'Social and Ethical Issue' in Nanotechnology? HYLE–International Journal for Philosophy of Chemistry, Vol. 11, No.1 (2005), pp. 5–18.

[25]Joy W. R. (Apr. 2000). Why the Future Doesn't Need Us. *Wired*, issue 8.04. Available at http://www.wired.com/wired/archive/8.04/joy.html.

[26]Kass L. R., *The Moral Meaning of Genetic Technology*, Commentary, Sept. 1999, pp. 34, 35.

such developments are grouped within the category of nanomedicine is more semantic than substantive, and reflects the ambiguity with which nonmedical (or, to be
more precise, nontherapeutic) uses of medical skill have been put in more
mundane technologies, and the unease with which they are seen as "medical" at
all—from physician participation in the execution of criminals, through cosmetic
plastic surgery, to the use of illegal steroids by athletes. Two of these three examples
offer current and somewhat rudimentary illustrations of putative "enhancements."
The shift from therapeutic to elective applications of medical skills, while today
on the fringe of healthcare practice, is destined to grow. One effect will be to
wrench the delivery of what we may term quasimedical applications further from
the professional tradition that even today offers a powerful context for the practice
of medicine. It may be expected that "nanoethics" will find a prime focus in quasi-
medicine that is developed from nanoscale research and development. This will
especially be the case with the "enhancement" of human intelligence.

This is noted emphatically in the European Commission HLEG report, in which
of the "converging technologies" of nanotechnology, biotechnology, information
technology, and cognitive science, it is the last named that raises quite the most
serious questions.[27] By the same token, it is singled out in the 2003 21st Century
Nanotechnology Research and Development Act of the U.S. Congress.[28] But the
matters of enhanced and artificial intelligence may not prove the most immediate
challenge to the integrity of the mind. It may be most starkly illustrated with reference to the prospect of the "pursuit of happiness" by means of cognitive "enhancements" that involve the manipulation of perception and memory, whether through
neuropharmacology (including what has been termed "cosmetic neurology") or
cognitive prostheses. A recent editorial in the journal *Neurology* discussed the challenge of nanomedicine in these terms:

> its presence is already beginning to be felt in neurology. Cochlear implants are the senti-
> nel example of mechanical interfaces providing sensory input to the human nervous
> system. Neural stimulators—for movement disorders and epilepsy—are other examples
> of technologies currently in (increasing) use. Some worry that these successes represent
> the beginnings of Cyborgs—individuals who are part human and part machine. For more
> than 50 years science fiction writers have imagined the potential for such human-robotic
> chimeras. Nanotechnology promises the potential of designing micromachines capable of
> dramatically advancing the potential of such interfaces.[29]

The initial development of such technologies will either be "dual use"—with
primary applications that are therapeutic in nature—or military. The ethics and
policy questions will, therefore, relate not primarily to the development of but
rather to particular applications of the technology for nontherapeutic or, in the
case of military application, civilian use.

[27]European Commission—High Level Expert Group (HLEG),(Alfred Nordmann, Rapporteur) (July 2004).
*Foresighting the New Technology Wave: Converging Technologies—Shaping the Future of European
Societies*, p. 12. Available at http://ec.europa.eu/research/conferences/2004/ntw/pdf/final_report_en.pdf.
[28]21st Century Nanotechnology Research and Development Act (2003), Pub. Law 107–314.
[29]Hauser S. L. (Sept. 2004). The Shape of Things to Come, *Neurology* Vol. 63: 948, p. 949.

THE QUESTION OF "NANOETHICS"

We have touched on the two distinct sets of issues raised as we look to the development of a nanoethics. The first concerns the nature of the exercise. Will it essentially be continuous with the "bioethics" that has come to dominate the public conversation about human values and biomedicine, adopting approaches that, in terms of ethical theory, may be generally described as utilitarian, though focusing strongly on (and somewhat in contradiction with) individual autonomy and "informed consent" as the fulcrum of the system? Is current "bioethics" to provide the basis of the "nanoethics" of the future? Any claim that it should, or expectation that it will, needs to engage the crisis of contemporary bioethics, which lies in its general withdrawal from substantive questions into a focus on procedural, decision-making, concerns; at the same time as it is still widely viewed by the public as a critical discipline pronouncing on the "rightness" or "wrongness" of procedure A or proposal B as the basis for private choices or public policy.

The evacuation from substantive moral engagement that has been characteristic of this generation's bioethics is partly the result of a commendable effort to find public traction for a discussion that has in the past largely been generated within particular communities—both within religious traditions, and also within a medical community that even today bears the lineaments of its Hippocratic antecedents. But the path of bioethics, instead of seeking consensus where it can be found and building on the commonalities evident across the diverse communities of contemporary culture, has led ironically in a direction that has emphasized diversity and led to a far more pluralistic understanding of Western (and, in this case, American) culture than the facts suggest. The cure has ended up exacerbating the disease. Nowhere is this more evident than in the secularism that underlies the bioethics of the early twenty-first century, in marked contract to that of a generation earlier when from both liberal and conservative directions (Paul Ramsey vs. Joseph Fletcher) the first generation of "bioethicists" sought to grapple in public language with the distinctions of the moral tradition that remains immensely powerful in Western, and especially American, civilization. The achievement of the UNESCO *Universal Declaration on Bioethics and Human Rights* (2005) lies not least in its refocusing the fragmented efforts of contemporary bioethics in the Enlightenment concepts of "human rights and fundamental freedoms." By the same token, a further corrective is offered by the *United Nations Declaration on Human Cloning* (2004), which urges all member states to prohibit "all forms of human cloning." While most U.S. bioethicists have been supportive of research or so-called "therapeutic" cloning, the United Nations vote shows how out of touch they are with opinion in the Western democracies, in which bioethics remains more firmly rooted in the distinctive religious and moral traditions of particular communities. (By way of illustration: only two European states currently permit "therapeutic cloning," and Canada has made it a felony.)

Thus, to question the fitness of current American bioethics to tackle the immense tasks raised by emerging technologies is, in fact, to raise a question of another kind, and ask whether current approaches will, in due course, collapse and be replaced by something that is more akin to the Enlightenment model underling the multilateral documents and more common in Europe (where, by an irony that should be noted, there are significantly higher levels of religious participation in public bioethics, despite equally significantly lower levels of personal religious observance than in the United States). It may be that the emergence of fundamental questions about the human good and the good of society in the context of emerging technologies will precipitate such a shift. The effort to "privatize" ethical decisions and assume an atomistic view of society is problematic enough in relation to "living wills" and, at the other end of life, preimplantation genetic diagnosis and fetal screening tests. Whether it can survive the challenge of radical efforts at the manipulation memory and perception, and human "enhancement," is another question.

The problem is brought into ready focus by the manner in which bioethics has essentially emerged as the conjoined twin of biopolicy. While in its emphasis on pluralism and privatization this has led mainstream bioethicists to oppose restrictive or regulatory approaches to technology, in the case of a re-emergent deontology, seeking commonalities across traditions and communities on the basis of the Enlightenment, the twinning of bioethics and biopolicy could readily work in the opposite direction, and offer the basis for a nanoethics that builds a vision for the common good—on the basis of shared convictions about "human rights and fundamental freedoms," and with a flipside in approaches to biopolicy that are not shy to encourage appropriate regulation.

Note that the second aspect lies in the substance of nanoethics. We have listed eight key areas where basic ethical issues are raised, including the enhancement question that transcends all others, not least since it suggests the possibility of an entirely fresh context in which to address the question of the human good, changing the conditions for anthropology in ways that could ultimately efface the *anthropos* by manipulating humankind, step by step, into some other kind of being.

In the near future, however, the task of nanoethics will be to aid in the proper informing of the various communities and traditions within our society, and to seek commonalities where otherwise secular and religious worldviews come to a common point—as the basis for shared vision that may be articulated across the various networks of civil society that make up this nation and all free nations. It is hard to see how such a task can be undertaken by "bioethics" in its current form, though not perhaps impossible to imagine the recrudescence of an older ethics that takes as its mainspring the Enlightenment tradition with its taproots in both pagan and Judeo–Christian worldviews that have shaped not simply the West, but the modern world. A focus on the integrity and givenness of human nature, and the assertion of human rights and fundamental freedoms, will span the gulfs of civil society and offer a substantive basis upon which to engage the greatest challenges to ethics and policy that have yet been faced.

CHAPTER 17

Anticipating the Impact of Nanoscience and Nanotechnology in Healthcare

DEBRA BENNETT-WOODS

(The master of it) anticipates things that are difficult while they are easy, and does things that would become great while they are small. All difficult things in the world are sure to arise from a previous state in which they were easy, and all great things from one in which they were small.

Lao Tzu[1]

INTRODUCTION

At first glance, the master philosopher Lao Tzu provides rather straightforward advice on the importance of planning. By anticipating problems and other untoward events early in the process, one has the time and clarity to plan for contingencies. An underlying assumption of the strategic planning process in business is that good planning provides better control and a more effective response to competition and other challenges of the business environment. Strategic planning generally includes an informed analysis and forecasting of the market, critical assessment of the assets and liabilities of the company, and agreement on clear goals and objectives in support of organizational mission, values, and culture. The planning imperative assumes the foundations of an enterprise are more easily built in the beginning than constructed after the fact. This insight is particularly true when the environment is increasingly fast-paced, turbulent, and complex.

If we apply the same basic thought process to planning for nanotechnology (NT), Lao Tzu's words suggest an even deeper level of meaning and cause for reflection.

[1]This translation from the Tao Te Ching by James Legge (1891) can be found in the *Sacred Books of the East, The Texts of Taoism,* Vol. 39. Available at http://www.sacred-texts.com/tao/sbe39/index.htm (retrieved on October 16, 2006).

Nanoscale: Issues and Perspectives for the Nano Century. Edited by Nigel M. de S. Cameron and M. Ellen Mitchell
Copyright © 2007 John Wiley & Sons, Inc.

NT becomes both the literal and metaphorical epitome of "the great" originating from the small and "the difficult" originating from what was previously easy. Engineering of the very small is being promoted as the next great thing in human technological dominance. Paradoxically, while the technical challenges are immense, the world around us has always been effortlessly engineered at the atomic and molecular scales. Similarly, current technologies may come to seem absurdly crude when compared with the simple elegance of manipulation at the molecular level; however, the issues raised by the power to engineer everything from consumer goods to the consumers themselves may prove far thornier and more difficult than concerns of the past.

The goal of this chapter is to marry these two levels of meaning in an effort to anticipate the systemic impact of nanoscience and nanoengineering on healthcare in the United States. Few industries provide more complex and fertile ground for strategic analysis,[2] and perhaps no industry represents as broad a potential for deep societal impact. So, what are the elements that must be addressed in a comprehensive strategic assessment, if we are to successfully plan for the societal impacts of NT on the healthcare system? How do the practical and metaphorical implications of NT itself inform the deeper analysis of what promises to be a paradigmatic shift in human ability?

I begin with a general discussion of the mandate for analysis, followed by a brief overview of current and anticipated NT-enabled applications in medicine. Concepts in frame analysis are introduced along with a popular model of organizational framing proposed by Bolman and Deal.[3] This model is then adapted and employed to illustrate a conceptual outline for strategically targeted and cross-disciplinary research, analysis and intervention.

A STRATEGIC MANDATE

Nanoscience and nanoengineering, which involve the study and manipulation of matter at atomic and molecular scales (typically 1–100 nm), are predicted to yield an abundance of enabling applications in biotechnology and medicine.[4] To the extent these applications live up to predictions, they are likely to pose a complex array of practical and ethical challenges. These challenges extend beyond the medically possible to cross into the spheres of social, cultural, political, and economic systems and understandings. The emerging field of research into the societal implications of NT abounds with calls for assessment of the likely impacts;[5] however, the task of conducting such a complex, prospective analysis is daunting.[6] It is as if we are being called upon to anticipate the manner and extent to which science and technologies that do not currently exist will affect a society that is

[2]Zuckerman, 1998.
[3]Bolman and Deal, 2003.
[4]Roco, 2005.
[5]Berube, 2006.
[6]Bennett-Woods, in press.

itself undergoing rapid, complex, and inherently uncertain political, economic, and social change.

While some societal disruptions can be envisioned based on technologies and trends of the past, NT suggests outcomes in biotechnology and medicine that are largely unprecedented, rendering prediction inherently difficult. At the same time, the current pace of research and development does not allow for more than a few years to assess, plan, and respond to evolving concerns. The general area of human health, and the field of medicine specifically, may prove particularly problematic to efforts in risk and impact assessment. The healthcare enterprise is a complex system. Insights into the behavior of complex systems found in chaos theory and complexity science show that even small changes can quickly cascade throughout a system.[7] Given the multitude of concerns expressed regarding the convergence of NT with biotechnology and related fields, a systematic approach that represents a multidisciplinary perspective is warranted. Toward this end, concepts in strategic frame analysis, from the field of organizational theory, will be introduced as a means of framing likely effects of NT in medicine and on the larger healthcare system.

SCOPE AND BACKGROUND

NT is projected to have a broad impact in both applied knowledge and the actual practice of medicine. Less attention has been paid to related and equally important effects on the healthcare system itself. How will the basic infrastructure of healthcare be altered in response to new medical realities? Will the general public question the safety of emerging medical technologies, and will the current healthcare work force be prepared to deliver them? How will limited resources be reallocated, and who is most likely to benefit? Will the prospect of radical human enhancement and life extension alter the very definition of human health and the goals of medicine? Although such questions can be broadly posed, the intent here is to focus narrowly on medical practice within the healthcare system of the United States, taking into account its unique infrastructure and cultural context.

In terms of background, even a brief overview of the potential medical applications of nanoscience and engineering is adequate to convey the magnitude of the impact it will have on the practice of medicine. While it is fair to say that working at the atomic and molecular levels is not new in medicine, the convergence of NT with other disciplines and technologies has already poised diverse areas of biomedical research on the brink of significant advancements. The term "nanomedicine" has been coined to represent a range of medical applications of nanotechnology. Areas of likely impact include pharmaceuticals, medical diagnostics, and medical devices and implants.

The pharmaceutical industry has a major stake in NT, which promises to enable methods that simplify, speed up, and reduce the costs of drug development and

[7]Coveney and Highfield, 1995.

testing, as well as increasing drug safety and efficacy.[8] Biocompatible nanoparticles may provide a new platform for drug delivery, including alternative routes of administration for existing drugs that will minimize drug degradation, allow site targeting, and reduce side effects.[9] In addition, nanoparticles can be designed so that drugs can be integrated into the body of the particle or attached to the surface, and then effectively delivered to specific organs or tissues, across membrane barriers, in a controlled release mode. For example, biodegradable polymer nanoparticles appear to be ideal candidates for cancer therapy, vaccine delivery, contraceptives, and targeted antibiotic delivery.[10] Future developments include "smart" delivery systems that are increasingly sensitive and responsive to changes in drug concentration and other factors.

Applications in pharmacogenomics will also allow for faster and more efficient discovery, while pharmacogenetics will enable better-targeted, more effective medicines and related therapeutic interventions.[11] Applications of nanoparticles are envisioned for the replacement or repair of defective or nonfunctional genes, as well as the possibility of genetic immunization with deoxyribonucleic acid (DNA) vaccines.

Much attention has also been given the area of medical diagnostics. Better image resolution, longer tissue retention, and tissue-specific targeting are being achieved through the use of nanoparticles in contrast agents for medical imaging.[12] Nano-enabled fabrication of high throughput screening microarrays of DNA, protein, carbohydrates, cells, and tissue will enhance rapid genotyping, genetic analysis, and DNA resequencing. Microarrays are also being developed for application in early diagnosis and monitoring of cancer, genetic epidemiology, tissue typing, microbial identification in infectious disease, and drug validation.[13] Together, the combination of enhanced diagnostics and therapeutics may lead to an increasingly predictive and preventative model of what has been termed personalized medicine.[14]

Another area of intense effort involves nanofabrication tools and techniques in the construction of medical devices and implants. Unique properties at the nanoscale will increase biocompatibility and enhance integration and longevity of implanted medical devices and prostheses.[15] Other projected developments include the creation of artificial cells, tissues, and organs. For example, artificial cells are currently being studied for treatment of diabetes and liver and kidney failure. Artificial blood, skin, and bone are also likely targets for development.[9]

Optimistic projections into longer-term nanomedical advances (15–30 years) are even more dramatic. J. Storrs Hall describes advanced surgical techniques that will perform repairs at the cellular level while the patient plays tennis in

---

[8]Ferrari and Downing, 2005.
[9]Kubik et al., 2005.
[10]Kayser et al., 2005.
[11]Lindpainter, 2002.
[12]Mazzola, 2003.
[13]Campo and Bruce, 2005.
[14]Weston and Hood, 2004.
[15]van den Beucken et al., 2005.

virtual reality rather than undergoing anesthesia.[16] He also projects a virtually endless supply of "spare parts" in the form of high-performance artificial organs, as well as molecular machines that can literally roll back the effects of aging. These nanorobots are envisioned to operate continuously within the body to detect and mitigate disease, and repair or replace cells, tissues, and entire organ systems damaged by trauma or age.[9]

Such powerful capabilities naturally open the door to projections of radical extension of the human lifespan, as well as radical enhancement of basic human capabilities. Why replace red blood cells with an exact replica when it might be possible to create an artificial cell capable of carrying enough oxygen to allow you to survive an extended loss of cardiac or respiratory function? If it is possible to create neural implants that preserve memory in a brain damaged by Alzheimer's disease, why not also provide enhanced capacity for general information storage and computational speed? Why stop at repairing a gene when you can select for or engineer genetic enhancements prior to birth?

FRAME ANALYSIS

> The system of nature, of which man is a part, tends to be self-balancing, self-adjusting, self-cleansing. Not so with technology.
>
> E. F. Schumacher[17]

Where does one begin to project the impact and potential concerns raised by a technological shift in the medical paradigm? As suggested by Schumacher above, we cannot assume the healthcare system will automatically self-correct as issues arise. The impacts are far-reaching and complex in their interrelationships. Any prospective analysis must assume a systems approach capable of anticipating effects at many levels. In addition, the standard academic approach in the social sciences is to observe what is, or wait for something to happen and then study it after the fact. What is needed here is a set of anticipatory tools and methods that account for the high levels of uncertainty inherent in paradigmatic change.

The origins of frame analysis are generally credited to sociologist Erving Goffman, and his seminal book *Frame Analysis: An Essay on the Organization of Experience.*[18] Goffman proposed a method for studying visual images and cultural representations. The concept of framing has since been loosely applied across a range of disciplines, including cognitive psychology, sociology and social movement theory, linguistics and discourse analysis, communication and media studies,

---

[16]Storrs Hall, 2005.

[17]This quotation is attributed to Ernest Friedrich. Shumacher (1911–1977), a British economist and advocate of sustainability. He proposed an "intermediate technology," that was affordable to poor people and could lead to higher productivity with less social dislocation. Available at http://www.brainyquote.com/quotes/authors/e/e_f_schumacher.html.

[18]E. Goffman, 1986.

political science, and policy studies.[19] While there is no generally accepted definition of a frame, all branches of frame analysis seek to explain how we process information to generate meaning.[20]

Frames are variously referred to as perspectives, mental models, lenses, windows, categories, or patterns that effectively focus our attention, organize our assumptions, and construct meaning. A simple example of framing in business practice is the SWOT analysis, in which an array of data and assumptions is used to generate lists of strengths, weaknesses, opportunities, and threats—four related, but also quite different, frames for business planning. While frame analysis may not always provide an immediate or obvious answer to a problem or concern, the process of framing reveals key questions about what has happened or what is likely to happen, as well as important intersections between frames and the potential for multiframe solutions. Furthermore, a model that naturally integrates disciplinary perspectives can begin to construct a common language that encourages a more efficient and collaborative approach.

A popular model that employs the basic concept of framing is found in organizational theory. Bolman and Deal propose a four-frame model that encompasses four major schools of thought in organizational theory and through which organizational realities can be analyzed and explained.[3] This model is of particular interest because it addresses the benefits of cross-disciplinary thought in complex analysis, and has been shown empirically to improve managerial effectiveness. The four frames are termed structural, human resource, political, and symbolic. Each frame represents a particular organizational reality, variously drawing on organizational perspectives from within sociology, psychology, political science, and anthropology.

The remainder of the chapter will employ a loose adaptation of the Bolman and Deal model to demonstrate the value of frame analysis when attempting to anticipate the impacts of NT on healthcare in the United States. The term organization should be read interchangeably with the healthcare system. The intent is illustrative rather than comprehensive, using the model simply as a brainstorming tool for foresighting and other planning activities to generate a more comprehensive and adaptive approach.

The Structural Frame

Rooted in economics and sociology, the structural frame is primarily concerned with maximizing the efficiency and performance of formal structures and processes in the pursuit of clearly understood and shared goals and objectives. Problems are viewed narrowly as structural deficiencies. They are assumed to be amendable to rational analysis, expertise, formal modes of communication, and subsequent forms of coordination and control, such as restructuring and implementing policies and procedures. In this frame, organized attempts are made to minimize and manage uncertainty using rational analysis and formal mechanisms of control. From the standpoint

---

[19]Benford and Snow, 2000.
[20]Fisher, 1997.

of the structural frame, the healthcare system represents a particularly broad, diffuse, and complex organizational structure with many related industries and delivery settings nested within the larger organization. As a result, many structural issues can be raised regarding the likely impacts of NT-enabled medicine.

Medical Efficacy. The history of medicine is one of discovery and increasingly effective responses to human disease, trauma, and aging. Between 1900 and 2002, the average U.S. life expectancy has increased from 48 to 75 years for men and from 51 to 80 years for women, due largely to the prevention and control of infectious disease.[21] Nano-enabled medicine and related biotechnologies are poised to substantially extend this general trend, initially through a combination of early diagnosis and innovations in treatment, later through prevention strategies related to genetic screening and manipulation, and eventually through body system repair and replacement.

In this subframe, NT is judged solely on the objective evidence of its ability to achieve the goals of healthcare as defined by reduced mortality and morbidity. The NT-enabled medical advances that replace and improve upon less effective approaches to diagnosis, treatment, and prevention are likely to achieve wide acceptance, so long as they prove safe. Eventually, medical efficacy may include various measures of human enhancement and the extension of the normal human lifespan beyond its currently observed natural limit of 120 years. These longer-term outcomes may prove more problematic for reasons raised in the other three frames.

Structure, Focus and Finance. Bawa identifies several structural barriers to commercial nanomedicine including large-scale production challenges, high production costs, a lack of safety and regulatory guidelines, scarcity of venture funds, and a limited number of near-term commercially viable products.[22] However, recent advances are making their way into medicine and early forecasts are favorable for NT commercialization as these structural barriers are overcome. The larger challenges may lie with the healthcare system itself.

Consider the current structure, focus, and financing of the U.S. healthcare system. The expenditures for healthcare in the United States totaled $1.7 trillion in 2003, 15.3% of gross domestic product, and more than any other country.[23] After a protracted struggle, which commenced in the early 1980s to control costs and create a more efficient and effective delivery system, expenditures continue to increase at more than double the rate of inflation. The use of acute inpatient services has declined overall as technological advances have allowed for more treatments to be delivered in outpatient settings or managed at home. However, inpatient care has become more complex and costly.[23] The fastest growing expenditure is for prescription drugs, increasing 11% in 2003.[23] Hospital care accounted for 31% of U.S.

---

[21]National Center for Health Statistics (NCHS), 2005.
[22]Bawa, 2005.
[23]NCHS, 2005.

health expenditures in 2003, followed by physician services at 22%, prescription drugs at 11%, and nursing home care at 7%.

The impact of NT on the structure and focus of healthcare will depend on how research and development is targeted, where the breakthroughs occur first, and how the market responds. The brief overview of current research directions emphasizes prevention through earlier and more effective screening, monitoring, and diagnosis, as well as the potential for less invasive surgical techniques, pre- and postnatal genetic interventions, and personalized approaches to drug design. Long-term potentials include organ and tissue replacement, as well as embedded systems for health monitoring and repair.

Each of these possibilities suggests a diminishing need for acute inpatient hospital care and far more technologically complex primary care and outpatient services. Betta and Clulow describe this as a fundamental shift from the hospital to the laboratory, in the form of predictive and manipulative services enabled by advances in biotechnology.[24] Although this trend has been evident in healthcare for a couple of decades now, the rapid dissemination of NT-enabled advances may accelerate the movement away from traditional inpatient hospital care as the centerpiece and anchor of the healthcare system. At the same time, the care that remains in the hospital setting will also reflect the new technological realities of advanced diagnostics and monitoring, individualized treatment capabilities, nanosurgery, and eventually the maintenance and repair of biocompatible implants and replacement organs.

Therefore, one central concern of the structural frame will be adapting the current infrastructure of healthcare toward a new primary focus on prevention. Both hospitals and alternate delivery settings will need to make substantial investments in retooling their material infrastructure to accommodate new equipment and services in the face of market pressures. The same will be true for the pharmaceutical industry and companies that provide outsourced laboratory and imaging services.

A second significant concern related to structure and focus is that of cost. Will NT-enabled technologies increase or decrease the overall cost of healthcare? Again, this will depend, in part, on how quickly and in what order NT-enabled technologies make their way into the healthcare system. While the term cost effective is occasionally applied to certain aspects of nanomanufacturing, the concept of personalized medicine intuitively suggests fewer economies of scale and higher costs on the front end of healthcare delivery. The need for companies to recover substantial investments in research and development also suggests that early adopters of NT-enhanced technologies will pay a premium for quite some time. There are well-established precedents in other industries for expecting the price of specific healthcare technologies to drop over time. In addition, effective prevention should theoretically save money in the long run. What is very difficult to predict at this juncture is the timespan over which these changes will occur, and at what point we will reach a critical mass of either budget neutrality or actual cost savings within the system as a whole.

---

[24]Betta and Clulow, 2005.

But what if the system never realizes any measurable cost savings? Radical success in preventing and treating common conditions, especially those of aging, will result in more and more patients living toward the end of the normal human life-span or beyond. Currently, more than 45 million Americans have no health insurance, costs continue to rise, and the fastest growing demographic is people 85 and older.[25] Employers, insurance companies, government programs, such as Medicare and Medicaid, and communities will face escalating costs of providing healthcare for a growing and aging population.

Workforce. Changes in the structure and focus of healthcare will be closely tied to impacts within the healthcare workforce. Little or nothing has been written on the competencies needed to prepare various providers for a shift in focus from curative to preventive services. In a recent article, Pruitt and Epping-Jordan argue for the need to prepare the workforce to shift its attention away from acute to chronic care,[26] but neglects to mention anything about the rapid increase in medical knowledge or the impact of emerging technologies on the nature and scope of treatment delivery. Likewise, only passing mention was found in any recent article regarding the need for substantive changes in the basic educational curricula for healthcare providers to accommodate the impact of nanoenabled genomics and proteomics on the delivery of healthcare.

Shortages within the healthcare workforce have been a focus of attention since the early 1990s and are projected to continue well into the next decade and beyond. Will shifts in the focus and structure of the delivery system reduce the necessary size of the workforce and counteract the current and projected shortages? Will it increase the size of the workforce as new technologies are adopted within existing systems that have to be maintained during the transition? Will it change the mix of the workforce, introducing new roles and targeted expertise, while reducing others? How will retraining of the existing workforce be accomplished? How quickly can universities adopt new curricula and programs to support a potentially rapid shift in focus and practice?

Regulatory Environment. Of some additional interest in the structural frame is the legal and regulatory environment that has begun to evolve in response to rapid advances in biotechnology, concerns about public safety, and controversy over issues related to patent law. The structural results of efforts to regulate NT technologies and products will manifest in this frame, but are most likely to originate in the human resource frame and be enacted via the political frame.

General questions have been raised by scientists and consumer advocates about the possible toxicity of nanoparticles and the environmental effects of nonbiodegradability.[5] However, attempts to set safety standards have been hampered by a lack of agreement on the classification and definition of NT[22] and the absence

[25]He et al., 2005.
[26]Pruitt and Epping-Jordan, 2005.

of established risk analysis and managements protocols for NT.[27] Existing regulatory entities, such as the Food and Drug Administration (FDA), will have to address testing and approval processes accordingly, and may need to expand its scope in terms of what warrants FDA approval.

A second regulatory concern deals with issues in patent law blocking viable efforts at commercialization, particularly for nanomedicine start-up companies.[5] Because NT represents clusters of technologies, medical applications may be caught up in the emergence of one or more "patent thickets" in which overlapping patent rights require parties wishing to commercialize a new technology to obtain multiple patent licenses from multiple patent holders. Various concerns include long delays in bringing nanomedical products to market, underutilization of new discoveries due to the unwillingness of parties to invest in the face of complex bargaining arrangements, and unreasonably high costs being passed on to the consumer and the larger healthcare system due to elevated royalty fees from acquiring multiple licenses.

Human Resource Frame

The human resource frame has foundations in psychology, with an underlying assumption that organizations and humans each have needs that the other is required to satisfy. Bolman and Deal's model narrowly focuses on the organizational workforce, with emphasis on employee development and empowerment and the management of interpersonal and group dynamics. However, it is not difficult to recognize the same basic issues between consumers and the healthcare system as a whole. The core challenges are creating mutual benefit and gaining trust. In this frame, actions and attitudes are driven more by subjective speculation and perceived self-interest than rational analysis. The response to uncertainty is often fear and resistance to change.

People have a tendency to become fearful and resistant when they feel threatened, and there are at least two common sources of threat. The first is safety, and the other justice. Threats to safety can appear in several forms. The most obvious is physical safety along with economic safety and general well being. While NT-enabled medicine and biotechnology hold open much promise, there are also risks. Minimally invasive surgery and narrowly targeted chemotherapy may enhance safety, while the potential toxicity and nonbiodegradability of nanoparticles may pose substantial risk. Public perceptions of the safety of nanomaterials will play a major role in acceptance, regardless of structural frame efforts in safety testing and protective regulations. This dynamic is particularly relevant if the general public does not trust the entities doing the testing and regulating, as is indicated in the work of Cobb and Macoubrie on public perceptions about NT.[28]

In general, American society has a technological bias that predisposes the public to trust and desire advances in healthcare. Healthcare is viewed as a social good with a majority of Americans generally supportive of medical research and technological

[27]Morgan, 2005.
[28]Cobb and Macoubrie, 2004.

advances. In fact, Macoubrie found the category of major advances in medicine to be the top anticipated benefit of NT (31%) among the private citizens surveyed.[29] It is notable, and completely consistent with the human resource frame, that medical advances also evoked some of the greatest concerns. With respect to NT in general, 95% of those surveyed did not trust either government or industry to effectively manage the risks of NT. In the healthcare arena, it does not take much to turn the tide on support if we look to examples, such as the impact of a few poor outcomes and research missteps on genetic therapy in the late 1990s.[30]

In a similar vein, acceptance of nanomedicine among the healthcare workforce will also be based on perceived benefit or harm. If changes in the delivery system dramatically effect the qualifications or reduce the need for segments of the current healthcare workforce, they will resist NT based on the potential for economic instability and job security.

Justice is another source of fear and resistance in this frame. The cost impact of nanomedicine is unknown at this point in time. Movement toward nanoenhanced diagnostics and treatments may, at least initially, add to escalating healthcare costs and further limit access for those without insurance or reliant on public assistance. If large public constituencies believe they are not likely to benefit equally from NT, then resistance is likely. Similarly, it may make a difference whether initial nanoenabled advances are targeted broadly toward prevention or narrowly toward treatment of certain diseases or end-of-life care. Whichever end of the prevention–treatment continuum is left out is likely to feel unjustly overlooked.

Political Frame

With obvious ties to political science, the political frame emphasizes the diverse and competitive nature of human systems. Organizations, and by extension social systems, are described as "coalitions of diverse individuals and interest groups"[3] competing for scarce resources and power in the face of conflicting interests. Particular foci of this frame include how decisions are made, resources allocated, and networks and alliances developed, all of which may lack the endorsement of either the structural or human resource frames. An important insight of this frame is that, although this frame is infused with narrow self-interest and the potential for abuse of power, nothing much gets done in organizations without a healthy and dynamic political frame. This frame sets the agenda, choreographs the decisions, and allocates the resources. The effects of uncertainty in this frame are to rev it up as stakeholders assess both threats and opportunities and begin to maneuver for power and advantage.

---

[29]Macoubrie, 2005.

[30]In 1999, in a highly publicized move that is generally considered to have been a major set-back to research in gene therapy, the FDA suspended all gene therapy operations following the death of 18-year-old Jesse Gelsinger from an immunological response to the virus that delivered the gene. Although the researchers were primarily faulted for not following their approved protocol and for failing to report prior adverse events, the backlash tended to focus on the therapy itself rather than potential research misconduct.

computational abilities, and large-scale replacement of organs and tissues with engineered biomaterials.

The Preamble of the WHO Constitution goes on to specify: "The enjoyment of the highest attainable standard of health is one of the fundamental rights of every human being without distinction of race, religion, political belief, economic or social condition." What is the highest attainable standard of health? Is this to be interpreted as a future moral imperative to provide whatever biotechnical enhancements are available to achieve an individual's own personal definition of health and well being?

Physician as Healer or Designer. The practice of medicine is defined as the "art and science of the diagnosis and treatment of disease and the maintenance of health."[33] For most of the history of medical practice, this has meant a limited set of interventions based on a crude understanding of human anatomy and physiology, herbal and other traditional remedies, crude surgical interventions, and comfort measures once a patient was injured or became symptomatic of a disease process. The role of the physician as healer was primarily limited to responding to medical conditions as they presented. As medical knowledge has increased, physicians have played an increasingly active role in prevention by proactively managing the potential for disease and disability.

The convergence of NT with biotechnology, information technology, and cognitive science introduces a new role, that of physician as designer. The widespread acceptance of assisted reproduction and the recent popularity of cosmetic surgery illustrate the possibility that consumers will readily accept and demand an array of enhancement technologies, including those associated with genetic selection and manipulation. Advances in assisted reproduction have already crossed the line between the preservation and the creation of human life. Sports doping and the Metabolically Dominant Soldier program of the Defense Advanced Research Projects Agency (DARPA) are clear examples of our willingness to exceed physiological boundaries in the pursuit of practical goals.[34] Increasing attention toward the creation of an "ageless body" challenges medicine to defy the assumption that death itself is inevitable.[35] How will the combination of this emerging power and external pressures to pursue its application affect the professional identities of physicians? Similar questions can be raised for other members of the healthcare team.

Whether explicit or implicit, any backlash against nano-enabled applications in medicine will likely be a statement on NT itself and the unprecedented power it gives science and technology. The phrase "playing God" has been used to describe many, prior advancements in healthcare from organ transplants, to advanced life support, to *in vitro* fertilization. Manipulation of human life at the level of our DNA, radical enhancements of human performance, and the potential to extend the normal human lifespan to 150 years or longer all extend this metaphor of divine power into truly uncharted waters.

---

[33]Miller-Keane, 1997.
[34]Garreau, 2005.
[35]President's Council on Bioethics, 2003.

The Boundaries of Social Justice. Healthcare has come increasingly to be viewed as a right in the United States. There is rising public pressure to respond to access problems in the current healthcare system by providing at least some minimum level of universal health coverage to every citizen. In 1983, the President's Commission for the Study of Ethical Problems in Medicine and Biomedical and Behavioral Research produced a report entitled *Securing Access to Health Care*.[36] The report argued that healthcare constitutes a social good of "special importance" based on its relationship to well being, opportunity, relief from concern, and the interpersonal significance of illness, birth, and death. Although intentionally stopping short of establishing a right to healthcare, the Commission concluded that society has an ethical obligation to ensure equitable access to healthcare for all, balanced by individual obligations, and without excessive burdens on society. The Commission's definition of equitable access as an adequate level of healthcare warrants reopening in light of NT-enhanced medicine.

What will constitute an adequate level of healthcare in light of potentially rapid advances in medicine? Does an adequate level of healthcare include expensive genetic screening as a preventive measure? Does an adequate level of healthcare include access to enhancement technologies that allow a person to compete effectively in a work environment in which the privileged already have routine access to available technologies? Do the boundaries of intergenerational justice render attempts at radical life extension too burdensome on society as a whole? In the United States, answers to these questions will be drawn from deep-seated cultural assumptions about social class, market-driven economics, and traditional values of fairness and due process.

Although the Commission explicitly chose to take a position of societal obligation rather than individual rights, the two are not necessarily different in the assumptions of the symbolic frame. If someone has an obligation, then someone must have a right. Who has a greater claim to scarce resource dollars—those in need of preventive medicine or those in need of curative medicine? How will individual responsibility for use of resources come to be defined in a future age of personalized medicine in which the effects of poor dietary and other lifestyle choices can be erased? Will the "duty to die" debate take a metaphorical about face as we discover a need to impose space and resource limitations on the human lifespan, rather than the human lifespan imposing natural limits on our use of space and resources? Will past intolerances and inequities tied to race and gender be transferred wholesale to the NT-enhanced and the -unenhanced?

Evolution or Devolution of the Human Person. The innate human appreciation of and drive for perfection is apparent in everything from Olympic

---

[36]Securing access to healthcare: a report on the ethical implications of differences in the availability of health services. United States. President's Commission for the Study of Ethical Problems in Medicine and Biomedical and Behavioral Research. Washington DC: President's Commission for the Study of Ethical Problems in Medicine and Biomedical and Behavioral Research: For sale by the Supt. of Docs., U.S. G.P.O., 1983. OCLC: 936584.

competition, to the Miss Universe pageant, to the membership of Mensa. Human aspiration lies at the core of human progress. Writers, such as James Hughes,[31] Ramez Naam,[37] Ray Kurzweil,[38] and Ronald Bailey,[39] speak longingly of the day human beings finally break free of our evolutionary bonds to "possess the power to guide our own development—to choose our paths, rather than allowing nature to blindly select for the genes that are best at spreading themselves."[40] Critics, such as Francis Fukuyama[41] and Bill McKibben,[42] caution against embracing these powerful technologies too quickly or at all. Popular culture is profoundly inconsistent in how it ascribes value to the human person in the face of technology. Literature and film are replete with the images of technology gone awry. From *Frankenstein* to *The Matrix* we are at once fascinated, entertained, frightened, and appalled with our own technological potential.

The symbolic frame loves drama, and what could be more dramatic than having the very nature of what it means to be human hanging in the balance? This is the symbolic issue that cuts the most deeply across cultural and spiritual assumptions. It threatens the viability of traditional cultural and religious beliefs regarding the origins, purpose, autonomy, and ends of the human species. As such, the prospect of a radical shift in the biomedical paradigm fuels everything from unbridled optimism, to knee-jerk resistance, to befuddled amusement and, disbelief.

Is there a point at which the technological manipulation of body and mind simply crosses a boundary to become something less than or more than human? Will society assign differential preferences and meanings to genetic, sensory, mechanical and cognitive enhancements? Will we be the beneficiaries of an entirely new level of human self-actualization and transcendence? Alternatively, will we be the victims of a technological determinism resulting from too much power to manipulate life at the molecular level, and too little wisdom to use it well? Does it matter?

To illustrate, Baylis and Robert examine moral arguments against genetic enhancement technologies, one of the likely beneficiaries of NT.[43] They offer a range of standard arguments against genetic enhancement, including: the transgression of divine and natural laws; unacceptable risks of harm; threats to genetic diversity, as well as our common genetic heritage; a paradox of short-term individual benefits with long-term societal harms; misuse of scarce resources; a widening of social and economic gaps, promotion of social conformity and homogeneity; erosion of free choice; and opposition to the means on moral grounds. They also explore a thesis of inevitability that defends genetic enhancement based variously on: opportunistic capitalism; heedless liberalism; human inquisitiveness; human competitiveness that seeks to maximize personal, social, and economic advantage; and an attitude of "just because we can." If read carefully, their analysis reveals

[37]Naam, 2005.
[38]Kurzweil, 2005.
[39]Bailey, 2005.
[40]Naam, 2005, p. 232.
[41]Fukuyama, 2002.
[42]McKibben, 2003.
[43]Baylis and Robert, 2004.

many complex connections between all four of the proposed frames as profound questions of meaning are filtered through imagined consequences that reside in one or more frame. Ultimately, these authors argue that the development and use of genetic technologies are inevitable for no one particular reason, but simply as a matter of destiny, a statement of profound symbolic relevance in its own right.

CONCLUSIONS

> Great things are not done by impulse, but by a series of small things brought together.
>
> Vincent Van Gogh[44]

The practical impact of technological convergences and NT-enabled technologies in medicine will have profound effects throughout the American healthcare enterprise. The Bolman and Deal model of organizational frames provides a useful tool for brainstorming issues from multiple disciplinary perspectives and mapping connections.[3]

The danger of focusing narrowly on issues in any one frame, is that the ability to effect change or control in the system may be compromised or lost. Two basic assumptions of the model have been alluded to both directly and indirectly. First, an action in any one frame is likely to ripple across and affect each of the other three frames. Second, what appears to be a problem in one frame, may actually originate in another frame, and the solution may lie predominantly in yet another frame. Therein lies the real strategic value of framing complex situations.

I propose that few structural, human resource, or political issues raised in this brief analysis will be effectively addressed without first coming to terms with the deeper issues of the symbolic frame. The European experience with genetically modified organisms (GMOs) is often cited as an analogy to NT challenges. However, the current debate about embryonic stem cell research, with its highly symbolic focus on the moral value of the human embryo, is perhaps a better exemplar of what might happen as NT-enabled medicine begins to roll out into the marketplace.

Rosalyn Berne posits a similar concern when she similarly argues for the importance of the symbolic in the conscientious development of NT.

> Those who are interested in the ethical development of nanotechnology, and in leading it towards humanitarian aims, will more likely achieve critical conscientiousness about this incredible and perhaps revolutionary enterprise, in identifying and exploring the roles of meaning-making, imagination, myth, metaphor and belief in nanotechnology development.[45]

She points out that current stakeholders are already engaged in meaning-making, and will be joined by consumers once NT products enter the marketplace. She is

---

[44]This quotation is widely attributed to the artist Vincent Van Gogh. Retrieved from the Open Encyclopedia Project Available at http://open-site.org/Society/Philosophy/Art/.
[45]Berne 2004, p. 637.

solidly rooted in the symbolic frame when defining the process of assigning meaning about NT "as an imaginatively and symbolically engaged process through which ideas about nature, matter, and control express beliefs about where this new evolution of technology might lead, how it may be used, and what purposes it may serve."[46]

With respect to medicine and healthcare, we must first tackle the question of how to define the scope of human health. This definition, along with a realignment of the purpose, meaning, and culture of medical practice, will drive efficient, effective, and consistent responses to many other issues. Effective action in the structural, human resource, and political frames is impeded by the lack of social consensus on questions dealing with healthcare as a basic right, justice in the face of limited resources, and limits on human enhancement and life span.

In the absence of meaningful guidance from the symbolic frame, any significant restructuring of the healthcare system will lack strategic and practical direction. Unrealistic expectations and unfounded fears among stakeholders will persist and likely increase over time as people struggle to come to terms with the deeper implications of unequal access to innovations, what may be sacrificed in the face of competing priorities, and the consequences of a radical transformation in human identity. In the absence of consensus or broad civil dialogue, political discourse is likely to remain highly polarized, with small-but-powerful interests setting agendas that may or may not represent either the short- or long-term public good. At best, knee-jerk regulatory responses to threats of global competition, corporate interests, or hype-generated public concern may come to dominate, shifting with the political winds.

This less-than-optimistic scenario is but one that must be considered when trying to foretell the future of the healthcare system in relation to NT. It is, by its very nature, incomplete. Many scenarios are possible; none are inevitable. If the social sciences are to be an effective adjunct to assessing the societal impacts of emerging technology and crafting a well-considered response, then we have to quickly develop new tools of inquiry that prospectively reflect the realities of a rapidly evolving future. Enhanced models of frame analysis may provide an effective option for such strategic collaboration.

BIBLIOGRAPHY

Bailey, R. (2005). Liberation biology: The scientific and moral case for the biotech revolution. Amherst, NY: Prometheus.

Bawa, R. (2005). Will the nanomedicine "patent land grab" thwart commercialization? *Nanomedicine*, 1(4), 346–350.

Baylis, F. and Rober, J. S. (2004). The inevitability of genetic enhancement technologies [Online version]. *Bioethics*, 18(1).

Benford, R. D. and Snow, D. A. (2000). Framing processes and social movements: An overview and assessment. *Annual Review of Sociology*, 26, 611–639.

---

[46]Berne 2004, p. 636.

Bennett-Woods (in press). A Framework for the Pragmatic Integration of the Ethical Considerations of Societal Impacts Into Funding Decisions for Emerging Technologies, *Journal of Nanotechnology Law and Business.*

Berne, R. W. (2004). Towards the conscientious development of ethical nanotechnology. *Science and Engineering Ethics,* 10(4), 627–638.

Berube, D. M. (2006). *Nano-Hype: The truth behind the nanotechnology buzz.* Amherst, NY: Prometheus Books.

Betta, M. and Clulow, V. (Spring 2005). Health care management: Training and education in the genomic era. *Journal of Health and Human Services Administration,* 465–500.

Bolman, L. G. and Deal, T. E. (2003). *Reframing organizations: Artistry, choice and leadership* (3rd ed.). San Fransisco: Jossey-Bass.

Campo, A. and Bruce, I. J. (2005) Diagnostics and high throughput screening. In Neelina H. Malsch (ed.) *Biomedical Nanotechnology,* Boca Raton, FL: Taylor & Francis.

Cobb, M. D. and Macoubrie, J. (2004). Public perceptions about nanotechnology: Risks, benefits and trust. *Journal of Nanoparticle Research: An Interdisciplinary Forum for Nanoscale Science and Technology,* 6(4), 395–405.

Covency, P. and Highfield, R. (1995). *Frontiers of complexity: The search for order in a chaotic world.* New York: Fawcett Columbine.

Ferrari, M. and Downing, G. (2005). Medical nanotechnology: Shortening clinical trials and regulatory pathways? *Biodrugs,* 19(4), 203–210.

Fisher, K. (1997). Locating frames in the discursive universe. [Online serial] *Sociological Research Online* 2(3). Available at http://www.socresonline.org.uk/2/3/4.html (retrieved on August 15, 2006).

Fukuyama, F. (2002). *Our posthuman future: Consequences of the biotechnology revolution.* New York: Picador.

Garreau, J. (2005). *Radical evolution: The promise and peril of enhancing our minds, our bodies – and what it means to be human.* New York: Doubleday.

Goffman, E. (1986) *Frame analysis.* Boston: Northeastern University Press.

Hall J. S. (2005). *Nanofuture: What's next for nanotechnology?* Amherst: New York: Prometheus Books.

He, W. et al. (2005). *65+ in the United States: 2005.* U.S. Census Bureau, Current Population Reports. Available at http://www.census.gov/prod/2006pubs/p23-209.pdf (retrieved on May 10, 2006).

Hughes, J. (2004). *Citizen cyborg: Why democratic societies must respond to the redesigned human of the future.* Cambridge, MA: Westview Press.

Kayser, O. et al. (2005). The impact of nanobiotechnology on the development of new drug delivery systems. *Current Pharmaceutical Biotechnology,* 6, 3–5.

Kubik, T. et al. (2005). Nanotechnology on duty in medical applications, *Current Pharmaceutical Biotechnology,* 6, 17–33.

Lindpainter, K. (2002). Pharmacogenetics and the future of medical practice. *Journal of Molecular Medicine,* 81, 141–153.

Macoubrie, J. (2005, September). Informed public perceptions of nanotechnology and trust in government. The Pew Charitable Trusts. Available at http://www.pewtrusts.com/pdf/Nanotech_0905.pdf (retrieved on July 14, 2006).

Mazzola, L. (2003). Commercializing nanotechnology. *Nature Biotechnology*, 21(10), 1137–1143.

Morgan, K. (2005). Development of a preliminary framework for informing the risk analysis and risk management of nanoparticles. *Risk Analysis*, 25(6) 1621–1635.

McKibben, B. (2003). *Enough: Staying human in an engineered age*. New York: Henry Holt and Company.

Naam, R. (2005). *More than human: Embracing the promise of biological enhancement*. New York: Broadway Books.

National Center for Health Statistics (2005). *Health, United States, 2005 with chartbook on trends in the health of Americans*. Available at http://www.cdc.gov/nchs/ (retrieved on July 30, 2006).

National Nanotechnology Initiative (NNI). (n/d). NNI budget by agency. Available at http://nnco5.nano.gov/html/about/nnibudget.html (retrieved on August 23, 2006).

O'Toole, M. T. (Ed.). (1997). *Miller-Keane encyclopedia and dictionary of medicine, nursing and allied health*. (6th ed.). Philadelphia: W.B. Saunders.

President's Commission for the Study of Ethical Problems in Medicine and Biomedical and Behavioral Research. (1983, March). *Securing access to healthcare*. [Online version.] Available at http://www.bioethics.gov/reports/past_commissions/securing_access (retrieved on July 2, 2006).

President's Council on Bioethics. (2003). *Beyond therapy: Biotechnology and the pursuit of happiness*. New York: Dana Press.

Pruitt, S. D. and Epping-Jordan, J. E. (2005). Preparing the 21st century global healthcare workforce. *BMJ*, 330, 637–639.

Roco, M. C. (2005). Converging technologies: Nanotechnology and biomedicine. In Malsch (Ed.), *Biomedical nanotechnology* (pp. xi–ixx). Boca Raton, FL: Taylor and Francis.

Van den Bueken, X. et al. (2005). Implants and prostheses. In Neelina H. Malsch (ed.) *Biomedical Nanotechnology*, Boca Raton: Taylor and Francis.

Weston, A. D. and Hood, L. (2004). Systems biology, proteomics, and the future of health-care: Toward predictive, preventative, and personalized medicine. *Journal of Proteome Research*, 3, 179–196.

World Health Organization (1948). *Preamble to the Constitution of the World Health Organization as adopted by the International Health Conference*. Available at http://www.searo.who.int/EN/Section898/Section1441.htm (retrieved on June 28, 2006).

Zuckerman, A. M. (1998). *Health care strategic planning: Approaches for the 21st century*. Chicago: Health Administration Press.

CHAPTER 18

Doing Small Things Well: Translating Nanotechnology into Nanomedicine

WILLIAM P. CHESHIRE, JR.

In life it is the little things that count.

Joseph Blount Cheshire[1]

INTRODUCTION

Enormous scientific discoveries sometimes concern very small things. The first glimpse of bacteria whirling within a drop of water positioned under finely ground microscope lenses set into motion a renaissance in medical science. Yet such microorganisms are like gargantuan mammoths when compared to entities at the nanoscale. Dimensions of Nature once unimaginably small have come within the reach of instruments of finer and finer precision. The emerging paradigm shift onto the nanoscale is yielding exciting new methods for visualizing and interacting with not only cells, but also the complex biomolecules that compose them. All manner of miniscule things can now be observed. Ever more intricate nanomarvels are evolving from design to construction to application.

The wave of nanomedicine draws swiftly near, bringing prospects for a new era of improving human health through efficient diagnosis and meticulous interventions that just a generation ago were the fancy of science fiction. The frontiers of nanomedicine will likely continue to unfold for decades to come, if not longer. There is for nanomedicine, as for nanotechnology in general, "plenty of room at the bottom."[2]

---

[1]London L.F. (1941). *Bishop Joseph Blount Cheshire: His Life and Work*. Chapel Hill: University of North Carolina Press, p. 81.
[2]Feynman R.P. (1960). There's plenty of room at the bottom. *Eng. Sci. (CalTech)* 23, 22–36.

Nanoscale: Issues and Perspectives for the Nano Century. Edited by Nigel M. de S. Cameron and M. Ellen Mitchell

Likewise the ethical implications of nanomedicine are enormous. Although scientific attention reduced to the nanoscale finds in the lining up of atoms an apparent simplicity of structure, the consequences of nanotechnology for human living and health are far from simple. Greater detail of knowledge about the nanoworld generates an explosive quantity of information. Information requires interpretation. The ability to act on that information demands care and creativity, and the multiplication of options for precise intervention entail choices.

Familiar ethical principles will be helpful in guiding decision making as we enter into the world of nanomedicine, just as established principles of engineering provide a basis for constructing nanodevices. Customary approaches may not, however, be fully translatable into the entirely new field of nanomedicine. The physical properties of nanoscale objects are unlike those of things around us in the everyday world. The ethical discussion, likewise, will borrow from past experience while also taking on new shape and proportions. As the mind zooms inward to consider the nanoworld, points of reference in the larger macroworld may at times seem distant or blurry. But they are there.

This discussion will consider the very small alongside the unfathomably immense. Whatever one's perspective, one thing is clear. Small things matter. Indeed, much wisdom is required to do small things well.

THE DISCOVERY OF CELLS AND GERMS

Today, it is nearly impossible to imagine what medicine would be like without awareness of the microbial world. Yet, throughout most of recorded history, knowledge of health and illness was confined to that part of nature visible to the human eye. The nineteenth century physician William Osler observed:

> The Greek physicians, Hippocrates, Galen, and Aretaeus, gave excellent accounts of many diseases; for example, the forms of malaria. They knew, too, very well, their modes of termination, and the art of prognosis was studied carefully. But of the actual causes of disease they knew little or nothing, and any glimmerings of truth were obscured in a cloud of theory.[3]

The invention of the compound microscope by Zaccharias and Hans Janssen in 1590 was a key development toward the transformation of medicine from a discussion of humors and miasmata into a study of cells and microorganisms underlying life and disease. Subsequent improvements on the design of the microscope enabled Anton van Leeuwenhoek in the late seventeenth century to observe and publish the first detailed descriptions of tiny "animalculæ" or organisms teeming in a drop of water, red blood cells coursing through capillaries, and living bacteria swimming in saliva.[4,5]

[3]Vallery-Radot R. (1928). *The Life of Pasteur.* (Translated from the French by R. L. Devonshire.) Garden City: Garden City Publishing Co.
[4]Dobell C. (1932). *Antony van Leeuwenhoek and His Little Animals.* New York: Harcourt and Brace.
[5]Schierbeek A. (1959). *Measuring the Invisible World: The Life and Works of Antoni van Leeuwenhoek.* New York: Abelard-Schuman.

The unveiling of the microbial world led to Theodor Schwann's nineteenth century discovery of the cell as the fundamental unit of biological organisms. In addition to describing the envelope of myelin surrounding nerve fibers that bears his name, Schwann explored the cellular origin of diverse bodily tissues, laying the groundwork for the modern field of histology.[6] The origin of the fields of cellular pathology and comparative pathology is credited to Rudolf Virchow, who recognized in leukemia that illness can arise from disturbances in specific types of cells. Virchow's law *Omnis cellula e cellula* (every cell originates from another cell)[7,8] foreshadowed the current interest in stem cells, the precursors of all specialized cells.[9]

From these first observations of the cellular realm came the germ theory of disease, now a cornerstone of medicine. Direct observation of disease-causing protozoa, bacteria, and spores through the lens of the microscope was integral to the development of the understanding that microorganisms are the actual causes of many types of infectious disease. Thus, Louis Pasteur famously demonstrated that fermentation of broth could be prevented by blocking access by airborne particles.[3] The medical establishment of the time was, however, slow to accept the germ theory, at first dismissing the handwashing practices of Ignaz Semmelweis as religious foolishness despite his remarkable success in curtailing the contagious spread of puerperal fever in the medical ward.[10,11]

Appreciating the microbial nature of infectious diseases has, over the last century, led to great strides in better health through hygienic practices, public sanitation measures, antibiotic drugs, and vaccines. Many a student has watched microbes dance under the microscope lens, and inspired by what Osler called "the infinitely little view of the nature of disease germs,"[3] he or she has embarked on a serious study of medicine.[12]

The greater the resolving power of a microscope, the finer is the level of detail that can be distinguished. The physics of light limits the maximum degree of resolution possible by light microscopy to 400×. At this maximum optical magnification, the closest that two distinct points can be resolved from one another is 0.2 μm (micrometer) or 2×10^{-7} m, which is about one-half of the wavelength of visible light. Optical microscopy is well suited to studying cellular structure, as cells in the human body are typically 50 μm in diameter. Human red blood cells, for example, are 6–8 μm in diameter, while neurons range from 4 to 100 μm in diameter.

[6]Schwann T. (1839). *Microscopical Researches into the Accordance in the Structure and Growth of Animals and Plants.* Berlin: Sander'schen Buchhandlung.

[7]Virchow, R.L.K. (1978) *Cellular pathology.* (1859 special ed.), London: John Churchill, pp. 204–207.

[8]Castiglioni A. (1941). *A History of Medicine.* (Trans by E. B. Krumbhaar) New York: Alfred A. Knopf, p. 696.

[9]Cheshire W.P. (2005). Small things considered: the ethical significance of human embryonic stem cell research. *New England Law Rev.* 39(3), 573–581.

[10]Nuland, S. B. (2003). *The Doctors' Plague: Germs, Childbed Fever and the Strange Story of Ignac Semmelweis.* New York: W.W. Norton & Co. Ltd.

[11]Best M. and Neuhauser D. (2004). Ignaz Semmelweis and the birth of infection control. *Qual. Saf. Health Care* 13, 233–234.

[12]Cushing H. (1925). *The Life of Sir William Osler.* Oxford: Clarendon Press, pp. 40–45, 56, 59–64.

PENETRATING THE SUBCELLULAR MATRIX

The story of molecular biology, of course, does not end at the optical resolution of cells. At the 0.2-nm (nanometer) level of resolution afforded by the electron microscope, the view of the living cell suddenly explodes into a bustling city, within which lies yet another astonishing world of continuously interacting macromolecules, receptors, mitochondria, ribosomes, membranes, proteins, electrolytes, and genes.

New imaging methods are making it possible to view biological structures at the molecular level. Scanning tunneling microscopes were the first to visualize individual atoms. Atomic force microscopy[13,14] and other new imaging technologies[15,16] are pushing further the boundaries of intricate detail subject to direct visualization.

Wrapped within the human cell nucleus, folded in a double helix, lies the molecular library encoding the genetic instruction book for all human physical traits. The complete sequencing of this 3 billion base pair (bp) deoxyribonucleic acid (DNA) molecule through the Human Genome Project[17] has opened up exhilarating new frontiers in medicine with incalculable implications for improving patient care. The eloquent DNA molecule is a mere 2.3 nm in width.

Directing the focus of investigation to smaller and smaller things has enlarged the detail of the knowledge gained to staggering proportions. Whereas scientists who peered into the first microscopes could keep track of their data with quill pen and paper, modern biotechnology relies on the fastest computers to process the vast volumes of new information added daily to the scientific literature.

THE NANOREALM

Nano refers to 1 billionth or 10^{-9} power. One nanometer (nm) is 1 billionth (10^{-9}) of a meter or 0.000000001 m. One nanometer is also 1 millionth of a millimeter (mm), or 1 thousandth of a micrometer. By comparison, one red blood cell is 7000 nm across. The human hair, which is among the smallest structures visible to the unaided eye, is 100,000–150,000 nm in thickness. Nanoscience consists of the scientific disciplines concerned with the study of molecules and atoms and objects on the scale roughly of 1–100 nm. Nanotechnology refers to the tools, methods, and procedures through which humanity acquires the ability to control Nature at the nanoscale. Nanomedicine refers to the application of nanoscience and nanotechnology to human health and disease.

[13]Gadegaard N. (2006). Atomic force microscopy in biology: technology and techniques. *Biotech. Histochem* 81(2), 87–97.

[14]Hansma H.G. et al. (1995). Applications for atomic force microscopy of DNA. *Biophys. J.* 68, 1672–1677.

[15]Rust M. et al. (2006). Subdiffraction limit imaging by stochastic optical reconstruction microscopy (STORM). *Nat. Methods* Aug. 9.

[16]Patton F.S. et al. (2006). Speckle patterns with atomic and molecular de Broglie waves. *Phys. Rev. Lett.* 97, 13202.

[17]Collins F.S. et al. (2003). A vision for the future of genomics research. *Nature* 422, 835–847.

THE TOOLS OF NANOMEDICINE

Nanoscale substances exhibit novel physical properties unlike their bulk counterparts. Their small size creates very high surface/volume ratios, which can facilitate molecular interactions. Some nanosubstances have chemically tailorable physical properties directly related to their size, shape, and molecular composition. Together, these properties render nanoscale probes exquisitely sensitive to target-binding molecular events, which can generate measurable effects. Nanoscale structures can be engineered to have considerable tensile strength and structural stability. Some nanoparticles have peculiar optical and magnetic properties.[18]

Nanotechnology proceeds in two directions. "Top-down" nanofabrication begins with existing structures and applications and scales them down to nanoproportions. For example, nanoscale imprinting utilizes high-resolution electron beam lithography to apply a controlled pattern of a molecular monolayer on a surface. Atomic force microscopy provides the means slowly to lay down individual atoms with nanometer precision. Top-down approaches are engendering "lab-on-a-chip" nanotechnologies.

Top-down nanofabrication methods could achieve nanofluidic systems in which tiny computer chips containing channels for processing nanoliter quantities of biological liquids could perform diagnostic tests rapidly, economically, and with unprecedented portability.

Nanocantilevers are microscopic flexible beams built by semiconductor lithographic techniques. When coated with molecular probes, for example, complementary DNA strands, molecular binding of the target of interest alters the mass of the mechanical oscillating cantilever. The slight change in its resonant frequency generates an electronic signal that can be conveyed to a computer. Neurocantilevers are starting to provide exquisitely sensitive real-time detection of genes, proteins, viruses, and bacteria.[19] Similar degrees of sensitivity in detecting biomarkers are also obtainable with nanowires measuring 20–250 nm in diameter. Nanowire conductivity changes when a target molecule binds to a probe attached to its surface.[20]

Nanoarrays can potentially reduce the size and sample volume requirements for biomolecule testing by orders of magnitude smaller than conventional microarrays.[21] Future nanoarrays may, for example, be capable of screening an individual's entire genome by analyzing samples arranged into 15-nm droplets within the space of a 2×2 cm^2 chip.[22]

"Bottom-up" nanofabrication begins with atoms or molecules and builds them up into more complex nanostructures. A notable example is the carbon sphere. Buckminsterfullerenes, named after R. Buckminster Fuller, the inventor of the geodesic dome, are 1-nm diameter spheres consisting of 60 carbon atoms arranged into an

[18]Leary S.P. et al. (2006). Toward the emergence of nanoneurosurgery: Part I—Progress in nanoscience, nanotechnology, and the comprehension of events in the mesoscale realm. *Neurosurgery* 57(6), 606–634.

[19]Leary S.P. et al. (2006). Toward the emergence of nanoneurosurgery: Part II—Nanomedicine: Diagnostics and imaging at the nanoscale level. *Neurosurgery* 58(5), 805–823.

[20]Patolsky F. et al. (2006). Nanowire-based biosensors. *Anal. Chem.* 78, 4260–4269.

[21]Lynch M. et al. (2004). Functional protein nanoarrays for biomarker profiling. *Proteomics* 4, 1695–1702.

[22]Available at http://www.azonano.com.

interlocking structure of 20 hexagons and 12 pentagons. These buckyballs condense to form a loosely bound, superconducting solid termed fullerite. Fullerenes can encapsulate atoms and are being investigated as potential drug delivery systems.

Carbon nanotubes are extremely stable, single-walled, 1-nm diameter structures thousands of a nanometer in length. Their material strength is comparable to that of diamond due to their carbon bond configuration. Their exotic conducting properties make them promising for the development of nanoscale electronic devices. The opening and closing of the free ends of nanotube pairs attached to electrodes in response to an electric current has made possible the construction of nanotweezers capable of grabbing and manipulating various other nanoscale structures.[23]

Quantum dots are fluorescent semiconductor nanocrystals composed of several hundred atoms. Most commonly, an inner core of cadmium is coated with an amphiphilic polymer, to which may be attached targeting molecules, such as monoclonal antibodies. Their size, shape, and composition can be systematically varied to produce materials with specific emissive, absorptive, magnetic, and light-scattering properties. When excited, for example, by ultraviolet (UV) light, they emit a single wavelength of light. When conjugated to proteins or nucleic acids, their brilliant colorful light emission renders them very powerful biological probes. When microinjected into living cells, they can track the cells' movement through the body.

Gold nanoshells are 100-nm spheres of gold-coated silica. Like quantum dots, they have size-dependent tunable optical properties.[24] Gold nanoshells can be conjugated with antibodies that recognize and bind to cancer cells.[25] The moment binding occurs, their absorption peak shifts to the near-infrared (IR) spectrum. By passing a beam of bright light harmlessly through normal tissue, researchers at Rice University were able selectively to heat up and destroy cultured cancer cells to which gold nanoshells had attached.[26]

Current research is also directed toward the development of "smart" nanoparticles capable of targeted drug delivery to cancer cells. Bioconjugated nanoconstructs would be customized to recognize cancer cells and ferry chemotherapeutic agents or therapeutic genes directly into malignant cells while leaving healthy cells untouched.[27] Nanoparticles are also being studied as potential vaccine delivery vehicles.[28]

---

[23]Kim P. and Lieber C.M. (1999). Nanotube nanotweezers. *Science* 286, 2148–2150.

[24]Lin A.W. et al. (2005). Optically tunable nanoparticle contrast agents for early cancer detection: model-based analysis of gold nanoshells. *J. Biomed. Opt.* 10, 64035–64045.

[25]Cuenca A.G. et al. (2006). Emerging implications of nanotechnology on cancer diagnostics and therapeutics. *Cancer* 107, 459–466.

[26]Hirsch L.R. et al. (2003). Nanoshell-mediated near-infrared thermal therapy of tumors under magnetic resonance guidance. *Proc. Natl. Acad. Sci. U.S.A.* 100, 13549–13554.

[27]Sinha R. et al. (2006). Nanotechnology in cancer therapeutics: bioconjugated nanoparticles for drug delivery. *Mol. Cancer Ther.* 5, 1909–1917; Salata O.V. (2004). Applications of nanoparticles in biology and medicine. *J. Nanobiotech.* 2, 3–8; Groneberg D.A. et al. (2006). Nanoparticle-based diagnosis and therapy. *Curr. Drug Targets* 7(6), 643–648; Jain K.K. (2005). Role of nanobiotechnology in developing personalized medicine for cancer. *Technol. Cancer Res. Treat* 4(6), 645–650; Mastrobattista E. et al. (2006). Artificial viruses: a nanotechnological approach to gene delivery. *Nat. Rev. Drug Discov.* 5(2), 115–121; Chowdhury E.H. and Akaike T. (2005). Biofunctional inorganic materials: an attractive branch of gene-based nanomedicine delivery for 21st century. *Curr. Gene Ther.* 5(6), 669–676.

[28]Koping-Hoggard M. et al. (2005). Nanoparticles as carriers for nasal vaccine delivery. *Expert Rev. Vaccines* 4(2), 185–196.

HEIR OF MICROMEDICINE

If micromedicine, in its investigation of life at the scale of roughly $1-1000$ μm, has revealed wonders important for medical progress, how much more might nanomedicine achieve?

Actually, forms of nanomedicine have been with us for some time, albeit by other names. The first quantitative descriptions of oxygen and its relationship to cellular respiration in the eighteenth century concerned gas exchange at the nanoscale. To administer oxygen by a face mask to a critically ill patient is to supply gaseous particles of approximately 1 nm in size. Alchemists of old and *doctours of physik* in Chaucer's day dabbled in nanoscale configurations of the elements—atoms, the sizes of which they could not measure—even before Mendeleev introduced his periodic table in 1869.[29]

Molecular biology has for decades investigated cellular structures at the nanoscale and has translated a wealth of discoveries into beneficial medical applications. In addition to elucidating the structure and function of membranes, lipids, receptors, microtubules, ribosomes, and mitochondria, molecular biology has delved into the mechanisms of proteins, immunoglobulins, neurotransmitters, and DNA. Greater understanding of the molecular basis of these intracellular systems, combined with progressively finer techniques of biochemical synthesis, has engendered antibiotics, anesthetics, anti-inflammatory agents, anticonvulsants, antidepressants, analgesics, and many more varieties of salubrious drugs.

As so often happens in the history of science, empirical discovery finds reality to be much grander than imagination had predicted. Molecular biology has for decades yielded hints of the magnificence of the yet-unexplored regions of the nanorealm at the next level of resolution within living organisms. Consider, for example, the brain. Hippocrates knew little more than that the brain was the seat of intelligence. In recent decades, microscopic studies utilizing sophisticated staining techniques have shown that the human brain comprises 100 billion neurons, each of which taps into its neighbors with $1-10,000$ synaptic connections.[30] This intricately designed labyrinth of neurons contains 1000 trillion synapses. Dozens of neurotransmitters and neuropeptides, the flow of which is very finely regulated, cross synaptic clefts 20 nm wide to signal adjacent neurons in the magnificent molecular symphony of intelligence.

Now that nanotechnology is providing the tools for more detailed exploration of the nanorealm, the ongoing empirical venture of investigating the molecular basis of life may look forward to ever expanding prospects for edifying scientific knowledge. Many such discoveries will undoubtedly benefit human health.[31]

---

[29]Mendeleev D.I. and Jensen W.B. (2005). *Mendeleev on the Periodic Law: Selected Writings, 1869–1905.* Dover: Dover Press.

[30]Swenson R.A. (2000). *More Than Meets the Eye: Fascinating Glimpses of God's Power and Design.* Colorado Springs: NavPress.

[31]Freitas R.A., Jr. (2005). What is nanomedicine? *Nanomedicine: Nanotechnology, Biology, and Medicine* 1, 2–9; Leary S.P. et al. (2006). Toward the emergence of nanoneurosurgery: Part III—Nanomedicine: Targeted nanotherapy, nanosurgery, and progress toward the realization of nanoneurosurgery. *Neurosurgery* 58(6), 1009–1026; Silva G.A. (2004). Introduction to nanotechnology and its applications to medicine. *Surg. Neurol.* 61, 216–220; Kubik T. et al. (2005). Nanotechnology on duty in medical applications. *Curr. Pharm. Biotechnol.* 6(1):17–33; Emerich D.F. (2005). Nanomedicine—prospective therapeutic and diagnostic applications. *Expert Opin. Biol. Ther.* 5(1), 1–5; see footnotes 13 and 19.

UNLIKE PREVIOUS MEDICINE

Whereas nanomedicine follows to a point the prosperous trajectory that micromedicine has taken, it is also exceeding that trajectory. Imagine that all of medical science until now has been like a series of projects in which skillful warriors have hurled projectiles at disease. With refinements in technology have come swifter and more accurate missiles. Our ancestors used stones and arrows in the warfare against disease, and now we work toward developing silver-bullet drugs, rocket-like surgical techniques, and smart-bomb targeted therapies. The space age of medical marksmanship is now upon us. Nanotechnology is poised to surpass the equivalent of gravitational escape velocity, taking medical science into brave new orbits of astronomical possibilities.

The terrain of nanotechnology viewed from the perspective of human eyes resembles in proportion the view of life on Earth from deep space. As scales of comparison soar from things far above to those deep within, I will chart some interesting analogies between the flight into space and the plunge into nanoscience. Triangulating these large and small scientific frontiers with creative human initiative will help to locate emerging trends in nanomedicine.

Nanomedicine will excel, for example, in resolution. Amazingly detailed and colorful images of extremely distant galaxies and nebulae from the Hubble telescope have repainted our view of the universe. Confronted by tremendous beauty, we are inspired to consider afresh our place within its magnificent expanse. Nanomedicine, similarly, will bring into view the incredible detail of the inner workings of human nature and other living creatures. Brilliant images of busy nanoworlds knit together within our own cells will testify how fearlessly and wondrously we have been made.[32]

Nanomedicine will excel in access to our most personal molecular signatures. Aerospace technology made possible space walks in which tethered astronauts could totter along the outer surface of their spacecraft, repair equipment, and pause breathlessly to behold the great blue marble below. Nanomedicine promises new ways to design probes that can walk along DNA strands, reading and activating or inactivating genes. Future nanoscale DNA sequencers could, in principle, capture an individual patient's entire genome to be stored on a compact disk—if not on something smaller. Nanotechnology is also moving toward the development of information storage devices that may surpass the density available on current digital media by orders of magnitude.

Nanomedicine will have access to nanoengineered materials of remarkable tensile strength. Successful space travel requires materials that are relatively lightweight, yet firm enough to withstand the forces of thrust and deceleration and reliably insulate the cabin from the cold, unforgiving vacuum of deep space. Nanotechnology promises to yield biocompatible structures of exceptional tensile strength, smoothness, and durability that may find application in artificial joints, replacement cardiac valves, and other prostheses for body parts lost to

---

[32]Brand P. and Yancey P. (1980). *Fearfully & Wonderfully Made*. Grand Rapids, MI: Zondervan.

disease or injury. Perhaps nanomedicine may also provide improved ocular lens implants to replace cataracts, pliable blood vessel linings to replace atherosclerotic plaques and aneurysms, or artificial vocal cords for patients who have undergone laryngectomy. Nanofabrication methods could envelop these materials with molecular coatings with nearly flawless atomic precision, achieving near-nanometer smoothness, thus reducing their bioactivity, leukocyte attraction, and fibrinogen absorption, allowing them to retain their structural integrity within the body.

Nanomedicine will find detours around some of the previous constraints of the laws of chemistry and physics. Astronauts in orbit experience the freedom of weightlessness and its attendant physiologic challenges to the circulation and the inner ear, as well as the basic physical challenge of moving about without falling into the walls or ceiling. Similarly, nanoparticles behave differently than their bulk counterparts. The construction of superconducting nanowires, through which electricity flows nearly without resistance, would be a step toward designing vastly improved magnetic resonance imaging (MRI) machines to analyze living tissue.

Nanomedicine is already offering unprecedented sensitivity of molecular detection. The detailed satellite imagery available as a result of space technology has become an invaluable resource to geography, geology, agriculture, land use and resource management, military surveillance, and planning for responses to natural disasters. Similarly, nanoparticles are improving the resolution of medical imaging.[33] One such technology takes advantage of the unusual magnetic properties of nanoscale substances. Metallic particles measuring 1–10 nm in size exhibit superparamagnetism, such that the energy required to change their direction of magnetic moment is comparable to ambient thermal energy. This means that, liberated from any fixed internal magnetic order, they will align with an externally applied magnetic field. This property permits enhanced contrast resolution of MRI[34] and is making possible the detection of cancer metastases not previously visible by clinical imaging methods.[35] Earlier detection of the spread of cancer means earlier treatment and a greater likelihood of remission or cure. Emerging detection technologies include nanocantilevers and nanofluidic chambers capable of detecting single virions, single bacteria, or just a few molecules of disease-identifying biomarkers.[36] Bio-bar-code amplification systems utilizing gold nanoparticles can detect specific DNA sequences present at zeptomolar concentrations consisting of only a few strands per sample.[37] These technologies will have important applications for

[33]Zhu D. et al. (2006). Biocompatible nanotemplate-engineered nanoparticles containing gadolinium: stability and relaxivity of a potential MRI contrast agent. *J. Nanosci. Nanotech.* 6(4), 996–1003.

[34]Muldoon L.L. et al. (2006). Imaging and nanomedicine for diagnosis and therapy in the central nervous system: report of the eleventh annual blood-brain barrier disruption consortium meeting. *Am. J. Neuroradiology* 27, 715–721.

[35]Harisinghani M.G. et al. (2003). Noninvasive detection of clinically occult lymph node metastases in prostate cancer. *N. Engl. J. Med.* 348(25), 2491–2499.

[36]see footnote 8; Zheng Y. et al. (2005). Rapid self-assembly of DNA on a microfluidic chip. *J. Nanobiotech.* 3, 2–12.

[37]Georganopoulou D.G. et al. (2005). Nanoparticle-based detection in cerebral spinal fluid of a soluble pathogenic biomarker for Alzheimer's disease. *PNAS* 102, 2273–2276; Keating C.D. (2005). Nanoscience enables ultrasensitive detection of Alzheimer's biomarker. *PNAS* 102(7), 2263–2264.

early and accurate diagnosis, disease prevention, genetic testing, pharmacogenomics, and bioterrorism defense, as well as many other needs.

Some applications require, not better detection, but camouflage. As the shape and
structural design of stealth aircraft render them invisible to radar detection, future nanomedicine may devise implantable biologic materials imperceptible to immune surveillance. Suppose one wishes to transplant a collection of functional cells, for example,
insulin-secreting pancreatic islet cells, from a healthy individual into someone with diabetes mellitus. Normally, the recipient's immune system would reject the transplanted
cells. For this reason, organ transplant recipients must take potent immunosuppressive
drugs, the many adverse effects of which include increased susceptibility to infectious
illness. Nanocages are microfabricated nanopore immunoisolation chambers that could
shield cellular implants from the host's immune system while permitting the passage of
smaller molecules, such as glucose.[38] Such a device could be designed as one component in an integrated, implantable drug delivery device combining nanosensors, drug
nanoreservoirs, nanopore pumps, nanochannels, and responsive microchips capable
of regulating drug delivery on demand.[39]

Nanomedicine may allow cells to boldly grow where no cell has grown before.
The utmost triumph of space programs to date has been the transport of men and
women beyond the Earth's atmosphere, to encircle the Earth in the space shuttle,
to inhabit the spacestations Skylab and Mir, to walk the surface of the Moon, and
then to return home safely. When someone suffers a serious spinal cord injury,
severed neurons do not regenerate. Their connections, a centimeter of scar tissue
away, might as well be as far off as the Moon. Researchers at Northwestern University have achieved one small step for mousekind by inducing mouse neurons to
regenerate along peptide-amphiphile nanofiber scaffolds.[40] Researchers at MIT in
collaboration with the University of Hong Kong using peptide nanofiber scaffolds
to reconnect a severed optic tract in hamsters achieved partial return of vision.[41]
This research holds promise in the search for neuroregenerative interventions for
victims of brain and spinal cord injury.[42]

Nanomedicine, as any new technology, is likely to have unforeseen environmental impact. Following 5 decades of space missions, more than 100,000 nuts,
bolts, and other fragments of artificial objects are whizzing through space at velo-
cities as high as 17,000 mph.[43] Until these bits of space junk orbiting the Earth

---

[38]Tao S.L. and Desai T.A. (2003). Microfabricated drug delivery systems: from particles to pores. *Adv. Drug Delivery Rev.* 55, 315–328.

[39]Gardner P. (2006). Microfabricated nanochannel implantable drug delivery devices: trends, limitations, and possibilities. *Expert Opin. Drug Deliv.* 3, 479–487; Sharma S. et al. (2006). Controlled-release micro-chips. *Expert Opin. Drug Deliv.* 3(3), 379–394; Santini J.T., Jr. et al. (2000). Microchip technology in drug delivery. *Ann. Med.* 32(6), 377–379.

[40]Silva G.A. et al. (2004). Selective differentiation of neural progenitor cells by high-epitope density nanofibers. *Science* 303, 1352–1355.

[41]Ellis-Behnke R.G. et al. (2006). Nano neuro knitting: peptide nanofiber scaffold for brain repair and axon regeneration with functional return of vision. *PNAS* 103, 5054–5059.

[42]Yang F. et al. (2004). Fabrication of nano-structured porous PLLA scaffold intended for nerve tissue engineering. *Biomaterials* 25, 1891–1900.

[43]Britt R.R. Space junk. Space.com, Accessed at http://www.space.com/spacewatch/space_junk.html. (Retrieved October 19, 2000).

eventually burn up upon entering the atmosphere, they pose a small risk to astronauts. Prior to the exploration of space, we had no experience with the consequences of leaving in orbit hardware no longer useful. Similarly, preliminary investigations of environmental fullerenes (buckeyballs) suggest the potential for cerebral neurotoxicity in some circumstances.[44] Inhaled manganese oxide ultrafine particles 30 nm in size created by welding have been detected in brain tissue.[45] On the other hand, some fullerene derivatives appear to have neuroprotective properties.[46] Much research has yet to be done to investigate the potential toxic and environmental effects of various nanoparticles and nanomaterials to ensure that nanoscale therapies will fall within an acceptable margin of safety.

Nanomedicine will deliver amazing new medical products directly to patients. Spin-offs from the space program included Tang, Velcro, Teflon, and many other products now available to the general public. Nanotechnology may provide the means to develop small-scale, sophisticated, and ultimately inexpensive biomedical devices to enable patients to participate more directly in their own healthcare. Possible examples might include over-the-counter genetic testing kits, nanodevices that can detect spoiled food by sampling the air, nanofilters for efficient water purification, extremely lightweight mosquito-repellent barriers that can be distributed to malaria-infested regions, and nanodevices that can continuously monitor cardiac impulses or blood glucose. Scaling the instruments of medical testing down in size could, if the cost of manufacture decreases, also improve access to medical technology in developing nations.

Nanomedicine holds almost endless opportunities for innovation. The Viking space missions landed mobile robotic vehicles on the surface of planet Mars to conduct scientific experiments. Nanomedicine may one day craft remotely controlled probes to penetrate, measure, and interact with living cells. Such probes might study normal cells, restore impaired cells, or destroy cancer cells.[47] Researchers at MIT have begun to lay the groundwork for radio-controlled biomolecules. By linking a metal nanocrystal to a DNA molecule, Hammad-Schifferli and colleagues were able reversibly to unwind a specific segment of DNA by applying an external magnetic field.[48] Further developments in such technology could lead to the capability of activating or deactivating desired genes by remote control.

Nanomedicine will open new channels of communication. Satellites launched into orbit around the Earth now relay electronic communications worldwide. For

---

[44]Oberdörster E. (2004). Manufactured nanomaterials (fullerenes, C_{60}) induce oxidative stress in the brain of juvenile largemouth bass. *Environ. Health Perspect.* 112, 1058–1062; Oberdörster G, et al. (2005). Nanotoxicology: an emerging discipline evolving from studies of ultrafine particles. *Environ, Health Perspect,* 113, 823–839; Hoet P.H.M. et al. (2004). Nanoparticles—known and unknown health risks. *J. Nanobiotech* 2, 12–26.

[45]Elder A. et al. (2006). Translocation of Inhaled Ultrafine Manganese Oxide Particles to the Central Nervous System. *Environ. Health Persp.* 114, 1172–1178.

[46]Dugan L.L. et al. (1997). Carboxyfullerenes as neuroprotective agents. *PNAS* 94, 9434–9439.

[47]Patel G.M. et al. (2006). Nanorobot: a versatile tool in nanomedicine. *J. Drug Target* 14(2), 63–67.

[48]Hamad-Schifferli K. et al. (2002). Remote electronic control of DNA hybridization through inductive coupling to an attached metal nanocrystal antenna. *Nature* 415, 152–155.

the first time in human history, video transmissions of breaking news around the globe are instantly available via television, anywhere serviced by electrical power. This network of satellites also make possible the Global Positioning System that accurately guides navigation by land and sea. Satellites also connect offices, homes, and students worldwide through the Internet. Nanomedicine may one day provide the means for brains to communicate directly with computers. Several notable developments are proceeding in this direction. Peter Fromherz and colleagues at the Max Planck Institute, in collaboration with Infineon Technologies, have succeeded in recording the electrical activity in cultured hippocampal slices by a high-resolution multitransistor array with more than 16,000 sensors/ mm^2. The hippocampus is the part of the brain that coordinates the laying down and retrieval of memories. The density of sensors on the chip ensures prolonged contact and interaction with each neuron on the tissue surface, making it possible to measure the flow of information through the neural network.[49] Fernando Patolsky, Charles Lieber, and colleagues at Harvard University have devised arrays of nanowires integrated with the axons and dendrites of live mammalian neurons. These hybrid structures permit exquisitely detailed real-time measurement of neuronal signals propagating along each neuron, which may have as many as 50 "artificial synapses."[50]

Further development of neural interfaces at the nanoscale may lead to refinements in brain–computer interface technology, which has already achieved a rudimentary phase.[51] Brown University's Gerhard Friehs, in collaboration with Cybernetics, implanted a 100-electrode array over the motor cortex of a C4 quadriplegic, allowing the paralyzed patient, with training, to move a computer cursor by thought, thereby sending simple instructions to a computer monitor and to an external prosthetic hand.[52] More sophisticated brain–computer interfaces are under development and could be used to restore communication to the severely paralyzed or locked-in patient.[53] Additionally, Theodore Berger and colleagues at the University of Southern California in Los Angeles are currently developing a computer chip that, once implanted into the brain, would function as a prosthetic hippocampus.[54] More sophisticated neural prostheses may eventually stretch their nanowire sensors in three dimensions (3D) to conform more intimately to the cytoarchitecture of the brain.[55]

[49]Hutzler M. et al. (2006). High-resolution multitransistor array recording of electrical field potentials in cultured brain slices. *J. Neurophysiol.* 96, 1638–1645.

[50]Patolsky F. et al. (2006). Detection, stimulation, and inhibition of neuronal signals with high-density nanowire transistor arrays. *Science* 313, 1100–1104.

[51]Leary et al. See footnote 18; Liopo A.V. et al. (2006). Biocompatibility of native and functionalized single-walled carbon nanotubes for neuronal interface. *J Nanosci. Nanotech.* 6(5), 1365–1374.

[52]Martin R. (2005, March). Mind control. *Wired*, Available at http: //www.wired.com/wired/archive/13.03/brain.html?pg=2&topic=brain&topic_set=.

[53]Kubler A. and Neumann N. (2005). Brain-computer interfaces—the key for the conscious brain locked into a paralyzed body. *Prog. Brain Res.* 150, 513–525.

[54]Graham-Rowe D. (2003, March 12). World's first brain prosthesis revealed. *New Scientist.*

[55]Berger T.W. and Glanzman D.L. (2005). *Toward Replacement Parts for the Brain: Implantable Biomimetic Electronics as Neural Prostheses.* Cambridge, MA: MIT Press.

These possibilities raise important questions about how emerging nanotechnologies ought to be harnessed and guided in the interest of human flourishing.

VISIONS OF MEDICAL NANOUTOPIA

The anticipated successes of nanotechnology have aroused optimism in the potential for medical science to deal definitively with human illness. Perhaps nanomedicine might accomplish for many diseases what microbiology, combined with vaccination technology and a committed public health effort, achieved for smallpox in 1979, when the disfiguring and deadly virus finally was globally eradicated.[56] Humanity often rises to the challenge in responding to dire circumstances. In keeping with this spirit of optimism, the National Cancer Institute has set the ambitious strategic goal of the elimination of suffering and death due to cancer by the year 2015.[57] The pursuit of preserving health, preventing and treating disease, and reducing suffering through nanotechnology is indisputably good. Even if only a portion of predicted nanomedicine applications prove possible, the benefit to patients would be enormous.

Less clear in the moral analysis, and the subject of considerable interest, has been the question of whether nanotechnology should be applied to projects intending to enhance human nature. Traditionally, the role of medicine has been to heal the sick, to restore what has gone awry, to respond to illness with therapy and to suffering with compassion.[58] Proponents of medical enhancement contend that the scope of medicine should extend also, if the patient requests it, to helping patients to be better than normal and to feel healthier than well. Some of the arguments advanced in favor of enhancement include respecting patients' autonomy, yielding to economic incentives, fulfilling the creative human impulse, and taking control over human destiny.[59]

The material efficiency of nanotechnology could render it a powerful tool for human enhancement. Once safe and affordable prosthetic nanotechnologies capable of joining with human tissue for the purpose of restoring lost function were developed, it might be difficult to prevent their use also for purposes of cybernetic augmentation of normal function. We continually upgrade our personal computers, electronic entertainment equipment, and cellular telephones with each phase of improvement in technology. If nanotechnology were to give us the opportunity, would we choose to upgrade our bodies and our minds?

The success of some forms of nanomedicine derives from the similarities of biological and mechanical systems. The study of cellular macromolecules, receptors,

---

[56]Fenner F. et al. (1988). Smallpox and Its Eradication. Geneva, Switzerland: World Health Organization.

[57]Available at http://strategicplan.nci.nih.gov/pdf/nci_2007_strategic_plan.pdf.

[58]Cheshire W.P. (2006). Drugs for enhancing cognition and their ethical implications: a hot new cup of tea. *Expert Rev. Neurotherap.* 6(3): 263–266.

[59]Parens E. (1998). *Enhancing Human Traits: Ethical and Social Implications*. Washington, DC: Georgetown University Press, Chatterjee A. (2004). Cosmetic neurology: the controversy over enhancing movement, mentation, and mood. *Neurology* 63, 968–974.

neural networks, and other highly developed natural biological systems is yielding clues useful to the design of efficient nanoscale structures and devices. Nanotechnology in some cases replicates, in other cases mimics, and in still other cases interfaces with human biomolecules at the subcellular level.[60] Some interpret these relationships of similarity, complementarity, and integration to reflect an essential interchangeability of living and artificially fabricated matter. The ability to merge the biological and the mechanical at the molecular level through nanotechnology begins, one atom at a time, to blur the distinction between patient and prosthesis.

The idea of "material unity at the nanoscale" lies at the heart of the Converging Technologies initiative of the National Science Foundation (NSF) and Department of Commerce.[61] Their 2002 multiauthored report sets a vision for combining *nano*technology, *bio*technology, *i*nformation technology, and *c*ognitive neuropsychology (NBIC) to improve human performance, and looks to nanodevices as the means to enhance mental function and connectivity. The beginning of the NBIC report reads: "We stand at the threshold of a new renaissance in science and technology, based on a comprehensive understanding of the structure and behavior of matter from the nanoscale up to the most complex system yet discovered, the human brain."[62] Mihail Roco and William Bainbridge continue: "Unification of science based on unity in nature and its holistic investigation will lead to technological convergence and a more efficient societal structure for reaching human goals."[62]

Nanotechnology-enabled brain-to-brain and brain-to-machine interfaces are among the stated NBIC research goals.[63] The NBIC agenda also highlights the prospect of reverse engineering the brain to enhance the function of computers.[63] Such a strategy would, the report speculates, increase our understanding of the scientific basis of biologically complex molecular systems and allow us to replicate them in software and nanoelectronics and to build interactive nanodevices having tens of billions of moving parts.[64] How far nanointerface microelectronics will advance remains a matter of speculation, but the comments of reputable researchers merit cautious scrutiny, and not just from the pragmatic perspective of what may be feasible. Two contributors to the NBIC report, Larry Cauller and Andy Penz, write, "We see this future in terms of a coming nano-neuro-cogno-symbiosis that will enhance human potential ... by opening direct channels of natural communication between body and artificial nervous systems for the seamless fusion of technology and mind."[65] Humanity would, according to the introduction of the NBIC report, "become like a single, distributed and interconnected 'brain' based in new core

[60]Pennadam S.S. et al. (2004). Protein-polymer nano-machines. Towards synthetic control of biological processes. *J. Nanobiotech* 2, 8–15.
[61]National Science Foundation and Department of Commerce. (June 2002). Converging Technologies for Improving Human Performance: Nanotechnology, Biotechnology, Information Technology, and Cognitive Science. Arlington, VA, pp. ix–x, 6, 162, 228, 256. Available at: http://www.technology.gov/reports/2002/NBIC.
[62]See footnote 61, p. 1.
[63]See footnote 61, p. xi.
[64]See footnote 61, p. 102–103.
[65]See footnote 61, p. 256.

pathways of society."[66] In the words of NBIC contributor Edgar Garcia-Rill, "the inescapable conclusion is that nanotechnology can help drive our evolution."[67] According to Roco and Bainbridge, "The twenty-first century could end in world peace, universal prosperity, and evolution to a higher level of compassion and accomplishment."[67]

Even allowing for the enthusiastic hyperbole that sometimes accompanies proposals written to solicit funding, these claims contain more than a nanogram of exaggeration. What inspires such zeal may be the degree of sheer control nanotechnology affords over material nature. If applied for enhancement purposes, nanomedicine could hold similar potential for commanding the shape of human nature, and for reconfiguring humankind, particularly if human nature were seen to be merely another form of material nature.

Nanotechnology has attracted a growing philosophical movement known as transhumanism. Nick Bostrom, the founder of the World Transhumanist Association, defines transhumanism as:

> the study of the means and obstacles to humanity using technological and other rational means to becoming posthuman, and of the ethical issues that are involved in this. 'Posthumans' is the term for the very much more advanced beings that humans may one day design themselves into if we manage to upgrade our current human nature and radically extend our capacities.[68]

Ray Kurzweil, whose book *The Age of Spiritual Machines*[69] gave momentum to the transhumanist movement, writes in his sequel *The Singularity is Near* of the hope of achieving a form of personal immortality by exploiting nanotechnology to scan and upload one's brain, and hence one's identity, onto a computer.[70]

Meanwhile, it is worth asking whether the transhumanists, in their quest for the perfection of humanity through nanotechnology, would unwittingly be leaving behind something good and true about human nature. To their credit, the transhumanists' dream is corrective, as well as perfectionistic. Transhumanists seek a way beyond human disease, misery, and disappointment. The reality of human suffering, however, is a sober reminder that the medical profession has a clear and compelling obligation to provide therapy for the sick, but only a weak obligation, if one at all, to promote enhancement for the well. In this world of limited resources, if medicine were to embark full speed down the

---

[66]See footnote 61, p. 6.
[67]See footnote 61, p. 228.
[68]Bostrom N. What is transhumanism? Available at http://www.nickbostrom.com/old/transhumanism.html. See also http://www.transhumanism.org; Hook C.C. (2004). Techno sapiens: nanotechnology, cybernetics, transhumanism, and the remaking of humankind. In Colson C.W. and Cameron N.M. de S., (2004). *Human Dignity in the Biotech Century*. Downer's Grove: Intervarsity Press, pp. 75–97.
[69]Kurzweil R. (1999). *The Age of Spiritual Machines: When Computers Exceed Human Intelligence*. New York: Penguin.
[70]Kurzweil R. (2005). *The Singularity is Near: When Humans Transcend Biology*. New York: Viking, pp. 199–200.

transhumanist path, it might do so at the cost of diverting attention from sick patients in need of care.

VISIONS OF MEDICAL NANODYSTOPIA

Utopian visions humanly implemented have a way of becoming dystopias. As for any developing technology, the first nanomedicine concerns to address are those of safety. Some forms of toxicity may be foreseeable, while others will be unforeseeable.

With regard to the environment, the high surface/volume ratio of nanoparticles gives some of them potent properties of chemical reactivity with biological molecules. Moreover, many types of nanostructures have never before existed. Their effects upon exposure to plant and animal life and on complex environmental ecosystems may be unpredictable. The precautionary principle suggests the need for tempering enthusiasm with due caution as the implementation of nanotechnology proceeds. More studies, and designated funding to enable those studies, will be needed to assess the environmental impact and potential toxicity of nanomolecules and nanostructures. Specific knowledge about toxicity will permit the development of appropriate guidelines and safeguards.

A special case is the "grey goo" nanomyth. In this chilling hypothetical scenario, first articulated by Eric Drexler in his book *Engines of Creation*[71] and popularized by Michael Crichton's novel *Prey*,[72] self-replicating nanorobots spread like locusts across the surface of the globe, efficiently and inexorably consuming all living matter. Grey goo is now believed to be an unlikely development, one reason being that building autonomous self-replicating nanomachinery would be unneccessary, as larger scale nanofactories would seem more efficient.[73] Although Drexler now dubs "grey goo" fears obsolete,[74] he advocates for prohibiting "the construction of anything resembling a dangerous self-replicating nanomachine."[74] Even aside from self-replicating devices, as a general principle, projects that lead to potentially irreversible consequences require more stringent safeguards.

Some nanotechnology dystopias, therefore, would be unintentional. Others might be intentional. If nanodevices can be designed to restore or enhance human tissue, they can also be constructed for harmful purposes. Nanodevices developed to be warriors against disease might be rearranged and reprogrammed to become nanoweapons in the hands of militant extremist groups. Their submicroscopic scale could render these tools of lethal molecular efficiency virtually undetectable. Particularly disturbing would be the prospect of malicious nanoweapons designed to profile genetic identity and selectively attack certain ethnic groups. It is thus possible to conceive of some types of nanodevices that should never be developed.

[71]Drexler K.E. (1986). *Engines of Creation: The Coming Era of Nanotechnology*. New York: Anchor.
[72]Crichton M. (2002). *Prey*. New York: Harper Collins.
[73]Phoenix C. and Drexler E. (2004). Safe exponential manufacturing. *Nanotechnology* 15, 869–872.
[74]Nanotechweb. Drexler dubs "grey goo" fears obsolete. Available at http://www.nanotechweb.org/articles/society/3/6/1/1 (retrieved June 9, 2004).

Other nanotechnology dystopias might exist only in the imagination. Enchanted by promotional claims of the powers of nanomedicine, and suspicious of the potentially unrecognized or unknown effects of nanoparticles on the human body, we may soon see occasional patients who have received nanomedical therapies proceed to develop an irrational fear of personal nanoparticle contamination. Similar fears of contamination have developed in relation to dental amalgams and other unvisualizable trivial exposures. While exceptional cases of toxic illness may exist, most complaints arise in the absence of a causal relationship between the artificial substance and the patients' symptoms.[75] Beset with unexplained vague somatic symptoms not due to traditionally defined disease, such patients focus their thoughts obsessively on the belief that their symptoms originate from artificial particles that have leaked into their tissues and are wreaking havoc on their health.[76] The invisibility of such small particles and the impossibility of detection by diagnostic equipment only intensifies their fear. Should this syndrome develop in the context of nanoparticle therapies, one or two affected patients with misguided initiative may publish webpages claiming that scientific studies disproving any pathophysiologic relationship reflect a medical or government conspiracy to withhold facts from the public. In such instances, the medical scientific community has an educational obligation to patients and the public. It may be possible to preempt such nanomyths with proper scientific studies and dissemination of reliable information.

Similar fear of contamination can burden rational patients if a genuine risk of toxicity is present from exposure to environmental substances too small or too widely scattered to be easily detectable. Such is the fear that the public would encounter if a "dirty bomb" containing radionuclides mixed with explosives were ever detonated in a populated area. If a highly toxic form of nanoparticle were to be invented, stringent safeguards would be needed, as for plutonium, to minimize the risk of large quantities of it or the means to its production falling into malevolent hands. Toxic types of nanoparticles might be more difficult to detect than radionuclides since they would not emit radioactive particles. Clearly, very powerful technologies require ongoing vigilance to ensure their safe use and storage.

Still other nanotechnology dystopias transcend questions of safety, toxicity, efficacy, and fair distribution of resources. For these scenarios, the human interests at stake are more subtle. These deeply human concerns have less to do with the physical effects of the technology itself and more to do with the motives of those who use it and the various ways the technology reshapes the lives of others.

Consider the potential to harness nanotechnology as a means of social control. It should not be surprising to observe that governments and political regimes around the globe have always differed in their views of freedom and individual privacy. Would some governments seek to develop nanotechnologies, or redirect existing nanotechnologies, for purposes of establishing omnipresent surveillance or imposing tight control over their citizens' lives and expression of speech? History is

[75]Binder L.M., Campbell K.A. (2004). Medically unexplained symptoms and neuropsychological assessment. *J. Clin. Exper. Neuropsychol.* 26, 369–392.
[76]Rachman S. (2004). Fear of contamination. *Behav. Res. Ther.* 42, 1227–1255.

replete with examples of technologies initially developed for good purposes becoming instruments of tyranny in the hands of less-benevolent regimes.

In a free and democratic society, professional, peer, and market pressures could exert subtle forms of coercion. In these instances, how people think about nanotechnology may be more powerful than what nanotechnology actually does. Consider, for example, the goal of human perfection. Some have viewed nanotechnology as the means not only to restore human tissue, but also to improve upon human nature by modifying, reconstructing, or augmenting the body or establishing interactive connections between the human body and sophisticated machines. If this were attempted, what standard of human nature should be sought? Who would decide which vision of humanity will be the goal to which the technology will be directed? Would all people be free to choose whether and how to use the technology?

Consider the prospect of the augmented physician. In a future decade, one might have the option of seeking treatment from a physician who has chosen to have his or her cognitive capacities enhanced through technology. Suppose that wakefulness-promoting drugs take the place of a morning cup of coffee to achieve maximum alertness and sustained mental concentration—such drugs are, in fact, already available.[56] Suppose, extending this hypothetical scenario a bit further, that advances in nanointerface microelectronics and nanowire fabrication make it possible to implant brain nanoelectrodes noninvasively, safely, and painlessly. Suppose that these nanoelectrodes can follow the subtle paths of axons in the brain and automatically match up with the 3D architecture of their target region, establishing functional connections with thousands or tens of thousands of neurons. A next-generation nanowire deep brain stimulator might, for example, steady a surgeon's hand during delicate operations.

A psychiatrist has a speech recognition chip implanted just beneath her scalp. The hidden device listens in at the bedside and quietly whispers key questions to guide the interview, or perhaps treatment suggestions that take into account all possible drug interactions, through a stimulating nanowire tipped with branching tentacles contacting her tympanic membrane. If the research that developed the device had been funded by the pharmaceutical industry, would its promptings be trustworthy?

A dermatologist has an implant reaching into his visual cortex that records images of skin conditions in every patient he encounters. The clinical images are stored in implanted nanoscale digital media for subsequent retrieval. Its software instantly compares incoming images with stored files and transmits suggested diagnoses directly to the physician's occipital cortex. He is able to visualize the automatically generated images as if watching a television screen. Because he does not notice them, he does not find the occasional subliminal flashes of images of cosmetic skin cream to be intrusive into his practice.

Another physician no longer spends time dictating clinical notes because a nanoscale voice detection system implanted in his auditory canal extracts information directly from his conversations with patients and constructs text that appears as a virtual image in his mind. By thought, he moves the cursor that edits the text.

Another physician, fatigued from long sleepless hours on night call, switches on a facial nerve stimulator that generates automatic smiles, dissociated from any hint of his true emotions.

Imagining such possibilities is fascinating, and, admittedly, it is tempting to be carried away by such fantastic ideas. Less plausible would be a nanointerface micro-electronic miracle chip connecting the physician's brain directly to the totality of the medical literature. Suppose information about rare diseases, drug interactions, and emerging techniques could be accessed on demand simply by thinking about them. It is difficult to imagine, however, given the complexity of medical data, how a direct brain-to-computer interface with a medical library could outperform manual and visual interfaces with external digital devices, even if future nanoengineering could achieve it. Of the more plausible feats, however, if any one of them could be accomplished, much could be achieved in terms of efficiency.

Enhancing professionals with nanotechnology has a seductive appeal and might be marketable. Wherein lies its appeal? Is efficient performance the supreme virtue worthy of being maximized? Does it matter what is performed, how it is performed, by whom, and for what purpose? Is systematic manipulation of matter by technical means the most important of human tasks? Or is there something else that makes a good doctor—something that a skilled automaton would be incapable of doing no matter its degree of mechanical precision, and that a human medical service provider augmented in efficiency by cybernetic accessories would not necessarily do better? It matters how we define *better*, that is, what values and purposes are implied.

Moreover, the gains in productivity and error prevention could provide strong incentives for physicians to opt for such augmentations. Patients might insist on the right to be seen by an *enhanced* specialist. Professional societies and physician employers might offer recommendations and implement incentives for physicians to comply. Legal liability risks and the increased remuneration possible by accelerating task throughput would also factor into decisions whether to plug into such technology. Even if safety factors were adequately addressed and physical harms minimized, the personal decision whether to participate in enhancing technology would thus not be fully autonomous. Were the use of enhancing technologies to become customary, it might be difficult for a physician to choose to remain unenhanced. Similar concerns would apply to the use of nanotechnology to augment human performance in other professions and vocations, in soldiers[77] and in students.[78]

The appetite for greater and greater personal enhancement might be insatiable. Cultivating the desire for more enhancement might also foster an attitude of despising plain, ordinary human nature as found in ourselves or in others.

Katherine Hailes paints a bleak picture of the path of technological posthumanism. "Humans can either go gently into that good night, joining the dinosaurs as a species that once ruled the Earth but is now obsolete, or hang on for a while longer by becoming machines themselves. In either case . . . the age of the human is drawing to a close."[79]

---

[77]Baard E. (2003, January 22–28). The guilt-free soldier: new science raises the specter of a world without regret. *Village Voice*.

[78]Sparks J.A. and Duncan B.L. (2004). The ethics and science of medicating children. *Ethical Human Psychology and Psychiatry* 6, 25–39.

[79]Hailes N.K. (1999). *How We Became Posthuman: Virtual Bodies in Cybernetics, Literature and Informatics*. Chicago: University of Chicago Press, pp. 2–3.

Medical nanodystopias, whether in fact or theory, awaken us to the far-reaching consequences of the decisions before us regarding nanotechnology, as well as to its profound implications for understanding and manipulating human nature. Among the potential applications of nanotechnology to society, the ones to medicine in particular press the question of what it means to be human. Even imagining theoretical and fanciful nanomedical technologies by thought experiments raises this timeless question in compelling new ways.

Technological dystopias are disturbing because of their embrace of nihilism. Tantalizing tales of better worlds in which efficient machines have supplanted human effort turn horrific once things start to go wrong. The irony is that things can go wrong even when the machines are working perfectly. The perfection of machinery for the sake of power, it turns out, tells us nothing about how such power should be used. Though wondrously efficient nanomachines may inspire admiration, ultimately they disappoint our longing for fulfillment because mechanistic efficiency has no purpose of its own. Technology's purpose derives from its human designers and wielders. In this respect, C.S. Lewis wisely observed that: "What we call Man's power over Nature turns out to be a power exercised by some men over other men with Nature as its instrument."[80]

Though optimistic in their portrayal of material performance, dystopias are pessimistic in their view of human nature. Nanodystopias tend to deepen that pessimism further by suggesting that efficient material power will inevitably triumph over all else. Humanity, vulnerably dwarfed by the glorious success of technology become autonomous, appears irrelevant. Resistance seems futile. Successful stories of dystopias, of course, always allow human nature to win in the end, but it is the premise of human nature in bondage or on the verge of defeat that sets the stage.

MEDICAL NANOREALISM

If our own human story is to succeed, with or without nanotechnology, we must have an accurate understanding of human nature. Much of the current discourse originates from two theories of human nature. For naturalists, human nature is the result of an unguided process of natural selection, whereby environmental hardships and competition from other life forms select for survival. Survival gives an opportunity to reproduce, such that chance genetic mutations lead to greater performance. The process of natural selection is, by this theory, ongoing. The potential for nanotechnology to accelerate our deciphering of the human genome and to alter DNA with surgical precision introduces the possibility of intentionally guiding or redirecting the otherwise blind process of human evolution. If in the natural order there is no fixed human nature, then human nature might be redesigned. Humanity, as the presumptive ultimate moral authority, would possess the right to remake itself and choose its own destiny. Once that assertion is accepted, the remaining ethical questions are mostly pragmatic and concern such matters as how to achieve germline

---

[80]Lewis C.S. (1978). *The Abolition of Man.* (originally published 1943) New York: Macmillan, p. 69.

intervention safely, who decides, and how to respect the thoughts and feelings of those about to become obsolete.

For theists, this author included, the naturalistic story is partly true, but there is also much more to reality than the eyes can see, than matter can model, or than determinism etched by chance can conceive. Theists perceive nature's elegant design, improbable origins, fragile persistence, and wondrous beauty to be the creative work of a transcendent Intelligence. The fact that the world is populated with persons is not an illusion conjured by our brains,[81] but rather evidence that we have been created for the purpose of entering into a relationship with a personal God. For theists, human nature is not accidental. Human nature is divinely ordained in the image of our Creator, wherein lies its intrinsic dignity. Human nature is also, as history repeatedly testifies, tragically fallen and in need of redemption. Theists hold to an eschatological hope in a new and perfected created order not of our own design. There is also within theistic traditions the understanding that humans as imagers of God are endowed with creativity and, in ministering to the needs of the world, have the capacity to be cocreators with God. Human creativity entails the responsibility of exercising it within universal moral principles provided for our welfare.

Nanotechnology offers a more detailed view of the material aspect of human nature. Although quite informative, the parts themselves are incapable of painting the complete picture of the whole. Just as prolonged close-up work can weaken the eye's ability to adjust to perceiving clearly objects at a distance, focusing exclusively on the improved resolution of human biology afforded by nanotechnology can lead to reductionistic thinking about human nature. The view of human beings through eyes of nanoscience may convey the sense of looking down into the well of human nature and, finding at its bottom simply an array of particles, mistakenly concluding that particles are all there is to know about human nature.

The apparent interchangeability of biological molecules and artificial machines at the nanoscale initially appeals to a reductionistic evaluation of living beings. A more thoughtful study of the Lilliputian landscape of nature, however, discovers evidence of stunning design surpassing the most clever examples of human engineering. While the naturalist may choose to interpret the intricate efficiency and irreducible complexity of biological molecules as an arbitrary resource, the theist will appreciate in nature's nanomachines the nimble fingerprints of an amazing Creator.

Nanotechnology also offers new methods of manipulating material nature, as well as the material aspects of human nature. The ability to alter matter at the atomic level confers greater leverage over things. While such nanoleverage may yield greater control, that does not ensure mastery. One reason is that human knowledge is incomplete. The consequences can be only dimly foreseen in a complex world. Furthermore, medical experience teaches that all interventions have potential adverse effects. The same will hold true for the most carefully conducted techniques in

---

[81]Farah M.J. and Heberlein A.S. (in press). Personhood and neuroscience: naturalizing or nihilating? *Am. J. Bioethics.*

nanomedicine. Human error, both unintentional and intentional, eventually tarnishes even the most cautiously laid out plans.

Nanomedicine offers promising new resources for serving patients' needs. Society should enthusiastically support the development and implementation of nanotechnology to improve human health and treat illness. Resources should also be allocated to exploring the ethical implications of nanotechnology and to educational programs to assist patients and professionals in understanding the impact of nanotechnology on our lives. As nanomedicine grows beyond its infancy and takes its first steps into hospitals and clinics, education will become increasingly important. The goals of education will include helping people to develop realistic expectations, exploring what nanoscience reveals about the nature of things and the limits of what it is able to describe, and understanding what nanomedicine can and cannot achieve.

Nanomedicine is upgrading the means, but does not change the primary goal, of medicine, which remains the care of patients. Nanomedicine may grant us the ability to conquer some diseases and for others to reduce more effectively their human injury. Some problems will be solved; some new problems will be created. The most efficient nanorobots, however, can neither conquer the curse of human frailty, nor compel physicians to care. Once that is acknowledged, we will have much for which to be thankful, and much to which to look forward.

> And the last great lesson is humility before the unsolved problems of the universe.
> William Osler[3]

CHAPTER 19

Nanotechnology and the Future of Medicine

C. CHRISTOPHER HOOK[1]

INTRODUCTION

When Eric Drexler resurrected and promoted Richard Feynman's vision[2] for nanotechnology in his doctoral thesis and subsequent technical volume, *Nanosystems*,[3] and popular-level book, *Engines of Creation*,[4] he predicted in the latter volume that one of the main uses of the technology would be in medical technology. Feynman himself claimed that his inspiration for molecular-level engineering was the combination of molecular machinery and synthetic processes in the living cell. Since that time nanoseers and promoters have been promising magnificent, even miraculous, treatments for all ailments, including cancer and all infectious diseases, repair of disordered and diseased tissues and organs, and even significantly prolonged life spans. A few go so far as to suggest that nanomedicine will bring about an end to the scourge of humankind, aging, and natural death itself.[5]

---

[1]Dr. Hook's comments are solely his own and do not necessarily reflect the views of the Mayo Clinic and Foundation.

[2]Feynman, R. (1960). There's Plenty of Room at the Bottom. *Engineering and Science* 23, 22–36. Available at http://www.zyvex.com/nanotech/feynman.html (retrieved October 17, 2006).

[3]Drexler, K. E. (1992). *Nanosystems: Molecular Machinery, Manufacturing and Computation.* New York: John Wiley & Sons, Inc.

[4]Drexler, K. E. (1986). *Engines of Creation.* New York: Anchor.

[5]Most vocally joining Drexler in proclaiming the coming nanoutopia is Ray Kurzweil (2005). *The Singularity Is Near: When Humans Transcend Biology.* New York: Viking. Kurzweil is so convinced that nanotechnology will enable humankind to achieve immortality, or at least dramatically extend life spans, that he is said to take more than 250 pills a day of vitamin, enzyme, and other supplements, and several times a week receives intravenous therapies so that he can live long enough to benefit from the cures nanotechnology will bring. In the event that he may die before the nanofountain of youth is developed, his body is to be frozen by the Alcore Foundation so that he can be resurrected once these tools are available.

Nanoscale: Issues and Perspectives for the Nano Century. Edited by Nigel M. de S. Cameron and M. Ellen Mitchell

Others see the development of molecular technologies for medical use as a road to technologies of human re-engineering, not just healing. Thus, in their projections and hopes, human beings will soon be able to not just heal, but also to "enhance" our bodies and minds, make ourselves stronger, smarter, . . . "better." These are amazing claims that include not just a little hyperbole, and that raise serious medical, social, and ethical questions. This chapter will provide a brief overview of the concept of nanomedicine and some of the ethical issues it raises, particularly so-called "enhancement" or human re-engineering projects. Hopefully, along the way, it attempt to bring the larger discussion of nanomedicine down from the stratospheric vapors of nanohype.[6]

According to Robert Freitas, one of the self-professed pioneers of nano-medicine, and the European Science Foundation, which adopted his definition almost verbatim, nanomedicine is "Most broadly, . . . the process of diagnosing, treating, and preventing disease and traumatic injury, relieving pain, and preserving and improving human health, using molecular tools and molecular knowledge of the human body. In short, nanomedicine is the application of nanotechnology to medicine."[7] The U.S. National Institutes of Health (NIH) Nanomedicine Initiative defines nanomedicine as "a new discipline that integrates nanotechnology with nanoscience of cellular processes and uses this information to diagnose and treat diseases."[8] Medical uses of microscopic, sub-cellular agents, or machines potentially include rational drug design, devices specifically targeting and destroying tumor cells,[9] or infectious agents, *in vivo* devices for at-the-site-of-need drug manufacture and release, tissue engineering, or re-engineering, early detection or monitoring devices, *in vitro* lab on a chip diagnostic tools,[10] devices to clear existing atherosclerotic lesions in coronary or cerebral arteries, biomimetic nano-structures to repair or replace deoxyribonucleic acid (DNA) or other organelles, provide artificial replacements for red blood cells and platelets,[11] augment or repair interaction between neurons in the brain, improve biocompatibility and the interface between brain tissue and cybernetic devices, and develop more

[6]A thorough treatment of all forms of nanohype, not just those related to medicine, may be found in Berube, D. M. (2006). *Nano-hype: The Truth Behind the Nanotechnology Buzz.* Amherst, New York: Prometheus Books.

[7]Freitas, R. A., Jr. (1999). *Nanomedicine, Volume 1: Basic Concepts.* Austin: Landas Bioscience.

[8]Available at http://nihroadmap.nih.gov/nanomedicine.

[9]McDevett, M. R. et al. (2001, November 16). Tumor Therapy with Targeted Atomic Nanogenerators. *Science 294*, 1537–40; Kumar, C., ed. (2006). *Nanomaterials for Cancer Therapy.* New York: Wiley-VCH.; Kawasaki, E. S. and Player, A. (2005). Nanotechnology, nanomedicine, and the development of new, effective therapies for cancer. *Nanomedicine: Nanotechnology, Biology, and Medicine* 1, 101–109.

[10]Park, So-Jung et al. (2002, February 22). Array-Based Electrical Detection of DNA with Nanoparticle Probes. *Science 295*, 1503–6.

[11]Freitas, R. A., Jr. Respirocytes: A Mechanical Artificial Red Cell: Exploratory Design in Medical Nano-technology. (Revised version). Available at www.foresight.org/Nanomedicine/Repirocytes.html.

durable prosthetic devices or implants.[12] Such tools have also been envisioned to provide new means of cosmetic enhancement, such as new forms of weight control, changing hair or skin color, removing unwanted hair, or producing new hair simulations.[13] If many of the potential therapeutic uses listed above become reality, producing effective treatment of life's greatest killers, such as cancer, infectious disease, and vascular disease, such treatments could also significantly extend lifespan, by forestalling nonaccidental death.

But what, if anything is truly different about nanomedicine from conventional medical science and therapeutic interventions? In many ways, not much. Medicine has long used pharmaceuticals that work at the molecular level. Aspirin, acetylsalicylic acid, is a molecule that binds irreversibly to cyclooxygenase 1, a critical component in the pathways that produce prostaglandins, important molecules in the pain and thermoregulatory mechanisms of the body, and thromboxane A2, a potent stimulator of platelet function. Aspirin is, thus, a molecular tool or machine that is used therapeutically to treat pain, fever, and to prevent unwanted clotting, such as a heart attack. In the past, aspirin was just a drug, but if it were developed now as a "molecular machine," rather than having been discovered as a component of willow bark, it would be claimed to be a nanotech breakthrough. Antibodies are molecular machines that recognize foreign materials in the body and activate a number of molecular and cellular processes of immunity. They are produced by complex molecular factories that receive their building instructions from a series of genetically based, yet environmentally stimulated and conditioned, instructions. Biotechnology has been coopting antibody mechanisms for many years to produce targeted drugs (e.g., Rituximab), which is used in the treatment of lymphoma and autoimmune disorders.

A number of other engineered molecules with unique physical, structural, and chemical properties have been used for many years, and more are rapidly being developed and brought to market (e.g., revlamid for the treatment of myelodysplastic syndromes, or gleevec for the treatment of chronic myelogenous leukemia). Essentially all nonbiological chemotherapeutic agents used to treat cancer work through molecular mechanisms either by the inhibition of specific enzymes critical to cell functioning, particularly DNA replication and cell division, or directly injuring the DNA molecules of the cells themselves.

[12]Drexler, K. E. *Unbounding the Future*; Freitas, Nanomedicine; Candell, B. C., ed. (1999). Nanotechnology: Molecular Speculations on Global Abundance. Cambridge: MIT Press; BECON (NIH Bioengineering Consortium). Nanoscience and Nanotechnology: Shaping Biomedical Research – June 2000 Symposium Report. Available at www.becon.nih.gov/poster_abstracts_exhibits.pdf and www.becon.-nih.gov/nanotechsymposiumreport.pdf; Greco, R. S., Prinz, F. B., and Smith, R. L. (2005). *Nanoscale Technology in Biological Systems.* New York: CRC Press; Pennadem, S. et al. (2004). Protein-polymer nanomachines. Toward synthetic control of biological process. *J. Nanobiotechnol.* **2**:8. Available at http://www.jnanobiotechnology.com/content/2/1/8. (Retrieved October 17, 2006.); Freitas, R. A., Jr., (2005). Current Status of Nanomedicine and Medical Nanorobotics. *J. Comput. Theor. Nanosci.* 2, 1–25; Freitas, R. A., Jr. (2005). What is Nanomedicine? *Nanomedicine: Nanotechnology, Biology, and Medicine* 1, 2–9; Urban, G. A., ed. (2006). *BioMEMS.* Dordrecht, The Netherlands: Springer.
[13]Crawford, R. (1999). Cosmetic Nanosurgery in Candell, *Nanotechnology.*

One of the methods that is used to produce some of these agents is recombinant DNA manufacturing in which the existing machinery of bacteria or animals (e.g., mice) are reprogrammed to produce the desired product in large quantities. The polymerase chain reaction (PCR), a tool used in molecular diagnostics for the detection of specific genes that has been with us since the late 1980s, is, in its fundamental essence, a type of molecular manufacturing. PCR takes a tiny quantity of DNA (actually one copy is all that is necessary) and expands the quantity of that gene or genetic material to levels more easily detected by traditional macromolecular methods.

These processes and products, while all working at the molecular level, have not traditionally been referred to as nanotechnology, but rather have been identified within the fields of biochemistry, pharmacology, molecular genetics, immunology, hematology, molecular cell biology, and biotechnology. Thus, much of "nanomedicine" is but a continuation of the understanding of physiological, biochemical, cellular, genetic, and disease processes that have been evolving over hundreds of years, but certainly accelerating dramatically during the past 4 decades. It is giving a new sexy label to the status quo to invite more public funding and private sector investment into the next new big thing. Since the Clinton administration was persuaded by the disciples of Feynman and Drexler to create a major research and development initiative in the United States, the National Nanotechnology Initiative in 1999, everything seems to be nano-this and nano-that. Findings that traditionally would have been published in traditional chemical journals are now being labeled nanochemistry, and a whole spate of new journals with "nano" in the title have appeared in the past 6 years and are continuing to appear at a rapid rate. It seems that a good way to be more competitive for grant funding these days, both in the hard sciences and in the area of science and technology studies (STS), or ethical, legal, and social implications (ELSI) projects is to include nano- or nanotechnology in the title of the proposal.

There are, however, potentially new and unique aspects of nanomedicine, as well. One area is the approach pursued primarily by Robert Freitas in his expanding multivolume work on theoretical nanomedicine.[14] Freitas, a disciple of Drexler, is primarily focusing of the creation of unique, nonbiological machines that would operate at the molecular, subcellular level. These devices would be made from carbon-based diamonoid materials, and would be, in essence, molecular analogs of traditional machines, with gears, levers, rotors, and so on. Freitas has developed conceptions of artificial red blood cells, respirocytes, that would carry oxygen, a tool that could be used in the field and would not require cross-matching or cryopreservation, and an artificial platelet, the clottocyte, both of which could have significant clinical impact if actualized. This project, however, has little to show in terms of practical or even actual results thus far.

Some groups have been working on using synthetic carbon nanotubes and spherical structures (Buckminster fullerenes, or buckyballs) in novel ways for

[14]Freitas, R. A., Jr. (1999). *Nanomedicine Volumes I, IIa, IIb, III.* Austin: Landes Bioscience (also 2003, and pending).

drug delivery,[15] but this work is hampered by findings of potentially serious toxicity to normal cells by the nanotubes and buckyballs.[16] Indeed, one of the major limitations to this so-called "top-down" approach to producing therapeutic agents is, and will be, biocompatibility and toxicity.

A few interesting projects by other researchers have made molecular motors, but used existing biological molecules for the majority of the device structure and function, rather than developing a truly unique, biologically independent mechanism.[17] This is not necessarily a weakness, in that any device that is to be used *in vivo* must be able to interact specifically and efficiently with the biological milieu. Good engineering practice would require using available energy sources, and so on.

Another unique approach to nanomedical methods is to reprogram existing biological agents (i.e., cells and viruses) to perform new functions.[18] Ehud Shapiro and his colleagues have been working on programmable cells.[19] Others researchers are also making strides in the area of cellular computing.[20] The idea here is to coopt the system of homeostasis and gene regulation in cells and reprogram these mechanisms to perform functions that clinicians and patients would desire to treat disease and/or maintain healthy functioning.

Shapiro's work begins by understanding that a computer is an apparatus (not necessarily electronic or silicon based) for making calculations or controlling operations that are expressible in numerical or logical terms. In this sense, a single bacterium of *Escherichia coli*, an ubiquitous organism, possesses roughly an equivalent degree of computing power as a Pentium II microprocessor. The difference is that its inputs come from biomolecules, signals from receptors on the surface of or internal to the cell, and that its outputs are also biomolecules, the quantity of which is regulated by gene expression, and produced by ribosomes and a complex system of enzymes. His "Doctor in a Cell" (which he projects could be available by 2020) could be trained to detect certain substances that produce or denote the presence of disease, and then to produce products to deal with the problem. For example, a cell might be trained to detect rising levels of prostate specific antigen (a protein that may indicate the presence of evolving prostate cancer), and, in response, produce and secrete an agent to inhibit the growth of or destroy the cancerous cells. One of the main differences between this approach for molecular engineering of the cell and Freitas' is that there have already been a number of reports of proof of concept with DNA and/or cellular computing (though nothing at the time of this writing is even close to a clinical trial).

---

[15]For example, see Luna Innovations Incorporated. Available at http://www.lunainnovations.com.

[16]Hoet, P. et al. (2004). Nanoparticles—known and unknown health risks. *J. Nanobiotechnol.* 2:12. Available at http://www.jnanobiotechnology.com/content/2/1/12 (retrieved October 17, 2006).

[17]Soong, R. et al. (2000, November 24). Powering an Inorganic Nanodevice with a Biomolecular Motor. *Science* 290, 1555–58. See also, Montemagno, C. et al. (1999). Constructing Biological Motor Powered Nanomechanical Devices. *Nanotechnology 10*, 225–31.

[18]For example, see Mao, C. et al. (2003). *Proc. Nat. Acad. Sci.* 100, 6946–6951.

[19] A number of Professor Shapiro's publications and presentations may be found. Available at http://www.wisdom.weizmann.ac.il/~udi.

[20]Amos, M., ed. (2004). *Cellular Computing.* New York: Oxford University Press.

NANOMEDICINE AND HUMAN RE-ENGINEERING

One of the most controversial claims for nanomedicine, is that these tools will also be able to re-engineer normal tissues to something even better, more durable, and less susceptible to disease or degeneration. Molecular tools could also introduce new devices to "enhance" function, such as improved eyesight with detectors perceiving wavelengths other than the visible spectrum; improved memory or mental function through molecularly based, artificial neurons or brain chips;[21] increased connectivity through neural implants that would allow the brain to be connected to the internet or other digitally based media.[22] In this way, nanomedical techniques join the list of other technologies that are being viewed as means to "human enhancement" or "transcendence through technology."

The most radical claims come from a small group of individuals who call themselves transhumanists[23] and/or prolongivists,[24] and who believe that through nanomedicine and other so-called "human performance enhancement technologies" (HPET) (e.g., genetic engineering, pharmacological enhancement, tissue engineering and replacement, and stem-cell and regenerative medical therapies), aging can be significantly slowed, and death forestalled, if not eliminated. Transhumanism is defined by philosopher and intellectual leader of the transhumanist movement, Nick Bostrom, as, "the study of the means and obstacles to humanity using technological and other rational means to becoming posthumans, and of the ethical issues that are involved in this. Posthumans' is the term for the very much more advanced beings that humans may, one day, design themselves into if we manage to upgrade our current human nature and radically extend our capacities."[25]

With these claims, come demands that these technologies be aggressively pursued by governments and research agencies around the world.[26] Further, the proponents of human performance enhancement (HPE) demand that the nature of medicine as a profession and activity must necessarily be altered to include HPETs.

---

[21]Urban, *BioMEMS* and Berger, T. and Glanzman, D., eds. (2005). *Toward Replacement Parts for the Brain: Implantable Biomimetic Electronics as Neural Prostheses.* Cambridge: MIT Press.

[22]Kurzweil, *The Singularity Is Near.*

[23]For a thorough introduction to transhumansim, visit the World Transhumanist Association website, particularly the FAQs. Available at http://www.transhumanism.org.

[24]The premier prolongevist is Aubrey de Grey. A large number of his articles and information regarding the Strategies for Engineered Negligible Senescence may be found. Available at http://www.sens.org.

[25]Bostrom, N. (2001). *What Is Transhumanism?* Available at http://www.nickbostrom.com (retrieved October 17, 2006).

[26]For example, on April 30, 2004, Ray Kurzweil, in testimony before a committee of the U.S. Congress, called for a nanobiotechnology research program for the goal of replacing the nucleus and ribosome machinery of each cell in the body with a nanocomputer and nanorobotics system to prevent diseases and aging, and enhance human capabilities. One of the more strident, and comical, demands comes from a chap who refers to himself as Elixxir in *The ImmorTalist Manifesto.* Bloomington, Indiana: Authorhouse. 2001.

Fortunately, the number of self-attributed transhumanists is low,[27] but they do tend to make a lot of noise, write many books,[28] have attained prominence in the bioethics community,[29] and seem to have the ear of significant policy makers within governmental and research organizations.[30] In fact, certain agencies of the U.S. government are explicitly promoting the transhumanist agenda. The Converging Technologies for Improving Human Performance Project [also known as NBIC, an acronym for Nano-, Bio-, Info- (or Information), and Cogno- (or Cognitive) Technologies] is sponsored by the National Science Foundation (NSF) and the Department of Commerce. Due to increased public criticism and scrutiny of the project since the publication of its first manifesto in 2002,[31] the project has continued to have annual conferences and promote its agenda of human re-engineering.

Although only a very small number of individuals presently would be willing to embrace the radical technoutopianism of the named transhumanists, ideas of the right to self-modification, the belief in transcendence through technology or technique,[32] and the ability of human beings to create a glorious future of happiness for all are commonplace throughout most of Western culture, particularly in the United States, despite all of the hard evidence produced by the twentieth century to the contrary.[33] These ideas are already corrupting the profession of medicine now and, thus, warrant our attention and analysis. A full-fledged critique of the transhuman delusion is beyond the scope of this chapter, but for our purposes

---

[27]The WTA website (www.transhumanism.org) claims 3766 members worldwide, most of whom reside in the North America and Europe. However, other radical utopian groups, that ultimately inflicted a huge amount of damage upon humanity, such as the National Socialists in Germany, started out with fairly low numbers, but were extremely aggressive in promoting their views, just as are the transhumanists.

[28]Naam, R (2005). *More Than Human: Embracing the Promise of Biological Enhancement.* New York: Broadway Books.; Kurzweil, *The Singularity Is Near*; Hughes, J. (2005). *Citizen Cyborg: Why Democratic Societies Must Respond to the Redesigned Human of the Future.* New York: Westview Press; Silver, L. M. (2006). *Challenging Nature: The Clash of Science and Spirituality at the New Frontiers of Life.* New York: HarperCollins Publishers. Silver probably would not use the term transhumanist as a self-descriptor, but his techno-utopianism, and pro-re-engineering views are indistinguishable from the more honest, card-carrying transhumanists.; Bailey, R. (2005). *Liberation Biology: The Scientific and Moral Case for the Biotech Revolution.* New York: Prometheus Books; Young, S. (2006). *Designer Evolution.* New York: Prometheus Books.

[29]Among the techno-utopian, prohuman re-engineering bioethicists are Art Caplan, Glenn McGee, Julian Suvalescu, John Harris, John Fletcher, David Magnus, Dan Brock, Peter Singer, and Ronald Cole-Turner.

[30]In addition to Ray Kurzweil frequently appearing before Congressional committees to discuss issues in technology, particularly nanotechnology, the Center for Responsible Nanotechnology, a nanotech advocacy group whose founders and spokespersons are full-fledged transhumanists, frequently speaks to governmental groups and leaders about policy issues.

[31]Roco, M. and Bainbridge, W. S., eds. (2002). *Converging Technologies for Improving Human Performance.* Available at http://wtec.org/ConvergingTechnologies (retrieved October 17, 2006).

[32]By technique, I am referring to the inclusion of bureaucratic and political systems as a form of technology in the manner of Jacques Ellul and Lewis Mumford.

[33]Johnson, P. (1992). *Modern Times.* New York: Harper Perennial.; Rummel, R. (1997). *Death by Government.* New York: Transaction Publishers.; and Delsol, C. (2006). *The Unlearned Lessons of the Twentieth Century.* Wilmington: ISI Books. All are excellent sources exploring this point. Indeed, the existentialist movement, and much of post-modernist philosophy has arisen as a critique of the failures of the politico- and technoutopian projects of the enlightenment.

now we will discuss: (1) the fundamental goals of medicine; (2) the distinction between treatment and legitimate medical activity and HPE, or re-engineering; (3) current trends in society and medicine that pervert the fundamental goals of medicine and the treatment—"enhancement" distinction; and (4) the practical realities of asking society and medicine to pursue the HPE/re-engineering project, and conclude with some medical observations about the dangers inherent to it. But, first to provide a context for this discussion, let us examine one transhumanist's suggestion for the alteration of the nature of medicine, Freitas' Normative Volitional Model. This model well articulates the unfortunate logical conclusion of some of the current trends and erosions of medicine that we will examine.

FREITAS' NORMATIVE VOLITIONAL MODEL[34]

Freitas has suggested that nanomedicine, and by implication other HPETs, requires a new concept of disease that transcends the classic model of disordered function. He calls this new model the Volitional Normative Model of Disease, which includes the following elements:

> Disease is characterized not just as the failure of "optimal" functioning, but rather as the failure of either (a) "optimal" functioning or (b) "desired" functioning. Thus disease may result from:
>
> 1. failure to correctly specify desired bodily function (specification error by the patient);
> 2. flawed biological program design that doesn't meet the specifications (programming design error);
> 3. flawed execution of the biological program (execution error);
> 4. external interference by disease agents with the design or execution of the biological program (exogenous error); or
> 5. traumatic injury or accident (structural failure).[34]

While encompassing traditional understandings of disease, this model additionally takes disease out of the context of an objective pathophysiological assessment and turns disease into whatever the patient defines it to be. Any limitation or undesired trait may now be declared a disease. This has little to do with the nature of genuine medicine. Some might see this model as simply continuing the contemporary trend of patient self-determination to a new level, but this course is fraught with both danger and injustice. To declare that a condition is a disease imposes a moral claim that medical services ought to be rendered for its modification, elimination, or amelioration. The balance between beneficence and nonmaleficence in weighing the benefits and risk of a proposed intervention may be inappropriately tipped in favor of what the patient desires, rather than needs. Physicians would be reduced to agents of wish fulfillment and technicians, rather than healers, and healing is an art and an

[34]Freitas, R. A., Jr. (1999). *Nanomedicine*, p. 20.

ethical commitment that goes well-beyond technical knowledge and physical acumen and dexterity. Further, claims to "treatment" for the purpose of wish fulfillment would unjustly deplete shared healthcare resources and funds, depriving those in real need of legitimate healing.

THE LEGITIMATE PURPOSES AND GOALS OF MEDICINE[35]

Medicine began when individuals suffering from injury, infectious diseases, or even genetic diseases, someone we now refer to as a patient, asked another individual, now known as the physician, to help them with their pain, weakness, or other limitations produced by the affliction. The goals of the treatments employed were to alleviate the pain, fever, and other manifestations of the malady, things we call symptoms. If possible, the treatments would also try to correct or eliminate the underlying problem so that the body could then heal itself, though only recently has this latter goal been practically achieved for a number of conditions. Even in contemporary medicine, much of what we do is palliative rather than curative: That is, we relieve symptoms, we help people live more successfully with their disease or malady, we push a disease like acute leukemia into a temporary remission, rather than put the disorder into permanent remission.[36] Cure when possible, care always.

Given this history and the realities of medical interventions, its successes and failures, the first goals of medicine were established as: (1) curing a disease when possible; (2) reducing suffering through attenuating disease activity, and/or symptom control; and (3) rehabilitation from the effects of injury and disease. As medical knowledge has expanded and the mechanisms of disorder have been better understood, it has also become possible, in some circumstances, to educate individuals, perform screening evaluations to detect some afflictions at a stage where cure could be achieved through intervention, encourage changes of behavior, and employ certain treatments (e.g., vaccinations) to forestall or prevent diseases, or, at least, lessen their severity. This has led to a fourth goal of medicine: preventing disease and/or injury.

With these focused goals, our society, and many societies throughout the world, consider medical care a valued resource in which there should be a shared commitment by all to provide the resource to a larger collective, through taxation and government healthcare programs, medical insurance programs, and so on. There is often the temptation to lump all of these concepts together under the term "health," making "health" the fundamental goal of medicine. While each of the

---

[35]For a very nice discussion of the goals of medicine, please see Callahan, M. and Callahan, D., eds. (1999). *The Goals of Medicine: The Forgotten Issues in Health Care Reform.* Washington, DC: Georgetown University Press.

[36]Some patients are cured of acute leukemia with chemotherapy and/or bone marrow transplantation, but, despite the fact that we may get the majority of patients into an initial remission, the majority will relapse and die from their disease or from a treatment-related complication.

concepts above do indeed flow into a larger concept of health, the definitions of health are far too amorphous, and potentially subjective, to be used as any practical guide. For example, the classic statement of the World Health Organization (WHO) defines health as "a state of complete physical, mental and social well-being and not merely the absence of disease or infirmity."[37] Clearly, this definition of health places health in its supposed totality well beyond what the art and science of medicine can address alone. Thus, the goals of medicine would be more accurately understood as the more specific issues listed in the previous paragraph.

What is not mentioned in that list of four is the promotion of happiness, or fulfilling the individual's aspirations for his or her life, aside, except, perhaps, as a secondary phenomenon to attending to physical malady and disability. However, guaranteeing happiness is what Freitas, the other transhumanists, and many in our culture seem to believe is owed them by medicine. This idea is the ultimate in folly, the product of a spoiled and overindulged baby-boomer generation, as no one's happiness can be guaranteed, or every aspiration fulfilled, especially in a culture, such as ours, in which fashions change daily, and whole industries exist to make people feel insecure, inadequate, deprived, lustful, envious, and so on, in order to generate false "needs" solely for the purpose of inducing sales. A few years ago, it was estimated that the average person in the United States is bombarded by more than 3000 advertisements each day,[38] none, or at least only a tiny portion of which, could possibly be understood as being published or broadcast for the actual benefit of the target.

Our culture has confused the right articulated in the Declaration of Independence, "to pursue happiness" (which is like most rights articulated in our founding documents a negative right—i.e., a right to be left alone to make one's own choices, rather than a positive right—claims imposed upon others to produce some sort of end or behavior), with a right to happiness. Further, it is a horrible mistake to impose a political aspiration on the concrete processes of medicine, which have more specific and realistic goals.

But what of mental health? Is that not part of medicine, and one that directly bears on an individual's happiness? Yes and no. First of all, mental health itself is a poorly defined concept. Are we referring to brain health or spiritual health? Are we referring to perceived unhappiness and stresses imposed by a pathological society, or individual social interactions, or genuine intrinsic functional problems within a specific individual. Despite what the pharmaceutical industry has been promoting since the development of serotonin uptake and release inhibiting (SSRI) antidepressants, medication is not the answer to a bad day. Medicine does not exist to fulfill the narcissistic whims of Western society, and it would destroy medicine, and further contribute to the social, moral, and spiritual degeneration of our society, to pervert it in the way the technoutopians and the moguls of the pharmaceutical industry demand.

---

[37]Preamble of the Constitution of the WHO.
[38]Quoted in the A&E television documentary, The Sell and Spin: A History of Advertising. 2002.

THE DISTINCTION BETWEEN LEGITIMATE TREATMENT AND "ENHANCEMENT"

Despite the claims made above, there is a growing debate in our culture over the difference between treatment and so-called "enhancements," with many bioethicists concluding that the distinction cannot be made.[39] Consequently, everything must be permitted. This is nonsense, and, in many cases, the result of poor thinking, laziness, or either a disingenuous or overt effort to promote wholesale support for and availability of enhancing technologies. Saying this is not to discount that there are challenging situations and gray areas, nor to declare that there is presently a complete definition, which will clearly distinguish between the two areas in every possible situation. But, if we look at most things honestly, we can conclude that those gray areas are actually a very small percentage of all possible interventions one could do to a human being for purported benefit. In the vast majority of cases, most of us can objectively state that a given intervention is either a treatment or an enhancement, and we should do so. It is a practical medical, social, and economic necessity to make these distinctions. Science and medicine, as opposed to law, politics, and philosophy, do not throw up their hands in the face of a seeming exception, a difficult case, or data that falls outside the current theory, stating that the exception invalidates everything. Rather, they continue to use the current system, where it has clear explanatory power, and works harder to understand the exception and refine the working theories to include and explain the challenging phenomenon. Though we know that Newtonian mechanics cannot explain all physical phenomena, particularly at the atomic and subatomic levels, they still enabled humankind to successfully transport human beings to the moon and back. As we continue to examine the borderline cases, it is the belief of this author that we will refine our definitions such that the border zone will become increasingly narrower. The key requirement for this to succeed, however, is that the medical and bioethics communities must want to create these lines of demarcation. Evidence suggests, however, that many of our most prominent bioethics commentators desire the opposite.

One of the first things I propose for future discussion of this issue is that the term "enhancement" be replaced by "re-engineering." Enhancement, as defined by the Oxford English Dictionary, is "to increase, intensify, raise up, exalt, heighten, or magnify."[40] Common parlance would also include the concept of improvement. The American Heritage Dictionary lists the definition of enhance as "to make greater, as in value, beauty, or effectiveness; augment."[41] To enhance is to make better, by some criteria. What is "better"? How is it defined, and by whom? Herein lies the real problem with so called HPETs—they necessarily assume in

---

[39]Exemplary of this intellectual surrender are many of the articles in the book: Parens, E., ed. (1998). *Enhancing Human Traits: Ethical and Social Implications.* Washington, DC: Georgetown University Press.

[40]Simpson, J. A. (ed.) and Weiner, E. C. (1989). *Oxford English Dictionary* (2nd ed. Vol. 1–20) London: Oxford University Press.

[41]Pickett, J. P. et al. (ed.) (2000). *The American Heritage Dictionary of the English Language* (4th ed.) Boston: Houghton Mifflin.

the designation "enhancing" that the change produced by the agent is indeed better. Better, however, by what standard? Just because one parameter might be improved, have the other necessary consequences of that change been factored in? The use of enhancement in the discussion, however, tends to weigh the argument in favor of the re-engineering side. After all, who wants to be against making something "better"?

Let us take SSRI antidepressants as an example of some of the consequences and the ambiguous nature of a supposed enhancement. Large direct-to-consumer marketing campaigns promoted a new "disease," social anxiety disorder,[42] for the commonplace, and normal, condition of shyness and/or inexperience in public speaking, and then promoted Prozac or another SSRI, as the cure. Always ending the ad with the phrase, "ask your doctor," thus lending a veneer of medical respectability and legitimacy to their propaganda, these campaigns contributed to excessive and inappropriate use of these potent agents, adding to the long list of activities and aspects of day-to-day life being medicalized for the real goal of promoting pharmaceutical sales (and also perhaps driving physicians crazy with a barrage of requests for unneeded pharmaceuticals).[43] Yes, these agents can reduce social anxiety and flatten out a borderline tendency to discouragement, dysthymia, but at a cost, at least to a significant number of users: weight gain, flattening of the personality, sedation (which will now, of course, need to be treated by another "enhancing" agent, Provigil), restless legs syndrome (which will of course need to be treated by Mirapex or other agents), loss of critical inhibitions and judgment because of the loss of healthy fears (which may lead to excessive spending, etc., adding additional stress to relationships), and, quite significantly, a physical and emotional dependency, which tends to keep people on these expensive agents sometimes indefinitely. Tapering off an SSRI, when successful, is a very prolonged process often fraught with emotional liability, anger, despondency, and other undesirable complications. This is not to say that SSRIs have not been a major advance in the treatment of depressive illness, for they have. For the appropriate patient with legitimate illness, SSRIs have often been life saving and/or restoring. But the wholesale marketing of these agents for less-than-serious depressive disease has been irresponsible and baited a gullible public into thinking that medicine can and should be available to it for mood engineering. When one understands what really is involved in this "treatment" for pseudo-social anxiety disorder, it is difficult to continue to describe it as an enhancement.

Enhancements are necessarily tied to individualized, subjective standards, and, by that fact alone, they should be excluded from the goals and operations of medicine. Human growth hormone used to "enhance" height to allow someone to try to achieve heights necessary to play professional basketball would have excluded

[42]There is a legitimate diagnosis of social anxiety disorder, which may exist independently or accompany another disorder, such as autism. However, the ad campaigns tended to show normal situations in which most people experience some degree of anxiety, with the degree varying on a wide, but normal, spectrum and attempted to turn that into a disease.

[43]Another term for medicalization is "disease mongering". A nice comment on disease mongering may be found at Moynihan, R. and Henry, D. (2006). The Fight Against Disease Mongering. *PLoS Medicine* 3(4), e191. Available at http://www.plosmedicine.org (retrieved October 17, 2006).

someone from being an astronaut during the first decades of the space program.[44] Breast implants to "enhance" the bust line of a young woman may also lead to significant pain and disfiguration if placed prior to the completion of her growth, can rupture, lead to infection, will probably require revision, and only seek to continue the assault against women's dignity by the fashion and entertainment industries that, despite their rhetoric to the contrary, necessarily reduce women to their physical attributes. Further, such modifications encourage the irresponsible and immature behavior of men to demean women, reduce male–female relationships to titillation and appearance, and, thus, promote superficial relationships rather than genuine love and commitment. Of course, the crushed psyche and hearts of the victims of such superficiality need to be "treated" by antidepressants, anxiolytics, and a whole industry of self-help gurus who parasitically continue to promote narcissistic self-absorption ever perpetuating the process.

The HPETs, in general, should be recognized for what they are: at least 99% scam. But, as our technology develops, we may soon be confronted by means to genuinely change the underlying substrate and function of our bodies, with the results being good or ill still very much in question. Thus, if we are to honestly label what is going on now, and may happen with nanotechnology, genetic engineering, and the host of other "enhancing" technologies it would be far more accurate to use the term re-engineering rather than enhancement. We are changing the structure and function of a being, but it is presumptuous to assume that the outcomes, when all weighed together, will necessarily be positive.

Even the simple "enhancements" available to us today, such as caffeine, are not without consequence, challenging the notion of overall betterment. Caffeine may lead to irritability, insomnia, headaches, tremors, gastric hyperacidity, and is slightly addictive (to which anyone who has suffered a caffeine-deprivation headache can well attest). Caffeine in coffee, tea, chocolate, and soda pop is ubiquitous and considered a relatively minor thing. But would we even need this enhancement if we had not created a society that increasingly forces us, or at the least encourages us, to get less than an adequate amount of sleep. The end result of chronic sleep deprivation and stimulant use to compensate is chronic neurotransmitter depletion, a situation we are only now beginning to understand through the evolving field of sleep disorders.

Next, we need to examine some of the cases that have purportedly made the distinction between treatment and re-engineering so difficult. Let us begin with the classic, oft-quoted canard of LeRoy Walters and Julie Palmer about vaccination: "In current medical practice, the best example of a widely-accepted health-related physical enhancement immunization against infectious disease. With immunizations against diseases like polio or hepatitis B, what we are saying is in effect, 'The immune system that we inherited from our parents may not be adequate to ward

---

[44]An excellent discussion of how the nontherapeutic use of human growth hormone is an example of how medical interventions become perverted and ultimately accepted for enhancement purposes can be found in Rothman, Sheila M. and Rothman, David J. (2003). *The Pursuit of Perfection: The Promise and Peril of Medical Enhancement*. New York: Pantheon Books.

off certain viruses if we are exposed to them.' Therefore we will enhance the capabilities of our immune system by priming it to fight against these viruses."[45] They then go on to use this factually and conceptually flawed statement to justify re-engineering, equating the two as similar in type and mechanism: "From the current practice of immunizations against particular diseases, it would seem to be only a small step to try to enhance the general function of the immune system by genetic means."[45]

Walters and Palmer have a fundamental misunderstanding of the nature and function of the immune system. We are not born with ready-formed antibodies to the specific antigens that occur in infectious agents. Our antibody immune surveillance structures evolve through environmental and infectious exposures. Unless one is born with a congenital defect in the immune system, such as common variable immunodeficiency, preventing us from doing so, each and every one of us must "train" our immune system over the course of our lives. Getting chicken pox, or mumps, or even polio stimulates the development of antibodies that prevent subsequent infections should we survive the infection. A vaccination is simply a means of training the immune system in a manner that hopefully avoids the morbidity and mortality that may accrue from an actual infection. Vaccinations are not "enhancements" or re-engineering, they are good, responsible stewardship that depend on normal, endogenous mechanisms of our body. So are sufficient rest, exercise, a healthy diet, education, good oral hygiene, and other things that have been labeled in these discussions, erroneously, if not dishonestly, as enhancements. Exercise and education, like vaccinations, depend on the underlying anatomy and physiology of the body to work as designed, and train them to optimal function. They do not re-engineer the underlying substrate, which, in contrast, is what nanotechnological replacement of normal cellular and extracellular structures, genetic engineering for other than repairing genes that do not support species-level norms of function, and other hard core HPETs would do.

Space limitations do not permit a further examination of the multitude of other interventions mistakenly claimed as enhancements (e.g., fluoridation, the use of aspirin for the prevention of heart disease, and others that the author has confronted in debate), but, suffice it to say, most when properly understood are either actually treatments of an underlying pathology or environmental threat, or are training underlying normal structures to achieve optimal, species-normal performance. They are not enhancements in the sense that the proponents of re-engineering would like us all to believe.

For the purpose of promoting more accurate terminology in the critical treatment versus enhancement discussion, I would like to propose that the definition of enhancement, or more accurately, re-engineering, be interventions that attempt to alter or "improve" a specific physiological or anatomical parameter when that parameter is already within or above species-normal boundaries, by means of changing

[45]Walters, LeR. and Palmer, J., quoted in Juengst, E. T. (1998). *What Does Enhancement Mean?* in Parens, Erik, ed. (1998). *Enhancing Human Traits: Ethical and Social Implications.* Washington, DC: Georgetown University Press.

the underlying biological substrate rather than employing normal physiological mechanisms for optimizing function, or by providing supra-normal levels of natural substances, hormones, or analogs thereof, than would be possible by normal physiological responses to the environment. "Enhancement" is necessarily a repudiation of the normal, as the pursuit of "enhancement," at least as so designated here, is a *de facto* declaration that normal function and normal levels of trained optimization are not good enough. With this definition, it may be easier to identify a category of activities that can appropriately be excluded from the activities and responsibilities of medicine. Whether others will be permitted to perform or access these interventions is a discussion for the larger society to engage, but the integrity of medicine, and the mutual demands placed upon medicine and the larger society supporting this shared resource, can remain intact and appropriate.

TRENDS CHALLENGING THE INTEGRITY OF MEDICINE

The challenge and problems caused by medicalization, that is, the conversion of a normal physical or trait variation, or social pathology, into a medical problem has already been mentioned in the conversion of shyness into a disease, "social anxiety disorder," requiring treatment by SSRIs. This is one of the latest examples of a long trend of "disorders" being created by industry to boost sales of a product that dates back at least to patent medicine hawkers trying to convince everyone that fatigue was always an indication of disease, and Listerine inventing the "disease" of "halitosis," with, of course, its product as the cure. What is different now, however, is that the process of medicalization is more and more encroaching upon the boundaries of trait and real organic disease and that the "cures" are restricted pharmaceuticals that require a physician's prescription. Medicalization threatens to erode the integrity of medicine because it cheapens medicine to the role of an instrument of industry and contributes to confusion about the understanding of normality, disease, treatment, and enhancement.

One of the most severe cancers within medicine has been the prostitution of plastic surgery to "aesthetic cosmetic" surgery.[46] Plastic surgery has a noble tradition of healing or ameliorating congenital defects and deformities produced by injury or disease. Some years ago, a debate raged within the subspecialty of plastic surgery regarding whether or not this group of specialized surgeons should perform nontherapeutic interventions for solely cosmetic and/or social purposes. The forces of integrity and commitment to medicine as a healing art lost out to the greed of those who knew they could make a bundle with these techniques, enthusiastically supported by healthcare administrators and executives more interested in profits than in the integrity of medicine, or the overall good of society.

Consequently, our culture is more and more becoming a spectacle of the unreal, the illusion, and the artificial image such that more and more individuals are feeling forced to submit to desecrating and mutilating procedures in order to feel accepted,

---

[46]Footnote 44, pp. 101–130.

or competitive in the workplace. One cannot go through a checkout line in a grocery or convenience store without headlines from the tabloids or glossy gossip magazines reporting on or speculating about some celebrity's "remodeling." The further plasticizing of the clueless and superficial corps of celebrities afflicting our culture would be comic were it not for the fact that legions of citizens who should know better want to jump on the band wagon and emulate their idols. We have glamorized these superficial transformations in such programs as *The Swan*, *Extreme Makeover*, and in China, *The Miss Artificial Contest*. What is never shown, however, is how much better, or worse, the frequently victimized and empty women who become the subjects of these extravaganzas actually do in their lives after their "transformations", and all the crews and cameras have gone on to another piece of modeling clay. The silence is suspicious and deafening. The problem here is that most of our happiness and contentment is a personal decision. Yet, we want to blame every possible external factor possible for our happiness or unhappiness, and we have been taught to believe this is how it must be done by a culture dominated by advertising and the sale of nonessential products, including the product that is our media. The fundamental problem, however, is spiritual, not medical, going to the very basis of our worldviews. No technology, medical or non-medical, can fix the spiritual crisis of our culture, and this physician strongly opposes perverting medicine to try to do so.

As an illustration of the extent of this particular problem we need but examine the number of citizens of the United States who pursue some form of cosmetic "enhancement." Statistics for 2002 from the American Society for Aesthestic Plastic Surgery revealed that there were 372,831 liposuctions, and 350,000 women had nonrestorative breast augmentations and reductions.[47] (Breast reductions may be for legitimate medical concerns such as neck and back strain, chronic headaches, and other maladies, but were a very small percentage of the breast interventions overall.) More than $7 billion was spent that year in the United States for cosmetic, that is nontherapeutic, plastic surgical interventions, which includes botulinum toxin, or Botox, injections. By 2004, the tab for these kinds of nontherapeutic procedures had increased to $12.5 billion.[48] So low have we descended into this nightmare that even the U.S. Army has begun to use cosmetic plastic surgery for soldiers and dependants as a recruitment tool. In the 2002 survey, 33% of male patients and 19% of female patients indicated that they pursued body modifications for employment-related reasons, particularly for those in sales or customer contact. One new, critically important medical procedure to enter the market is the designer vagina.

So obsessed with the superficial is our culture that in a survey,[49] almost a decade old as of this writing, 15% of women and 11% of men said they would sacrifice more

[47]The American Society for Aesthetic Plastic Surgery, 2004 Statistics. Available at http://www.surgery.org/press/statistics-2002.php (retrieved October 17, 2006).
[48]The American Society for Aesthetic Plastic Surgery, 2004 Statistics. Available at http://www.surgery.org/press/statistics-2004.php (retrieved October 17, 2006).
[49]1997 Psychology Today Poll of 4000 respondents.

than five years of their lives to be able to achieve their weight goals. The survey revealed that, at that time, there was a growing number of women actively proselytizing other women not to have children because to do so would ruin their figure and appearance. Well over one-half of all women and roughly one-half of the men surveyed were dissatisfied with their bodies. Two-thirds of the women dissatisfied with their bodies indicated that fashion models made them want to loose weight, while one-half also said that the models made them angry and resentful. Most concerning was that in the 13–19-year-old age group of women, the group least likely to be overweight at all, roughly two-third were dissatisfied with their weight. The glitzy, artificial world of fashion is inducing a mass neurosis of envy and self-loathing, and pushing women and men to subject themselves to the knife. This is a deep and serious social pathology that medicine cannot fix, and perverting medicine into an accomplice to this insanity is a betrayal of the legacy of the healing arts.

The images of fashionable "beauty" are carefully crafted by industry in such a way that only a scarce few do not require, by the plans of the technocrats, significant alterations in hair color, skin tone, body, or feature shape. For example, in the 1980s, one cosmetic company ran a series of television and print ads using an attractive actress who cooed: Don't hate me because I'm beautiful. The obvious message was, of course: You should make yourself beautiful too by using our products. However, the real message of the campaign was, You should hate yourself for being less attractive than our air-brushed model. Marketing companies know that, by inducing this self-loathing, this envy of another, the consumer will buy the technology of aesthetic transformation. There appears to be a mass blindness to the reality that current ideals of beauty and acceptable attire are not established democratically, but created by people whose job it is to sell products and services, and then pushed unilaterally upon an obedient and unquestioning public.

In addition, it is only getting worse. It is not just the thin forms of anorexic models that is the problem, but the increasing use of morphing technologies to alter the images plastered on billboards, commercials, the covers of magazines, and other forms of advertising to induce people to strive for looks that simply are no longer humanly feasible short of re-engineering. An *Esquire* magazine article once ran a piece on the actress Michelle Pfeiffer entitled "What Michelle Pfeiffer Needs . . . Is Absolutely Nothing." The implication was that this beautiful actress was the perfect physical model, the standard of aesthetic perfection. However, the pictorial image of this paragon of physical beauty had to be significantly modified in order to produce the purported perfection, including complexion clean-up; softening of the eye, smile, and ear lobe lines; addition of color to the lips and hair; trimming the chin; removing neck lines; and adding hair on the top of the head.[50] But, at least, there was a real person as a base in this instance.

---

[50]The Michelle Pfeiffer cover shot in *Esquire* (December 1996) is discussed by Bob Andelman in *Nothing Except $1,525 Touch Ups*, at Adbusters. Available at http://www.andelman.com/mrmedia/95/7.4.95.html.

The future, however, will be even less generous to mere *Homo sapiens* as one can now see in the Miss Digital World contest found on the Internet. The site effuses: "Every age has its ideal of beauty, and every age produces its visual incarnation of that ideal: the Venus de Milo, the Mona Lisa, the 'divine' Greta Garbo, Marilyn Monroe ... Miss Digital world is the search for a contemporary ideal of beauty, represented through virtual reality."[51] As one examines the contestants, it becomes clear that few, if any, mere mortal human woman will be able to attain the morphology of these avatars without surgical modification.

As another example, we can look to the world of sports, which is constantly being rocked by scandal after scandal of individuals willing to violate the spirit of athletics to utilize illegal forms of "enhancing" interventions, from anabolic steroids to crythropoietin doping. In each case, there has been a physician who has been willing to violate his or her professional responsibilities and engage in the enhancement dance by providing prescriptions for the agents used.

I have belabored these points regarding cosmetic and athletic re-engineering to point out that our society is already pursing, to some extent, the folly of the transhumanist belief in re-engineering, and doing so in ridiculously high numbers. It is highly doubtful that this culture will be less susceptible to other forms of HPET whether they involve re-engineering cognitive abilities, tissue re-engineering, and genetic re-engineering, with or without the assistance of nanotechnology. The Genetics and Public Policy Center of Johns Hopkins University, for example, revealed in a survey of more than 4000 adults in the United States that around one-third of the respondents supported using preimplantation genetic diagnosis for the selection and/or augmenting of traits, such as intelligence and strength.[52]

All of these examples illustrate why it is critical that we establish lines of demarcation between treatment of genuine disease and disability, and re-engineering interventions, and excluding by policy the latter from the responsibility and practice of medicine, particularly as powerful technologies (e.g., nanotechnology), may be used to pursue the re-engineering project. Our society cannot afford, whether it be financially, socially, or spiritually, to permit or pursue the medicalization of the worst aspects of our corrupt hearts: lust, hatred, envy, greed, avarice, and narcissism.

SOME CONSEQUENCES OF THE RE-ENGINEERING PROJECT

So what is so wrong with getting a better memory, having faster, endogenous access to the internet or other databases through implants, and replacing easily destroyed tissues with tissues supported by stronger and more resilient substrates? Perhaps nothing; perhaps everything. As I have already illustrated, no change to our underlying structures and physiology comes without some price, or consequence, and that

[51]For information on this "virtual beauty contest," see Miss Digital World's Web site. Available at http://www.missdigitalworld.com/MDWContest/showpage/6.
[52]Kalfoglou, A. et al. (2004). *Reproductive Genetic Testing: What America Thinks*. Washington, DC. Genetics and Public Policy Center. Available as a pdf at http://www.dnapolicy.org/pub.reports.php.

it is inappropriate to describe these things as true enhancements when we have no idea of what that price will be. Our existing restorative efforts are not perfect, without some consequence or potential toxicity, and it is the height of ignorance, and arrogance, to presume that the proposed re-engineering projects will not carry significant burdens with each supposed benefit they may produce.

Our bodies are extraordinarily complex systems that we still only partially understand. In my experience as a physician, there is little, if anything, that we ever do to perturb these systems that does not, or at least has the potential to, produce some consequence, many times unexpectedly. (Does anyone remember Vioxx?) Fortunately, the majority of the time these consequences are slight, manageable, or compensated for by the flexibility and resilience of the larger system. But sometimes the consequences are life threatening, debilitating, and costly.

We know next to nothing about the toxicity of nanoscale materials, though early data raise serious concerns.[53,54] Assuming we could, at some point, ensure that the nanotech "enhancements" would have little direct toxicity, we still have no clue as to how they might otherwise effect function and vitality of the organism as a whole. Brain chips might possibly supplement memory, but how will those devices also effect personality and thought processes? Memory is laid down by parts of the limbic system, specifically the hippocampus, a part of the brain that is also critically involved in emotion. Any device that is to participate in the creation and retrieval of memories will necessarily do so through the hippocampus. Recognizing the sensitivity of brain function to perturbations created by a multitude of chemicals, and structural changes produced by tumors and strokes, even small ones, it is inconceivable that these brain devices would not produce many unexpected, and potentially disturbing and harmful, consequences.

Similarly, will nanostructures used to increase the strength and resiliency of connective tissues and bones also interfere with normal processes of healing when injury does occur? Drexler's, Freitas', and Kurweil's nanobots are still very much science fantasy, but anyone who has ever really dealt with the human body and medical intervention will tell you that there is likely going to be some sort of rejection response to the foreign product, especially given the necessary size of such a complex device. Some of the most frequent and severe toxicities of engineered molecules like Rituximab are allergic reactions. The nanoseers all like to claim that nanotherapeutics will be without traditional toxicities, but such claims reveal a profound ignorance of real-life physiology and medicine.

Another disturbing aspect of the claims for nanomedicine is that, through nanotechnology, we will be able to cure every disease because we can address it and fix it at the molecular level. At this point, we need to step back, take a deep breath and remember the history of medicine during the past hundred years. As the twentieth century began, with genetics and evolution being the new kids on the scientific block, eugenics was all the rage, claiming that, through proper breeding and

[53]Oberdörster, G. et al. (2005). Nanotoxicity: An Emerging Discipline Evolving from Studies of Ultrafine Particles. *Environmental Health Perspectives* 113, 823–839.
[54]Ross, P. (2006, May). Tiny Toxins? *Technology Review.*

genetic cleansing, humankind could usher in a glorious future of a stronger, smarter, and healthier race. Eugenics, however, proved to be little more than a vehicle for tyranny and prejudice, and the cause of no small amount of pain, suffering, and death.

Next, nuclear physics and the power of the atom captured the scientific and public imagination, and soon radiation therapy was being used to treat everything from tonsillitis and hangnails to impotence. It soon became clear that radiation was an extremely toxic treatment, which should only be used for certain types of neoplastic disease and thyroid disease. It is unknown how many cases of leukemia and other forms of secondary malignancy, tissue sclerosis, coronary artery disease, and other organic dysfunction ultimately resulted from this radioactive enthusiasm.

Then, in the 1970s President Nixon declared war on cancer with the expectation that, given recent developments in cytotoxic chemotherapy, we would soon cure all cancer. When traditional cytoxic agents proved incapable of eliminating cancer, the discovery of the antineoplastic effects of interferon and interleukin 2 during the 1980s became the next cure-all for cancer. Yet, horrendous toxicity and disappointing results have relegated interferon to treating a few hematological malignancies and melanoma, usually as a second-line agent.

More recently, we have been subjected to an immense propaganda campaign, dependent more on posturing, patent seeking, prospective profits, and politics than on any real scientific evidence, which declares that embryonic stem cells will be the salvation of humankind, curing every possible disease, and prolonging life dramatically through tissue replacement therapies. Already, the scientific community is engaging in major back-pedaling on these predictably outrageous claims. No human artifact has "unlimited potential," except perhaps the human imagination to delude itself.

I think it is not unfair to say that nanotechnology, and more specifically nanomedicine, will ultimately fall short of its projected miracles. It will be found to produce significant complications, and it will not be able to cure every disease or repair every injury. Life and its processes just do not work like that. This is not to demean nanotechnology, but rather, simply to face reality—something that the technoutopians have a hard time doing. Utopians of every stripe, whether it is the political utopians of the twentieth century who produced the ghastliest and bloodiest century in human history due to their wholesale misunderstanding and/or denial of human nature, or the technoutopians who would destroy humanity to supposedly save it, are sad and potentially dangerous fanatics. To the degree that we attempt to replace real life with our technological analogs to pursue a doomed to fail quest to achieve control over every prospect for disease and deterioration (and it will require a complete substitution of the whole biological system to avoid the processes of decay and death), we will loose something immeasurably valuable and irreplaceable—our very nature and being.

Personally, as a physician, and particularly as a hematologist–oncologist, I look forward to many legitimate tools that nanotechnology can bring to medical practice, and I believe these are worth pursuing. But I just as strongly believe that we must constrain the uses, and also the potential liabilities, of biomedical nanotechnology to genuine healing, rather than use them to pursue the re-engineering–enhancement project. Potential harm has long been accepted as a possible consequence of medical interventions through the principle of double effect. A similar calculus, however, cannot be performed when the intervention is being applied to a normal individual whose motivation is being created by social pathology, propaganda, and/or his or her own internal demons.

PART IV

NANO AND SOCIETY: THE NELSI IMPERATIVE

The 2003 21st Century Nanotechnology Research and Development Act underlines the importance of research into NELSI (nanotechnology's ethical, legal, and societal implications). Basing its thinking on the parallel ELSI program of the human genome project, Congress stated that the National Nanotechnology Initiative should not be limited to technical work, but include engagement with a range of related questions. In addition to NELSI, they include environmental, health, safety (EHS), workforce implications, and education.

We turn to the NELSI agenda in this final section. Michele Mekel and Nigel Cameron, from the Center on Nanotechnology and Society at Illinois Institute of Technology (IIT), survey the NELSI agenda and review the initial response of the National Science Foundation (NSF) to the Congressional mandate. They also survey the various NELSI-related projects in the United States. David Guston from Arizona State University, Principal Investigator on the NSF's major funded NELSI project, the Center for Nanotechnology in Society, discusses both its agenda and the key method of Real-Time Technology Assessment that it is bringing to bear in its analysis. Kristen Kulinowski from Rice University describes the work of ICON, the International Council on Nanotechnology, a project that seeks to bring together all parties in the nanotechnology conversation.

Vivian Weil, an engineering ethicist who is involved in both the IIT and ASU centers, reviews ethical principles and pitfalls the process of taking innovation from the lab to the marketplace. Finally, Nigel Cameron looks to the future, suggesting NELSI-related "points to consider" in respect of the future management of the National Nanotechnology Initiative to ensure the integrity of the NELSI task, as well as looking ahead to the fundamental ethics and policy decisions that await policymakers as the "enhancement" potential of nanotechnology becomes more evident.

Nanoscale: Issues and Perspectives for the Nano Century. Edited by Nigel M. de S. Cameron and M. Ellen Mitchell
Copyright © 2007 John Wiley & Sons, Inc.

359

CHAPTER 20

The NELSI Landscape

MICHELE MEKEL and NIGEL M. DE S. CAMERON

> Now nanotechnology had made nearly anything possible, and so the cultural role in deciding what *should* be done with it had become far more important than imagining what *could* be done with it.
>
> Neal Stephenson[1]

INTRODUCTION

In the sci-fi world of Neal Stephenson's *The Diamond Age*, the nanoscale science and technology challenges being grappled with at present had long been resolved and far surpassed. In this futuristic world, however, nano's ethical, legal, and social implications (NELSI) were still beleaguering humanity. His novel serves as a cautionary tale for today's "nanoethicists," science and technology studies (STS) scholars, nanoscale scientists and technologists, nanoentrepreneurs, nanoinvestors, policymakers, regulators, and society as a whole.

At this nascent stage of nanoscale science, technology, and engineering, there is still time (how much, we do not know) to proactively and openly tackle NELSI in tandem with, or even one step ahead of, nanoinnovation. The success of such efforts will depend on a constellation of concomitant factors, including resources, restraint, public engagement, political will, international risk governance, creativity, and a healthy dose of realism. While this is a tall order, numerous efforts—albeit rather disparate and disjointed—have commenced. This chapter will analyze those efforts, with a particular focus on those based in the United States, and put forward a framework for NELSI with an eye toward avoiding

---

[1]N. Stephenson, The Diamond Age or, a Young Lady's Illustrated Primer 37 (1996) (hereinafter Stephenson) (emphasis added).

Nanoscale: Issues and Perspectives for the Nano Century. Edited by Nigel M. de S. Cameron and M. Ellen Mitchell
Copyright © 2007 John Wiley & Sons, Inc.

nanoethics[2] minefields, such as those presented in Stephenson's futuristic, nano-"enhanced" realm.

NELSI AND THE NANOSPHERE: SETTING THE STAGE

Just as nanotechnology's origins are often attributed to Richard Feynman's much-cited 1959 talk, *There's Plenty of Room at the Bottom,*[3] in hindsight, NELSI's earliest inklings can be traced to an unlikely source—K. Eric Drexler, the foremost proponent of the so far largely theoretical "molecular manufacturing"[4] branch of nanotechnology, who in his 1986 book, *Engines of Creation: The Coming Era of Nanotechnology,* warned of what has become dubbed the "grey goo" scenario.[5] Opening that chapter of his book, Drexler postulated that:

> Replicating assemblers and thinking machines [could] pose basic threats to people and to life on Earth . . . . We cannot hope to foresee all the problems ahead, yet by paying attention to the big, basic issues, we can perhaps foresee the greatest challenges and get some idea of how to deal with them.[5]

With impressive foresight, he went on to prognosticate that:

> Entire books will no doubt be written on the coming social upheavals: What will happen to the global order when assemblers and automated engineering eliminate the need for most international trade? How will society change when individuals can live indefinitely? What will we do when replicating assemblers can make almost anything without human labor? What will we do when AI systems can think faster than humans?[5]

---

[2]Currently, the term "nanoethics" is starting to be viewed as a synonym for NELSI (and its variants) to encompass the full spectrum of nontechnical issues related to nanotechnology. This regrettable phenomenon seemingly discounts the specific listing of ethical issues alongside others in the 21st Century Nanotechnology Research and Development Act, and other official documentation from the National Nanotechnology Initiative. While we acknowledge that all human activities give rise to ethical considerations, it is vital to avoid assimilating the ethics agenda into critical, but separate, regulatory and related matters. See Chapter 16 for a further discussion of nanoethics. For a discussion of NELSI's variants, see M. Mekel, *Small Science, Big Potential, and Potentially Bigger Issues,* 49 Development 47 (Dec. 2006) (hereinafter Mekel).

[3]R. P. Feynman, *There's Plenty of Room at the Bottom,* Dec. 29, 1959, Annual Meeting of the American Physical Society, California Institute of Technology, available at http://www.zyvex.com/nanotech/feynman.html (last visited Oct. 13, 2006).

[4]The notion of molecular manufacturing is explained on e-drexler.com as follows: "By holding and positioning molecules, nanomachines will control how the molecules react, building up complex structures with atomically precise control." e-drexler.com, *Molecular Manufacturing Will Use Nanomachines to Build Large Products with Atomic Precision.* Available at http://www.e-drexler.com/p/04/03/0325molManufDef.html (last visited Oct. 13, 2006).

[5]K. E. Drexler, Engines of Creation: The Coming Era of Nanotechnology (1986). Available at http://wfmh.org.pl/enginesofcreation/EOC_Chapter_11.html (last visited Oct. 13, 2006) (hereinafter Drexler).

Then, after warning of the future potential for nano- and/or artificial intelligence-(A.I.) triggered environmental disasters, military misuses, and state-sponsored human rights abuses, he articulated, with axiomatic insight, a truism inherent in all disruptive technologies throughout history: "[I]n the future, as in the past, new technologies will lend themselves to accidents and abuse."[5]

Today, exactly two decades later, alongside the exponential advances in nano-scale science and engineering, there has been a considerable and sudden upsurge in NELSI interest and activity, with the advent of the National Nanotechnology Initiative (NNI), the U.S. research and development (R&D) program "established to coordinate the multiagency efforts in nanoscale science, engineering, and technology."[6] In fact, the number of entities claiming a stake in the NELSI arena has become nearly as prolific as the number of nanoinnovations during the same period.

The creation of the NNI in 2000, the concomitant federal funding—to the tune of $1 billion a year[7]—for all things nano under the NNI's auspices, and the 21st Century Nanotechnology Research and Development Act's[8] NELSI-focused mandate (discussed below) are, without a doubt, the predominant motivations behind NELSI's current draw. Nevertheless, other incentives also are part of the calculus of factors—albeit some more cynical than others—that contribute to NELSI's present-day popularity, including: nanotechnology's emergence as the techno buzzword *du jour*;[9] an economic interest in defining and managing both nano's real and perceived risks,[10] allied, as some would argue, with a healthy dose of social engineering;[11,12]

---

[6]National Nanotechnology Initiative, *About the NNI*. Available at http://www.nano.gov/html/about/home_about.html (last visited Oct. 13, 2006).

[7]"In the U.S. alone the funding for nanotechnology initiatives that flows through the NNI is $1 billion per annum." National Research Council, A Matter of Size: Triennial Review of the National Nanotechnology Initiative, S-3 (2006) (hereinafter National Research Council).

[8]The Act was passed in 2003.

[9]For a discussion of nano's commoditization as a brand for marketing purposes, see Mekel, footnote 2. For an in-depth look at the hyperbole surrounding nanotechnology, see David M. Berube, Nano-Hype: The Truth Behind the Nanotechnology Buzz (2006) (hereinafter Berube).

[10]Swiss Re, the world's largest reinsurer, was the first to seriously flag these issues and their significance for the commercial viability of nanotechnology:

While its commercial utilisation has triggered debate in specialist circles and the term "nanotechnology" itself is rapidly becoming a media buzzword, there is still no universal assessment of the opportunities and hazards of this new scientific discipline. The word "nano-technology" itself actually connotes less a technology than a generic term for a large number of applications and products which contain unimaginably small particles and demonstrate special properties as a result .... Too little is known about risks of this kind, and the paucity of data gives rise to a host of fears and alarmist scenarios.

A. Hett (ed.), Swiss Re, Nanotechnology: Small Matter, Many Unknowns 3 (2004). Available at http://www.swissre.com/ (last visited Aug. 8, 2006) (hereinafter Hett).

[11]"In political terms, social engineering refers to top-down efforts by governmental bodies and other entities to guide public opinion about a specific thing or phenomenon, such as nanotechnology, in a particular direction." M. Mekel, *Civil Engagement Versus Social Engineering: What Can be Learned from NanoJury UK*, 2 Nanologues 5, n. 5 (2005) (hereinafter Mekel 2).

[12]See D. Berube, Presentation at the Nano and Bio in Society Conference, Mar. 29, 2006, Chicago, Illinois (hereinafter Berube 2); Berube. See footnote 9, p. 309.

and, even, *bona fide* concerns about potential health, environmental, societal, and ethical hazards posed by nano as an emerging, disruptive, and enabling technology. Reflecting the variation in the underlying motives, the NELSI efforts undertaken to date differ significantly in scope, jurisdiction, purpose, and target audience.

When Congress Talks: The NELSI Mandate

To provide a proper framework within which to view these nanoethics efforts, particularly those funded under the auspices of the NNI, one first must look to the language of the actual NELSI mandate contained within the Act. This legislation states, in pertinent part:

SEC. 2. NATIONAL NANOTECHNOLOGY PROGRAM.

* * *

(b) PROGRAM ACTIVITIES.—The activities of the Program shall include—

* * *

(10) ensuring that ethical, legal, environmental, and other appropriate societal concerns, including the potential use of nanotechnology in enhancing human intelligence and in developing artificial intelligence which exceeds human capacity, are considered during the development of nanotechnology by—

(A) establishing a research program to identify ethical, legal, environmental, and other appropriate societal concerns related to nanotechnology, and ensuring that the results of such research are widely disseminated;

(B) requiring that interdisciplinary nanotechnology research centers established under paragraph (4) include activities that address societal, ethical, and environmental concerns;

(C) insofar as possible, integrating research on societal, ethical, and environmental concerns with nanotechnology research and development, and ensuring that advances in nanotechnology bring about improvements in quality of life for all Americans; and

(D) providing, through the National Nanotechnology Coordination Office established in section 3, for public input and outreach to be integrated into the Program by the convening of regular and ongoing public discussions, through mechanisms such as citizens' panels, consensus conferences, and educational events, as appropriate.[13]

Do People Listen?: The NNI's Performance on the NELSI Front

In addition to reviewing the congressional mandate, it is also critical to consider the NNI's execution of its responsibilities, which includes funding NELSI activities under the umbrella of the NNI's programmatic rubric.[14] The NNI's Societal

---

[13]15 U.S.C. § 7510 (2)(b)(10) (2004).

[14]Approximately two-dozen U.S. governmental agencies and departments participate in the NNI. Of those, approximately one-half oversee nano-related funding. In terms of the federal funding for programmatic NELSI-related efforts, the lion's share is managed by the NSF. National Research Council. See footnote 7, pp. 1-3–1-5.

Dimensions Program Component Area (PCA), which is one of seven PCAs, consists of three themes:

(1) research to characterize environmental, health, and safety (EHS) impacts of the development of nanotechnology and assessment of associated risks; (2) education-related activities such as development of materials for schools, undergraduate programs, technical training, and public outreach; and (3) research directed at identifying and quantifying the broad implications of nanotechnology for society, including social, economic, workforce, educational, ethical, and legal implications.[7]

The NNI spending under the Social Dimensions PCA, which, as noted above, includes much more than just NELSI, has been reported at $82 million, out of the $1.1 billion in nano-related funding, for 2006.[15] That $82 million was divided among NNI-participating entities as follows:

- The NSF awarded $60 million.
- The Department of Health and Human Services oversaw $8 million under the auspices of the National Institutes of Health (NIH) and handled $3 million through the National Institute for Occupational Safety and Health.
- The Environmental Protection Agency managed $4 million.
- The Department of Defense (DOD) and the Department of Justice each directed $2 million.
- The Department of Energy (DOE), the Department of Commerce under the National Institute of Standards and Technology, and the U.S. Department of Agriculture each controlled $1 million.[15]

A Matter of Size: Triennial Review of the National Nanotechnology Initiative (NRC Report) had just been released, as of this writing, by the National Research Council. This review is also specifically called for in the Act, which provides:

SEC. 5. TRIENNIAL EXTERNAL REVIEW OF THE NATIONAL NANOTECHNOLOGY PROGRAM.

(a) IN GENERAL.—The Director of the National Nanotechnology Coordination Office shall enter into an arrangement with the National Research Council of the National Academy of Sciences to conduct a triennial evaluation of the Program, including—

(1) an evaluation of the technical accomplishments of the Program, including a review of whether the Program has achieved the goals under the metrics established by the Council;

(2) a review of the Program's management and coordination across agencies and disciplines;

(3) a review of the funding levels at each agency for the Program's activities and the ability of each agency to achieve the Program's stated goals with that funding;

---

[15]See footnote 7, pp. 1–5.

(4) an evaluation of the Program's success in transferring technology to the private sector;

(5) an evaluation of whether the Program has been successful in fostering interdisciplinary research and development;

(6) an evaluation of the extent to which the Program has adequately considered ethical, legal, environmental, and other appropriate societal concerns;

(7) recommendations for new or revised Program goals;

(8) recommendations for new research areas, partnerships, coordination and management mechanisms, or programs to be established to achieve the Program's stated goals;

(9) recommendations on policy, program, and budget changes with respect to nanotechnology research and development activities;

(10) recommendations for improved metrics to evaluate the success of the Program in accomplishing its stated goals;

(11) a review of the performance of the National Nanotechnology Coordination Office and its efforts to promote access to and early application of the technologies, innovations, and expertise derived from Program activities to agency missions and systems across the Federal Government and to United States industry;

(12) an analysis of the relative position of the United States compared to other nations with respect to nanotechnology research and development, including the identification of any critical research areas where the United States should be the world leader to best achieve the goals of the Program; and

(13) an analysis of the current impact of nanotechnology on the United States economy and recommendations for increasing its future impact.

(b) STUDY ON MOLECULAR SELF-ASSEMBLY.—As part of the first triennial review conducted in accordance with subsection (a), the National Research Council shall conduct a one-time study to determine the technical feasibility of molecular self-assembly for the manufacture of materials and devices at the molecular scale.

(c) STUDY ON THE RESPONSIBLE DEVELOPMENT OF NANOTECH-NOLOGY.—As part of the first triennial review conducted in accordance with subsection (a), the National Research Council shall conduct a one-time study to assess the need for standards, guidelines, or strategies for ensuring the responsible development of nanotechnology, including, but not limited to—

(1) self-replicating nanoscale machines or devices;

(2) the release of such machines in natural environments;

(3) encryption;

(4) the development of defensive technologies;

(5) the use of nanotechnology in the enhancement of human intelligence; and

(6) the use of nanotechnology in developing artificial intelligence.

(d) EVALUATION TO BE TRANSMITTED TO CONGRESS.—The Director of the National Nanotechnology Coordination Office shall transmit the results of any evaluation for which it made arrangements under subsection (a) to the Advisory Panel, the Senate Committee on Commerce, Science, and Transportation and the House of Representatives Committee on Science upon receipt. The first such evaluation shall be transmitted no later than June 10, 2005, with subsequent evaluations transmitted to the Committees every 3 years thereafter.[16]

Clearly, carrying out such a comprehensive charge is a lofty and laborious task. In executing its duties, the review committee provided an excellent and detailed exposition of the NNI's convoluted structure. The NRC Report, however, is as much an appeal for continued and increased nano R&D funding as it is an assessment of the NNI's performance. Moreover, some of the specific provisions contained within the legislative directive were given short shrift. Those particular provisos include:

- "evaluation of the extent to which the Program has adequately considered ethical, legal, . . . and other appropriate societal concerns;"[17]
- "use of nanotechnology in the enhancement of human intelligence;"[18] and
- "use of nanotechnology in developing artificial intelligence."[18]

In the preface to the NRC Report, the committee hastily dispensed with the enhancement of human intelligence and the development of A.I., stating:

> As a result of its reflections on and discussions of what is regarded as the more futuristic aspects of nanotechnology—such as the use of nanotechnology in developing artificial intelligence, and similar topics popularized by science fiction—the committee decided that an assessment of such topics in the context of a need for standards and guidelines would be premature and speculative at best. Therefore, the committee chose to address potential real risks rather than perceived risks pertaining to nanotechnology.[19]

Yet, an entire section of the report was devoted to molecular manufacturing,[4] which is currently more in the realm of fiction than A.I. and human cognitive augmentation. Moreover, the committee found mechanisms by which to appraise the United States' position in nano R&D, despite the facts that: (1) a lack of standards presently plagues nano and is presently the focus of much effort by various standard-setting bodies;[20] and (2) differences in how the definitions of and funding for nano R&D vary from nation to nation greatly complicate comparisons.[21] Even though the committee noted an absence of accepted reporting protocols for tracking investments in nano across the NNI's participating agencies and departments, as well as a dearth of metrics for measuring the NNI's economic impact, it, nonetheless, somehow found a way to comment on and make recommendations for these mandated areas of focus.[7]

[16]15 U.S.C. § 7510 (5) (2004).
[17]15 U.S.C. § 7510 (5)(a)(6) (2004).
[18]15 U.S.C. § 7510 (5)(c)(5) (2004).
[19]See footnote 7, p. viii.
[20]See footnote 7, pp. 4-12–4-13.
[21]See footnote 7, pp. 2-1–2-2.

While not dispatched in quite as cursory a manner as A.I. and human cognitive enhancement, NELSI was clearly an afterthought in the review process. The NRC Report explicitly states: "Although they were *not* a central issue for its deliberations, the committee recognized that addressing ethical and societal concerns pertaining to the emergence of nanotechnology *will be* an important part of responsible development."[22] That is all they have to say on the subject.

In summary, the NRC Report appears preoccupied, first and foremost, with maintaining U.S. leadership in nano R&D, funding, and commercialization. A second prominent area of concern emphasized in the report is nano EHS. Rounding out the most pressing nano issues, as determined by the review committee, were nano education with the end goal of workforce development, and public perception management (under the guise of "public education" and "public engagement") to ensure consumer acceptance of nano.

AN OVERVIEW OF U.S. NELSI INITIATIVES

The entities engaged in the NELSI arena in the United States are generally affiliated with academic institutions or are NGOs.

Academic-Based NELSI Initiatives

As a rule, the academic-based entities conducting NELSI initiatives are dependent on grant funding to support their efforts—and, oftentimes, their very existence. As detailed earlier in the chapter, to date, such funding has come primarily through federal agencies participating in the NNI—especially the NSF.

Among those entities perhaps the most significant is a broad, NSF-funded nanotechnology in society network, described in Chapter 21 by one of its principals, David Guston. This network includes the Center for Nanotechnology in Society at Arizona State University; the Center for Nanotechnology in Society at University of California, Santa Barbara; the nanoCenter at the University of South Carolina; and other collaborating universities.[23] From Guston's chapter, it is clear that the diffused network has numerous prongs and functions with differing purposes and target audiences—including the network partners, themselves, which is critical for coordination given the involvement of approximately 100 researchers who are geographically dispersed across various sites.[24] In fact, the network's multinodal nature, which was a decision made by the NSF in awarding the funding, has caused nanocommentators, such as Berube, to express trepidation about the network's effectiveness: "A network rather than a [singular] center seems problematic .... While spreading the wealth seems the egalitarian thing to do, it is questionable whether it is good for society."[25] Nevertheless, it boasts

---

[22]See footnote 7, p. 4–13 (emphasis added).
[23]See Chapter 21 for further details.
[24]See Chapter 21.
[25]Berube, footnote 9, pp. 423–324.

an impressive semblance of interdisciplinary STS scholars, and it is still very early days for the network, which was officially announced in October 2005, and which has at least another 4 years, as of this writing, to go under the initial grant term.

Another NELSI initiative, which, as noted above, has ties to the NSF's nanotechnology in society network, and which was one of the earliest NELSI initiatives, is the nanoScience & Technology Studies project (nSTS) within the University of South Carolina's nanoCenter. With the mission of conducting "research and education about the societal, epistemological, and ethical dimensions of nanotechnologies,"[26] nSTS-affiliated investigators have edited and authored various books[27] and other publications. Additionally, nSTS has implemented "The South Carolina Citizens' School of Nanotechnology," the purpose of which is to introduce the lay-public to nanotechnology through a basic, survey-styled course.[28]

Among the other initiatives in the field, which are also at least partially supported by NSF grants, are The Initiative on Nanotechnology and Society at the University of Wisconsin-Madison,[29] which, through its affiliated scholars, was involved in the first, small-scale citizens' consensus conference on nanotechnology;[30] and the Agrifood Nanotechnology Project within Michigan State University's Institute for Food and Agricultural Standards, which focuses solely on nanotechnology's applications and implications for the agrifood sector.[31] Additionally, the NSF funds the National Nanotechnology Infrastructure Network, which includes the SEI-Nano Portal to "foster understanding of the societal and ethical components of nano R&D" at Cornell University.[32]

Another university-based NELSI project is the Center on Nanotechnology and Society (Nano & Society), at Chicago-Kent College of Law within Illinois Institute of Technology—with which both authors of this chapter are affiliated. Established in the spring of 2005, Nano & Society's mission is to catalyze informed interdisciplinary research, education, and dialogue on the ethical, legal, policy, business, and broader societal implications of nanoscale science and technology—with a special focus on the human condition. In doing so, it brings together scholars, researchers, and policy leaders in law, ethics, science, social sciences, and industry. Nano & Society also offers a number of resources to aid the national conversation on NELSI, including: its website;[33] NELSI Global, a web-based policy document archive, which serves as a unified clearinghouse that enables users to identify and

---

[26]Available at http://nsts.nano.sc.edu/.

[27]See, for example Berube, footnote 9; Joachim Schummer and Davis Baird (eds.), Nanotechnology Challenges: Implications for Philosophy, Ethics and Society (2006).

[28]Available at http://nsts.nano.sc.edu/outreach.html.

[29]Available at http://www.lafollette.wisc.edu/research/Nano/.

[30]See Report of the Madison Area Citizens Consensus Conference on Nanotechnology (2005). Available at http://www.lafollette.wisc.edu/research/Nano/nanoreport42805.pdf. For more information about consensus conferences, see Michele Mekel, *Civil Engagement Versus Social Engineering: What Can be Learned from NanoJury UK*, 2 Nanologues 2, 5 (2006).

[31]Available at http://www.msu.edu/~ifas/index.htm.

[32]Available at http://sei.nnin.org/.

[33]Available at http://www.nano-and-society.org.

access key public policy documents from across the world;[34] publications, including academic and popular press articles, an electronic newsletter, and a print series; and hosting and participating in numerous national symposia, including its Annual Nanopolicy Conference in Washington, DC, and its webcast Nano Forum series.

NGOs and Other Entities with a NELSI Focus

The most visible and vocal of the NGOs that have thrown their hats into the nano ring are the environmental groups, including ETC Group (ETC), Friends of the Earth U.S. (FOE–US), the International Center for Technology Assessment (ICTA), and Environmental Defense (ED). Their approaches run the gamut from: ETC's call for a moratorium;[35] to the ICTA and FOE–US's filing of the first-ever legal petition with the Food and Drug Administration, demanding nano-specific review and regulation of nano-infused products falling with in the agency's purview, as well as a recall of nano-enhanced sunscreens currently on the market;[36] to ED's joining forces with industry in calling for greatly increased funding for EHS research.[37]

Additionally, the U.S. nanosphere includes a hybrid entity of note, which is not exactly an NGO because it receives some government funding. This organization is Woodrow Wilson International Center for Scholars' Project on Emerging Nano-technologies (WWICS PEN) in Washington, DC.[38] The project, which is predominantly supported by a grant from the Pew Charitable Trusts Foundation, commenced in the spring of 2005. Since then, WWICS PEN has released several reports on nano, addressing public perception, regulation, and EHS, and it has also unveiled a publicly accessible nano consumer products inventory via its website.[39]

GOVERNANCE IN A NANOWORLD

One of the most widely recognized multinationals, the United Nations Educational, Scientific and Cultural Organization (UNESCO), began exploring nanotechnology

[34]Available at http://www.nano-and-society.org/NELSI/.

[35]Available at http://www.etcgroup.org/en/issues/nanotechnology.html. Note that the ETC Group is headquartered in Canada; however, it maintains a U.S. office. The full name of the ETC Group is the Action Group on Erosion, Technology and Concentration. Available at http://www.etcgroup.org/en/about/. The NGO adopted this name in 2001, as it was formerly known as the Rural Advancement Fund International (RAFI).

[36]See International Center for Technology Assessment et al., Petition Requesting FDA Amend Its Regulations for Products Composed of Engineered Nanoparticles Generally and Sunscreen Drug Products Composed of Engineered Nanoparticles Specifically, No. ____, (FDA filed May 16, 2005). Available at http://www.icta.org/doc/Nano%20FDA%20petition%20final.pdf (last visited May 22, 2006). For more information on the filing and on these NGOs, see Chapters 6 and 13.

[37]See F. Krupp and C. Holliday, *Let's Get Nanotech Right*, The Wall Street Journal, June 14, 2005, at B2. Available at http://www.environmentaldefense.org/documents/5177_OpEd_WSJ050614.pdf (last visited Oct. 20, 2006). See also http://www.environmentaldefense.org.

[38]Available at http://www.nanotechproject.org/.

[39]Available at http://www.nanotechproject.org/inventories.

through its World Commission on the Ethics of Scientific Knowledge and Technology (COMEST) in late 2003, and, in mid-2005, it convened a nanotechnology and ethics expert group with an interest in mapping NELSI from a global perspective.[40] The most notable result of this initiative thus far is a 2006 report entitled *The Ethics and Politics of Nanotechnology.*[41]

While clearly not as prominent as UNESCO, another international organization, the International Council on Risk Governance (IRGC), has been very active in early nanotechnology governance efforts on a global scale. The IRGC is a private–public partnership founded in 2003 and based in Switzerland, engages governments, industry, and academia in multidisciplinary assessment of emerging risk issues.[42] The projects' funders include the Swiss government and Swiss Re, the world's largest reinsurer.[43] Among IRGC's undertakings is a project on nanotechnology, which "addresses the need for adequate risk governance approaches at the national and international levels in the development of nanotechnology and nanoscale products."[43] Chaired by Mihail Roco, who is credited with being the architect of the NNI, this IRGC initiative has issued a number of white papers, including one that proposes a "global framework for the risk governance of nanotechnology."[43]

The International Council on Nanotechnology (ICON), described by Kristen Kulinowski in Chapter 22, is an international, multistakeholder organization with representation from academia, government, industry, and NGOs. Under the mission of developing and communicating "information regarding potential environmental and health risks of nanotechnology" in order to "foster risk reduction while maximizing social benefit," ICON grew out of an affiliates program at the Center for Biological and Environmental Nanotechnology, an NSF-funded Nanoscale Science and Engineering Center at Rice University.[44] Unveiled in the fall of 2004, ICON, which derives its funding in large measure from its industrial partners, and has multiple risk-management initiatives underway.

BIG ISSUES FROM SMALL SCIENCE: FORMULATING A NELSI FRAMEWORK

Building on the agenda for nanoethics set forth in the context of nanomedicine in Chapter 16, this chapter strives, in part, to offer a framework for the broader and more comprehensive NELSI.[2] Specifically, there appear to be four overarching

---

[40]UNESCO, The Ethics and Politics of Nanotechnology 4, 13 (2006), available at http://unesdoc.unesco.org/images/0014/001459/145951e.pdf. (Last visited Oct. 20, 2006) (hereinafter UNESCO). See also http://portal.unesco.org/shs/en/ev.php-URL_ID=6314&URL_DO=DO_TOP IC&URL_SECTION = 201. html. A process is also underway at OECD.
[41]UNESCO, footnote 40.
[42]Available at http://www.irgc.org/irgc/about_irgc/.
[43]Available at http://www.irgc.org/irgc/projects/nanotechnology/.
[44]See Chapter 22.

categories into which the issues presently fall and are likely to fall in the future: (1) nanoethics; (2) nanogovernance and nanopolicy; (3) risk management, socially responsible development, and sustainability; and (4) public engagement. Each area will be discussed individually to highlight the types of considerations anticipated to arise under each.

Nanoethics

The ethical agenda items set forth in Chapter 16 are broadly relevant to many of nanotechnology's predicted applications. The most germane of these expansive nanoethics concerns are:

- Creation of enormous and unfounded expectations for the benefits to be derived from the nascent technology, which can only lead to public disappointment and distrust (as discussed in Chapter 19 by C. Christopher Hook).
- Need for nonsensationalized, public information and a transparent mechanism for disseminating such information, so that it can be vetted and assessed on its merits.
- Articulation of the essence of humanness, coupled with a recognition of and respect for human rights—including individual rights, and procedural and substantive due process—and human dignity in the development, testing, commercialization, and application of the resulting technological advances—especially given how some frenetically tout nanotechnology as the "magic bullet" for all that ails humankind and as the mechanism that will "liberate" *Homo sapiens* from the vagaries of biology and perhaps even death itself, marking the advent of human engineering and "enhancement."[45]
- Safeguards on privacy and confidentiality, as balanced against security, in an era of envisaged nanosensor ubiquity.
- Development of parameters for military applications, national security, and associated public disclosure, which is of particular significance as the DOD is second only to the NSF in providing nano R&D funding.[46]
- Equity and justice—on both a present-day and intergenerational basis—so as to mitigate a "nanodivide," which is likely to create a widening, on perhaps the most dramatic scale yet, of the technologically magnified gulf between the "haves" and the "have nots," domestically and globally.[47]

As a cautionary note, however, it bears mentioning that, at present, the term "nanoethics" is beginning to be used interchangeably with NELSI to encapsulate the complete range of nontechnical issues related to the nascent technology and its applications. This is both unfortunate and counter to the Act's express

[45]For broader discussions of human enhancement and transhumanism, see Chapters 3, 10, and 19.
[46]National Research Council, footnote 7, p. 1–5.
[47]As noted in Chapters 5 and 24, a prime example of how we choose to draw the line between "treatment" and "enhancement" will have a substantial bearing on this subcomponent.

listing of ethical issues alongside other areas of concern. The Act specifies, in relevant part:

SEC. 2. NATIONAL NANOTECHNOLOGY PROGRAM.

* * *

(b) PROGRAM ACTIVITIES.—The activities of the Program shall include—

* * *

(10) ensuring that *ethical*, legal, environmental, and other appropriate societal concerns . . . by—

(A) establishing a research program to identify *ethical*, legal, environmental, and other appropriate societal concerns related to nanotechnology, and ensuring that the results of such research are widely disseminated;

(B) requiring that interdisciplinary nanotechnology research centers established under paragraph (4) include activities that address societal, *ethical*, and environmental concerns; [and]

(C) insofar as possible, integrating research on societal, *ethical*, and environmental concerns with nanotechnology research and development, and ensuring that advances in nanotechnology bring about improvements in quality of life for all Americans . . . .[48]

While there are ethical dimensions to all human activities, it is essential to avoid amalgamating the ethics agenda into important-but-distinct deliberations about regulatory and related considerations.[49]

Nanogovernance and Nanopolicy

As nanotechnology is seen to be disruptive within and across industries and nations, the development—and hopeful harmonization—of nanopolicy on the national, international, and multinational levels will prove both extremely critical and inordinately challenging. Nevertheless, from the questions that abound—within the United States alone—as to the application of existing laws, the roles of various regulatory agencies, the sufficiency of voluntary reporting programs and self-regulation, and the need for nanospecific laws and regulations, it is clear that nanogovernance is already behind the eight ball in the fast-paced world of nanoinnovation.[50] Europe faces the same questions.[51] On the broader scale, industry-specific and international efforts toward nanostandards development, including nomenclature and metrology, have commenced.[52] How these efforts will sync with each other, as well as with

[48]15 U.S.C. § 7510 (2)(b)(10) (2004) (emphasis added).
[49]For an overview of those regulatory issues, see Chapters 11 and 13.
[50]See Chapters 11 and 13.
[51]See Chapter 12.
[52]See M. C. Roco and E. Litten (eds.), International Risk Governance Council, Survey on Nanotechnology Governance: Vol. A. The Role of Government (2005); Mekel, footnote 2.

various national and international approaches to nanopolicy, remains one of nano's many unknowns.

Risk Management, Socially Responsible Development, and Sustainability

Two types of risk impact business: (1) real risk, such as toxic torts rising from substances such as asbestos; and (2) perceived risk, such as the market impact of widescale European rejection of genetically modified food.[53] Both can have a devastating economic impact on the bottom line if not proactively addressed and managed. With emerging, broadbased innovations, such as nano, the potentially destructive effect of such hazards is particularly pronounced. Today, nano is starting to wrestle with these risks. Currently, the real risks, those related to nanotoxicity under the EHS rubric, are garnering the most attention.[54] Nevertheless, the dangers of perceived risk have started to cross the radar screen. This is due, in part, to the widely publicized, recent recall of a "nanoposer" product in Germany, a bathroom sealant called Magic Nano, which caused respiratory distress in a number of consumers who used it.[2] Upon investigation, however, it was determined that, despite the appearance of "nano" in the product's name, it contained no nanoparticle ingredients.[2] This nanoscare shined a spotlight on how nano, which crosses numerous industry sectors as an enabling technology, could easily become severely tainted by a "nanofaker" that is simply free riding on the nanowave.[2]

Public Engagement

Vigorous, open public participation is a core tenet of democratic society. This tenet is founded on the concept that citizens are entitled to a say in all matters that impact them and the society in which they live, and it presupposes that members of the public are capable of understanding complex issues and of making and articulating informed decisions about such issues.

To that end, awareness of a phenomenon is the most basic prerequisite for such participation. Yet, with regard to nanotechnology, it has been shown time and again that public awareness of nano, itself—let alone the complex NELSI it brings to bear—is low.[55] But even when such awareness reaches a critical mass, the question lingers whether it will prove not only necessary, but also *sufficient* to engender public engagement in NELSI.

Some scholars have already answered this question with a resounding No. For example, David Berube has postulated that, even with heightened awareness, citizen engagement in NELSI, at least in the United States, will remain lackluster

---

[53]See Chapter 24 for more about the genetically modified food debacle in Europe.
[54]The Swiss Re report, entitled *Nanotechnology: Small Matters, Big Unknowns*, brought theses issues to light early on in the context of insurability. See Hett, footnote 10.
[55]See Jane Macoubrie, Informed Public Perceptions of Nanotechnology and Trust in Government, 8 (2005); D. A. Scheufele and B. V. Lewenstein, *The Public and Nanotechnology: How Citizens Make Sense of Emerging Technologies*, 7 J. Nanoparticle Res. 662 (2005).

because "public opinion in such matters has found few venues for expression because America does not have a viable public sphere over science and technology."[56] Additionally, there is concern that, "[t]o be real participants in democracy, people need to understand the terms in the debate," and, given the complexities of today's science and technology dialogue, to the extent it exists, "[t]he threshold of understanding for true public participation is [already] relatively high and will need to be even higher."[57] Others, however, remain more optimistic. Davis Baird and Tom Vogt have written that "science is being questioned and challenged by a public sphere in an unprecedented manner" because of a cultural shift regarding science that has resulted in a "stronger conceptualization and embedding of science in society."[58]

Moreover, STS scholars and nanoethicists have expressed uneasiness with the potential for social engineering in the nano discussion. "In political terms, social engineering refers to top-down efforts by governmental bodies and other entities to guide public opinion about a specific thing or phenomenon, such as nanotechnology, in a particular direction."[11] In Chapter 21, Guston eloquently warns that: "Indeed, the longer we travel down the paths of the various nanotechnology roadmaps that [Mihail] Roco and various agencies have construed, the more it will seem as if our choices about nanotechnology were never choices but were foreordained." Similarly, the allegation has been made that the NNI's NELSI-focused initiatives are "a grand effort in perception management by the U.S. government."[59] Furthermore, in his recent book, Berube stated that the embrace of NELSI by those in government "may simply be perception management [by such actors] to distance themselves from culpability should anything disastrous ensue."[60]

But, as with all things nano, the sufficiency of public awareness, the public's ability to and interest in engaging in the dialogue, and the role of social engineering in framing nano remain a mystery.

LESSONS IN NELSI FROM *THE DIAMOND AGE*

> We must have technology to live, . . . but we must have it with our own *ti*.
>
> Neal Stephenson[61]

As this chapter began with a quote from Neal Stephenson's *The Diamond Age*, so it shall end. While, today, the nanodevices in Stephenson's novel seem to be purely science fiction, their realization may not be that far off into the future, given the

---

[56]Berube, footnote 9, p. 340.

[57]J. D. Miller, Panel Presentation at "Why Should We Care About Genetics," Illinois Humanities Council, Museum of Contemporary Art, Chicago, Illinois.

[58]D. Baird and T. Vogt, *Societal and Ethics Interactions with Nanotechnology ('SEIN')—An Introduction*, 1 Nanotechnology Law Business 391–401 (2004).

[59]Berube. footnote 12.

[60]Berube, footnote 9, p. 309.

[61]Stephenson, footnote 1, p. 457.

push to move nano R&D forward at an exponential pace. The choice of what nanoin-novations come about and how they are deployed, however, is the province of humankind, and in exercising that election, we must do so with an eye toward pre-serving "'our own *ti*'"—our underlying essence, our values, and our humanity. To achieve that fundamental, but deceptively difficult, directive, NELSI, as part of a robust and vital public dialogue, must be part and parcel of the nano R&D equation at every stage.

In Europe, this process of ensuring the coevolution of science and technology has recently begun under the guise of nano-enabled "converging technologies," as discussed at length in Chapter 3. The European Commission (EC) commissioned analysis of these very issues. Based on the European "*ti*," to use Stephenson's word, the reports generated by this activity have set forth a normative structure for guiding the development and application of such transformative emerging technologies.[62] The touchstones are European hallmarks of identity, social ideals, and values.[63] As such, the EC framework articulates, for example, that such technol-ogies "should be dedicated to engineering *for* the mind" rather than "pursu[ing] engineering *of* the mind."[64]

In concluding, the EC-commissioned initiative acknowledged: "Confronted by deeply transforming and potentially disruptive changes in relation to nature, society and individuals, citizens and governments shoulder grave responsibilities."[65] That is indisputably fact rather than fiction.

---

[62]See European Commission, High Level Expert Group, Foresighting the New Technology Wave: Converging Technologies Shaping the Future of European Societies (2004). Available at http://www.ntnu.no/2020/final_report_en.pdf. (Last visited Oct. 20, 2006) (hereinafter Hleg Report); European Commission, Special Interest Group II, Foresighting the New Technology Wave Expert Group Sig 2 Final Report, V3.7 (11.7), Sig II — Report on the Ethical, Legal and Societal Aspects of the Converging Tech-nologies (NBIC) (2004). Available at http://ec.europa.eu/research/conferences/2004/ntw/pdf/sig2_en.pdf (last visited Oct. 20, 2006) (hereinafter SIG II Report).
[63]See Sig II Report, footnote 62.
[64]HLEG Report footnote 62, p. 42. See also SIG II Report, footnote 62.
[65]HLEG Report footnote 62, p. 51.

CHAPTER 21

The Center for Nanotechnology in Society at Arizona State University and the Prospects for Anticipatory Governance

DAVID H. GUSTON[1]

INTRODUCTION

In the spring of 2005, I found myself in a taxi cab, being driven from my home in the desert foothills of suburban Phoenix to Sky Harbor Airport. Despite the early hour, the following conversation between the driver and me ensued

"Where ya' headed?"
"Washington, DC."
"Business or pleasure?"
"Business."
"Oh, whaddaya do?"
"Teach over here at ASU."
"Whaddaya teach?"
"Political science."
"Ah, really? What in particular?"

---

[1] Any opinions, findings, and conclusions or recommendations expressed in this material are those of the author and do not necessarily reflect the views of the National Science Foundation (NSF).

Nanoscale: Issues and Perspectives for the Nano Century. Edited by Nigel M. de S. Cameron and M. Ellen Mitchell
Copyright © 2007 John Wiley & Sons, Inc.

After the previous, perfunctory responses, I was left with a decision about the kind of detail with which to describe my work. I chose the straightforward:

"I study the politics of science and technology."
"Yeah, so what's in DC?"
Here it got more complicated, I thought and then answered:
"Well, I'm trying to get a grant from the government to study the social, ethical, and political implications of nanotechnology."
"So, like, what do you think of the situation with quantum computing and security?"

This anecdote is true, and I proceeded to find out what more my interlocutor knew about quantum computing (which was more than me!) and how he came about to know it. I discovered he was not an underemployed aerospace engineer. He was a cab driver who had happened to have taken a couple of community college courses, had a girlfriend who had an ex-boyfriend who knew something about cryptography, had read a couple of issues of *Scientific American*, and so on. He was just one citizen with untutored information about an emerging area of science and technology, and he wanted to talk about what it meant for an issue to which he felt some connection.

The anecdote provided me with a wonderful lead in for my purpose in Washington, which was the reverse site visit portion of the competition, sponsored by the NSF, for a Center for Nanotechnology in Society to study just such issues. This chapter describes the intellectual rationale for and activities of the Center for Nanotechnology in Society at Arizona State University (CNS–ASU), which are hewn from an ensemble of social science and humanities research perspectives that my colleagues and I call "real-time technology assessment."[2] The cab driver and those like him, however, provide CNS–ASU with its normative rationale. Ordinary people do care about what novelty in science and technology mean for them, and they should have a way of effectively engaging with how decisions that mean something for them—even in science and technology—are made.

STUDYING NANOTECHNOLOGY IN SOCIETY

My trip to Washington turned out to be fruitful. The NSF decided to create CNS–ASU with a $6.2-million grant, spread over 5 years. Beyond our center, NSF created a broader nanotechnology in society network, including a $5-million center at the University of California, Santa Barbara (UCSB), and $1.5-million-projects at the University of South Carolina and at a collaboration including the University of California, Los Angeles (UCLA), Harvard, and Illinois

[2]Guston, D. H. and Sarewitz, D. (2002). Real-Time Technology Assessment. *Technology in Society*, 24, 93–109.

Institute of Technology (IIT). In the same October 2005 press release in which NSF announced these awards to study the societal implications of nanotechnology, it also announced a $20-million Nanoscale Informal Science Education Network (NISE Net) to promote nanotechnology education among the general public through science museums. Although a complete catalogue of its activities in the societal implications of nanotechnology is beyond the scope of this chapter, NSF has funded other project and team-level activities including: a CAREER award to Rosalyn Berne of the University of Virginia to investigate the perspectives of nanotechnology researchers on the ethical aspects of their work[3]; a small number of Nanoscale Undergraduate Education (NUE) awards dealing with social or ethical issues; and Nanoscale Interdisciplinary Research Team (NIRT) awards to the same South Carolina and UCLA/Harvard/IIT groups that later received project funding in the CNS competition.

Although NSF had identified for itself the societal implications of nanotechnology as an important area of inquiry,[4] a direct impetus for NSF to fund research in the societal implications of nanotechnology was the 21st Century Nanotechnology Research and Development Act,[5] which Congress passed in 2003 to authorize large pieces of the National Nanotechnology Initiative (NNI). The NNI, led by the entrepreneurial activities of Mihail Roco,[6] had been underway since President Clinton rolled it out in 2000 near the very end of his administration. In the Act, Congress mandated "establishing a research program to identify ethical, legal, environmental, and other appropriate societal concerns related to nanotechnology"[5] and "insofar as possible, integrating [such] ... concerns with nanotechnology research and development."[5] Congress further authorized the integration of "public input and outreach ... through mechanisms such as citizens' panels, consensus conferences, and educational events."[5] In calling for such a participatory approach, Congress was taking an unprecedented step in its crafting of science and technology policy.[7] In short, Congress also wanted my cab driver, and others like him, to have their say in how nanotechnology was going to develop.

The remainder of this chapter will describe how the CSN–ASU envisions its roll in research, education, and outreach on the societal implications of nanotechnology. The first section below focuses on what we mean by "anticipatory governance," such that CNS–ASU is an effort toward it. Descriptions of the center's research programs

[3]Berne, R. W. (2005). *Nanotalk: Conversations with Scientists and Engineers about Ethics, Meaning, and Belief in the Development of Nanotechnology.* Mahwah: Lawrence Erlbaum Associates.

[4]Roco, M. C. and Bainbridge, W. S. (Eds.). (2001). *Societal Implications of Nanoscience and Nanotechnology.* New York: Springer.

[5]21st Century Nanotechnology Research and Development Act, Pub. L. No. 108–153 (2003), 15 U.S.C. § 7510 (2004).

[6]McCray, W. P. (2005). Will Small Be Beautiful? Making Policies for Our Nanotech Future. *History and Technology*, 21(2), 177–203.

[7]Fisher, E. and Mahajan, R. L. (2006). Contradictory Intent? U.S. Federal Legislation on Integrating Societal Concerns into Nanotechnology Research and Development. *Science and Public Policy*, 33(1), 5–16.

and its education and outreach programs, constitute the second and third sections, respectively. The fourth section below describes in more detail how we envision this ensemble of activities coming together to create the prospect for changes in science and in society that would represent the anticipatory governance of nanotechnologies.

CNS–ASU AND ANTICIPATORY GOVERNANCE

In the 2003 Act, the deliberations of public officials converged with the insights of social science on the understanding that nanotechnology research, considerations of its societal implications, and public engagement should be addressed jointly in the knowledge production process.[8] These social science insights come from a synthesis of two bodies of knowledge that have, unfortunately, been somewhat estranged from one another—at least in the American context: innovation studies,[9] which demonstrated that innovation arises from continual interactions among a variety of actors in a variety of institutions; and the social studies of science and technology,[10] which have demonstrated that knowledge and society are "coproduced" through the interactions of scientists and nonscientists in a variety of institutional settings.

The CNS–ASU responds to this convergence by attempting to build a new institutional capacity for understanding and governing the transforming power of science and technology. The approach we take in this response is what we call "real-time technology assessment" (RTTA). RTTA is an ensemble of fairly traditional social scientific methods arranged to perform not only the traditional tasks of technology assessment, that is, tracking and evaluating emerging nanotechnologies, but also two more innovative tasks that we encourage called "reflexivity" and "anticipatory governance." Reflexivity in this context is the ability of nanoscale science and engineering (NSE) researchers to be more aware of the kinds of decisions they are making, on behalf of society, in their research. Anticipatory governance is the ability of a variety of stakeholders and the lay-public to prepare for the issues that NSE may present before those issues manifest or reify in particular technologies.

Using "anticipatory," we distinguish this concept from the more reactionary and retrospective activities that follow the production of knowledge-based innovations rather than emerge with them. The anticipation comes with the capacity to: (1) understand, beforehand, the political and operational strengths and weaknesses of such tools; and (2) imagine sociotechnical futures that might inspire their use. With nanotechnology, we like to think that we have a good shot at anticipatory

[8]Bennett, I. and Sarewitz, D. (2006) *Too Little, Too Late? Research Policies on the Societal Implications of Nanotechnology in the United States*. Manuscript submitted for publication.

[9]Mowery, D. C. and Rosenberg, N. (1991). *Technology and the Pursuit of Economic Growth*. Cambridge: Cambridge University Press.

[10]Jasanoff, S. (2004). *States of Knowledge*. New York: Routledge; Latour, B. (1988). *Science in Action: How to Follow Scientists and Engineers Through Society*. Cambridge: Harvard University Press.

governance because those of us concerned with its societal implications have gotten into the game both a little bit earlier than with other knowledge-based innovations and in a manner in which both we and our technical target audience have learned from recent, and not fully satisfactory, histories of ethical, legal, and social implications research on the human genome project, and experiences with genetically modified organisms. Yet, when we recognize that the NNI was planned as a technical endeavor prior to any major consideration of its societal implications other than speculations on the part of its advocates, and that more than one dozen Nanoscale Science and Engineering Centers were created before the nanotechnology-in-society network, one begins to get a sense of how the prospects for anticipation should not be hyped.

Using "governance," we mean a capacity that is broadly distributed through society and not lodged merely in government or in any other single sector or group. The range of activities implicated by the concept of governance is broad. On one side, they are bounded by the kind of technological determinism of the 1933 Chicago World's Fair, "Science Finds, Industry Applies, Man Conforms"— an attitude often apparent in the nanoadvocacy that Langdon Winner effectively cited in his testimony before the House Science Committee in the run-up to the 2003 Nanotechnology R&D Act.[5] Policy makers have adopted this rhetoric, for example, then Undersecretary of Commerce Phillip Bond in his contribution to the 2003 NNI Workshop on the Societal Implications of Nanotechnology, insisted that nanotechnology was coming and that we, as a society, had better get used to it.[11] On the other end of the governance spectrum is the opposite expression of technological choice, "the ban." In this stem-cell era, "the ban" has gotten a lot of play. Earier technological eras had their bans, as well—"ban the bomb", for example. Society effectively bans all sorts of research for broadly consensual reasons (e.g., the safety of human research subjects), even if such experiments could provide useful information and might plausibly be consented to (e.g., effects of pesticides on humans). In nanotechnology, we have heard calls not for a ban as such, but for the ban's temporary cousin, "the moratorium." The ETC Group made an early splash by recommending a halt in nanotechnology research until environmental issues are settled.[12] Friends of the Earth has called for a similar moratorium until the safety of nanoproduction for workers and consumers can be ensured.[13]

Between adapting and banning are a wide range of governing options, including—in no particular order—licensing, civil liability, insurance, indemnification,

[11]Bond, P. J. (2003). Nanotechnology: Economic Opportunities, Societal and Ethical Challenges— Remarks delivered December 9, 2003. Available at http://www.technology.gov/speeches/ PJB_031209.htm (retrieved on October 19, 2006).

[12]ETC Group. (April 7, 2006). Nanotech Product Recall Underscores Need for Nanotech Moratorium: Is the Magic Gone? Available at http://www.etcgroup.org/en/materials/publications.html?id = 14 (retrieved on October 19, 2006).

[13]Archer, L. (May 16, 2006). Nanomaterials, Sunscreens and Cosmetics: Small Ingredients, Big Risks. Available at http://www.foe.org/new/releases/may2006/nanostatement5162006.html. Accessed on October 16, 2006 (retrieved on October 19, 2006).

testing, regulation, restrictions on age or other criteria (rather than on ability to pay), labeling, and on and on. Some tools, like labeling and life-cycle analysis, complement private-sector governance by providing more complete information necessary for market efficiency. Some, like civil liability and indemnification, may distort markets for important reasons of justice or critical technology development. Anticipatory governance means laying the intellectual foundation for (any of) these approaches early enough for them to be fully effective.

CNS–ASU AND RTTA

The RTTA hopes to being to lay such an intellectual foundation. As initially described,[2] RTTA consisted of four functions: (1) understanding what NSE may become, both socially and technically, by developing historical case studies of previous knowledge-based innovations (e.g., genetically modified organisms, nuclear power, space exploration) for systematic comparison; (2) mapping the NSE research program through bibliometric and patent analysis to understand its dynamics and geographies; (3) communicating NSE advances, and societal perspectives on them, between NSE researchers and various publics, in part to help identify, beforehand, particularly problematic research thrusts or applications; and (4) assessing and choosing particular directions for NSE development based not only on the ideas and preferences of the researchers but also on the ideas and preferences of various publics. This initial version of RTTA aspired to document changes in how NSE researchers imagined the implications of their work and conducted their research with that new understanding.

Implemented as the programmatic core of a research center, the RTTA agenda is somewhat altered, but essentially the same as initially conceived. It is organized around four research programs, all of which comprise important and related activities:

RTTA 1: Research and Innovation Systems Analysis (RISA), which characterizes the scope and dynamics of the NSE research enterprise, public and private, and the plausible linkages between it and public values and outcomes. The working assumption for the RISA activities is that a clear empirical understanding of the specific research enterprise, including its promise and specific achievements, is a critical component to sound governance of NSE. The RISA consists of three activities: Research Program Assessment, which develops empirically based insights about the dynamics of the NSE enterprise, as indicated by publications, patents, and other data sources; Public Value Mapping, which assesses the social outcomes or "public value" of NSE research activities by comparing the societal goals articulated for NSE with actual performance; and Workforce Assessment, which conducts supply-and-demand analyses for regional labor markets in nanotechnology.

The principal outcome from the RTTA 1 activities to date is a sophisticated bibliographic definition of nanotechnology, developed through a process that included significant collegial consultation and a small survey (19 expert respondents of 75) of

NSE researchers.[14] The Research Program Mapping team, based at Georgia Institute of Technology, will conduct three types of analyses with the data: (1) mapping emerging US and international NSE developments in research and early commercialization; (2) identifying and probing the drivers of leading regional clusters in NSE research and commercialization worldwide; and (3) assessing the extent to which NSE is emerging as a convergent general purpose technology.

RTTA 2: Public Opinion and Values (POV) monitors the changing values of the public and researchers regarding NSE over time, and examines the role of the media in reflecting and influencing those values. The working assumption is that a clear understanding of the values that both lay-citizens and researchers themselves bring to bear is also a critical component of sound governance for NSE. POV also consists of three activities: (1) Public Opinion Polling, which conducts a national random-digit dial telephone survey ($N = 1200$) to understand the knowledge of and attitudes toward nanotechnology by the US public; (2) Media Influence, which explores the role of the media in public opinion around NSE through experimental interventions involving the award-winning web site, The Why Files (www.whyfiles.org); and (3) Researchers' Values, which surveys NSE researchers, in parallel with the public opinion survey, to understand their knowledge of attitudes toward nanotechnology in society.

As of this writing, POV activities were still in various stages of research framing or data collection. We expect the public opinion survey to be an important contribution because it will provide the first significant longitudinal data with earlier surveys and comparative data with Eurobarometer surveys. The Researchers' Values project is critical because of the guiding observations of RTTA, drawn from both the innovation and the science and technologies studies literatures, that value-laden choices made during knowledge creation are critical for research outputs and, ultimately, outcomes. To hone this perspective, we are also extending the frame of the Researchers' Values inquiry to conduct telephone and face-to-face interviews with Hispanic and Latino/a NSE researchers in an attempt to understand if values derived from their ethnic background provides them with any distinctive perspective on nanotechnology in society. To cultivate additional perspectives from communities holding potentially different sets of values, CNS–ASU is also collaborating with the Hispanic Research Center at ASU and with CNS–UCSB and other nanotechnology-in-society grantees to sponsor a conference on nanotechnology for undergraduate students from underrepresented perspectives, to be held in spring 2007.

RTTA 3: Deliberation and Participation (D&P), which engages researchers and various publics in deliberations and participatory forums about NSE and its plausible futures. The working assumption in D&P activities is that, through iterative and interactive deliberations, we can build a greater capacity to anticipate potentially troublesome issues in the development of NSE and steer away from such trouble. D&P consists of four activities: (1) Scenario Development, which

[14]Porter, A. et al. (2006). Refining Search Terms for Nanotechnology. White paper available at http://cns.asu.edu/cns-library/author.htm (retrieved on October 19, 2006).

develops a variety of plausible, technically validated nanotechnological futures to serve as a substrate for planning and analysis by other CNS–ASU activities; (2) InnovationSpace, which creates a NSE track within a transdisciplinary undergraduate course that joins students from the ASU schools of design, engineering, and business, in which students working in cross-functional teams develop new venture proposals for nanotechnology products; (3) CriticalCorps, which helps illuminate the social significance and consequences of nanotechnological design concepts by applying techniques of cultural criticism; and (4) the National Citizens' Technology Forum, which organizes six networked groups of lay-citizens to deliberate on questions about the societal implications of NSE that they themselves frame.

The scenarios that we have developed in our first round of activity (see Table 1) are drawn largely from examples in the open literature—scientific, political, and fictional. To validate them, we are creating an online, wiki-like site through which various communities of experts and the public can comment on and construct more elaborate scenarios in a quasiexperimental design. These more elaborate scenarios will serve as important inputs for other CNS–ASU activities. For example, InnovationSpace students draw on the scenarios to imagine NSE-based products that they might attempt to develop. CriticalCorps will similarly use the scenarios, which have been self-consciously written in a fashion that is "naïve" to their societal implications, as grist for cultural criticism. The National Citizens' Technology Forum will use the scenarios as examples of what citizens in the future might expect from NSE research and development.

Finally, **RTTA 4:** Reflexivity Assessment and Evaluation (RAE), which assess the impact of the information and experiences generated by CNS–ASU, and to assess CNS–ASU activities more generally. The working assumption of RAE is that we can build a successful, long-term collaboration with NSE researchers, as envisioned by the 2003 legislation, and in the process create better outcomes for researchers themselves and for society at large. The RAE has two activities: (1) Reflexivity Assessment, which documents over time any changes in the identity, knowledge, or practice of the NSE researchers with whom we work; and (2) Evaluation, which attempts to understand what the role of CNS–ASU is in the larger context of nanotechnology-in-society, how well its programs are conceived, and how well they are operating.

The RAE activities assess the success of CNS–ASU at a variety of levels. The interviews on identity, knowledge, and practice suggest what kind of impact the center's engagement is having on individual NSE researchers. For example, one senior NSE researcher at ASU attributes his success in 2006 in getting a NIRT award from NSF after failing in the previous cycle to his reconceptualization of his work following interactions with CNS–ASU. The evaluation activities investigate the kind of success that CNS–ASU, as a larger organization and as an agent of institutional change, might have. For example, during the 2005–2006 academic year, a doctoral student under CNS–ASU funding at the University of Colorado conducted his research as an "embedded humanist" in a mechanical engineering

Table 21.1. Topics of NSE Scenarios Currently Under Development at CNS–ASU

Implantable chips that relay information directly into the brain through nanoscale wires
A floor that tracks the movements of everything on it with nanosensors in the floor material
A semiautonomous robot that climbs and cleans the sides of buildings, using nanohairs similar to those on the toes of a gecko
A contraceptive and STD-protective sponge with nanoparticles that target sperm and disease organisms
Computer processors built from neurons wired with nanoscale transistors
Rapid, high-sensitivity DNA sequencing using nanoscale templates to screen large volumes of throughput (e.g., wastewater) for target DNA
Nanoscale "tagents" to stick to, label, and track anything and any exposure situation
Self-cooking food packaging that uses piezoelectric nanowires to turn the energy from shaking the package into heat
Programmable tattoos with metallic nanoink that responds to magnetic fields
"Bar-free" prisons using nanoscale drugs in prisoners' bodies that radio frequencies can trigger to incapacitate potential escapees

laboratory to test a working assumption of RTTA that choices that researchers make in the laboratory are subject to what the student called "mid-stream modulation."[15] Another kind of an evaluative project brings CNS–ASU into comparative study as a boundary organization[16] that, like other systems that encourage the integration of technical, local, and lay-knowledge, may be a particularly helpful strategy for "inclusive management."[17] A final example is the comparison of CNS–ASU and its RTTA activities with the constructive technology assessment activities of our European, and particularly Dutch, counterparts through collaborative activities, including the International Nanotechnology and Society Network (INSN)[18] and through personnel exchanges and international writing projects.[19]

In addition to its RTTA research programs, CNS–ASU also maintains two, more traditional, research thrusts in areas of substantive interest for nanotechnology: Freedom, Privacy, and Security (FPS); and Human Identity, Enhancement, and Biology (HIEB).

The goal of FPS is to develop theory and explore cases of surveillance and nano-sensing technologies, including issues of effectiveness, potential ubiquity and embeddedness, and impacts on practices of surveillance and on the individuals and

[15]Fisher, E. (2006). Midstream Modulation of Technology: A Case Study in US Federal Legislation on Integrating Societal Considerations into Nanotechnology. A doctoral dissertation completed at the University of Colorado, Boulder.

[16]Guston, D. H. (1999). *Social Studies of Science.*

[17]Feldman, M. S., Khademian, A. M., Ingram, H., and Schneider, A. (2006). Ways of Knowing and Inclusive Management Practices.

[18]See International Nanotechnology and Society Network (INSN). Available at www.nanoandsociety.org (retrieved on October 19, 2006).

[19]For example footnote 21; van Merkerk, R., Guston, D. H., and Smits, R. (November, 2006). *An International Comparison of Recent Technology Assessment Approaches: Bypassing Collingridge.* Presented at the Annual Meeting of the Society for the Social Studies of Science, Vancouver.

communities subject to surveillance. One current project assesses the assumptions and values underlying the design of nanosensors, especially those designed for use on or in the human body. The design of such sensors represents a strategic site for potential collaboration between CNS social scientists and NSE researchers to consider how and when to embed safety and access controls into the devices being envisioned or produced, and to identify areas, where such safeguards cannot be ensured.

The goal of HIEB is to explore the historical, philosophical, cultural, and political dimensions of the interactions between human biology and human values in the context of new nanotechnologies. One current project targets our understanding of the prospects for human nanobiotechnology in the context of ongoing bioethics research into chimeras and hybrids, that is, biological entities that join the characteristics of more than one species. Akin to other CNS–ASU projects, HIEB thrives on close collaboration with NSE laboratories, particularly those in ASU's Biodesign Institute. This collaboration is facilitated by a CNS graduate student with a biology background who, in addition to her ethics work with the center, also works 10 hours a week in a collaborating wet lab.

To find a highly qualified team to conduct top-notch research across such a breadth of social science and humanities fields, CNS–ASU assembles a research team of more than 80 individuals (including faculty, postdoctoral fellows, graduate students, and undergraduate students) at six universities across the country: Arizona State University; University of Wisconsin, Madison; Georgia Institute of Technology; North Carolina State University; Rutgers University; and the University of Colorado, Boulder. The center also has narrower collaborations with individuals and small groups at the University of Georgia, the University of New Hampshire, and elsewhere.

EDUCATION AND OUTREACH

Drawing its mandate from the 2003 Act, which, as described above, spoke to public engagement in nanotechnology research and development, CNS–ASU educational and outreach activities are both critical to our agenda and tightly integrated with our research activities. Our formal education activities cover the range from precollege teacher training to postdoctoral training. This scope of coverage is critical if only because the benefits of nanotechnologies—like the benefits from other major technological changes—may preferentially accrue to those who are educationally positioned to take advantage of them, and the costs may be incurred by everyone—but perhaps particularly by those who are least well off and least prepared for change.

In precollege education, CNS–ASU collaborates with the Center for Research on Education in Science, Mathematics, Engineering, and Technology (CRESMET) with the latter's ongoing work in Project Pathways, a 5-year education research project funded by NSF through the Math and Science Partnership program. In Project Pathways, ASU researchers collaborate with high school math and science teachers in five Phoenix metropolitan school districts, aiming to develop a new model to support teachers' continuing education and professional development. The CRESMET's original design included four graduate courses for teachers, coupled with ongoing professional learning communities. The collaboration between CNS and CRESMET is enabling the development of a fifth course, "Nanotechnology & Society: How Will We Guide and Be Guided by the New Science of Very Small Things?" This course will integrate the natural science that underlies nanotechnology with an inquiring look at how nanotechnology products might affect people, social systems, and the environment. The centerpiece of the new module is a pair of three-week experiences in which teachers will follow a nano-scale phenomenon from an idea in a laboratory to a product marketed to the public. In the process, they will pursue questions ranging from research mechanics (Should we publish or patent?) to engineering (Does this work? Safely?) to marketing (What price point will attract buyers?). At every stage, the course instructors will prompt the teachers with questions that encourage them to think about and discuss the societal consequences of their products and who is responsible for anticipating and governing them. The teachers will then return to their classrooms with lesson planning for delivering an adapted version of the course to their high school students.

In addition to traditional coursework, at the undergraduate level, CNS–ASU is developing two innovative activities to train undergraduates in transdisciplinary ways of learning about and using nanotechnology. The first is a learning community in which students take nine credits' (three courses') worth of material around a single theme—nanotechnology in society—but approached from three separate-but-integrated perspectives. In this learning community, beginning in the spring semester of 2007, students will receive instruction on foundational aspects of NSE, on the social aspects of NSE, and on the politics and policy of NSE. The second is the InnovationSpace course, discussed briefly above, in which senior-level students from design, business, and engineering engage in hands-on activities to develop user scenarios, define new product offerings, build or conceive of engineering prototypes, and create business plans and visual materials to communicate the end results. In its first year of operation in academic year 2006–2007, Innovation-Space will draw on scenarios developed through the RTTA activities, with a focus on the thematic area of Freedom, Privacy, and Security.

At the graduate level, in addition to coursework, CNS–ASU hopes to implement a trans-disciplinary program referred to as "PhD plus," in which NSE doctoral students will include, as an element of their dissertations, a chapter on the societal context or implications of their work. Drawing, in part, on relationships already established through the shared support of graduate students—CNS cofunds three graduate students with laboratories in the Biodesign Institute—and through other activities to expose NSE research students to societal perspectives, CNS will match such students with a social science or humanities mentor and serve as a member of their thesis committee. NSE students who are already engaged in CNS activities may pursue such related "PhD plus" inquiries as the use of "science cafés" to engage the public on nanotechnology and the question of why NSE researchers are less likely to talk about the potential negative consequences of their work than the potential positive consequences.

Outreach and engagement of the public at CNS–ASU is centered on one very large-scale activity and a variety of smaller-scale ones. The large-scale activity, referred to briefly above, is the National Citizens' Technology Forum (NCTF). A citizens' technology forum is an extensive and intensive form of public deliberation akin to a consensus conference or citizens' jury in which a group of 15 or so ordinary citizens inform themselves and the deliberate on a matter of social import, as well as scientific and technical complexity. Many such panels have been conducted in Europe, often as part of a government-related technology assessment as in the case of the Danish Board on Technology, which pioneered the public version of consensus conferences, but they have less frequently appeared in the United States and at no time under the aegis of a government activity. To date, there have been at least two local citizens' forums on nanotechnology in the United States (in North Carolina and in Wisconsin), but CNS–ASU plans to make the scope of its forums national by conducting six forums simultaneously across the country and having electronically mediated deliberations across the six groups, as well as face-to-face deliberations within them. Given the novelty and ambition of the project, and the specific call for such mechanisms of public engagement in the 2003 Act, we hope not only to demonstrate the capacity to engage the public extensively and intensively in this way, but also to provide useful input to public and private sector decision makers in nanotechnology.

CNS–ASU has and will continue to create smaller-scale outreach and engagement activities as well, including its Science Café series in which NSE researchers at ASU head off-campus to speak with a small group of citizens in an informal environment, like a coffee shop or café. CNS–ASU organized three cafes in the Spring of 2006 and, to date, has organized an additional two in the Fall of 2006— one of which was a Spanish language cafe held at a community center in a Hispanic community. Although not formally linked to the broader movement, the CNS–ASU Science Café series is akin to any one of a number of efforts now emerging across the globe to create informal cultural dialogue around scientific issues. The venue for the cafes have been absolutely critical to its purpose of interaction with community, as opposed to university, members. The early cafés were held in a coffee shop near campus, and the audience was almost exclusively university-related. Upon moving the café to a bookstore about 5 miles from campus, the attendance changed to almost entirely nonuniversity, but without any drop-off in the number of attendees. The CNS–ASU cafés have also tweaked the format to introduce greater dialogue: Instead of having a single speaker presenting a small lecture and then responding to questions, these cafés instead pair a NSE researcher with a CNS social scientist or humanist, and the resultant exchange between the two speakers creates a more dialogic exchange with the audience, as well.

CNS–ASU AND THE PROSPECTS FOR ANTICIPATORY GOVERNANCE

The research program of CNS–ASU, laid out here, represents a relatively faithful elaboration of the original vision for RTTA. In CNS–ASU, it is supplemented by an ambitious set of education, outreach, and engagement activities meant to expand the reach of science studies beyond the social sciences, and the reach of the center beyond the campus.

In addition to laying out this program, Guston and Sarewitz also laid out three challenges to implementing it.[2] First, the interdisciplinary collaborations necessary to the success of RTTA are profoundly difficult when much of the academic enterprise is still carved up into isolated disciplines and separated into natural and social sciences. CNS has had a much easier time of it at ASU than others have elsewhere, as interdisciplinarity is one of the university's new watchwords. Under the presidency of Michael Crow, himself a trained student of science policy, ASU is attempting to transform itself into "a New American University" in which most of the articulated institutional "design aspirations" are directly applicable to the mission of CNS (see Table 21.2).[20] One indicator of the receptivity to collaborations between social and natural scientists at ASU is the response of an NSF review panel to a proposal submitted by CNS–ASU personnel in response to the solicitation in Ethics Education in Science and Engineering. The panel indicated that the admittedly innovative proposal did not meet the program requirement for transferability to other universities well enough because, while the proposal's interdisciplinary nature may work at ASU, it would be too challenging to implement at other universities.

A second challenge earlier identified is that broad public knowledge of and stakeholder interest in nanotechnology lag behind the need to explore which values are embedded in emerging nanotechnologies. This challenge has perhaps eased, as many stakeholder and interest groups now actively inform and agitate on nanotechnology issues ranging from the environmental health and safety of nanoparticles to state-based NSE initiatives. Alas, polls continue to show modest portions of the broader public knowledgeable of or engaged with nanotechnology, and massive outreach to the public is not part of the mission of the nanotechnology-in-society network. Rather, it is the specific mission of the Nano-scale Informal Science Education Network (NISE Net), referred to at the beginning of this chapter. Although each of the "nodes" of the nano-in-society network has its own connections with NISE Net, more formal connections that could encourage the presence of a more robust societal perspective in its activities are still lacking at the time of this writing.

The third challenge is the scale of the NSE enterprise. Even at the time that RTTA was being formulated as a way to engage it, NSE was larger than a single RTTA project could ever hope to engage with or assess, let alone steer. This challenge has only become more difficult because the NSE enterprise, as a whole, has grown much larger and faster than the societal implications work that might engage it: a $1-billion/per year NNI in the United States overwhelms the $3-million/year nanotechnology-in-society network. As Bennett and Sarewitz (forthcoming) describe, research on the societal implications of nanotechnology

[20]See Center for Nanotechnology in Society at Arizona State University website. Available at http://cns.asu.edu/index.htm (retrieved on October 16, 2006).

Table 21.2. Design Aspirations for a New American University

Leveraging place	Embracing our cultural, social, economic, and physical setting
Transforming society	Becoming a force, and not only a place
Academic enterprise	Serving as a responsible knowledge entrepreneur
Use-inspired research	Improving the human condition through the appropriate application of knowledge
Focus on the individual	Outcome-determined excellence; commitment to intellectual and cultural diversity
Intellectual fusion	Disciplinary/interdisciplinary/transdisciplinary/postdisciplinary
Social embeddedness	Social enterprise development through direct engagement
Global engagement	Transnational–transcultural focus and impact

got off to an exceptionally slow start. This happened for a variety of reasons, and there is enough blame to go around—to the technocrats, nanoenthusiasts, and compliant politicians who started the NNI with insufficient attention to how the societal implications research was going to be funded, competed, and connected to the policy process; and to the societal implications research community, which was tardy in identifying the profound opportunity for inquiry and impact that nanotechnology offered. It is ironic of course, that the NNI is touted, not wrongly, as that major R&D initiative, which has best confronted its societal implications, most welcomed its potential critics, and most actively sought out public involvement. But no matter how much better prepared we are for nanotechnology's societal implications, we are not well-enough prepared. Thus, while CNS–ASU has been pleased to note that in presentations the government's original nano-impressario Mihail Roco (2006) has taken to citing "anticipatory governance" of nanotechnology as a goal,[21] we require both a more systematic approach and a way to make up for lost time.

While CNS–ASU may have overcome the first of these challenges in its local context at ASU, all three remain major roadblocks to the prospects for anticipatory governance. Indeed, the longer we travel down the paths of the various nanotechnology roadmaps that Roco and various agencies have constructed, the more it will seem as if our choices about nanotechnology were never choices, but were preordained by the nature of NSE itself. Nevertheless, there are resources for change that are already available. As this chapter has made clear, the federal authorizing legislation for nanotechnology R&D itself not only legitimates, but also demands the kind of activities that feed anticipatory governance. To the extent that Congress takes its words seriously in overseeing the NNI, anticipatory governance will benefit. A second resource is the requirement at NSF that proposals and reviews address not only the intellectual merits of the project at hand, but also the broader impacts of the project—including its societal implications. To the extent NSF takes its review criteria seriously, then scientists and engineers who are

[21]Roco, M. (2006, May). "Keynote Address." Center on Nanotechnology and Society First Annual Nano-Policy Conference, National Press Club, Washington, DC.

more collaborative with social scientists and more sophisticated in thinking about the broader impacts of their projects will benefit. A third resource is that many educational programs, including the high school guidelines in the State of Arizona, have adopted contextual learning goals for science, as well as other subjects. To the extent that the societal implications of knowledge-based innovation, including nanotechnology, can be brought into the classroom along with the critical basics of science, technology, engineering, and mathematics, society will benefit from educating a broader base of citizens who, like my cab driver in Phoenix, are independently engaged in such issues and who find the time and—with the success of the congressional goals of public engagement—the outlets to participate.

BIBLIOGRAPHY

Berne, Rosalyn. 2005. *Nanotalk: Conversations with Scientists and Engineers about Ethics, Meaning, and Belief in the Development of Nanotechnology.* Mahwah, NJ: Lawrence Erlbaum Associates, Publishers.

Bond, Philip. 2005. "Preparing the Path for Nanotechnology." p. 16–21 in *Nanotechnology: Societal Implications—Maximizing Benefit for Humanity.* Roco, Mihail and Bainbridge, William S., eds. Arlington, VA: National Science Foundation; Available at www.nano.gov/nni_societal_implications.pdf.

Crow, Michael M. 2006. "Vision and University Goals: 2002–2012." Available at http://www.asu.edu/president/vision-goals/index.html.

ETC Group. 2003. "Size matters: No small matter, II — The case for a global moratorium." Occasional Paper Series 7(1).

Feldman, Martha S., Khademian, Anne M., Ingram, Helen, and Schneider, Anne. 2006. "Ways of knowing and inclusive management practices."

Fisher, Erik. (2006). Midstream Modulation of Technology: A Case Study in US Federal Legislation on Integrating Societal Considerations into Nanotechnology. A doctoral dissertation completed at the University of Colorado, Boulder.

Fisher, Erik and Mahajan, Roop. 2006. "Contradictory intent? US federal legislation on integrating societal concerns into nanotechnology research and development." *Science and Public Policy* 33(1):5–16.

Guston, David H. 1999. *Social Studies of Science.*

Guston, David H. and Sarewitz, Daniel. 2002. "Real-time technology assessment." *Technology in Culture* 24:93–109.

Jasanoff, Sheila, ed. 2004. *States of Knowledge.* New York: Routledge.

Latour, Bruno. 1988. *The Pasteurization of France.* Cambridge: Harvard University Press.

McCray, Patrick. 2005. "Will small be beautiful? Making policies for our nanotech future." *History of Technology* 21(2):177–203.

Mowery, David and Rosenberg, Nathan. 1991. Technology and Economic Growth. New York: Cambridge University Press.

Porter, Alan, Youtie, Jan, and Shapira, Phil. (2006). "Refining Search Terms for Nanotechnology." White paper Available at http://cns.asu.edu/cns-library/author.htm (retrieved on October 19, 2006).

Roco, Mihail and Bainbridge, William S. 2001. *Societal Implication of Nanotechnology.* Boston: Kluwer Academic Publishers.

Roco, M. (2006, May). "Keynote Address." Center on Nanotechnology and Society-First Annual NanoPolicy Conference, National Press Club, Washington, DC.

van Merkerk, Rutger, Guston, David H., and Smits, Ruud. 2006, November. "An international comparison of recent technology assessment approaches: bypassing Collingridge." Presented at the Annual Meeting of the Society for the Social Studies of Science, Vancouver.

Winner, Langdon. 2003. "Testimony to the Committee on Science of the US House of Representatives on the Societal Implications of Nanotechnology." Available at http://www.house.gov/science/hearings/full03/apr09/winner.htm.

CHAPTER 22

The International Council on Nanotechnology: A New Model of Engagement

KRISTEN M. KULINOWSKI

INTRODUCTION

History will record the early years of the new millennium as an extraordinarily active time for the discussion of nanotechnology in the public domain. One would be hard-pressed to find a historical precedent in which the debate over a technology began so early and became so active in relation to its commercial development. It has usually been the case that society has been forced to try to catch the techno-horse after it has left the barn. But, for nanotechnology, the debate coalesced during the span of only a few years and at a time when most people were unable to define the term or point to how it may impact them personally. Setting aside societal concerns about corporate control and social justice, the debate about deleterious impacts has evolved during the past 5 years from one centered almost exclusively on apocalyptic scenarios of fictional nanobots reducing the planet to dust to a more-focused discussion on how best to assess and manage the impacts of passive nanoparticles incorporated into existing commercial products. One agent of change, among many other notable contributors, is the International Council on Nanotechnology (ICON), along with its parent organization the Center for Biological and Environmental Nanotechnology (CBEN).

ICON is an international, multistakeholder organization whose mission is to develop and communicate information regarding potential environmental and health risks of nanotechnology, thereby fostering risk reduction while maximizing societal benefit. The Council has evolved into a network of scholars, industrialists, government officials, and public interest advocates who share information and perspectives on a broad range of issues at the intersection of nanotechnology and

Nanoscale: Issues and Perspectives for the Nano Century. Edited by Nigel M. de S. Cameron and M. Ellen Mitchell
Copyright © 2007 John Wiley & Sons, Inc.

environment, health, and safety. ICON's development is coincident with the evolution of nanotechnology policy during the last five years, and, therefore, the story of its genesis necessarily chronicles the early years of the National Nanotechnology Initiative (NNI), the creation of the first federally funded nanotechnology research centers, and the first stirrings of civil debate about nanotechnology's impacts on society.

THE NNI AND THE GENESIS OF CBEN

On January 21, 2000, President Clinton announced the creation of the NNI, a coordinated federal program to fund research and development at the nanoscale. The NNI was the outcome of more than 4 years of growing interest among staff members in several federal agencies and the culmination of efforts by Clinton science advisor Neal Lane and others to elevate nanotechnology's prominence in the federal research portfolio.[1] Promoting the individual agency efforts in nanoscale Research and Development (R&D) to the level of a national initiative raised the profile of nanotechnology in the eyes of the media, academia, state governments, foreign governments, and the private sector.

Among the biggest beneficiaries of this new initiative was the National Science Foundation (NSF), which received the largest portion ($150 million) of FY 2001 NNI money, amounting to a 55% boost in its nanotechnology budget over the previous year.[2] The NSF directed a portion of this FY 2001 funding to the creation of six new Nanoscale Science and Engineering Centers (NSECs).

> NSECs will address opportunities that are too complex and multi-faceted for individuals or small groups of researchers to tackle on their own. They will bring together researchers with diverse expertise, in partnership with industry, government laboratories, and/or partners from other sectors, to address complex, interdisciplinary challenges in nanoscale science and engineering, and will integrate research with education both internally and through a variety of partnership activities (National Science Foundation).

The first class of NSECs (there are now 16) included CBEN, established at Rice University in 2001 by a grant from NSF's Division of Engineering Education and Centers. The CBEN proposal addressed three of the six research themes identified in the 2000 NSF program solicitation: (1) *biosystems at the nanoscale*; (2) *nanoscale processes in the environment*; and, to a lesser extent, (3) *societal and educational implications of scientific and technological advances on the nanoscale*.[3] At this early stage, NSF's program goals in biosystems and the environment were directed

[1]Lane, N. and Kalil, T. "The National Nanotechnology Initiative: Present at the Creation." *Issues in Science and Technology* Summer (2005).
[2]National Research Council. *Small Wonders, Endless Frontiers: A Review of the National Nanotechnology Initiative*. Washington, DC: The National Academy of Sciences, 2002.
[3]National Science Foundation. "FY 2001 Nanoscale Science and Engineering Solicitation (NSF 00-119)." 2000.

toward developing fundamental understanding of interactions and creating new nanobased tools for societal improvement. One would have to read between the lines of the solicitation to tease out an intention to explore potential negative outcomes arising from toxicity or environmental contamination. "Societal implications" was defined as "ethical, legal, social, economic and workforce implications of nanotechnology," including, for example, public understanding, ethical and legal ramifications of nanoscience, barriers to commercialization, and lifecycle assessment of manufacturing processes.[3] No explicit attention was paid in the solicitation to potential negative outcomes.

Rice University's proposal was probably unique in two respects: (1) its emphasis on *both* biological and environmental systems; and (2) its acknowledgment that introduction of large quantities of nanomaterials into the environment could have unforeseen and deleterious impacts.[4] The proposal's anticipation of large-scale production of nanomaterials was, in itself, seen in 2001 as much too premature by some reviewers and appropriately foresighted by others. In the end, the proposal was recommended for funding and the first federally funded program focused on exploring nanotechnology's impacts on human health and the environment commenced in September of that year.

CBEN's mission is to develop sustainable nanotechnologies that improve human health and the environment. The Center's research portfolio focuses on the interface between intentionally engineered nanoparticles and systems based in water, which is a common feature of biology and larger ecosystems. There is an explicit recognition that engineered nanomaterials have great potential to solve existing societal challenges in medicine and environmental remediation (*applications* of nanotechnology), as well as cause new problems due to potential toxicity or environmental contamination (*implications* of nanotechnology). That is, if these small particles can have an impact, we will try to harness that for societal good while being mindful of unintended consequences. In contrast to an NSEC that focuses on, for example, nanoelectronics or device architectures, CBEN's emphasis on human health and the environment has created an organic connection between the Center and groups exploring nanotechnology's implications for society from a theoretical or policy perspective. This has defined and shaped the Center's knowledge transfer activities in profound and exciting ways.

The mission of major NSF research centers goes beyond conducting cutting-edge research to advance the frontiers of knowledge. The NSF requires that the results of the discovery transcend the boundaries of the academy, through educational programming, public outreach, and strategic partnerships with industry, government laboratories, and/or other users of research outcomes. The goal of industrial interactions is to ensure that knowledge created in the research laboratory is developed and applied for the benefit of society. In many centers, partnerships with industry take the form of an affiliates program wherein established companies pay a membership fee or provide in-kind support to the center in exchange for campus visits,

---

[4]Access to confidential proposals submitted to the NSF is restricted to NSF staff and external reviewers participating in the formal review process.

involvement in collaborative research projects, student internships, first right of refusal for intellectual property generated by the center, access to faculty for consultations, and other benefits appropriate to the individual center. The relationship is meant to be mutually beneficial: The industrial participant develops a close connection to or becomes a partner in the development of new, cutting-edge knowledge, and center researchers better understand how their work might be applied outside the academy.

SOCIETAL DEBATE HEATS UP

CBEN began a modest industrial affiliates program in 2002, but the creation of start-up companies spun out of the research labs was emphasized more heavily in the Center's early years as the primary vehicle for knowledge transfer as established companies had yet to make major commitments to in-house nanotechnology R&D at that time. Early participants in CBEN's industrial affiliates program included a major producer of carbon particles and a start-up company that spun out of pre-Center work on carbon nanotubes. But CBEN's establishment coincided with the rise of public controversy over nanotechnology's impacts on society. Inspired in part by Bill Joy's *Wired* convergence nightmare[5] and Drexler's grey-goo scenario,[6] Michael Crichton penned the ultimate fictional nanothriller, *Prey*.[7] For many in the lay public, this was the first time they had heard the term nanotechnology. Journalists jumped on the story of nanobots run amok, and a national, if not international, conversation about nanotechnology's risks began in earnest. In contrast to the generally positive media coverage that nanotechnology had enjoyed up to that point, companies were now confronted with the prospect that their burgeoning investments in nanotechnology might be derailed by real or perceived risk factors.

It was also around this time that the first stirrings of civil society discord over nanotechnology began to emerge. First out of the gate was the Action Group on Erosion, Technology and Concentration (ETC) in 1999,[8] though the seeds had been planted a decade earlier. The ETC Group, which had been active in campaigns against genetically modified organisms (GMO) under the name Rural Advancement Foundation International (RAFI), made a practice of scanning the horizon of new technology to see where science and industry were heading next. According to ETC executive director Pat Roy Mooney, science will always seek to control the basic building blocks of nature and, as atoms are more basic than genes, science would soon seek control over these through the emerging field of nanotechnology.[9] Mooney spotted some reporting on nanotechnology in trade journals as early as 1988,

[5]Joy, B. "Why the Future Doesn't Need Us." *Wired Magazine* April 2000.
[6]Drexler, K. E. *Engines of Creation*. New York: Anchor, 1986.
[7]Crichton, M. *Prey*. New York: HarperCollins Publishers, Inc., 2002.
[8]Mooney, P. R. Development Dialogue. The ETC Century: Erosion, Technological Transformation and Corporate Concentration in the 21st Century: Dag Hammerskjöld Foundation, 1999.
[9]Mooney, P. R. "Origins of Etc Interest in Nanotechnology." Personal communication, 2006.

which merited a mention in RAFI's 1989 board report, but he was not sure if nanotech would pan out to be more than futurist imaginings and pushed it to the back burner.[9] In 1999, as RAFI was undergoing a change in direction and transforming itself into the ETC Group, he took another look and was surprised to find that a lot had happened in 10 years. With growing federal and private investments and increasing patent activity, nanotechnology became worthy of more focused scrutiny, especially because it appeared that other civil society organizations were not doing so.[9]

The ETC Group's first major report on nanotechnology came in response to early research indicating that nanoparticles might have harmful impacts on health and the environment. This report ruffled more than a few nanofeathers when it called on governments to "declare an immediate moratorium on commercial production of new nanomaterials and launch a transparent global process for evaluating the socio-economic, health and environmental implications of the technology."[10] ETC recently renewed its call for a moratorium in the immediate aftermath of the Magic Nano bathroom sealant recall.[11] While acknowledging that the product might not contain any nanotechnology, (as German regulators eventually confirmed),[12] ETC nonetheless reiterated its concern that "no government anywhere regulates nanoscale materials if the same chemical substance has been vetted at the macroscale."[13] All told, ETC Group has published more than 2 dozen reports[14] on nanotechnology, covering everything—toxicity, intellectual property, regulation, international governance, and agriculture—and remains the most relentless and prolific civil society critic of nanotechnology.

Other civil society organizations began to take issue with nanotechnology, albeit slowly. In May 2002, Greenpeace UK convened with *New Scientist* a debate about the societal impacts of nanotechnology, artificial intelligence, and robotics in which a science fiction writer, a professor of science and society, a "futurologist," and a professor of science communication were asked to answer the following questions: What will be the meaning of these developments?; What are the dangers to ourselves and to the environment?; and Will we even notice that we have handed over power and control of our lives to our creations?[15] The lack of specific information on nanotechnology products and their potential implications prompted

---

[10]ETC Group. No Small Matter! Nanotech Particles Penetrate Living Cells and Accumulate in Animal Organs. 2002. Available at http://etcgroup.org/documents/Comm_NanoMat_July02.pdf (June 30, 2002).
[11]BfR (Federal Institut for Risk Assessment). *Exercise Caution When Using "Nano-Sealing Sprays" Containing a Propellant.* 2006. Available at http://www.bfr.bund.de/cms5w/sixcms/detail.php/7699 (March 31, 2006).
[12]von Bubnoff, A. *Study Shows No Nano in Magic Nano, the German Product Recalled for Causing Breathing Problems.* 2006. Small Times. Available at http://www.smalltimes.com/document_display. cfm?document_id=11586 (May 26, 2006).
[13]ETC Group. Nanotech Product Recall Underscores Need for Nanotech Moratorium: Is the Magic Gone? 2006. Available at http://etcgroup.org/documents/NRnanoRecallfinal.pdf (April 7, 2006).
[14]ETC Group. 2006. Available at http://etcgroup.org/search2.asp?srch=nano (August 30, 2006).
[15]Greenpeace UK. *Technology: Taking the Good without the Bad?* 2002. Available at http://www.greenpeace.org.uk/contentlookup.cfm?CFID=191866&CFTOKEN=43159399&ucidparam=20020522151023 (May 30, 2002).

Greenpeace Environmental Trust to commission a report from Alexander Huw Arnall at Imperial College.[16] Published in 2003, *Future technologies, today's choices: Nanotechnology, artificial intelligence and robotics; A technical, political and institutional map of emerging technologies* acknowledged nanotechnology's potential to *improve* the environment while warning against deleterious outcomes and the lack of public input into determining research directions.[17] Since then, Greenpeace UK has collaborated with the *Guardian*, the Interdisciplinary Research Collaboration (IRC) in Nanotechnology of the University of Cambridge, and Policy, Ethics and Life Sciences at University of Newcastle to sponsor a community jury on nanotechnology called NanoJury UK in which a group of West Yorkshire residents quizzed experts and engaged in discussions about nanotechnology's impacts. Ultimately, the jury recommended greater access to information about nanotechnology, greater public input into decision making, and market incentives for developing nanotechnologies with proven health and environmental benefits.[18] Greenpeace UK chief scientist Doug Parr has supported a moratorium "until the hazards are characterized and understood."[19] ETC's writings and the 2003 Greenpeace report generated much publicity that had nanotechnology enthusiasts scrambling to respond.

The most provocative enthusiast to rise to nanotech's defense was F. Mark Modzelewski, founder and former executive director of the NanoBusiness Alliance, the first industry trade association created "to advance the emerging business of nanotechnology and microsystems."[20] In a December 2002 interview, Modzelewski went on the offensive against civil society critics of nanotechnology, characterizing them as "Luddites" and "enemies."[21] He responded to the Greenpeace UK report with equally withering criticism, attacking it as "industrial terrorism" and accusing Greenpeace simultaneously of using nanotechnology to "raise funds and pretend they care about something," and of achieving their ultimate Luddite objective of "creating a choke point on the development of industry

[16]Brown, D. *Greenpeace Wades into Nano Debate with Report That Calls for Caution.* 2003. Small Times. Available at http://www.smalltimes.com/articles/article_display.cfm?ARTICLE_ID=268886& p=109 (July 24, 2003).

[17]Arnall, A. H. *Future Technologies, Today's Choices: Nanotechnology, Artificial Intelligence and Robotics; a Technical, Political and Institutional Map of Emerging Technologies.* London, England: Greenpeace Environmental Trust, 2003; *Without a Reality Check, Claims of Nanotech's Benefits Are a Con.* 2003. Small Times. Available at http://www.smalltimes.com/document_display.cfm?document_id=6706 (September 26, 2003).

[18]Parr, D. *Nanojury Uk–Reflections and Implications of Recommendations.* 2005. Greenpeace UK. Available at http://www.greenpeace.org.uk/MultimediaFiles/Live/FullReport/7332.pdf (November 2005).

[19]Parr, D. *Without a Reality Check, Claims of Nanotech's Benefits Are a Con.* 2003. Small Times. Available at http://www.smalltimes.com/document_display.cfm?document_id=6706 (September 26, 2003).

[20]NanoBusiness Alliance. *About the Alliance.* 2002. Available at http://www.nanobusiness.org/ (September 2006).

[21]Wolfe, J. *Thinking Small: Mark Modzelewski.* 2002. Forbes/Wolfe Nanotech Report. Available at http://www.forbesinc.com/newsletters/nanotech/public/samples/nano_mark_mdec2002.pdf (December 2002).

and technology."[22] Six months later, he touched off an online firestorm by referring to molecular manufacturing proponents as "bloggers, Drexlerians, pseudopundits, panderers and other denizens of their mom's basements."[23] Without another, more moderate voice weighing in, Modzelewski's remarks were the *de facto* industry perspective. The polarization of opinion over nanotechnology's impacts on society had begun to take shape, with the rhetoric on both sides throwing off more heat than light. Many watched with increasing dismay as nanotechnology began going down the same road of civil division that GMOs had, with "industry" vociferously defending the technology and deriding the concerns of its critics, and the civil society organizations digging in for a long and protracted battle. To avoid the GMO backlash scenario, a different model of exploring nanotechnology's impacts was needed.

GENESIS OF ICON

As the only academic research center at the time with a focus on nanotechnology's impacts on human health and the environment, CBEN was poised to play a major role in shaping the conversation on risk. By early 2003, after 18 months of operation, Center leaders had met with more than 30 companies and given in excess of two dozen public talks about nanotechnology environment, health, and safety (EHS). The message was always the same: nanotechnology has enormous potential to improve our lives, but the risks of engineered nanoparticles must be explored in conjunction with the technology's development. For many of these audiences, CBEN was the first academic group to broach the subject of potential risk, and its aggressive public outreach and balanced message secured the Center's position as a thought leader in this growing area of interest. This led to an invitation to testify before the U.S. House of Representatives Committee on Science hearing on the 21st Century Nanotechnology Research and Development Act of 2003. In her testimony, Vicki L. Colvin, CBEN director and Rice University professor of chemistry, argued for increased funding of risk-related research in the pending bill as a way of removing a potential barrier to nanotechnology's success.[24] By being proactive in anticipating and creating processes and policies to deal with potential nano negatives, Colvin argued that the industry would be healthier and could avoid the kind of backlash that met the societal introduction of GMOs. This would, however, require funding for more research to answer the many outstanding questions about nanomaterial impacts on health and the environment, and willingness by

---

[22]Brown, D. *Greenpeace Wades into Nano Debate with Report That Calls for Caution*. 2003. Small Times. Available at http://www.smalltimes.com/articles/article_display.cfm?ARTICLE_ID=268886&p=109 (July 24, 2003).
[23]Modzelewski, F. M. *Industry Can Help Groundbreaking Nanotech Bill Fulfill Its Promise*. 2004. Small Times. Available at http://www.smalltimes.com/articles/article_display.cfm?ARTICLE_ID=269225& p=109 (January 26, 2004).
[24]Colvin, V. L. *Testimony before the Hearing on the Societal Implications of Nanotechnology*. U.S. House of Representatives Committee on Science, April 9, 2003, Washington, DC.

government and industry to focus attention on these questions. The technical arguments behind these policy recommendations were outlined later that year in an article in *Nature Biotech.*[25]

Early implications research, such as that reported at CBEN's first NanoDays symposium,[26] increasing level of social commentary on nanomaterials' health and environmental implications, shifting tone of media coverage from optimistic to cautionary,[27] and lack of clear guidance from regulators created a climate of uncertainty for industrial investments in nanotechnology. Many in the industry were hungry for context and uncertain of where to go to get it. This need, combined with a stated desire to address both real and perceived risks, motivated some companies to redefine how they interacted with a variety of stakeholders. There were, at the time, few venues for companies, government regulators and researchers, academics, and nongovernmental organizations to explore these issues in a safe space. One notable exception was the Woodrow Wilson International Center for Scholars, which began a series of invited talks and dialogues on environmental and regulatory issues in nanotechnology in late 2002 under its Foresight and Governance Project.[28] These events brought together diverse groups of people from government, NGOs, industry, and, to a lesser extent, academia to discuss issues surrounding nanotechnology and served to highlight many of the outstanding questions facing society. For many attendees, the Wilson Center events underscored the need to move beyond dialogue and begin taking action to address the issues, the question being, how? These pressures seemed to call for novel models of engagement capable of addressing risk issues proactively and with multiple stakeholder input.

The genesis of ICON can be traced back to the summer of 2003 when scientists and managers from the E. I. du Pont de Nemours and Company (DuPont) approached CBEN to discuss a collaborative research project on toxicity of nanoscale particles that had begun in February of that year. A team of researchers and managers visited Rice University in August 2003 to review the toxicology project, tour the research facilities, and get an update on recent policy activity in the United States. During these discussions, the concept was first broached of broadening DuPont's engagement beyond a technical research project undertaken within a bilateral academic–industrial center affiliates program. Motivating this conversation was a growing awareness that nanotechnology had the potential to develop into a

[25]Colvin, V. L, "The Potential Environmental Impact of Engineered Nanomaterials." *Nature Biotech* 21.10 (2003): 1170.
[26]Center for Biological and Environmental Nanotechnology. *Nanodays Symposium.* 2002. Rice University. Available at http://cohesion.rice.edu/centersandinst/cben/events.cfm?docid=5650 (October 2002).
[27]Berger, E. "Nanotech Encounrters New Barrier: Environmental Risks Rise as Costs Decline." *Houston chronicle* December 12 2001, Metfront ed., sec. A: 31; Moore, Julia. "The Future Dances on a Pin's Head. Nanotechnology: Will It Be a Boon—or Kill Us All?" Commentary. *Los Angeles Times* November 26, 2002, California Metro ed., sec. B: 13.
[28]Colvin, V. L. "Avoiding the 'Wow' to 'Yuck' Trajectory." *Nanotechnology and the environment.* Washington, DC: Foresight and Governance Project of the Woodrow Wilson International Center for Scholars, 2002.

socially contentious issue of interest to multiple stakeholders and that the solutions to the most critical issues might best be developed with a diverse group of people working together toward a common goal. The Nanotechnology Policy Center (NPC), as the concept was called then, was proposed as a multi-party coalition that would inject sound scientific data into the public debate about nanotechnology's risks and benefits. Key roles of the NPC were to quickly publicize new risk-related data, respond to concerns of civil society organizations with factual information from the technical literature, and serve as a resource for the public on nano-EHS. From the beginning, the group was intended to maintain a neutral stance on nano-technology (engaging in neither boosterism nor fearmongering), have an international composition, and remain open to discussion with all stakeholders. This last point was especially important to several prospective industrial sponsors, who sought a way to share information and perspectives with critical voices in society outside the glare of the media spotlight.

By October, the NPC had not yet been formed, but our expanded thinking had already made its mark on CBEN when L'Oréal became the first company without a clearly defined interest in CBEN's technical research program to join the Center's affiliates program. L'Oréal was most interested in getting updates on U.S. policy activity and understanding potential public concerns about nanotechnology. By the time DuPont returned to Rice in January 2004 for another research update, the concept for NPC had evolved to include a list of proposed activities that spanned technical research in nano–cell interactions, policy projects such as development of nanomaterial standards and terminology, and social studies of risk perception and communication. These were all areas in which we perceived existing gaps in knowledge and little coordinated international action to close them. A timeline was established for start-up and proposals for potential first-round projects were solicited from a number of social scientists and nanotechnologists. The name of the group was changed to the International Council on Nanotechnology (ICON).

Focused recruiting of participants from the four stakeholder groups began in earnest, and the concept for a rough governance structure was hammered out in the first half of 2004. ICON was conceived as a program within CBEN, a sort of industrial affiliates program in that companies pay membership fees to join, but distinct from a traditional, bilateral program in several significant ways. First, balance in membership is sought from four distinct stakeholder groups: academics, government officials, industry, and non-governmental organizations. Second, the companies sign membership agreements that effectively cede control over the use of their money to an executive committee whose membership is balanced among the four groups. Third, governance of the activities is managed by a steering committee according to a charter document. The initial governance model for the organization had an advisory council composed of representatives from the four stakeholder groups, ideally balanced by type and geographic location, headed by a smaller executive committee that made funding recommendations to the director. In addition to making final budgetary decisions, the director was charged with overseeing the staff and any grantees awarded funding to implement specific projects. The intent was to distribute the authority for decision making equally to all stakeholder groups and, more significantly, to offset the potentially dominating influence of industry, whose participants at that time outnumbered any other single stakeholder group and whose contributions were the primary source of support. This governance structure is still in place today except that the responsibilities originally assigned to the director now reside with an executive director (Fig. 22.1). The council is run by its volunteers who are supported by a small staff. The volunteers populate working groups that advance the activities of the council and also constitute the steering committee, which is chaired by the director and charged with revising the strategic plan and making funding recommendations. Funding decisions rest with the smaller executive committee, which is chaired by the executive director and has one member each from industry, academia, government, and non-governmental organizations. In this way, no one stakeholder group can dominate the discussions and direction of the Council.

The public launch of the group was held in October 2004 at a leadership meeting at Rice University facilitated by the Meridian Institute. By that time, four companies had joined at the Founding Sponsor level (DuPont, Procter & Gamble, Swiss Reinsurance Company, and Intel) and L'Oréal and Mitsubishi Corporation came on board during the next few months. In addition to representatives from these companies and Rice personnel, meeting participants included two academics, two U.S. government officials, representatives from nonaligned companies, a local Houston environmental lawyer, and five NGO representatives (two of these by telephone). Not everyone who participated in the meeting had joined the advisory board at that time. Some of the government officials had yet to receive formal approval to join, and several of the NGO participants were openly skeptical about the proposed

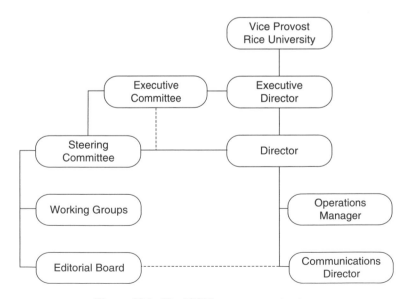

Figure 22.1. The ICON governance structure.

model of engagement. (One NGO representative stated that he was at the meeting to "kick the tires").[29] However, a *Washington Post* article overstated the lack of engagement by reporting that none of the invited NGO groups had agreed to join the meeting.[30] One NGO participant said wryly that he was "surprised to read in the paper that I'm not here today."[31] Subsequent reporting corrected that error, noting that the NGO groups had participated as guests.[32]

The launch meeting helped to refine some of the rough concepts for ICON's priorities and structure. Discussion focused on the purpose and scope of ICON, particularly with respect to how broad the council's mission should be. Some favored keeping it focused on EHS issues, which were perceived as more manageable in the near-term, while others felt that a focus on EHS to the exclusion of larger social and economic concerns, such as ownership and control was too narrow and threatened to take nanotechnology down the path trodden by GMOs. Ultimately, the group decided to focus on EHS issues in the early stages as the Council became established. There was a general sense that there was increasing activity within the nanorisk space, but very little connection among the various threads. It was on coordinating existing efforts and filling in the gaps that many thought ICON should focus. Most of the rest of the meeting was taken up with the development of specific action items for the participants to accomplish in the near future, several of which developed into the present-day working groups. The ICON launch can best be described as "rocky," which may not be too surprising given that it is, as one participant put it, "a kind of social experiment,"[30] but, after the meeting, Mooney captured the mood in his comment that "all of the parties have a significant amount of good will and want to talk. That's not a bad starting point."[33] Moreover, the launch served to focus the attention of a diverse group of highly motivated individuals, most of whom are still active in ICON today.

ICON really began to coalesce after the development of a detailed strategic plan in August 2005. The plan is updated on a semiannual basis and has four primary objectives, which coincide with the Council's four working groups.

ICON Objectives:

1. Create a multistakeholder, international and neutral forum for exploring health and environmental risk issues (Governance Working Group).
2. Establish and maintain a globally recognized, credible knowledge base (contextual summaries of research publications, access to experts, roadmap for future research needs and capabilities) of nanotechnology EHS peer reviewed information (Knowledge Base Working Group).

[29]Walsh, S. "Personal Communcation at Icon Launch Meeting." Houston, TX, 2004.
[30]Weiss, R. "Nanotech Group's Invitations Declined." *Washington Post* 2004: A04.
[31]Mooney, P. R. "Personal Communcation at Icon Launch Meeting." Houston, TX, 2004.
[32]Powell, K. "Green Groups Baulk at Joining Nanotechnology Talks." *Nature* 4 November 2004: 5; Service, Robert F. "Nanotech Forum Aims to Head Off Replay of Past Blunders." *Science* 5 November 2004: 955.
[33]Service, R. F. "Nanotech Forum Aims to Head Off Replay of Past Blunders." *Science* 5 November 2004: 955.

3. Facilitate and develop activities producing or using information in support of nanotechnology health and environmental risk assessment and management (Best Practices Working Group).

4. Broadly communicate the results of ICON's efforts and translate complex scientific concepts into material that can be understood by a broad cross-section of stakeholders (Communications Working Group).

Currently, ICON has working groups on the topics of: Environmental Health and Safety Knowledge Base; Best Practices; Communications; and Planning. The Knowledge Base activities include: producing, maintaining, and enhancing the database of citations to peer-reviewed publications in nanotechnology EHS, producing a series of backgrounders on key issues in nano EHS; and developing a research needs assessment for critical gaps in nano EHS knowledge. Best Practices activities include: producing a survey of current practices for nanomaterial handling in the workplace; and development of consensus best practices. The Communications group produces documents that describe ICON and its activities, maintains an information portal for nano EHS, and develops a media alert service for journalists writing about nano EHS.

EHS DATABASE

The EHS database published by ICON marks the first effort to integrate the vast and diverse scientific literature on the impacts of nanoparticles. Many nanoparticles exhibit unique chemical, electrical, optical, and physical properties by virtue of their size, shape, or surface characteristics. The great diversity of nanoparticle types that have already been created has made it difficult for scientists to make general statements about the potential safety hazards that nanoparticles might pose to living organisms. This problem is exacerbated by the limited scientific data on the topic. While there is a significant body of research on the impacts of incidental nanoparticles—a class of particles that are the unintentional byproduct of another process, such as combustion, and are often referred to as ultrafine particles—the specific effects of only a few engineered nanomaterials have been studied. This shortfall in scientific knowledge is beginning to be addressed through targeted research funding programs and other initiatives. However, nanotechnology's breadth poses unique challenges in this regard, and knowing which questions new research monies should be targeted toward requires an understanding of what is already known.

Currently, nano-EHS papers are scattered throughout the literatures of biomedical application developers, toxicologists, environmental engineers, and nanomaterials scientists. One is just as likely to find these papers in *Nano Letters* as in *Toxicology Letters*. By gathering them in one place, researchers, government funding agencies, and public advocates can better see the big picture and identify potential gaps in knowledge. In addition, the free service makes the information available to those who do not have access to expensive, specialized citation indexes.

This need to collect currently available knowledge on EHS issues of nanoscale materials was recognized by the Environmental Safety and Health working group of the National Nanotechnology Initiative Chemical Industry Consultative Board for Advancing Nanotechnology (CBAN), which was established by the Vision 2020 Chemical Industry Technology Partnership. The CBAN working group, which includes EHS specialists at several chemical companies, myself, and contacts from multiple government agencies, commissioned Dr. Tim Borges and Ms. LeeAnn Wilson at Oak Ridge National Laboratory to begin compiling a database through a Chemicals Plus project of the Industrial Technologies Program of the Department of Energy's Office of Energy Efficiency and Renewable Energy. In April 2005, a database of 1347 records dating from 1967 through early 2005 was handed over to ICON to web-publish and maintain.

Work on publishing the database came in three phases. In phase one, the raw database was converted into a web-accessible format and capability was added to search the records by author, year, keywords, or assigned category. The first assigned categories to be added were nanoparticle production method, nanoparticle type, and paper type (i.e., peer-reviewed study, review paper, commentary, or public policy report). With these enhancements in place, the database was launched in August 2005 with more than 1200 records.[34] In phase two, the database was updated to include records through the present and with the addition of six new assigned categories in the search function: (1) exposure pathway (oral–ingestion, dermal, etc.); (2) method of study (in vitro, in vivo, environmental, etc.); (3) exposure or hazard target (mammalian, aquatic, etc.); (4) risk exposure group (industrial or research worker, etc.); (5) publication type (peer reviewed journal article, etc.); and (6) target audience (technical, public, policy, etc.) Usage statistics for the month of July 2006, which are representative of totals for the previous 6 months, reveal that the database has been accessed nearly 200 separate times with 600 individual page loads, indicating multiple searches per visit. Between 20–40% of database users are from outside the United States.

A recent enhancement to the ICON database is the function that allows one to search two other external databases—the National Institute for Occupational Safety and Health (NIOSH) Nanoparticle Information Library (NIL)[35] and the Project on Emerging Nanotechnologies' (PEN) Inventory of Nanotechnology Environment, Health and Safety Research[36]—with the same search terms. In this way, if one wants to get an idea of, for example, the state of knowledge of impacts of carbon nanostructures, one can determine what has already been published on toxicity and environmental impact from ICON, what users or producers

[34]International Council on Nanotechnology. *Environmental Health and Safety Database*. 2005. Rice University Center for Biological and Environmental Nanotechnology. Available at http://icon.rice.edu/research.cfm. August 15 2005; Leslie, Mitch. "Gauging Nanotech Risks." *Science* 309 (2005): 1467.

[35]National Institute for Occupational Safety & Health. *Nanoparticle Information Library*. 2005. Available at http://www2a.cdc.gov/niosh-nil/ (August 25, 2006).

[36]Project on Emerging Nanotechnologies. *Inventory of Nanotechnology Environment, Health and Safety Research*. 2005. Woodrow Wilson International Center for Scholars. Available at http://www.nanotechproject.org/18/esh-inventory (August 25, 2006).

report about their properties from NIL, and what work is currently underway to further assess their impacts from PEN. The third phase of the project will create a virtual journal interface to the database and post commentaries on key papers and backgrounders on hot topics.

BEST PRACTICES FOR NANOMATERIAL HANDLING

Worker safety has emerged as one of the most pressing concerns in nano EHS. It makes sense to focus on the people who are producing and using nanomaterials as a first priority, as workers are likely to be exposed to nanoparticles in higher quantities over longer periods of time than the average consumer. If fixed or bound nanoparticles represent a lower exposure potential for consumers, the same may not be true for the worker who handles them during their incorporation into their matrices. Moreover, workers in a research setting may be dealing with exotic or unique nanoparticles whose hazard profiles have not been firmly established. NIOSH has been out in front of this occupational health issue and recently published an updated version of its *"Approaches to safe nanotechnology: An information exchange with NIOSH,"*[37] which seeks input from affected stakeholders so that its guidance can be refined. Yet little is known about how workers are handling nanomaterials, what hazard communication is occurring, and where employers are going to get information about safe practices. To answer these questions, ICON put out a call for proposals to survey the industry regarding their current practices and awarded the project to a team at the University of California at Santa Barbara (UCSB).[38] Stage One of the project involves an analysis of ongoing and planned activities by others to summarize current nanotechnology industrial safety practices. Stage Two will involve conducting interviews of nanotechnology companies in Europe and Asia via telephone and the web regarding their current industrial safety practices.

RESEARCH NEEDS ASSESSMENT

In 2006, ICON received a grant from the NSF to develop an International Nanomaterial Environmental Health and Safety (nanoEHS) Research Needs Assessment document in collaboration with a wide range of stakeholders. The goal of this assessment is to establish a science-based assessment of nanoEHS hazards of different classes of nanomaterials in their native form and in expected formulations with

[37]National Institute for Occupational Safety and Health. *Approaches to Safe Nanotechnology: An Information Exchange with NIOSH.* 2006. Department of Health and Human Services. Available at http://www.cdc.gov/niosh/topics/nanotech/safenano/pdfs/approaches_to_safe_nanotechnology.pdf (July 31, 2006).

[38]International Council on Nanotechnology. *Current Practices for Working with Nanomaterials.* 2006. Rice University Center for Biological and Environmental Nanotechnology. Available at http://icon.rice.edu/projects.cfm?doc_id=4388 (March 15, 2006).

research needs prioritized to validate the classes of nanomaterials and the principles that relate properties to predicted hazards. There have been several efforts aimed at summarizing the current state of knowledge about the interactions of nanomaterials with living systems.[39] However, the ICON effort is the first international, multistakeholder effort engaging governments, universities, industry, and non-governmental organizations in prioritizing specific research needed for different classes of nanomaterials. In contrast, the Royal Academy, NIOSH, HSE, SCENIHR, and SRC & Chemical Vision 2020 studies were not international, and, while issues of nanoparticles risk were identified and nanoparticle interactions with living systems were highlighted, there was little prioritization of research needs. Of the international studies, the Swiss Re and the OECD reports provided high-level summaries of the critical issues, while the *Particle and Fibre Toxicology* study provided an in-depth assessment of the issues and research needs, but there was no participation from Asia and no prioritization of research needs. Furthermore, many of these studies were completed by 2004 and significant advances have been made in our understanding of the key issues during the past two years. The ICON study is unique in providing an international, multistakeholder forum for the prioritization of nano-EHS research to establish and validate interaction principles for different classes of nanomaterials.

Two meetings are proposed to accomplish these objectives. In the first meeting, a team of approximately 30 stakeholders will identify classes of nanomaterials based on the physical properties that enable prediction of their behavior and fate throughout their life cycle, and identify points in the lifecycle at which nano-EHS research is needed in order to validate the behavior of nanomaterials and their potential hazards.

[39]Hett, A. *Nanotechnology: Small Matter, Many Unknowns*. 2004. Swiss Reinsurance Company. Available at http://www.swissre.com. (November 2004); HM Government. *Characterising the Risks Posed by Engineered Nanoparticles: A First Uk Government Research Report*. 2005. Department for Environment, Food and Rural Affairs. Available at http://www.defra.gov.uk/environment/nanotech/research/pdf/nanoparticles-riskreport.pdf. (December 30, 2005); National Institute for Occupational Safety and Health. *Strategic Plan for Niosh Nanotechnology Research: Filling the Knowledge Gaps*. 2005. Department of Health and Human Services. Available at http://www.cdc.gov/niosh/topics/nanotech/pdfs/NIOSH_Nanotech_Strategic_Plan.pdf. (September 28, 2005); Organisation for Economic Cooperation and Development. *Report of the Oecd Workshop on the Safety of Manufactured Nanomaterials: Building Co-Operation, Co-Ordination and Communication*. 2006. OECD Environment, Health and Safety Publications Series on the Safety of Manufactured Nanomaterials. Available at http://appli1.oecd.org/olis/2006doc.nsf/linkto/env-jm-mono(2006)19. (April 28, 2006); Scientific Committee on Emerging and Newly Identified Health Risks. *Opinion on the Appropriateness of Existing Methodologies to Assess the Potential Risks Associated with Engineered and Adventitious Products of Nanotechnologies*. 2005. European Commission Health & Consumer Protection Directorate-General. Available at http://ec.europa.eu/health/ph_risk/committees/04_scenihr/docs/scenihr_o_003.pdf. (September 2005); The Royal Society and The Royal Academy of Engineering. *Nanoscience and Nanotechnologies: Opportunities and Uncertainties*. 2004. Available at http://www.nanotec.org.uk/finalReport.htm. (July 29, 2004). Vision 2020 Chemical Industry Technology Partnership and Semiconductor Research Corporation. *Joint NNI-CHI CBAN and SRC CWG5 Nanotechnology Research Needs Recommendations*. 2006. Available at http://www.chemicalvision2020.org/pdfs/chem-semi%20ESH%20recommendations.pdf. (January 1, 2006); Oberdörster G. et al. "Principles for Characterizing the Potential Human Health Effects from Exposure to Nanomaterials: Elements of a Screening Strategy." *Particle and Fibre Toxicology* 2.8 (2005).

This will require a review of the known chemical and physical properties that contribute to nano-EHS hazards and risk, and the development of classifications that are correlated to the International Standards Organization/American Society for testing and Materials International terminology, validating these against existing and research nanomaterials. The team will develop a matrix of material class, interaction principles, and risk factors for different applications. They will also review when knowledge will be needed to validate the classes, and hazards and risk of the classes based on an informed understanding of the timeframe of potential use in different applications.

In the second meeting, a broader range of researchers will review the results of the first meeting, proposed classes of nanomaterials, the matrix of class versus application, the proposed hierarchy of risk assessment, and the research timeframe. This meeting will validate or modify the proposed classes of nanomaterials, and will identify the critical research needs for different nanomaterial classes for applications and lifecycle assessments. The researchers will also identify the toxicology, toxicokinetic, ecotoxicological research and metrology, and monitor capabilities needed to establish interaction principles for different classes of nanomaterials. These research and capability needs will be mapped against the proposed timing of research needs for different applications.

CONCLUSIONS

Ultimately, ICON seeks to identify and close gaps in knowledge about nanotechnology's risk factors in a science-based, inclusive manner. By engaging broadly with multiple groups from diverse perspectives, ICON's activities may help to reduce the polarization this emerging technology has already inspired and enable society to benefit from its considerable promise. If this "social experiment" proves to be successful, ICON may become a model for the introduction of emerging technologies yet to come.

BIBLIOGRAPHY

Arnall, Alexander Huw. *Future Technologies, Today's Choices: Nanotechnology, Artificial Intelligence and Robotics; a Technical, Political and Institutional Map of Emerging Technologies.* London, England: Greenpeace Environmental Trust, 2003.

Berger, Eric. "Nanotech Encounrters New Barrier: Environmental Risks Rise as Costs Decline." *Houston Chronicle* December 12 2001,Metfront ed., sec. A: 31.

BfR (Federal Institute for Risk Assessment). *Exercise Caution When Using "Nano-Sealing Sprays" Containing a Propellant.* 2006. Available at http://www.bfr.bund.de/cms5w/sixcms/detail.php/7699 (March 31, 2006).

Brown, Douglas. *Greenpeace Wades into Nano Debate with Report That Calls for Caution.* 2003. Small Times. Available at http://www.smalltimes.com/articles/article_display.cfm?ARTICLE_ID=268886&p=109 (July 24, 2003).

Center for Biological and Environmental Nanotechnology. *Nanodays Symposium.* 2002. Rice University. Available at http://cohesion.rice.edu/centersandinst/cben/events. cfm?doc_id=5650 (October 2002).

Colvin, Vicki L. "Avoiding the 'Wow' to 'Yuck' Trajectory." *Nanotechnology and the environment.* Washington, DC: Foresight and Governance Project of the Woodrow Wilson International Center for Scholars, 2002; "The Potential Environmental Impact of Engineered Nanomaterials." *Nature Biotech* 21.10 (2003): 1170; *Testimony before the Hearing on the Societal Implications of Nanotechnology.* U.S. House of Representatives Committee on Science, April 9, 2003, Washington, DC.

Crichton, Michael. *Prey.* New York: HarperCollins Publishers, Inc., 2002.

Drexler, K. Eric. *Engines of Creation.* New York: Anchor, 1986.

ETC group. 2006. Available at http://etcgroup.org/search2.asp?srch=nano (August 30, 2006).

ETC Group. *Nanotech Product Recall Underscores Need for Nanotech Moratorium: Is the Magic Gone?* 2006. Available at http://etcgroup.org/documents/NRnanoRecallfinal.pdf. (April 7, 2006); *No Small Matter! Nanotech Particles Penetrate Living Cells and Accumulate in Animal Organs.* 2002. Available at http://etcgroup.org/documents/ Comm_NanoMat_July02.pdf (June 30, 2002).

Foresight and Governance Project. *Dialogue Series on Nanotechnology and Federal Regulations.* 2003–2004. Woodrow Wilson International Center for Scholars/Meridian Institute. Available at http://www.merid.org/showproject.php?ProjectID=9233.0 (August 1, 2004).

Greenpeace UK. *Technology: Taking the Good without the Bad?* 2002. Available at http:// www.greenpeace.org.uk/contentlookup.cfm?CFID=191866&CFTOKEN=43159399&- ucidparam=20020522151023 (May 30, 2002).

Halford, Bethany. "Nano Database Goes Online." *Chemical and Engineering News* 83.42 (2005): 33.

Hett, Annabelle. *Nanotechnology: Small Matter, Many Unknowns.* 2004. Swiss Reinsurance Company. Available at http://www.swissre.com (November 2004).

HM Government. *Characterising the Risks Posed by Engineered Nanoparticles: A First UK Government Research Report.* 2005. Department for Environment, Food and Rural Affairs. Available at http://www.defra.gov.uk/environment/nanotech/research/pdf/ nanoparticles-riskreport.pdf (December 30, 2005).

International Council on Nanotechnology. *Current Practices for Working with Nanomaterials.* 2006. Rice University Center for Biological and Environmental Nanotechnology. Available at http://icon.rice.edu/projects.cfm?doc_id=4388. (March 15, 2006.) *Environmental Health and Safety Database.* 2005. Rice University Center for Biological and Environmental Nanotechnology. Available at http://icon.rice.edu/research.cfm (August 15, 2005).

Joy, Bill. "Why the Future Doesn't Need Us." *Wired Magazine* April 2000.

Lane, Neal, and Thomas Kalil. "The National Nanotechnology Initiative: Present at the Creation." *Issues in Science and Technology* Summer (2005).

Leslie, Mitch. "Gauging Nanotech Risks." *Science* 309 (2005): 1467.

Modzelewski, F. Mark. *Industry Can Help Groundbreaking Nanotech Bill Fulfill Its Promise.* 2004. Small Times. Available at http://www.smalltimes.com/articles/article_display. cfm?ARTICLE_ID=269225&p=109 (January 26, 2004).

Mooney, Pat Roy. *Development Dialogue. The ETC Century: Erosion, Technological Transformation and Corporate Concentration in the 21st Century*: Dag Hammerskjöld Foundation, 1999; "Origins of ETC Interest in Nanotechnology." Personal communication, 2006; "Personal Communcation at Icon Launch Meeting." Houston, TX, 2004.

Moore, Julia. "The Future Dances on a Pin's Head. Nanotechnology: Will It Be a Boon—or Kill Us All?" Commentary. *Los Angeles Times* November 26, 2002, California Metro ed., sec. B: 13.

NanoBusiness Alliance. *About the Alliance.* 2002. Available at http://www.nanobusiness. org/ (September 2006).

National Institute for Occupational Safety and Health. *Approaches to Safe Nanotechnology: An Information Exchange with NIOSH.* 2006. Department of Health and Human Services. Available at http://www.cdc.gov/niosh/topics/nanotech/safenano/pdfs/approaches_to_safen_anotechnology.pdf. (July 31, 2006); Nanoparticle Information Library. 2005. Available at http://www2a.cdc.gov/niosh-nil/. (August 25, 2006); *Strategic Plan for Niosh Nanotechnology Research: Filling the Knowledge Gaps.* 2005. Department of Health and Human Services. Available at http://www.cdc.gov/niosh/topics/nanotech/pdfs/NIOSH_Nanotech_Strategic_Plan.pdf (September 28, 2005).

National Research Council. Small Wonders, Endless Frontiers: A Review of the National Nanotechnology Initiative. Washington, DC: The National Academy of Sciences, 2002.

National Science Foundation. "FY 2001 Nanoscale Science and Engineering Solicitation (Nsf 00-119)." 2000.

Oberdörster G, et al. "Principles for Characterizing the Potential Human Health Effects from Exposure to Nanomaterials: Elements of a Screening Strategy." *Particle and Fibre Toxicology* 2.8 (2005).

Organisation for Economic Cooperation and Development. *Report of the Oecd Workshop on the Safety of Manufactured Nanomaterials: Building Co-Operation, Co-Ordination and Communication.* 2006. OECD Environment, Health and Safety Publications Series on the Safety of Manufactured Nanomaterials. Available at http://appli1.oecd.org/olis/2006doc.nsf/linkto/env-jm-mono(2006)19 (April 28, 2006).

Parr, Douglas. Nanojury UK—Reflections and Implications of Recommendations. 2005. Greenpeace UK. Available at http://www.greenpeace.org.uk/MultimediaFiles/Live/FullReport/7332.pdf. (November 2005.); Without a Reality Check, Claims of Nanotech's Benefits Are a Con. 2003. Small Times. Available at http://www.smalltimes.com/document_display.cfm?document_id = 6706 (September 26, 2003).

Powell, Kendall. "Green Groups Baulk at Joining Nanotechnology Talks." *Nature* 4 November 2004: 5.

Project on Emerging Nanotechnologies. *Inventory of Nanotechnology Environment, Health and Safety Research.* 2005. Woodrow Wilson International Center for Scholars. Available at http://www.nanotechproject.org/18/esh-inventory (August 25, 2006).

Scientific Committee on Emerging and Newly Identified Health Risks. *Opinion on the Appropriateness of Existing Methodologies to Assess the Potential Risks Associated with Engineered and Adventitious Products of Nanotechnologies.* 2005. European Commission Health & Consumer Protection Directorate-General. Available at http://ec.europa.eu/health/ph_risk/committees/04_scenihr/docs/scenihr_o_003.pdf (September 2005).

Service, Robert F. "Nanotech Forum Aims to Head Off Replay of Past Blunders." *Science* 5 November 2004: 955.

The Royal Society and The Royal Academy of Engineering. *Nanoscience and Nanotechnologies: Opportunities and Uncertainties.* 2004. Available at http://www.nanotec.org.uk/finalReport.htm (July 29, 2004).

Vision 2020 Chemical Industry Technology Partnership and Semiconductor Research Corporation. *Joint NNI-CHI CBAN and SRC CWG5 Nanotechnology Research Needs Recommendations.* 2006. Available at http://www.chemicalvision2020.org/pdfs/chem-semi%20ESH%20recommendations.pdf (January 1, 2006).

von Bubnoff, Andreas *Study Shows No Nano in Magic Nano, the German Product Recalled for Causing Breathing Problems.* 2006. Small Times. Available at http://www.smalltimes.com/document_display.cfm?documentid=11586. May 26, 2006.

Walsh, Scott. "Personal Communcation at Icon Launch Meeting." Houston, TX, 2004.

Weiss, Rick. "Nanotech Group's Invitations Declined." *Washington Post* 2004: A04.

Wolfe, Josh. *Thinking Small: Mark Modzelewski.* 2002. Forbes/Wolfe Nanotech Report. Available at http://www.forbesinc.com/newsletters/nanotech/public/samples/nano_markm_dec2002.pdf (December 2002).

CHAPTER 23

From the Lab to the Marketplace: Managing Nanotechnology Responsibly

VIVIAN WEIL

INTRODUCTION

A vision of managing nanotechnology responsibly evidently animated the National Nanotechnology Initiative (NNI) from its beginnings in the year 2000. Proposals to the National Science Foundation (NSF) responding to that initiative were to include attention to ethics and societal implications. In September 2000, NSF itself provided a boost to this dimension of the NNI by sponsoring a large conference to consider ethics and societal implications of nanoscience and nanotechnology. The volume produced from that conference contained the earliest essays on these topics by philosophers, social scientists, scientists, and engineers.

The following year, an assessment of the coverage of ethics and societal implications in early funded nano-projects received wide circulation in a volume from the National Research Council titled *Small Wonders, Endless Frontiers: A Review of the National Nanotechnology Initiative*.[1] A second NSF-sponsored conference on ethics and societal implications took place in December 2003 on the very day that President Bush signed the nanotechnology legislation approved by the Congress. That conference produced reports and analysis that advanced discussion and were later published in print and online.[2] Encouraging these efforts was a sense that lessons learned from mistakes in managing earlier radically innovative

---

[1]Committee for the Review of the National Nanotechnology Initiative, Division on Engineering and Physical Sciences, National Research Council. Washington, DC: National Academy Press, 2002.

[2]Nanotechnology: Societal Implications—Maximizing Benefits for Humanity. Report of the National Nanotechnology Initiative Workshop December 3–5, 2003, Arlington, VA. Sponsored by National Science Foundation.

Nanoscale: Issues and Perspectives for the Nano Century. Edited by Nigel M. de S. Cameron and M. Ellen Mitchell

technologies, such as biotechnology in agriculture, could be applied to manage nanotechnology responsibly.

One important conclusion drawn from negative reactions to these earlier technologies was that it is a mistake to protect dramatically new technologies from public scrutiny and engagement. Combined with simultaneously extolling the benefits of these innovative technologies, the protective approach seems to increase the likelihood of public distrust, disenchantment, and rejection when problems arise.

Insights that illuminate the context for responsible management of emerging innovative technologies come from two fields of scholarly study that have flourished in recent decades, Science and Technology Studies (STS), and Engineering and Scientific Research Ethics. STS studies shed light on complex interconnections between and among science, technology and society, showing that technologies are embedded in society and are shaped by society. STS scholars have analyzed in detail how society shapes technological developments and how technologies shape society.[3] A new technology option, the cell phone, for example, requires a social infrastructure. Societal arrangements that provide that infrastructure influence ongoing technological development.

Engineering and scientific research ethics investigation yields understanding of responsibilities of engineers and scientists in the intricacies of organizations and institutions in which they do their work. It identifies and examines impediments to responsibility in organizational contexts, such as microscopic vision—the precise but very limited field of vision that misses the big picture.

ON-THE-GROUND NANODEVELOPMENTS

Framed by the foregoing observations, an overview of actual developments in nanotechnology will set the stage for considering requirements for responsible management of nanoenterprises. Two striking features of nanotechnology developments in the period since the NNI's launch bring responsibility issues and societal implications to the foreground. One is that technical nanospecialists and business people proceed in the face of great ignorance. Knowledge is lacking about short, as well as long-term, consequences, and obstacles to risk identification and management are widely noted. Understanding of the distinctive properties observed at the nanoscale appears to advance slowly in comparison to the pace of commercialization. A second striking feature is that a "wish list" of nano developments has emerged from the vigorous promotion of nanotechnology during the NNI's first 5 years. That "wish list" has yet to be carefully assessed from an ethical perspective. Dramatic new capabilities anticipated in medicine, for example, are associated with sophisticated diagnostics and therapeutics. Disease prevention and public health, two major areas of intense concern, in developed countries, as well as in developing countries, are near the periphery and seldom mentioned. Because explicit social policy drives formation of nanoenterprises and public funding supplies approximately $1 billion of support each year, thoughtful sorting of priorities is called for.

---

[3]See, for example, Chapter 21.

Start-up companies and nanoendeavors of large, established companies develop rapidly, and promptly release products containing nanomaterials into the marketplace. In fact, more than 200 products containing nanomaterials are already on the market.[4] On December 2, 2005, a company received federal approval to sell catheters coated with a compound of nanoscale silver particles for use in wounds produced by surgery. Before the end of the month, the company began shipping the devices. It is anticipated that this product will provide an alternative to antibiotics to which bacteria are developing resistance. Since ancient times, physicians have had knowledge about the antiseptic powers of silver. But when the product was released, it was not yet clear to scientists how the surprisingly low concentration of silver in these new nanocoatings kills so many bacteria, nor was it clear how the remarkable capacity of these coatings to adhere to glass and plastic arises.[5]

Also moving quickly are universities. Supported by public and private funds, they erect buildings, and establish educational and research programs. They create capabilities and mechanisms for rapid translation of research products to the marketplace. For example, in early 2006, at a cost of almost $60 million in largely private funds, Purdue University opened a new facility devoted to nanoenterprises. The building provides state-of-the-art facilities not only to conduct research, but, as importantly, to develop innovations from research for the marketplace.

The products of this activity have barely been noticed by the public.[6] Nanodevelopments receive little coverage in mainstream news media and the popular press. Attention comes chiefly from business and science reporters in specialized sections and periodicals. Yet, on the World Wide Web, commercial nano enterprises, including companies, commercial research and promotional associations, conferences, and newsletters of various kinds, have rapidly become ubiquitous.

RATIONALE FOR CONCENTRATING ON RESPONSIBLE MANAGEMENT: PUBLIC TRUST

These on-the-ground characteristics of nanotechnology development generate specific concerns about the prospects for public trust, a matter of wide concern

---

[4]Woodrow Wilson International Center for Scholars, Project on Emerging Technologies. Inventory of Nanoparticle-Containing Products. Available at http://www.nanotechproject.org/44.
[5]Feder, B. J. "Old Curative Gets New Life at Tiny Scale," The New York Times, December 20, 2005, p. D5.
[6]Research findings released September 19, 2006, from the first major national poll on nanotechnology in more than 2 years indicated that, while more Americans are now aware of the emerging nanoscience, the majority of the public still has heard little to nothing about it. The poll was commissioned by the Project on Emerging Nanotechnologies at the Woodrow Wilson International Center for Scholars and was conducted by Peter D. Hart Research Associates in August 2006.
The poll also found that the public looks to the federal government and independent parties to oversee nanotechnology research and development. These results, according to experts, necessitate increased education and stronger oversight as a means to increase public confidence in nanotechnology.

since the dramatic exposure of lying by government officials in the 1960s.[7] Subsequent surveys tracked a continuing decline of public trust in government. The loss of confidence reached to other institutions, to people running major institutions of medicine, and to people in charge of companies.[7]

Continuing evidence of erosion of public trust in the U.S. government makes concern about trust a salient issue of the new millennium. Studies show greater trust in scientists than in government. They also show that "most Americans seem to be distrustful of business leaders in the nanotechnology industry and their ability and willingness to minimize potential risks to humans."[8] This is to say that commercialization is rushing forward lacking a reservoir of public trust.

Why is public trust important? Trust is a social good, a social resource. Words such as "reservoir" used in connection with material resources (e.g., water) are associated with the term "trust." Trust is essential for sustaining personal relationships, and for maintaining the social cohesion that supports ongoing institutions and new social endeavors, such as the NNI. Trust is as essential to organizations and to institutional initiatives as it is to personal relationships.[9]

Trust is an attitude. It includes confidence and a positive inclination to act on that confidence. In regard to the nano area, public trust would mean that members of the public have confidence that actors involved in developing, producing, overseeing, and regulating nanotechnology are taking appropriate precautions for the public's welfare. It would also mean that members of the public have a positive attitude toward the purchase of nano products. As for the actors in all of these nanotechnology endeavors, they must take reasonable care to avoid harming the public. In other words, those who are trusted must be trustworthy.

There is a tight link between trust and trustworthiness. The concern with responsible management arises from the need to foster and assure trustworthiness. This is an important element of the rationale for attending to responsible management and the organizational context of trust. The concern of avoiding adverse consequences is another important element of the rationale.

Reports of recent studies of public attitudes toward nanotechnology reinforce the importance of directing attention to responsible management. The research indicates that members of the public have an initially favorable attitude toward nanotechnology.[10] Those surveyed indicated, however, that they approved on the

---

[7]The year 1960 brought the revelation that President Dwight Eisenhower had lied in response to a question about the U-2 incident, in which an American spy plane had been forced to land in the Soviet Union. Bok, Sissela, *Lying: Moral Choice in Public and Private Life.* 1978. Pantheon Books: New York. p. xviii.

[8]NSF *Science and Engineering Indicators 2006*, Chapter 7 Science and Technology: Public Attitudes and Understanding.

[9]Weil, V., "Introducing Standards of Care in the Commercialization of Nanotechnology," *Inter. J. Appl. Philos.*, (to be published).

[10]Macoubrie, J., "Informed Public Perceptions of Nanotechnology and Trust in Government, "The Pew Charitable Trusts, Project on Emerging Technologies, Woodrow Wilson International Center for Scholars, September 2005, p. 5; Weil, V., "Ethics and Nano: A Survey" in *Nanotechnology: Societal Implications—Maximizing Benefit for Humanity*, Available at www.nano.gov 2005 and forthcoming in volume by Springer Science and Business Media.

assumption that "usual levels of government regulation and control are in place."[11] The public counts on companies to test new products adequately before releasing them into the marketplace, and it relies on the federal government to impose regulations as needed.[11] However, Clarence Davies's systematic and thorough examination of current government regulation of nanomaterials and nanoproducts and the prospects for regulation in the future indicates that government regulation and control over this area are at a very early stage of development.[12]

Moreover, the scientific studies needed to inform regulation are similarly at an early stage.[13] Surveys of nanoscientists and investigation of public attitudes show that concerns about health and environmental effects rank near the top among concerns about adverse effects.[10] Yet we know very little about adverse effects. The federal government investment in environmental and health research is shockingly low. The President's science advisor, John H. Marburger, III, observed in 2005 that current toxicity studies in progress through the NNI are "a drop in the bucket compared with what needs to be done."[14] The amount planned for 2006 amounts to less than 4% of the $1 billion budgeted for nanoscience and nanotechnology.

The examination of responsible management that follows springs from recognizing the public's expectation of adequate, informed oversight and regulation, as well as the need for the trustworthiness of actors in nanoenterprises. The discussion is premised also upon the need to protect public health and the environment in view of the lack of knowledge about both short- and long-term effects and the formidable challenges of identifying and managing risks. In the meantime, it is possible to address one source of protection for the public, responsible management.

RESPONSIBILITY IN RESEARCH

Specialists in science and ethics began to give serious attention to ethics and responsibility in scientific research in the 1980s in the wake of well-publicized cases of scientific misconduct. The field of scientific research ethics received a significant boost in 1989 when the National Institutes of Health (NIH) began to require ethics in the education of graduate students supported by NIH training grants. By the time the National Academy of Sciences produced its volume on responsible conduct in 1991, specialists in science and ethics had identified key

[11]Footnote 8, p. 36.
[12]Davies, C. *Managing the Effects of Nanotechnology*. Jan. 2006. Project on Emerging Nanotechnologies. Washington, DC.:Woodrow Wilson International Center for Scholars & Pew Charitable Trusts.
[13]Service, R. F., "Priorities Needed for Nano-Risk Research and Development," *Science*, 314, 5796, (Oct. 6, 2006):45; Stone, V. and Donaldson, K., "Nanotechnology: Signs of Stress." *Nature Nanotechnol.* 1.1 (2006): 23–24.
[14]Remark in a private conversation with permission to quote.

issues specific to science.[15] They included management, ownership, and sharing of data, reporting of research results, authorship and credit, and communication in research groups.

Empirical research across fields showed that standards were often unstated, hence rarely discussed or assessed, and norms and practices varied across disciplines.[16] Since then, activity in scientific professional societies has produced discipline-specific codes of ethics. The NSF awards have supported research and education projects in scientific research ethics and prepared a younger cohort of research ethics specialists. These endeavors have produced considerable resources for responsible management of scientific enterprises. An important conclusion that emerges from these activities is the need for research groups and communities to formulate and discuss their standards.

This conclusion applies, of course, to research in nanoscience and engineering. The key issues are no less important in nano research. Moreover, the emphasis in nano undertakings related to convergence (e.g., of nanotechnology, biotechnology, information technology, and cognitive science) and on collaboration across disciplines intensifies the need to address differences among norms and standards along with other key issues. That effort has to begin with research groups and communities undertaking critical discussion of their standards. The critical examination should lead to crafting standards for collaboration with researchers from other communities and disciplines. Some professional societies have already done this.[17] In actual practice in multidisciplinary research settings, it becomes necessary to make adjustments among standards from different research communities to produce agreement needed for allocating credit, for example.

Scientists and engineers conduct nanoresearch in a range of settings. Predictably, they include laboratory facilities in universities and companies. In addition, facilities to be shared by academic and commercial researchers have been developed, primarily through the National Nanotechnology Infrastructure Network (NNIN). Because of the extreme costliness of nano instrumentation, the NSF supports this geographically distributed network of facilities for regionally based use by paying academic and commercial customers. The NNIN and other facilities for such shared use are largely based in universities. National scientific laboratories, such as Argonne, also enter into such arrangements.

In all these settings, translation becomes an important issue in two distinct senses. In one sense, it is a matter of forging a common language among specialists from different disciplines. In another sense, the term is used to refer to the process of carrying research from the bench to implementation in products that enter the

---

[15]Panel on Scientific Responsibility and the Conduct of Research, National Academy of Sciences, National Academy of Engineering, Institute of Medicine. Washington, DC: *Responsible Science, Volume I: Ensuring the Integrity the Research Process.* National Academies Press, 1992.

[16]Swazey, J. P. et al. (1993). Ethical problems in academic research: A survey of doctoral candidates and faculty raises important questions about the ethical environment of graduate education and research. *Am. Sci.*, 81, 542–553.

[17]International Medical Informatics Association. Undated. IMIA Code of Ethics for Health Information Professionals. Available at http://www.imia.org/English_code_of_ethics.html.

marketplace. At times, researchers insist that they find no language barriers within multidisciplinary nanoprojects. However, in view of empirical evidence of variation in norms and standards across disciplines, the likelihood of miscommunication and misunderstandings is too high to ignore. Language differences reflect differences in technical fields and are embedded in cultural differences among disciplines.

Historian of science, Peter Galison, uses the metaphor of "trading zones," borrowed from anthropology, to refer to areas of interaction. His examination of multidisciplinary undertakings in the physical sciences shows the development of a kind of "pidgin" language that over time can evolve into a "creole" language.[18] Individuals who have mastery of the different disciplinary languages in "trading zones" can become specialists, experts who can facilitate effective interchange. In light of these considerations, explicit attention to forging common understandings becomes an essential component of responsible management.

The second sense of translation, transferring research advances from the bench to the marketplace, raises responsibility issues at important junctures along the transfer route. This linkage of a highly innovative area of research to the marketplace forces associated research engineers and scientists to confront an additional set of translation responsibilities. For example, at the bench level of research, it may be necessary to anticipate worker safety issues in implementation. That will likely require communication with others in other specialties along the transfer route, and perhaps additional research. Nanospecialists may need to prompt researchers in other specialties, such as toxicology, to initiate research.

Those oriented toward a view of research more independent of practical application may initially struggle with such demands for taking a wider perspective and looking ahead on the route toward implementation. The imperative for scientists to take an expanded view of their role and responsibility comes not only from the nano area. From the perspective of the STS emphasis on the interactive shaping of science, technology, and society, it is a call to scientists to become more self-conscious agents in social endeavors. Workers are on the front line of exposure to results of nano innovations at the bench level. By bearing that in mind when shaping their research or by taking early steps toward the protection of workers, researchers respond to societal expectations.

Realizing that mechanisms are needed to help nano researchers expand their outlooks and become proactive, specialists in engineering and scientific research ethics and STS test innovative approaches. With support from the NSF, this author tried out a model for stimulating attention to responsibility and the social context in a nano research setting. At a regional nano facility (not included in NNIN), this author and the facility manager supervised an advanced philosophy of science graduate student (pre-doc) who performed as a participant observer. His three tasks were: (1) to stimulate ethical discussion among researchers in the

---

[18]Galison, P. (1997) *Image & Logic: A Material Culture of Microphysics*. Chicago: University of Chicago Press. M. E. Gorman (2004) "Collaborating on Converging Technologies: Education and Practice," In Roco, M. C. and Montemagno, C. D. (Eds.) *The Evolution of Human Potential and Converging Technologies* (Vol. 1013, pp. 25–37), New York: The New York Academy of Sciences.

facility; (2) to test in the facility a questionnaire designed to raise awareness of ethical issues; and (3) to produce a report on the participant/observer effort as a kind of case study.

The underlying idea is that a new model for research, with ethics or STS specialists working in tandem with technical researchers, is called for in the development of emerging technologies. With the help of the required protective clothing, head-gear, and goggles, the pre-doc overcame the obstacle of appearing to be an outsider. He learned to use downtime to start shop-talk about science. Possessing a good foundation in biology, he could exchange scientific information with people in physics. After a time, he initiated ethics discussions and soon found technical people bringing in articles and questions relating to their discussions. Eventually even technical people from companies entered the discussion. This empirical ethics research took place over a period of 7 months.

Toward the end of this period, the trainer in the facility, a Ph.D. physicist, sent this author an email describing how his conversations with the pre-doc had "changed his perspective on nanotechnology." The trainer said he had found the ethical and societal issues interesting and important. The pre-doc had succeeded with his first task, stimulating ethics discussion. He has ended his time in the facility, but the second task, the testing of the questionnaire, is ongoing. While the facility has a new manager, the questionnaire remains a component of the orientation package for newcomers to the facility.

The flow of responses continues and is under study. Nevertheless, it is safe to say that responses from respondents who had no personal interaction with the pre-doc showed less effort. One might have hoped that the questionnaire by itself would spark interest in the social context and ethical issues, but that seems not to be the case.[19]

Other models of research bringing together technical nanospecialists and STS researchers and ethics specialists are being tried and evaluated. Testing of a model of Real-Time Technology Assessment is underway at Arizona State University through its NSF-funded Center for Nanotechnology in Society.[20] The NNIN at Cornell University made possible a short-term participant–observer effort involving a student in a nano research setting and the production of a video report.[21] The NSF-supported models of collaborative research involving STS and nanospecialists were implemented within engineering contexts at the University of Virginia starting from a relatively early point in the first 5 years of the NNI.

The innovative idea of in-tandem research clearly inspires an array of efforts with prospects for wider use. This is another dimension of multidisciplinarity. The well-tested insight that it is useful to formulate explicit guidelines within research settings to address ethics and responsibility issues should also inspire wider activity, including regular discussion of guidelines and their revision as needed.

---

[19]The report–case study is scheduled for presentation at the 2006 annual meeting of the STS professional society.

[20]Website, Center for Nanotechnology in Society, Arizona State University. Available at http://cns.asu.edu/.

[21]Available at http://sei.nnin.org/activities.html.

TRANSLATION TO THE MARKETPLACE

Early in the transition to implementation, the issue of how to engage the public has to be confronted. The need to be engaged with the public is a central insight from experience with earlier radically innovative technologies—"disruptive technologies," as some define them. Companies in agricultural biotechnology, for example, plunged ahead without input from the public about expectations, desires, or concerns about risks. Companies later ran afoul of product bans and closed markets. The public should be included in the beginning of product development not only later in dealing with adverse effects that have already come about. That means bringing in the voices of many publics "upstream" in settings that give access to decision making. This is to recognize the need for radically new interchange and interaction between nanospecialists–nanoinsiders and members of the public.

Introducing multiple perspectives into decision making at an early stage, including those of outsiders, would involve fashioning a more open, transparent, and deliberative decision-making process. Those are daunting tasks, but they are not impossible. To show the feasibility of fashioning such a process, I will focus on the prospect of formulating standards of care in nanoenterprises that are associated with producing nano products for the marketplace.

One favorable indication for the enterprise of formulating standards of care is the evidence that companies are uncertain about standards for testing products before releasing them to the marketplace. David Rejeski, Director of the Project on Emerging Nanotechnologies at the Woodrow Wilson International Center for Scholars, reported in mid-January 2006 on National Public Radio that he gets calls every week from start-ups and large companies seeking guidelines and criteria for adequate testing. As his colleague, J. Clarence Davies, remarked: "There is a lot of insecurity in the industry."[22] This evidence of the realization within companies of the need for voluntary standards of care makes it plausible to suggest instituting processes to formulate standards. It presents an important opportunity for companies to work with stakeholders and members of the public in a collaborative process to establish standards of care across a product's full life cycle.[23] This would be one form of public engagement with access to decision making and a strategy for managing responsibly.

It is not far-fetched to envision groups including outsiders with diverse perspectives taking part in determining standards. They would have models to examine and draw from, for example, voluntary standards (not government imposed) adopted by the chemical industry in 1990. The Chemical Manufacturers Association undertook an initiative that produced the program, "Responsible Care: A Public Commitment."[24] To meet a major objective of the program and respond to public concerns

---

[22]Balbus et al., "Getting Nanotechnology Right the First Time," *Issues in Science and Technology*, Summer 2005, p. 68.
[23]See footnote 22, p. 66.
[24]Harris, C., Pritchard, M. and Rabins, M. *Engineering Ethics: Concepts and Cases*. 3rd ed. Thomson Wadsworth: Belmont, CA. pp. 221–222.

it established a public advisory panel that included 15 nonindustry representatives of the public.[25]

The public reasonably expects organizations engaged in development or production of nano products to adopt appropriate standards of care. It seems that companies' own self-interest should motivate them to do this. Yet, it must be acknowledged that very often in the past, companies have failed to take measures to protect the public even when it seemed to be in their self-interest to do so. And many would anticipate that companies would resist the recommendation to include outsiders in a company activity, such as formulating standards. A perceived need to protect proprietary information, such as trade secrets, would presumably motivate that reaction.

It is evident that self-interest cannot be counted on to induce companies to take the suggested measures. The U.S. system of government regulation developed out of recognizing that fact. Furthermore, litigation in civil courts has, over the last 40 years, turned out to be another necessary form of government control. Yet, there is a reason for expecting self-interest to motivate companies, especially with regard to standards for testing nanoproducts for acute effects. Many in the nanocommercial world recognize that an outcome that comes to look like a disaster could be very damaging to continuing rapid advance in the nanodomain.

The capacity for instituting an inclusive process for devising appropriate standards of care is greater in large, established companies than in start-ups. The former have procedures for orderly attention to many areas of corporate concern. They have the motivation of having billions of dollars at stake. They may also have a memory of products that caused them trouble, in the way Teflon created problems for DuPont. However, there is a worry about start-ups. Their number is significant and growing. The business posture of start-ups is to move ahead adventurously and quickly, taking risks, while lacking established practices. Yet in the face of great ignorance, they have an interest in taking such steps as adapting the orderly procedures of large, established companies.

In racing forward, managers in start-ups may simply not have learned about procedures that large oil and chemical companies have put in place to avoid disasters. Without massive effort or financial investment, managers in start-ups can become informed and, in a deliberative process, adapt such procedures to their own circumstances.[26] If, as I argue, start-up companies have the capability to learn about and take lessons from chemical companies' well-tested procedures for meeting a reasonable standard of care, then they have an obligation to do so. They have to be made aware of the need for the suggested procedures and of the advantages of making those processes inclusive. It is reasonable to think an inclusive process of determining standards would eventuate in standards within the capabilities of companies.

---

[25]This paragraph and the one below draw from discussion in V. Weil, "Introducing Standards of Care in the Commercialization of Nanotechnology," to be published in the *Inter. J. App. Philos.*

[26]I am indebted to my colleague, Jay Fisher, Director, Ed Kaplan Entrepreneurial Studies Program at Illinois Institute of Technology and a veteran of 30 years at the Amoco Corporation, for making the point that ignorance and absence of orderly procedures existing in large, established companies may be the important issue with start-ups. He holds that the main problem is making these companies aware of what they need.

The concern that companies would resist including outsiders in determining standards of care for fear of revealing proprietary information also can be countered. Companies might well pull back initially. However, they might be persuaded that it is possible to consider appropriate standards without impinging on proprietary information. They might come to see that they can adequately restrict certain information viewed as proprietary while making available enough information to devise relevant standards of care. This expectation is based on assuming that generic features of company activities would figure most importantly in determining standards of care.

Furthermore, this is not the only arena in which companies have to balance other needs and benefits against the need to protect information viewed as proprietary. In the promotion of new products, for example, they must strike a balance between providing information about new and attractive features of their products and holding back information that they deem crucial to their competitive edge. In recruiting and contracting with new employees who are publishing researchers, they must make similar accommodations. Companies can fail to think through these issues carefully and make almost automatic protective responses. However, when they come to weigh in the benefits from establishing voluntary standards of care in open deliberation, they may be persuaded.

EXPERIMENTS IN PUBLIC ENGAGEMENT

The desirability of public involvement "upstream" in nanoenterprises has led to experimentation with a number of models for bringing citizens and nanospecialists together. Citizen juries, consensus conferences, and variations on these forms of citizen participation (generally referred to as citizen panels) are neither interest groups nor civic associations. They are groups of interested citizens brought together to represent a range of perspectives. The groups generally meet several times over a defined period to discuss developments of an emerging technology with specialists. For nanospecialists these discussions are opportunities to share information and get feedback from a diverse group of citizens. Citizens have an opportunity to learn about nanotechnology and share their perceptions and concerns with specialists. Citizen panels are a methodology for promoting dialogue between experts and the public that can integrate public responses into processes of technological innovation.[27]

Other discussion formats in other settings are springing up, as well. At the University of Wisconsin in Madison and at Arizona State University in Phoenix, nanoscientists and engineers in NSF-supported research projects, are experimenting with science cafés to engage with members of the public.[28] Admittedly, the cafés do not start out with nano specialists and members of the public on equal footing. The transition from scientists imparting information and taking questions from lay

[27]Brown, M. B., "Survey Article: Citizen Panels and the Concept of Representation," *J. Polit. Philos.*, Vol. 14, No. 2, pp. 203–225.

[28]University of Wisconsin NSEC and Arizona State University (ASU CNS) are NSF-supported projects.

people to genuine dialogue does not begin at once.[29] From these experiments, insight may be gained about how a transition to genuine dialogue develops.

Other efforts are underway to foster dialogue that can bring citizens' expectations and concerns to bear on processes of innovation. In the United Kingdom, noted STS specialists who study citizens' understanding of science and engineering have formed The Nanotechnology Engagement Group. Under the rubric, "small talk," they experiment with discussions to allow nanoscientists to discuss their work and aspirations with lay people. They too report that the question-and-answer sessions tend toward traditional top-down science communication although they are intended to be more participative.[30]

Still another experiment in scientists' interacting with citizens to promote citizen understanding about nanotechnology is a Citizens' Nanotechnology School at the University of South Carolina.[31] The school offers an innovative short course that brings together scientists and citizens. For some time, specialists at the University of South Carolina in behavioral science and philosophy of science have been engaged in discussion with nanoscientists and engineers. Together, they devised the earliest models of transferring the experience of discussion in academe between scientists and non-scientists to models of engagement with the public.

CONCLUSIONS

The various measures for fostering responsible management of nanotechnology enterprises are radically innovative. They do not get implemented quickly. The challenge to promoting activity in companies to develop standards of care, even without citizen participation, is daunting. The experimentation with models to advance public engagement is growing, but not rapidly enough to match a rush to market, which is accelerating at a very rapid pace. The implementation of government regulation the public counts on faces challenges from the underfunding and understaffing of regulatory agencies, as well from conceptual problems (e.g., definitional issues) and the dearth of scientific findings on toxicity, for example, to inform regulation. The most recent news, reports of nano food products already in the marketplace, underlines the need for the voluntary measures examined above.[32] Near the end of the report, the author, Barnaby Feder, noted, "F.D.A. officials say companies like Kraft are voluntarily but privately providing them with information about their activities. But many independent analysts say the level of disclosure to date falls far short of what will be needed to create public confidence."[32]

---

[29]E-mail message from David Guston, Director of the ASU CNS and author's observation from report of participating scientist at NSEC site visit, April 27, 2006.

[30]The Nanotechnology Engagement Group, Policy Report 1, March 2006. Available at www.involving. org, p. 13.

[31]Toumey, Chris, *Nature Nanotechnol.*, 1,(1), pp. 6–7.

[32]Feder, B. "Engineering food at the level of molecules." *New York Times*, October 10, 2006 C-1. This news report depended on information volunteered by companies.

CHAPTER 24

Nanotechnology and the Global Future: Points to Consider for Policymakers

NIGEL M. DE S. CAMERON

SCIENCE POLICY AND NANOTECHNOLOGY

The impetus for enhanced federal spending on nanoscale research and development, presaged by President Clinton's 2000 Caltech speech[1] and shaped in the $3.7-billion 21st Century Nanotechnology Research and Development Act signed by President Bush in December 2003, has been generated by excitement about nanotechnology's potential to revolutionize many areas of the economy, to contribute to human health and well being, and to enhance national security—all on a scale hitherto unimaginable.

Work at the nanoscale will have the effect of effacing standard distinctions between scientific disciplines, and granting humankind a new magnitude of manipulative capacity over matter. It, therefore, presents the human community with challenges that are, in many aspects, unique. One sees this when comparing the National Nanotechnology Initiative (NNI) (and its siblings in Europe and Japan) with other large-scale transformative science projects. Analogies may be drawn with the space program and the human genome project, yet these vast enterprises offer only partial parallels with the claims and challenges already emerging in the nano-arena. A further, sobering, analogy may be found in agricultural biotechnology—specifically, the promise and problematics of "genetically modified" (GMO) crops; for while the development of GMO foods was not chiefly funded through public science, its advocates have made transformative claims on a par with these other initiatives. The search for a context for nanotechnology among these parallel projects is instructive. The space program was, of course, funded entirely from the

---

[1]President Clinton's Address to Caltech on Science and Technology (Jan. 21, 2001). Available at http://pr.caltech.edu/events/presidential_speech/.

Nanoscale: Issues and Perspectives for the Nano Century. Edited by Nigel M. de S. Cameron and M. Ellen Mitchell
Copyright © 2007 John Wiley & Sons, Inc.

public purse, as was the genome project until late interventions by commercial interests. The development of GMO foods, by contrast, was a largely commercial undertaking. Nano, however, spans the public–private divide: the U.S. NNI and its parallels in Europe and Japan are feeding around $4 billion per annum into nano research and development, and an equivalent sum is being invested by the private sector. One reason, needless to say, is high expectations for economic return.

The most obvious difference between nano and these other enterprises is that, however great the financial investment and number of investigators involved, the other initiatives were each narrowly targeted toward a singular goal: to get a man on the moon, to enable therapeutic genetic interventions, and to increase crop yield. The underlying motives deployed in pursuit of these goals were, of course, diverse: private profit, politics, the healing of the sick, and the quest for knowledge, in varying proportions.

Partly as a result, a striking contrast lies in the fact that, unlike the two most clearly comparable federal science initiatives—the space program and the human genome project—there is only very limited public awareness of the NNI. Even in the policy community, and among lawmakers themselves, levels of awareness and understanding are remarkably low. This is a double-edged sword. On one hand, it has meant little public and political scrutiny for substantial new spending; on the other hand, it has also meant little public and political buy-in. Confidence in the project, and its technology, could, therefore, prove fragile. This is potentially troubling, not least for those focused on the substantial economic expectations of nanotechnology. The experience of the European GMO debacle, which is surprisingly unfamiliar in the United States, stands as a potent cautionary tale, and the stakes in nano are far higher. Just as the NNI lacks the clearly defined, narrow goal of the other major science projects we have considered, its effects are expected to be pervasive and to have greater—perhaps much greater—transformative and disruptive effects. Already, the use of nanointegrated products cuts across key industry sectors.

This public disinterest can be explained by such factors as low levels of public attention to science policy in general, and recent dominance of the science space by the stem-cell and cloning debates. Moreover, the wide variety of applications of nanoscale research has meant there is no single encapsulation of the technology (if, in fact, one views nanotechnology as a technology, rather than a brand or an enabling platform). Nor is there—yet—a significant and distinctive political constituency with an interest in critiquing, or even monitoring, what is taking place in the nanosphere.

Nevertheless, given the scope of nanoscale research and development, it should be a matter of serious concern that public understanding is so low. For nanotechnology, as is generally agreed by its advocates and critics alike, is potentially the most disruptive technology ever developed, with dramatic implications for human society. While at one level its significance is understood chiefly in economic terms, it is agreed by all who have weighed its implications that it may—many would say *will*—bring about fundamental, disruptive, change in almost every aspect of human culture.

The fundamental lesson of the costly European GMO fiasco lies in the need to build public confidence in new technology in order to ensure a market for the resulting products. At the height of the GMO controversy, Monsanto, the company with the greatest exposure, lost more than $5 billion in capitalization in a fast rising market—its profitable agricultural biotechnology business being at one point valued by that market at below zero. After disaster struck, Monsanto's chairman took the unprecedented step of apologizing to the conference of the environmentalist group Greenpeace with the phrase, "We forgot to listen."[2] As nanoproducts are already used in a wide range of industry sectors, the threat of market resistance based on the GMO model, is in economic terms, significantly more serious than in the case of agricultural biotechnology products. Aside from other nano-related concerns, it is vital, from the perspective of risk management, to address transparently and systematically the two basic sets of questions that determine public confidence in a new technology: (1) safety, and (2) social and ethical implications.

Moreover, as my colleague Vivian Weil has argued in her paper in the wide-ranging National Science Foundation (NSF) volume *Societal Implications of Nanotechnology*,[3] it is important that those engaged in the technology (whether as researchers, investors, or public science administrators) do not assume that their own vision for the technology and its transformative possibilities for society is shared by the public. This caveat is central to the healthy development of public science. Educational efforts are under way to raise awareness of nanotechnology among both adults and children. It is vital that these approaches do not degenerate into propaganda exercises, efforts to convince people that the projected outcomes of nanotechnology are benefits to be welcomed and not to be questioned. Aside from the unethical nature of such an approach, it risks backfiring when the public realizes what is actually happening.

Lack of public scrutiny during the early stages of nanotechnology's development may explain the sometimes strange and provocative language in which the significance of nanoscale work has been framed by some of the nanotechnology leaders. The focus on the "convergence" of nanotechnology, biotechnology, information technology, and cognitive science (nano-bio-info-cogno or NBIC) is the theme of the most substantial and far-reaching nano document issued to date by the NSF, the nearly 500-page report on *Converging Technologies for Improving Human Performance* published in 2002. The immediate intent of such "converging technologies" language is to demonstrate how at the nanoscale these areas of science and engineering come together, and the NNI thrust to develop synergies between and among them, therefore, has merit. The chief difficulty raised by this language is that the goal of "convergence" is specified not in terms of health, economic growth, and the advance of knowledge, but, rather, as that of "improving human performance."

[2]Vidal J. (Oct. 7, 1999). We Forgot to Listen, Says Monsanto: GM Company Chief Takes Blame for Public Relations Failures and Pledges to Answer Safety Concerns. *The Guardian*. Available at http://www.guardian.co.uk/print/0,3909773-103528,00.html.
[3]National Science Foundation. Roco, M.H. and Montemagno, C.D., eds. (2004). *Societal Implications of Nanoscience and Nanotechnology* 244–251.

While that phrase is, in principle, capable of several interpretations, the first NSF NBIC symposium makes plain that those responsible have in mind a larger-than-life reading that incorporates a vision for fundamental change in human capacities.

Among the goals and anticipated results: "enhancing individual sensory and cognitive capacities... improving both individual and group creativity... communication techniques including brain-to-brain interaction, perfecting human-machine interfaces including neuromorphic engineering ...."[4] It asks: "How can we develop a transforming national strategy to enhance individual capacities and overall societal outcomes? What should be done to achieve the best results over the next 10–20 years?"[5] At the end of one list of long-term implications, it specifies: "Human evolution, including individual and cultural evolution."[6] Then, it goes on to state:

> Technological convergence could become the framework for human convergence. The twenty-first century could end in world peace, universal prosperity, and evolution to a higher level of compassion and accomplishment. ... [I]t may be that humanity would become like a single, distributed and interconnected "brain" based in new core pathways of society.[7]

While the document includes the usual disclaimers, these extreme, and sometimes sophomoric, claims (e.g., world peace) are presented as key ideas in the thinking of the leaders of the NNI. It is influenced at many points by the assumptions of the fringe philosophical futurist movement known as "transhumanism," which sees new technologies as mechanisms for achieving fundamental changes in human nature. It is also laced with glorified claims about the hypothetical benefits of the technology.

As required by Congress, the NNI is funding nano ELSI (NELSI) research alongside its technical work. Funding is being disbursed chiefly through the NSF, and the emphasis is twofold: (1) incorporating NELSI aspects into technical projects; and (2) establishing a national center for nanotechnology and society, which is now based largely at Arizona State University and University of California-Santa Barbara. Both of these approaches have merit, although they also leave the strategic decisions behind them open to criticism, especially if a key goal is to generate critical assessment of the possibilities of the new technology and, thereby, to instill public confidence. In the case of embedding NELSI into technical projects, there is a natural danger that NELSI will be an afterthought added on to predominantly technical projects in order to obtain funding. In respect of the creation of a national NELSI center, its establishment as an NSF-funded project creates the impression of a conflict of interest. Whatever the merits of its work, it will have to fight an uphill battle to establish its independence in the public mind.

[4]Roco, M. and Bainbridge, W. S., eds. (2002). *Converging Technologies for Improving Human Performance*, p. ix. Available at http://wtec.org/ConvergingTechnologies (retrieved October 17, 2006).
[5]See footnote 4, p. x.
[6]See footnote 4, p. 4.
[7]See footnote 4, p. 6.

POINTS TO CONSIDER

The Administration of the National Nanotechnology Initiative

As this ground-breaking work goes forward, it is vital that NELSI projects be energetically developed with an immediate twofold intent. First, they need to assess the ethical, legal, and social entailments of new technological possibilities in order to anticipate problems and opportunities that may lead to regulatory, legislative, or other governance interventions. Second, although it is recognized that NELSI assessment is key to securing public confidence that a continuing, robust, and transparent critique of the technology is in hand, there is the wider consideration that NELSI projects serve as bridgeheads into the public's awareness of the societal significance of the technologies. This second prong of educating the public is especially critical if, as many observers believe is likely, the questions raised by developments in emerging technologies become a dominant (perhaps *the* dominant) theme of public and political debate in the twenty-first century. Thus, NELSI should be seen as an exercise in strategic communication. Plainly, its goal must be to introduce the public to the facts and make the public aware of informed opinion in respect to these facts, and this must be achieved in a transparent and self-critical fashion—or else it will run the risk of being seen as mere propaganda by enthusiasts for the technology or those who wish to embrace it to develop particular applications, or both. That is to say, while some see NELSI as a means of embedding public support for the technology and others seek to use it to critique some, perhaps all, of its possibilities, a true, honest, unsensationalized, and grounded NELSI function is central to the success of democractically developing science policy in a century that will, on any accounting, witness vast shifts in our technological capacities. Far from being an add-on, NELSI is the very nub of democratic accountability.

The discussion that follows assumes the particular structure and current operation of the United States NNI, and offers a series of points that will need to be considered in order to ensure democratic accountability for the transformative potential of the technology, and, in the process, to provide its best chance for success.

Public Science, Transparency, and Ideology. We have noted the importance of not taking for granted the existence of any social consensus on the significance of nano as an agent for social transformation. If technology offers new social options, their implications should be made plain and submitted transparently to public scrutiny and democratic evaluation at the earliest possible stage. This emerges as a key responsibility of public officials charged with the funding and/or overseeing the technology, in collaboration with those engaged in NELSI research. Their special knowledge grants them unique insight into the likely direction of research and its social implications.

Therefore, public science officials must take great pains not to permit their enthusiasm for the technology in question, or their personal, political, philosophical, or wider social views to influence their discharge of the public trust and the manner in which they articulate the significance of their work.

Public Confidence. High economic hopes for the technology depend entirely on the maintenance of public confidence as public awareness of nano grows. As such awareness in the United States is presently very low, this is an issue to be handled with particular care. The European experience with GMO foods demonstrates that economic effectiveness requires there to be confidence—both in terms of safety, and in terms of the nature of the technology in relation to human identity and broad social goods. In the GMO case, failures in both of these respects fed each other and generated an economic catastrophe for the industry that observers, such as Deutsche Bank, feared could readily spread to the U.S. market. As economic benefit is sought from the technology, it is necessary to go to great lengths on both these fronts. As to the latter, expectations of the socially transformative effects of the new technology are, at this stage, strongly counterindicated. Buy-in to any such agenda is required from major cultural stakeholders through the democratic process. Any attempt to short-circuit this process, especially at a stage when there is very limited public awareness of the technology, will inevitably raise the risk profile of the technology. Moreover, it is not the role of public science and its administrators to serve as advocates for any particular transformative social agenda. That is a political task, and one of the virtues of democracy is that the political process will always tend to ground new thinking in wide public understanding and assent, so as to produce a stable social and economic context.

Public Education. One of the early features of the NNI has been a focus on education, both in postsecondary contexts and K-12, as well as more broadly within the culture. All such educational initiatives need to inculcate a perspective that both informs as to current and potential future technologies and develops a critical disposition toward their possibilities so they are open to assessment by a free society. The two commitments are concomitant and, ironically, may offer the best chance of securing durable public acceptance of a technology's potential. Any use of public education for purposes of advocacy, whether of the technology as such or of any ideological vision stemming from it, must be eschewed as inappropriate and potentially counterproductive. This approach contrasts with the advocacy approach that has already been in evidence with regard to nanotechnology and that has helped to generate radical critique from those unconvinced of its merits. Educational initiatives, especially those focused on K-12 and in formal contexts, such as undergraduate teaching, should not have as their prime aim commending the technology or any particular implications or allaying public fears on issues, such as safety, but rather cultivating among Americans, young and old, an awareness of the basic questions raised by all technologies (including potential benefits, hyped promises, unintended consequences, problems of public accountability, and misuse). Only a public able to critique technology can come to stable acceptance of one with such transformative implications as nano.

In general, the United States has tended to accept technology less critically than Europe. If the twenty-first century is to be marked by rapid and fundamental technological change, how we handle emerging technology issues will become a dominant feature of our culture. The goal of education must, therefore, be the development of a public disposition skilled in critical assessment as a basis for discrimination and acceptance. Education must, at all costs, avoid being or being seen as advocacy.

NELSI Components in Technical Projects. Recognition of the central place of NELSI questions has led to an approach that seeks to co-locate NELSI components (perhaps focused on ethics, perhaps broader) within technical research and development projects. Various benefits may be identified in such an approach: alerting investigators to NELSI questions as research is designed and pursued; educating them in NELSI concerns; alerting the institution and broader community to NELSI issues raised by the research; reviewing the NELSI implications of research for public funding bodies; and, if all else fails, whistle-blowing.

At the same time, this approach also has built-in disadvantages that need to be carefully noted. There will be an inevitable tendency for NELSI investigators to be co-opted into the technical aims of the project and to serve as apologists, both for nano in general and for specific research and development goals. They will, by definition, be junior partners in such projects, and dependent on the goodwill of the principal investigators and other technical colleagues for their participation. They are unlikely to be added to the team in the first instance if they are known to hold deep-seated reservations about the technology, or if they tend to be particularly critical of new technological developments in general. Moreover, it may be expected that they are less likely to critique research on their own campus, especially research that, to some extent, funds their own, than developments elsewhere. To say this is not to impugn the integrity of investigators, but rather to draw attention to elements of this research model, which could readily lead to pressures on NELSI researchers, and potential conflicts of interest. Moreover, it is not clear how NELSI research can usefully relate to narrowly defined research projects in the lab. The tendency will be for such projects to generate broad-based commentary on the technology, and for it to be favorable.

So, while benefits for this approach may be acknowledged, it is necessary to design clear protocols to mitigate potential problems that will be applied to all technical projects that include NELSI components, in the interests of good practice. They will include the following:

- The NELSI component should be substantial in proportion to the overall scope of the project.
- It should include senior investigators (at co-PI level).
- It should be structured in such a manner as to free NELSI investigators from close dependence on technical colleagues for funding approvals, and so on.
- It should, as a rule, include substantial participation by NELSI collaborators from at least one other institution.
- It should result in publications that are distinct from the technical work of the project.
- The NELSI component projects and investigators should be publicly accountable on the national scale, through funded conferences, publications, and site visits that focus their efforts and ask hard questions about their role within particular institutions.

NELSI Audit. As nanotechnology is not a defined field of study comparable, say, to human genetics, and as it is anticipated that sweeping and disruptive social applications may result from nanotechnology research and development, the NELSI task needs to be addressed both in individual projects and key areas of the field.

Therefore, a NELSI audit should be regularly conducted in each basic research area, and individual technical research proposals should include provisions for such an audit of their ethical, legal, and social implications. Research proposals should include a "NELSI impact statement," which will serve as the basis for NELSI review of the research as it progresses—either within the institution (as outlined under the NELSI Audit Section) or by arrangement with a separately funded NELSI project.

The intention is both to identify emerging NELSI issues, and—perhaps even more importantly—to alert funding bodies to the possible implications of the research, both as a project seeking support and when it is underway. This, in turn, will aid public oversight via the democratic process and ensure that public debate is well-informed about the work of publicly funded investigators.

Dissenting Voices. Because of the radical implications posed by nanotechnology's possibilities, and the associated need to allay public concerns through transparent assessment, it is important to encourage and showcase, through funding and in other ways, the work of informed critics of the technology, as a key element in the wider NELSI assessment. This is one mechanism by which NELSI funding activities will be seen to maintain impartiality on controversial questions.

Risk Management. A basic caveat that we have noted lies in the GMO food story in Europe, in which failure to assess realistically public perception of and reaction to new technology resulted in a comprehensive failure to manage risk. Any perception that those funding and researching the technology are serving as its advocates and playing down criticisms on safety or broader social grounds can lead to disaster. A prudent risk management approach should be applied consistently at all levels, in contrast to an advocacy approach. The high social and economic stakes of this technology require nothing less.

The Development of Nanotechnology Policy

As the 2003 Act makes plain, the response of the U.S. Congress to the nascent NNI was both enthusiastic and anxious. The Act, and subsequent congressional follow-up, reflects serious concerns in respect of the implications of a technology that is seen as exciting and potentially crucial to the economy of the future. The Act achieves a careful and commendable balance. It is notable that the one specific NELSI question picked out for attention is that of intelligence—both the development of "artificial intelligence" (AI), and the enhancement of human intelligence.[8] This concern is also reflected as the key response in the European Commission's High Level

---

[8]21st Century Nanotechnology Research and Development Act (2003), Pub. Law 107–314, § 2 (10), § 9 (b)(10).

Expert Group (HLEG) established to review the U.S. approach to "converging technologies."[9] The HLEG seeks to remove the "C" (for cognitive science) from the NBIC formula, and to focus on beneficial effects on the human brain from technologies on the outside, not the inside—engineering "for" the brain, not "of" the brain.

While, unlike the European HLEG, the 2003 Act does not take a view as to the wisdom of the artificial enhancement of human intelligence, its focus on the importance of the question should lead to serious reflection about the implications of such developments. It is to be regretted that the National Research Council's (NRC) triennial review of the NNI, required by the Act, entirely fails to tackle this question.[10] Indeed, it dismisses such concerns as the subject of "popular science fiction." As the congressional mandate to the NRC specifies the enhancement of human intelligence and development of artificial intelligence, as two particular matters of concern, this response is, to say the least, bizarre. Serious research is in progress on both of these fronts, and it is closely connected with developments in nanotechnology.

While there are other areas in which nanotechnology-related efforts geared toward the "enhancement" of human capacities may emerge (e.g., in muscle development), it is in the possibility of the engineering of the brain that the most consequential of all technological interventions lies. A basic question that may soon be confronted in technology policy is whether interventions that are developed for therapeutic purposes (e.g., to aid recovery of function by those who have suffered strokes or other disabling conditions of the brain) should be able to be applied for nontherapeutic purposes (e.g., the superhuman enhancement of intelligence or memory). It could be argued that this is the most significant policy question ever confronted by the human community, as one answer could set us upon a path that leads toward a cyborg-esque future in the melding of human and machine. The central question that needs to be addressed is the degree to which such choices will enhance—or degrade—not particular capacities, but our general capacity to experience our humanness.

A prime source for this discussion is the President's Council on Bioethics' report, *Beyond Therapy*, which sets out a comprehensive reflection on the move from "therapy" to "enhancement" in the new technologies. The Council's point of departure is the "therapy/enhancement" dichotomy, though it recognizes the inherent problems. The Council sets out the problem thus:

> We want better babies—but not by turning procreation into manufacture or by altering their brains to give them an edge over their peers. We want to perform better in the activities of life—but not by becoming mere creatures of our chemists or by turning ourselves into tools designed to win or achieve in inhuman ways. We want longer lives—but not at the cost of living carelessly or shallowly with diminished aspiration for living well, and not by becoming people so obsessed with our own longevity that we care little about the next generations. We want to be happy—but not because of a drug

[9]See Chapter 3.
[10]See Chapter 20.

that gives us happy feelings without the real loves, attachments, and achievements that are essential for true human flourishing.[11]

There is an intentional ambivalence in each of these statements. While something in each of us would seek the end without regard to the means, in most of us there is a stronger intuition that declares the means to be central to the proper attainment of the end. We reflect on the stories of the heroic and the defiant that we wish our children to read, on the lives of courage and accomplishment that we seek for them. We muse on the accolades that we covet for ourselves. We discover that whatever our religious or non-religious understanding of the world, whichever location we find for ourselves on the cultural spectrum, and whether we tend to favor or suspect the latest in technology, there is in most of us a solid core of commonality. We admire striving; we despise those who cheat; we applaud the extraordinary achievements of those who triumph over adverse and desperate circumstances; we seek an understanding of our own lives in valiant terms, as those who might one day be said to have fought the good fight and to have kept the faith—whatever that faith may have been. We touch bottom in a common acknowledgement of what it means to be human, and, for all our, diversity we grasp human greatness when we see it. We hold Martin Luther King, Winston Churchill, Abraham Lincoln, and Mother Teresa of Calcutta among our heroes.

The President's Council on Bioethics go back to Aldous Huxley as their point of reference, with their intuition that the naïve predictions of bliss that will result from an unfettered application of these new technologies will come unstuck in "the humanly diminished world portrayed in Aldous Huxley's novel *Brave New World*, whose technologically enhanced inhabitants live cheerfully, without disappointment or regret, 'enjoying' flat, empty lives devoid of love and longing, filled with only trivial pursuits and shallow attachments."[12]

To speak simply in terms of "therapy" versus "enhancement" is difficult, as one person's therapy becomes another's enhancement (whether growth hormone, or neuroprosthesis). Yet, the line is fundamental in sketching the point at which the human condition begins to come under threat. One way in which we may articulate the question of human nature that recognizes the blurring of therapy and enhancement is in terms of analogy. Technological interventions, if they are to sustain and not compromise the human condition, need to retain congruence with human nature. They must not trespass upon its analogy. The analogy of human nature offers a means of construing the given-ness that we inherit as biological, psychosocial beings who are members of the species *Homo sapiens*. While a comprehensive definition of what it means to be human escapes us, that does not render us unable to address the question. We may not fully comprehend, but we may seek to apprehend, what it is to be human. Just as we can identify the essential canine-ness of dogs, and felineness of cats, without which we would cease to recognize our

---

[11]The U.S. President's Council on Bioethics Report *Beyond Therapy: Biotechnology and the Pursuit of Happiness* (Washington, DC: President's Council on Bioethics, 2003), p. xvii.
[12]See footnote 11, p. 7.

pets for what they are, we share substantial intuitions as to what it is to be human. While they may not amount to the tight definition that would be required in legislation, our stories of heroism and tragedy—from Shakespeare to news reports in *The New York Times* and the human quirkiness of the cartoons of *The New Yorker*—afford us powerful defining marks of our common humanity. This central recognition on our part, bounded on one side by our shared notions of heroism and achievement, and on the other by the ambiguity of such subhuman exigents as steroid use in sports or Viagra for sexual performance helps frame the human question.

This same point can be expressed in terms of convergence. The model of "converging technologies," with its concept of enabling technologies that catalyze change in others, needs to go further if we are to grasp the final significance of these developments for the human good. Alongside technological convergence must be placed the convergence of technology with the humanities and the arts; not the convergence of the human and the technological, but the convergence of our spheres of knowledge, of what have been famously called the "two cultures."[13] The final objects of convergence are not the respective technologies that are mutually enabled, rather they are humankind and the impact convergence, and converging technologies, will have on our nature and that of our communities. Technology ultimately enables humans. We must, therefore, find the primary context for our assessment of converging technologies not simply in "the human condition," abstractly conceived but also in: the warp and woof of the humanities and the arts; and our induction of the given-ness of human nature and the conditions needed for its flourishing. This is the most important of all conversations, and it must frame every consideration that touches upon the role of these technologies, if we are to find a human future that is not that of Tennyson's poem *The Lotos Eaters* or, perhaps, the Borg from *Star Trek*.

In fact, our human nature is susceptible to the ambiguous technological benefits of enhancement in two fundamental ways. One would focus on the consciousness, through the manipulation—mechanical or pharmaceutical—of memory and the mind, generating intrinsic "enhancements" to our nature that have the effect of enhancing our experience while leaving the outside world untouched. As we know from the use and abuse of Prozac and other mood influencing drugs, "enhancements" in the subjective world can lead to powerful effects through behavioral changes in the world at large. Yet their focus is subjective. The second is the extrinsic, the objective, in which what is "enhanced" is not our affect but our capacities for perception, reason, and action. Between them, they would seem to encompass the universe of possibility.

Such concerns may, perhaps, be most starkly illustrated with reference to the "pursuit of happiness" by means of cognitive "enhancements" that involve the manipulation of perception, memory, and mood or emotion, whether through

---

[13]Snow C.P., *The Two Cultures*, Cambridge: University Press, repr. 1993.

neuro-pharmaceuticals or cognitive prostheses. It is these possibilities that have lead the European HLEG to speak of technologies to be engaged "*for* the mind," rather than "*of* the mind."[14] The President's Council report avers that "the emotional flourishing of human beings in this world requires that feelings jibe with the truth of things, both as effect and as cause."[15] The report continues:

> We do not really want the pleasure without the activity: we do not want the pleasure of playing baseball without playing baseball, the pleasure of listening to music without the music, the satisfaction of having learned something without knowing anything . . . . We embrace neither suffering nor self-denial by suggesting that disconnected pleasure (or contentment or self-esteem or brightness of mood) produced from out of a bottle is but a poor substitute for happiness.[16]

THE LEWIS PARADOX: THE ABOLITION OF MAN?

As policy debates increasingly focus on the purpose for which these transformative technologies should be applied, the lesson of C.S. Lewis' 1943 essay "The Abolition of Man," will take on special relevance.

His argument opens with the claim that technology, which is said to extend the power of the human race, is in fact a means of extending the power of "some men over other men." As a result, he continues, "From this point of view, what we call Man's power over Nature turns out to be a power exercised by some men over other men with Nature as its instrument." He hastens to add that while it can be easily said that "men have hitherto used badly, and against their fellows, the powers that science has given them," that is not his point. He is not addressing "particular corruptions and abuses which an increase of moral virtue would cure," but rather "what the thing called 'Man's power over Nature' must always and essentially be." For "all long-term exercises of power, especially in breeding, must mean the power of earlier generations over later ones."

What Lewis invokes is the biological equivalent of inter-generational economics. In the nature of the case, the genetic accounting is of an even higher level of significance than economic relationships running through time, although the principle is the same: the impact of one generation's decisions on subsequent generations. So Lewis states: "We must picture the race extended through time from the date of its emergence to that of its extinction. Each generation exercises power over its successors: and each, in so far as it modifies the environment

---

[14]European Commission, High Level Expert Group. (2004). *Foresighting the New Technology Wave: Converging Technologies—Shaping the Future of European Societies* 13. Available at http://www.ntnu.no/2020/final_report_en.pdf (retrieved on October 19, 2006).
[15]See footnote 11, p. 264.
[16]See footnote 11, pp. 264–265.

bequeathed to it and rebels against tradition, resists and limits the power of its predecessors."[17] There can be no net "increase"

> in power on Man's side. Each new power won *by* man is a power *over* man as well. Each advance leaves him weaker as well as stronger. In every victory, besides the general who triumphs, he is a prisoner who follows the triumphal car. . . . *Human* nature will be the last part of Nature to surrender to Man. The battle will then be won. We shall have 'taken the thread of life out of the hand of Clotho' and be henceforth free to make our species whatever we wish it to be. The battle will indeed be won. But who, precisely, will have won it?

Because "the power of Man to make himself what he pleases means, as we have seen, the power of some men to make other men what *they* please. . . . Man's final conquest has proved to be the abolition of Man." While much of Lewis' analysis is directed at the possibility of inheritable genetic interventions, his thesis is of general application to the dynamic relation between technology and human nature. And his key perception is that the employment of radical manipulative powers upon our own selves, the seeming triumph of technological ingenuity, entails in truth the turning of human nature into one more manufacture, another artifact of human design.[17]

While, therefore, the therapy–enhancement distinction is problematic, the line to which it draws attention—between the medical model and the manufacturing model—is central to distinguishing humane technological interventions from the ultimately inhumane, in which the transformative potential of technology could result in the fundamental reshaping of the human condition.

The Challenge to the Policy Community

The European High-Level Expert Group report's response to the National Science Foundation's first conference report on *Converging Technologies for Enhancing Human Performance*, has demonstrated the volatility and far-reaching significance of the fundamental questions raised by nanotechnology for human society and human nature itself. It is notable that while these questions have not yet the subject of legislation or regulation, they have been discussed at high levels in both the United States and Europe. The concern articulated by the then Under-secretary for Technology at the U.S. Department of Commerce, Phillip J. Bond, that the technology be rapidly developed and yet that, also, the human condition be respected, suggests exactly the dilemma that will long occupy policy leaders in jurisdictions around the globe. It us mirrored in the focus of the U.S. Congress on the vital significance of the NELSI agenda, and the special importance of questions affecting enhanced human intelligence and the development of superhuman artificial intelligence.

[17]Lewis C.S. (1943). The Abolition of Man. Chapter 3. Full text available at http://www.columbia.edu/cu/augustine/arch/lewis/abolition1.htm.

Extraordinary benefits are expected to be realized from the development of nano-technology. The policy task is to ensure that they are realized while not shying way from the disbenefits that may also accrue. At the same time, it is crucial to secure and carry public confidence in the technology, no small task in light of the sobering history of the European GMO experience and the "science fiction" scenarios that both boosters and critics of nanotechnology suggest may flow from its unfolding. While their wilder extremes may be discounted, it can hardly be assumed that in the course of the next generation these hopes and fears will come to nothing. Sooner rather than later, nanotechnology policy concerns as diverse as privacy, safety, economic inequity and "enhancement" should take their place high on the agenda of nations and the international community. Foresight on the part of policy-makers may prove crucial in guiding the human community through the most dramatic technological revolution in history.

Bibliography

1. KEY DOCUMENTS RELATED TO NELSI

This listing includes reports, academic studies, articles, and fiction.

21st Century Nanotechnology Research and Development Act, Pub. L. No. 108–153 (2003), 15 U.S.C. § 7510 (2004).

21st Century Nanotechnology Research and Development Act, 15 U.S.C. § 7510 (2004). Available from http://www.nano-and-society.org/NELSI/documents/21stcenturynanor&dact.pdf (retrieved October 14, 2006).

Archer, L. (2006, May 16). Nanomaterials, Sunscreens and Cosmetics: Small Ingredients, Big Risks. Available at http://www.foe.org/new/releases/may2006/nanostatement5162006.html. Accessed on October 16, 2006 (retrieved on October 19, 2006).

Bennett, I. and Sarewitz, D. (2006). *Too Little, Too Late? Research Policies on the Societal Implications of Nanotechnology in the United States.* Manuscript submitted for publication.

Berube, D. M. (2006). *Nano-hype: The Truth Behind the Nanotechnology Buzz.* New York: Prometheus Books.

Bond, P. J. (2003). *Nanotechnology: Economic Opportunities, Societal and Ethical Challenges*—Remarks delivered December 9, 2003. Available at http://www.technology.gov/speeches/PJB_031209.htm. (Retrieved on October 19, 2006.)

Crichton, M. (2002). *Prey.* New York: Harper Collins Publishers.

Drexler, K. E. (1996). *The Engines of Creation: The Coming Era of Nanotechnology.* London: Fourth Estate, Available at http://www.e-drexler.com/d/06/00/EOC/EOC_Cover.html.

ETC Group. (April 7, 2006). *Nanotech Product Recall Underscores Need for Nanotech Moratorium: Is the Magic Gone?* Available at http://www.etcgroup.org/en/materials/publications.html?id=14 (retrieved on October 19, 2006).

Nanoscale: Issues and Perspectives for the Nano Century. Edited by Nigel M. de S. Cameron and M. Ellen Mitchell
Copyright © 2007 John Wiley & Sons, Inc.

European Commission. (2004). *Converging Technologies—Shaping the Future of European Societies.* Available from http://ec.europa.eu/research/conferences/2004/ntw/pdf/final_report_en.pdf (retrieved October 15, 2006).

European Commission. (2004). *Foresighting the New Technology Wave SIG I–Quality of Life.* Available from http://ec.europa.eu/research/conferences/2004/ntw/pdf/sig1_en.pdf (retrieved October 15, 2006).

European Commission. (2004). *Foresighting the New Technology Wave—Expert Group SIG II—Report on the Ethical, Legal and Societal Aspects of the Converging Technologies (NBIC).* Available from from http://ec.europa.eu/research/conferences/2004/ntw/pdf/sig2_en.pdf (retrieved October 15, 2006).

European Commission. (2004). *New Technology Wave: Transformational Effect of NBIC Technologies on the Economy—SIG III Report on Economic Effect.* Available from http://ec.curopa.eu/research/conferences/2004/ntw/pdf/sig3_en.pdf (retrieved October 15, 2006).

Feynman, R. P. *There's Plenty of Room at the Bottom: An Invitation to Enter a New Field of Physics* (Speech, December 29, 1959). Annual Meeting of the American Physical Society at California Institute of Technology. Available from http://www.zyvex.com/nanotech/feynman.html (retrieved October 15, 2006).

International Risk Governance Council. (2005). *Survey on Nanotechnology Governance: Volume A. The Role of Government.* Mihail C. Roco & Emily Litten. Available at http://www.irgc.org/irgc/projects/nanotechnology/_b/contentFiles/Survey_on_ Nano-technology_Governance_—Part_A_The_Role_of_Government.pdf (retrieved October 15, 2006).

International Risk Governance Council. (2006). *White Paper No. 1: Risk Governance—Towards an Integrative Approach.* Ortwin Renn. Available at http://www.irgc.org/irgc/projects/risk_characterisation/_b/contentFiles/IRGC_WP_No_1_Risk_Governance_(reprinted_version).pdf (retrieved October 15, 2006).

International Risk Governance Council. (2006). *White Paper No. 2: Nanotechnology Risk Governance.* Ortwin Renn & Mihail C. Roco. Available at http://www.irgc.org/irgc/_b/contentFiles/IRGC_white_paper_2_PDF_final_version.pdf (retrieved October 15, 2006).

Joy, B. (2000). "Why the Future Doesn't Need Us." *Wired* vol. 8.04. Available from http://www.wired.com/wired/archive/8.04/joy.html (retrieved October 15, 2006).

Krupp, F. and Holliday, C. (2005, June 14). "Let's Get Nanotech Right". *Wall Street Journal.*

National Nanotechnology Initiative. (2005). *The National Nanotechnology Initiative at Five Years: Assessment and Recommendation of the National Nanotechnology Advisory Panel.* Available at http://www.nano-and-society.org/NELSI/documents/NNIpcastreport0505.pdf (retrieved October 14, 2006).

National Nanotechnology Initiative. (2004). *The National Nanotechnology Initiative Strategic Plan.* Available at http://www.nano-and-society.org/NELSI/documents/NNIreport1204.pdf (retrieved October 14, 2006).

National Research Council. (2006). *A Matter of Size: Triennial Review of the National Nanotechnology Initiative.* Available at http://newton.nap.edu/catalog/11752.html (retrieved October 15, 2006).

National Science Foundation and Department of Commerce. (2002). *Converging Technologies for Improving Human Performance*: *Nanotechnology, Biotechnology, Information*

Technology, and Cognitive Science. Mihail C. Roco and William Sims Bainbridge eds. Available at http://2100.org/Nanos/NSF.pdf (retrieved October 15, 2006).

National Science Foundation and Department of Commerce. (2002). *Converging Technologies for Improving Human Performance: Nanotechnology, Biotechnology, Information Technology, and Cognitive Science.* Mihail C. Roco and William Sims Bainbridge eds. Available at http://2100.org/Nanos/NSF.pdf (retrieved October 15, 2006).

National Science Foundation. (2001). *Societal Implications of Nanoscience and Nanotechnology.* Available at http://www.cns.ucsb.edu/filestore/2001 Societal Implications.pdf (retrieved October 15, 2006).

President's Council of Advisors on Science and Technology. (2005). *President's Council of Advisors on Science and Technology on the National Nanotechnology Initiative.* Available at http://www.nano-and-society.org/NELSI/documents/NNIpcastreport0505.pdf (retrieved October 14, 2006).

See Center for Nanotechnology in Society at Arizonal State University website. Available at http://cns.asu.edu/index.htm (retrieved on October 16, 2006).

See International Nanotechnology and Society Network (INSN). Available at www.nanoandsociety.org (retrieved on October 19, 2006).

Stephenson, N. (1995). *The Diamond Age: Or, A Young Lady's Illustrated Primer.* New York: Bantam Books, Inc.

Swiss Re. (2004). *Nanotechnology: Small Matter, Many Unknowns.* Available at http://www.swissre.com/INTERNET/pwswpspr.nsf/fmBookMarkFrameSet?ReadForm&BM=../vwAllbyIDKeyLu/ulur-5yaffs?Open (retrieved October 15, 2006).

ten Have, Henk A. M. J. (ed.) (2007). *Nanotechnologies, Ethics, and Politics.* Paris: UNESCO.

The Royal Society and the Royal Academy of Engineering. (2004). *Nanoscience and Nanotechnologies: Opportunities and Uncertainties.* Available at http://www.nanotec.org.uk/finalReport.htm (retrieved October 15, 2006).

United Nations Educational, Scientific and Cultural Organization. (2006). *The Ethics and Politics of Nanotechnology.* Available at http://unesdoc.unesco.org/images/0014/001459/145951e.pdf (retrieved October 15, 2006).

2. ONLINE RESOURCES ADDRESSING NELSI

This select listing includes government, academic, and NGO websites from a wide range of perspectives.

Entity Name: National Nanotechnology Initiative
Link: www.nano.gov/
Description: The National Nanotechnology Initiative (NNI) provides a multiagency
             framework for nanoscale science and engineering research and
             development conducted within the United States. One of the NNIs
             program component areas addresses the societal dimensions of
             nanotechnology.

Entity Name: European Commission Nanotechnology Home Page
Link: www.cordis.europa.eu/nanotechnology/home.html
Description: Carries news and updates as well as links to the various European Union
             nanoscale projects.

Entity Name:	Center on Nanotechnology and Society at Chicago-Kent College of Law, Illinois Institute of Technology
Link:	www.nano-and-society.org/
Description:	The Center on Nanotechnology and Society (Nano & Society) was created to catalyze informed interdisciplinary research, education and dialogue on the ethical, legal, policy, business, and broader societal implications of nanoscale science and technology, with a special focus on the human condition

Entity Name:	Center for Nanotechnology in Society at Arizona State University
Link:	www.cns.asu.edu/
Description:	The Center for Nanotechnology in Society (CNS–ASU) is funded by the National Science Foundation to study the societal implications of nanotechnology. As a boundary-spanning organization at the interface of science and society, CNS–ASU conducts research on reflexiveness and social learning with the hope of signaling emerging problems, enabling anticipatory governance, and guiding knowledge and innovation toward socially desirable outcomes, and away from undesirable ones.

Entity Name:	International Council on Nanotechnology at Rice University
Link:	http://icon.rice.edu/
Description:	The International Council on Nanotechnology (ICON), based at Rice University, is an organization whose activities engage industry, academia, nonprofit foundations, and government. ICONs mission is to assess, communicate, and reduce nanotechnology environmental and health risks while maximizing its societal benefit.

Entity Name:	Center for Nanotechnology in Society at University of California, Santa Barbara
Link:	www.cns.ucsb.edu/
Description:	Center for Nanotechnology in Society at University of California, Santa Barbara (CNS–UCSB) is funded by the National Science Foundation, and its mission is to serve as: a research and education center; a network hub for researchers and educators concerned with nanotechnologies' societal impacts; and a resource for studying these impacts.

Entity Name:	^{nano}Science and Technology Studies at the University of South Carolina
Link:	www.nsts.nano.sc.edu/
Description:	The ^{nano}Science and Technology Studies at the University of South Carolina (STS) is a group of researchers trained in a variety of disciplines who are engaged in research and education related to the societal, epistemological, and ethical dimensions of nanotechnologies.

Entity Name:	Initiative on Nanotechnology and Society at the University of Wisconsin, Madison
Link:	www.lafollette.wisc.edu/research/Nano/

Description: The Initiative on Nanotechnology and Society at the University of Wisconsin-Madison is network that supports multidisciplinary research and projects on the ethical, legal, social, and broader policy issues posed by nanotechnology.

Entity Name: National Nanotechnology Infrastructure Network's Societal and Ethical Issues in Nanotechnology at Cornell University
Link: www.sei.nnin.org/index.html
Description: The National Nanotechnology Infrastructure Network's Societal and Ethical Issues in Nanotechnology component (NNINs SEI) is located at Cornell University and, as part of the NNIN, is funded by the National Science Foundation. The SEI examines social and ethical implications of nanotechnology, and operates within the rubric of the larger NNIN.

Entity Name: Woodrow Wilson International Center for Scholars' Project on Emerging Nanotechnologies
Links: www.wilsoncenter.org/nano
Description: The Woodrow Wilson Center for International Scholars' Project on Emerging Nanotechnologies, which is primarily sponsored by the Pew Foundation, collaborates with researchers, government, industry, policymakers, and others to identify and develop strategies for closing gaps in knowledge and regulatory processes related to nanotechnology.

Entity Name: International Risk Governance Council
Link: www.irgc.org/irgc/projects/nanotechnology/
Description: The International Risk Governance Council's (IRGC) project on nanotechnology addresses the need for adequate risk governance approaches at the national and international levels in the development of nanotechnology and nanoscale products. The primary aim of the project is to create a forum that enables a wide range of stakeholders to collaborate, using and adapting the risk governance approach established by IRGC, both to determine deficits in the risk governance of nanotechnology, and to establish recommendations for the management of these deficits.

Entity Name: ETC Group (or Action Group on Erosion, Technology and Concentration)
Links: www.etcgroup.org/en/ www.etcgroup.org/en/issues/nanotechnology.html
Description: ETC Group, or Action Group on Erosion, Technology and Concentration, evaluates emerging technologies, including nanotechnology, giving consideration to their potential impacts on the poor and other marginalized populations.

Entity Name: Friends of the Earth Nanotechnology Project
Link: www.nano.foe.org.au/
Description: The Friends of the Earth Nanotechnology Project aims to catalyze debate on nanotechnology and its impact on health, the environment, and society, with a goal of ensuring that new nanotechnologies are developed in the public interest.

Entity Name: Foresight Nanotech Institute
Link: www.foresight.org/
Description: Foresight Nanotech Institute, founded by K. Eric Drexler and Christine
                  Peterson, has pressed the transformative possibilities of nanotechnol-
                  ogy and also developed draft policy statements to ensure its responsible
                  development. It was one of the earliest voices in the field.

Entity Name: Center for Responsible Nanotechnology
Link: www.crnano.org/
Description: The Center for Responsible Nanotechnology (CRN) is a project focused
                  on promoting visionary approaches to nanotechnology.

INDEX

Nanoscale: Issues and Perspectives for the Nano Century. Edited by Nigel M. de S. Cameron and M. Ellen Mitchell
Copyright © 2007 John Wiley & Sons, Inc.

445

Bailey, Ronald, 249, 310, 343
Bainbridge, William Sims, 27, 33,
    131, 283, 328, 329, 343, 379, 428
Baird, Davis, 369, 375
Balbus, J., 88, 421
Balch, W. R., 108
Baluch, A., 247
Bandura, A., 59
Bawa, R., 259, 301
Bayertz, K., 65
Bayh–Dole Act of 1980, 87, 276–277
Bayliss, F. T., 310–311
Beauchamp, Tom, 127
Beauty, 353
Behkam, B., 253
Beliefs, science and, 56–59
Belkin, Aron, 100
Benford, Robert D., 300
Benigeri, M., 55
Bennett, I., 380
Bennett, P., 24
Bennett-Woods, Debra, 296
Benson, H., 45
Berger, Eric, 400
Berger, Theodore W., 326, 342
Bergeson, L. L., 173, 233
Bergin, A. E., 45
Berne, Rosalyn W., 311–312, 379
Bertilson, Margarita, 132
Berube, David M., 47, 58, 296, 306, 338,
    363, 374–375
Best, M., 317
Best practices, for handling nanomaterials,
    406. See also Safety
Betta, M., 302
Biases, prediction and planning,
    105–109
Binder, L. M., 331
Bioconservatives, 66, 67
"Bioethics," 284–285, 285–288, 293
Blair, R. D., 254
Bleeker, R. A., 242
Bluedorne, A. C., 102, 104
Boehlert, Sherwood, 3, 10
Bok, Sissela, 416
Bolman, L. G., 296, 300, 311
Boltzmann, Ludwig Eduard, 120
Bond, Phillip J., 283, 381, 437
Boninger, David S., 105
Boorstin, Daniel, 124
Borg, 435. See also Cyborgs
Borges, Tim, 405
Borkovec, T. D., 100

Bostrom, Nick, 138, 147, 148, 329, 342
Boyd, J. N., 103
Brand, P., 322
Brave New World (Huxley), 7, 434
Breast augmentations/implants, 349, 352
Bretthorst, G. Larry, 147
Breyer, Stephen, 156, 157, 186, 246
Brin, David, 96
British Standards Institution (BSI), 192.
    See also United Kingdom (UK)
Britt, Robert Roy, 324
Brock, T. C., 107
Brown, Douglas, 398, 399
Brown, M. B., 423
Bruce, I. J., 298
Brumfiel, G., 219
Brunner, T. J., 122
BSE (bovine spongiform
    encephalopathy), 25, 132
Buckyballs, 76. *See also* Fullerenes
Buehler, R., 109
Buller, David, 63
Bush, George W., 67, 282, 425

CAA. *See* Clean Air Act (CAA) of 1963
California Institute of Technology (Caltech),
    President Clinton's address to, 425
Callahan, D., 345
Callahan, M., 345
Calman, K., 24
Cameron, Nigel M. de S., 286, 359
Campbell, K. A., 331
Campo, A., 298
Canadian National Research Council, 28
Cancer, war on, 356
Canton, J., 33, 43, 51
Caplan, Art, 131
Carbon fullerenes, 219. *See also*
    Fullerenes
Carbon nanotubes, 320
Carl, Robert F., Jr., 76
Carnot, Sadi, 119, 120
Carstensen, Laura L., 104
Castells, M., 28
Castiglioni, A., 317
Cauller, Larry, 328
Causal models, 48–50
CBAN. *See* Consultative Board for
    Advancing Nanotechnology (CBAN)
CBEN. *See* Center for Biological and
    Environmental Nanotechnology (CBEN)
CDC. *See* Centers for Disease Control and
    Prevention (CDC)